Large-Scale Motions
in the Universe:
A Vatican Study Week

Large-Scale Motions in the Universe: A Vatican Study Week

Edited by

Vera C. Rubin and George V. Coyne, S.J.

Princeton University Press

Princeton, New Jersey

**PONTIFICIA
ACADEMIA
SCIENTIARVM**

Copyright © 1988
Pontifical Academy of Sciences

Distributed by: in Italy and Vatican City State: Libreria Editrice Vaticana
V-00120 Vatican City State
Elsewhere: Princeton University Press
Princeton, New Jersey 08540
USA

CONTENTS

II. LARGE-SCALE VELOCITY FIELDS

V.   THEORY: *AB INITIO*

MESSAGE

Leiden, February 25, 1988.

If it is true that putting the right questions is already half the solution of a problem, the Organizing Committee of the Vatican Study Week on *Large-Scale Motions in the Universe* may have gone a long way towards the realization of the aim of the symposium prior to the meeting itself. Those who have read the questions but, like myself, were not part of the small group attending will be anxious to learn the answers reached which are described in the present book. Judging from previous experience with these Study Weeks I am confident that the discussions will have have been extremely fertile. The principal question I am left with myself is: Have we perhaps been unable to formulate the prime illuminating question?

J. H. OORT

LIST OF PARTICIPANTS

N.A. BAHCALL, Space Telescope Science Institute, 3700 San Martin Drive, Baltimore, MD 21218, USA.

J.R. BOND, Institute for Theoretical Physics, 60 St. George St., Toronto, Ontario M5S 1A1, Canada.

D. BURSTEIN, Department of Physics, Arizona State University, Tempe, AZ 85287, USA.

G.V. COYNE, S.J., Specola Vaticana, V-00120 Vatican City State.

M. DAVIS, Department of Astronomy, Campbell Hall, The University of California, Berkeley, CA 94720, USA.

A. DEKEL, Racah Institute of Physics, The Hebrew University, Jerusalem, 91904 Israel.

G. EFSTATHIOU, Institute of Astronomy, Madingley Road, Cambridge CB3 OHA, United Kingdom.

S.M. FABER, Lick Observatory, The University of California, Santa Cruz, CA 95064, USA.

M.J. GELLER, Harvard-Smithsonian Center for Astrophysics, 60 Garden St., Cambridge, MA 02138, USA.

M.P. HAYNES, Department of Astronomy, Space Sciences Building, Cornell University, Ithaca, NY 14853, USA.

J.P. HUCHRA, Harvard-Smithsonian Center for Astrophysics, 60 Garden St., Cambridge, MA 02138, USA.

N. KAISER, Institute of Astronomy, Madingley Road, Cambridge CB3 OHA, United Kingdom.

D.C. KOO, Lick Observatory, The University of California, Santa Cruz, CA 95064, USA.

A.N. LASENBY, Cavendish Laboratory, University of Cambridge, Madingley Road, Cambridge CB3 OHE, United Kingdom.

D. LYNDEN-BELL, Institute of Astronomy, Madingley Road, Cambridge CB3 OHA, United Kingdom.

J. MOULD, Division of Physics, Mathematics and Astronomy, California Institute of Technology, Pasadena, CA 91125, USA.

P.J.E. PEEBLES, Joseph Henry Laboratories, Princeton University, P.O. Box 708, Princeton, NJ 08544, USA.

V.C. RUBIN, Department of Terrestrial Magnetism, Carnegie Institution of Washington, 5241 Broad Branch Road, N.W., Washington, D.C. 20015, USA.

W.R. STOEGER, S.J., Specola Vaticana, V-00120 Vatican City State.

A. SZALAY, Department of Atomic Physics, Eötvös University, H-1088 Budapest, Hungary.

R.B. TULLY, Institute for Astronomy, University of Hawaii at Manoa, 2680 Woodlawn Drive, Honolulu, HI 96822, USA.

N. VITTORIO, Istituto Astronomico, Università degli Studi di Roma, La Sapienza, Via Lancisi, 29, I-00161 Rome, Italy.

A. YAHIL, Department of Earth and Space Sciences, State University of New York at Stony Brook, Stony Brook, NY 11794, USA.

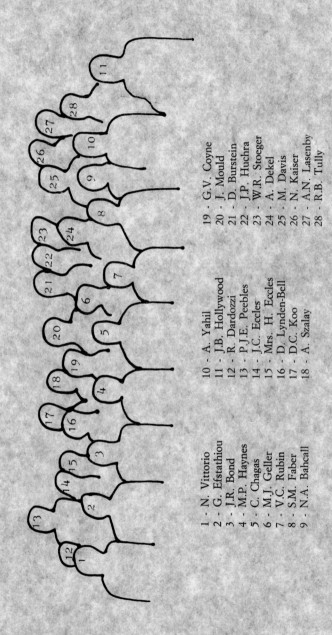

1 - N. Vittorio
2 - G. Efstathiou
3 - J.R. Bond
4 - M.P. Haynes
5 - C. Chagas
6 - M.J. Geller
7 - V.C. Rubin
8 - S.M. Faber
9 - N.A. Bahcall

10 - A. Yahil
11 - J.B. Hollywood
12 - R. Dardozzi
13 - P.J.E. Peebles
14 - J.C. Eccles
15 - Mrs. H. Eccles
16 - D. Lynden-Bell
17 - D.C. Koo
18 - A. Szalay

19 - G.V. Coyne
20 - J. Mould
21 - D. Burstein
22 - J.P. Huchra
23 - W.R. Stoeger
24 - A. Dekel
25 - M. Davis
26 - N. Kaiser
27 - A.N. Lasenby
28 - R.B. Tully

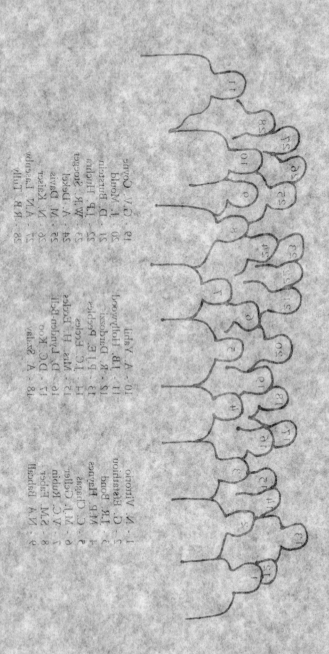

PREFACE

It is only 200 years since Sir William Herschel used the proper motions of 13 stars to discover the motion of the sun. During the week of 9 to 14 November, 1987, twenty-three astronomers met for a Study Week in Vatican City at the invitation of the Pontifical Academy of Sciences to discuss a contemporary equivalent of that study, *Large-Scale Motions in the Universe*. For the participants, the week was a gift; a gift of time during which we could describe our recent work to our colleagues. Unlike most scientific meetings in which only positive forward progress is reported, we hoped to include in our discussions also the uncertainties inherent in these initial steps in untangling the structure and the motions of our region of the universe. What's unclear, what's hidden, what's misinterpreted, what's misunderstood? All of these questions were to be raised.

To guide our deliberations many of the participants had used the occasion of the IAU Symposium 130 at Lake Balaton, Hungary in June, 1987 to identify some thought-provoking questions which would be relevant to the Vatican Study Week. These questions were distributed to the participants well in advance of the Study Week, along with a preliminary program devised by the Organizing Committee. For each of the nine sessions, a topic was identified, as well as a chairperson to introduce the subject, and a summarizer to act as a good listener and to remind us where difficulties still lay.

Within this framework and the limits of time, all participants were free to offer contributions. The preliminary questions are reprinted here; the program of the sessions can be recovered approximately by examining the Table of Contents of this volume. Written texts were not required, or even expected, at the time of the meeting. It was hoped that texts prepared immediately following the Study Week would better reproduce the flavor of discovery and uncertainty characteristic of the sessions and of the discussions. For the final session, each participant presented a less-than-five-minute statement (view-graphs not permitted) of anything on his/her mind related to the subject matter of the meeting. These statements and Avishai Dekel's summary contribution were recorded and are reproduced here after minor editing. Among this group of independent astronomers, those who violated the time and viewgraph restriction can now be identified.

We missed the participation of Marc Aaronson, who had accepted an invitation to be with us, but whose tragic death intervened. We also missed Yakov Zeldovich who was unable to attend. We note with sorrow that Prof. Zeldovich died a few weeks after our meeting. We acknowledge our scientific debt to both of them; they were missed.

All of the participants wish to thank President Carlos Chagas of the Pontifical Academy of Sciences for offering us the opportunity to exchange ideas in such a pleasant setting, and to the Director of the Chancery of the Academy, Ing. Renato Dardozzi, for administering the meeting arrangements. We also thank George V. Coyne, S.J., Director of the Vatican Observatory, for taking care of all the details which made the meeting go so well, and for his hospitality at Castel Gandalfo where we spent a delightful afternoon and evening. Jim Peebles, in his after dinner toast, correctly identified George as the "prime mover" of the Study Week. We thank also Rita Callegari, Secretary of the Vatican Observatory, and the staff of the Academy for their roles in making problems seem small scale. And I thank the other members of the Organizing Committee; Sandra Faber, Donald Lynden-Bell, and Alex Szalay, for their help with all stages in the planning for the Study Week. I also note with thanks that Jim Peebles offered valued advice as an ex-officio member of the Organizing Committee. To Janice Dunlap who helped in many ways including preparing the index I am grateful.

Often in planning this Study Week I was reminded of an old Peanuts cartoon. Peanuts is a popular American cartoon, apparently equally popular all over the world as well. In this cartoon Lucy is instructing her fall guy, Charlie Brown, as follows. On the oceans of the world are many ships, and some of these ships carry passengers. One of the activities the passengers like most is to sit on the deck and watch the ocean. Some of these passengers arrange their deck chairs to face where they are going; some of the passengers arrange their chairs to see where they have been. Lucy asks, "On the great ship of life, Charlie Brown, which way are you going to place your chair, to see where you are going, or to see where you have been?" And Charlie Brown answers, "I can't even seem to get my chair unfolded."

Some of us are trying to find out where we are going, some are looking at where we have been. For many of us, the Vatican Study Week offered an incomparable opportunity to attempt to unfold our chairs.

VERA C. RUBIN, CHAIRPERSON

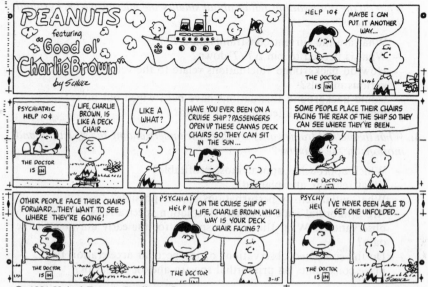

© 1981 United Feature Syndicate, Inc.

SOME QUESTIONS CONCERNING
THE STRUCTURE AND MOTIONS IN THE UNIVERSE

1. *What is a good description of the 3-dimensional structure of the Universe,*
   $V < 10,000$ *km s^{-1}? at largest distances?*

   — What is the convincing evidence that some structures are strings,
     chains, filaments, sponges, bubbles, sheets, voids?
   — Do structures exist which exceed 100 Mpc?
   — What is the amplitude of structures on the largest scale?
   — Did these structures originate in the very early universe?

2. *What is the spectrum of peculiar velocities as a function of scale?*

   — Can details of the velocity field of the Local Supercluster be map-
     ped and understood in the context of the distribution of galaxies?
     Can gravitation account for the magnitude and direction of the
     motion?
   — How reliable is our current knowledge of the distribution and mo-
     tions of clusters and superclusters, and how seriously should we
     take the present conclusions based on the data?
   — How large are the internal motions in the 3-dimensional structures?

3. *What do the observations tell us about small-scale microwave background fluctuations?*

   — What are the current constraints on $\Delta T/T$?
   — What is needed to pin $\Delta T/T$ down?
   — How do background fluctuations relate to the velocity fields in various models?
   — Could background fluctuations have been wiped out by re-ionization of the intergalactic medium?

4. *What theoretical frameworks are there for understanding the structures and motions in the universe?*

   — Which approaches appear most profitable?
   — What remains unexplained?
   — What initial density perturbations appear possible?
   — Are primordial scale invariant fluctuation spectra ruled out by the observations? What fluctuation spectra for the early universe fit the data?
   — Do the combined observations of the peculiar acceleration field place constraints on primordial spectra and/or scenarios?

5. *What are realistic prospects for N-body simulations?*

   — What do we learn from a comparison of simulations and observations, both in density and in velocity?
   — How can we quantify the spatial distribution of galaxies in theories and observations?
   — Are there alternatives to ξ, the correlation function?
   — What are the observational and theoretical constraints on biasing?

6. *What caused the formation of superclusters, clusters, and galaxies?*

   — Did this process not generally occur much before $z = 3$, and can we understand why?
   — Why did quasar formation stop essentially after $z = 3$ or so?
   — What can we learn from the clustering of high redshift objects: galaxies, clusters, quasars, Ly-α clouds?
   — What do galaxies look like at $z = 3$? 2?
   — What limits can be placed on interactions of galaxies since $z = 1$?
   — How big are galaxy halos?
   — What empirical handles do we have on the mass distribution?

7. *What do we have to observe to understand the universe?*

   — What are the most definite testable predictions of the competing cosmological scenarios: cold dark matter, hot dark matter, strings, explosions?
   — How compelling is the evidence that the density of the universe equals the critical density?

8. *Ten years hence, what will we wish we had discussed in November, 1987?*

I

TWO-DIMENSIONAL AND
THREE-DIMENSIONAL STRUCTURE

GALAXY AND CLUSTER REDSHIFT SURVEYS

MARGARET J. GELLER and JOHN P. HUCHRA

Harvard-Smithsonian Center for Astrophysics, Cambridge

ABSTRACT

We discuss the status of galaxy and cluster redshift surveys. We concentrate on the CfA redshift survey and a deep Abell cluster redshift survey.

Five strips of the CfA redshift survey are now complete. The data continue to support a picture in which galaxies are on thin sheets which nearly surround vast low density voids. In this and similar surveys the largest structures are comparable with the extent of the survey. Voids like the one in Boötes are a common feature of the galaxy distribution and present a serious challenge for the models. We suggest some statistics which may be useful for comparing the data with the models.

The deep cluster survey of Huchra *et al.* (1988) is nearly independent of the sample analyzed earlier by Bahcall and Soneira (1983). For this new sample the amplitude of the correlation function is a factor of ~ 2 less than for the earlier sample. However, the difference may not be significant because the cluster samples are sufficiently small that they may be dominated by single systems. The deeper sample also fails to support the claim by Bahcall, Soneira, and Burgett (1986) of ~ 2000 km s^{-1} peculiar motions; the limit for the deep sample is ≤ 1000 km s^{-1}.

1. INTRODUCTION

In 1978 Jôeveer, Einasto, and Tago suggested that the large-scale distribution of galaxies has a "cellular" pattern in which rich clusters are connected by "filamentary" structures. The data at that time were incomplete and could only hint at such structure. The discovery of the void in Boötes (Kirshner *et al.* 1981, KOSS hereafter) and the 21-cm survey of the Pisces-Perseus chain (Haynes and Giovanelli 1986, Giovanelli *et al.* 1986) soon lent support to this picture. These first surveys gave no clear message about the frequency of the structures.

It has become increasingly clear that large-scale features in the galaxy distribution are ubiquitous. The deep surveys of Koo, Kron, and Szalay (1986) show that voids are common even at high redshift. Because surveys are one-dimensional, the constraints on the sizes of the voids are poor. The AAT surveys also reveal voids along with thin structure perpendicular to the line-of-sight (Peterson *et al.* 1986). The continuing Arecibo survey delineates nearby voids and appears to support the interpretation of the Pisces-Perseus chain as a filamentary structure.

The extension of the Center for Astrophysics (CfA) redshift survey (discussed in Section 2) indicates that bright galaxies are distributed on thin sheets — two-dimensional structure— which surround (or nearly surround) vast voids. Recent completion of a southern hemisphere survey (da Costa *et al.* 1988) gives some support to this picture although the survey is not sufficiently deep or dense to be directly comparable with the recent CfA results. The 21-cm data (Haynes and Giovanelli 1986) also reveal sheet-like structures in the Perseus-Pisces region. The message of all surveys is now clear: large structures are a common feature of all surveys big enough to contain them.

Because redshift surveys of individual galaxies and redshift surveys of clusters of galaxies do not overlap substantially, the relationship between the large-scale distribution of rich clusters and the large-scale distribution of galaxies remains unclear. Analyses of existing surveys indicate that clusters of galaxies and individual galaxies are not equivalent tracers of the large-scale matter distribution. Section 3 is a discussion of a recently completed deep survey (Huchra *et al.* 1988) which is largely independent of the nearby sample analyzed by Bahcall and Soneira (1983).

We take $H_o = 100h$ km s^{-1} Mpc^{-1}, with $h = 1$ unless otherwise specified.

2. THE LARGE-SCALE GALAXY DISTRIBUTION

Over the last few years each new approach to mapping out the distribution of individual galaxies has uncovered unexpectedly large structures. Surveys like the one of the Boötes (KOSS 1987) region are an efficient method of finding large voids. Surveys like the CfA redshift survey extension (Huchra and Geller 1988) which are complete over a region of large angular scale are less efficient for identifying individual large structures, but they are necessary for quantitative characterization of the distribution of galaxies over a range of scales. Table 1 is a list of existing surveys.

The goal of the Center for Astrophysics redshift survey extension is to measure redshifts for all galaxies in a merge of the Zwicky *et al.* (1961-1968)

TABLE 1
GALAXY AND CLUSTER REDSHIFT SURVEYS

FORMAT

Name	Authors	N_{obj}	Date
A. LARGE AREA COMPLETE			
GALAXIES			
HMS	Humason *et al.*	500	1956
RSA	Sandage and Tammann	1300	1981
RC1 + RC2	de Vaucouleurs *et al.*	1200	1964, 1976
CFA1	Huchra *et al.*	2400	1982
UGC Gals	Bothun *et al.*	4000+	1985
CFA Slice	de Lapparent *et al.*	1100	1986
SSRS	da Costa *et al.*	1800	1987
Nearby Gals	Tully and Fisher	3000	1987
CFA2	Huchra and Geller	15000	1990
ABELL CLUSTERS			
D4	Hoessel *et al.*	116	1980
SAO	Karachentsev *et al.*	100+	1983, 1986
D5 ($m_{10} \leq$ 16.5)	Postman *et al.*	561	1985, 1988
Abell-Corwin	Corwin, Olowin *et al.*	500+	1989
B. LARGE AREA INCOMPLETE			
GALAXIES			
Markarian	Arakelian *et al.*	1300	1970's
Böotes	KOSS	400	1983, 1987
Arecibo	Giovanelli and Haynes	4000+	1985+
IRAS Gals	Strauss *et al.*	2600	1988
C. SMALL AREA (Usually Deep)			
GALAXIES			
AAT	Ellis, Shanks, Fong *et al.*	260	1985, 1988
KPNO	Koo and Kron	250+	1987
Cor Bor	Postman *et al.*	250	1987
Century	Geller *et al.*	2500	1990
Southern strip	KOSS	2000+	1990
Southern AAT	Efstathiou *et al.*	5000+	
ABELL CLUSTERS			
Superclusters	Ciardullo *et al.*	5	1985
Deep Abell	Huchra *et al.*	145	1988

and Nilson (1973) catalogs which have $m_{B(0)} \leq 15.5$ and $|\, b \,| \geq 40°$. There will be about 15,000 galaxies in the complete survey; more than half of these already have measured redshifts. About 2,500 of the galaxies with measured redshifts lie in three "slices" where the survey is now complete: (1) a slice with $8^h \leq \alpha \leq 17^h$ and $26.5° \leq \delta < 32.5°$ (de Lapparent, Geller, and Huchra 1986; dLGH hereafter); (2) a slice with $8^h \leq \alpha \leq 17^h$ and $32.5° \leq \delta < 38.5°$; and (3) a slice with $8^h \leq \alpha \leq 17^h$ and $38.5° \leq \delta < 44.5°$. In the southern Galactic hemisphere 2 slices covering $6° \leq \delta < 12°$ and $18° \leq \delta < 24°$ with $0^h \leq \alpha \leq 4^h$ and $20^h \leq \alpha \leq 24^h$ are also complete. More than 60% of the redshifts were measured with the 1.5-meter telescope and the MMT at Mt. Hopkins. The mean external error in the redshift measurements is ~ 35 km s^{-1}.

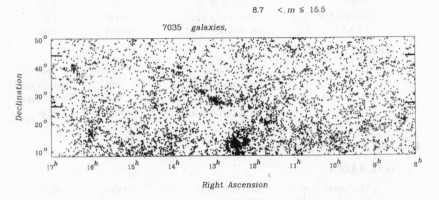

FIG. 1. Positions of galaxies in the merged Zwicky-Nilson catalogs with $m_{B(0)} \leq 15.5$ in the northern galactic cap. The bold ticks indicate the limit of the complete strips.

Figure 1 shows the positions of 7035 galaxies in the Zwicky-Nilson merge which have $m_{B(0)} \leq 15.5$ and $8^h \leq \alpha \leq 17^h$ and $8.5° \leq \delta < 50.5°$. The grid is Cartesian in α and δ. The deficiency of galaxies west of 9^h and east of 16^h is caused by galactic obscuration. The bold ticks indicate the limits of the three complete strips of the survey. The Coma cluster is the dense knot at $\alpha = 13^h$, $\delta = 30°$. Figure 2 is a similar map for the region of the southern galactic hemisphere covered by the CfA survey. Again the bold ticks indicate the limits of declination strips which are complete.

Figure 3 is a plot of the observed velocity versus right ascension for the strip centered at $29.5°$ (dLGH): the strip is $6°$ wide in declination. The plot includes only the 1065 galaxies with redshifts $\leq 15,000$ km s^{-1}. A galaxy of

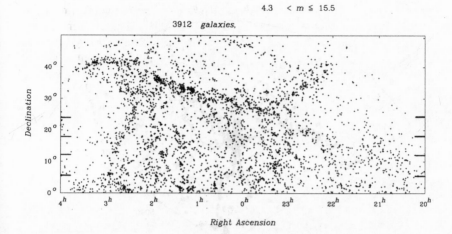

FIG. 2. Position of galaxies in the merged Zwicky-Nilson catalogs with $m_{B(0)} \leq 15.5$ and in the southern galactic cap. The bold ticks indicate the limits of the complete strips.

characteristic luminosity, M^*, is at roughly 10,000 km s^{-1} in this survey. Nearly every galaxy in this slice lies in an extended thin structure. The boundaries of the empty regions are remarkably sharp. Several of the voids are surrounded by thin structures in which the inter-galaxy separation is small compared with the radius of the empty region. The edges of some of the largest structures may be outside the right ascension limits of the survey. The only pronounced velocity finger in this slice is the Coma cluster at $\sim 13^h$.

The first slice alone demonstrates that the thin structures in the distribution of galaxies are cuts through two-dimensional sheets, not one-dimensional filaments. If the ~ 150 Mpc long structure which extends across the entire survey (from 9^h to 16^h between 7,000 km s^{-1} and 10,000 km s^{-1}) is a filament, a thin linear structure should be visible on the sky. This statement is particularly strong because the structure lies near the survey limit. The required filamentary structure is absent from Figure 1. Because structure on the sky can be caused by patchy obscuration and/or by inhomogeneities in the galaxy catalog, structure on the sky cannot provide complete proof (or disproof) of the filamentary nature of a structure in redshift space. A second argument against the filamentary nature of the structures in Figure 3 is that several thin, elongated structures lie in this single survey slice: the intersection of a slice with a three-dimensional network of filaments is *a priori* unlikely to be a two-dimensional network of filaments.

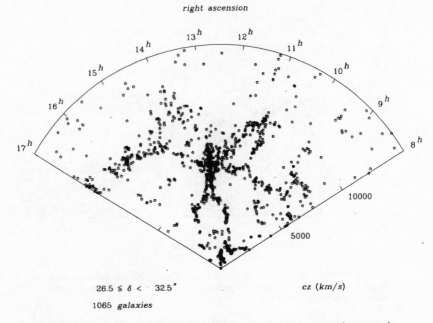

FIG. 3. Observed velocity versus right ascension for the survey strip centered at $\delta = 29.5°$. The strip is $6°$ in declination. Only the galaxies with velocities $\leq 15,000$ km s^{-1} are shown.

A geometric structure in which thin sheets surround or nearly surround voids accounts for the data. Examples include "bubble-like" and "sponge-like" geometries. We use the word "bubble" to convey the image of a structure dominated by thin sheets and holes. Note that the "bubbles" are not necessarily round. In this picture the 150 Mpc "filament" is made up of portions of adjacent "bubbles" and the richest clusters like Coma lie in the interstitial regions (where several "bubbles" come together).

In this interpretation of the data, we assume that the maps are similar in redshift space and in real physical space. Simulations of both the cold dark matter (White *et al.* 1987) and the adiabatic models (Centrella *et al.* 1988) suggest that this assumption is reasonable, i.e. distortions caused by large scale flows are minimal on these scales (but see Kaiser 1987).

Maps of the adjacent slices support the qualitative, homogeneous picture suggested by the first slice. Figure 4 shows the slice centered at 35.5°, just to the north of the slice in Figure 3. Once again the galaxies are in thin structures. Furthermore these structures are natural extensions of the structures in Figure 3; the structures are highly correlated in the two slices. The two closed structures at ~ 11^h (9000 km s^{-1} $\leq$ cz $\leq$ 11,000 km s^{-1}) and at 14^h (7000

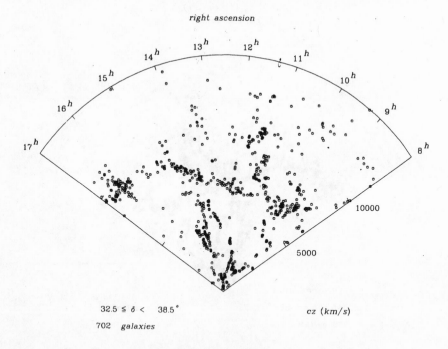

right ascension

$32.5 \leq \delta < \quad 38.5°$

702 galaxies

cz (km/s)

FIG. 4. Same as Figure 3 for a slice centered at $\delta = 35.5°$.

km s^{-1} $\leq$ cz $\leq$ 11,000 km s^{-1}) are not so clearly delineated in Figure 4 as in Figure 3.

Figure 5 shows the cone diagram for the two slices taken together. The distribution remains remarkably inhomogeneous with empty voids outlined by thin structures. Because the surfaces are curved or inclined relative to the plane of the slice, the structures are thicker here than in Figures 3 and 4. The largest low density region in the survey is located between 13^{h}20^m and 17^h with 4000 $\leq$ cz $\leq$ 9000 km s^{-1}. The diameter of this void is ~ 5000 km s^{-1} or 50 Mpc in the absence of large-scale flows. Recent infra-red Tully-Fisher measurements indicate that the diameter of the void is the same in physical space as in redshift space to within ~ 300 km s^{-1} (Geller *et al.* 1988a). The underdensity in the region ($\leq$ 20% of the mean) and the scale of the structure are comparable to the corresponding parameters for the void in Boötes.

Note that voids are not empty: they are regions which are underdense relative to the global mean. The galaxies inside the large void have normal properties and their infra-red Tully-Fisher distances put them at the relative distances indicated by their redshifts (Geller *et al.* 1988a). These galaxies may form a tenuous structure which would not be detected in a sparse survey like

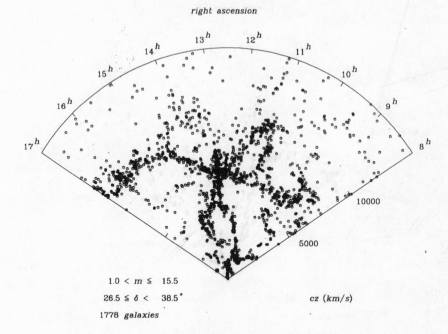

right ascension

1.0 < m ≤ 15.5

26.5 ≤ δ < 38.5°   cz (km/s)

1778 galaxies

FIG. 5. Cone diagram for the slices in Figures 3 and 4 taken together. This is 12° wide in declination and is centered at 32.5°. The sample contains ~ 1700 galaxies.

the KOSS survey of Boötes. Their spectroscopic properties are similar to those in other portions of the survey.

Figure 6 shows the cone diagram for the third slice centered at $\delta = 41.5°$. At first glange this slice gives a somewhat different visual impression from the first two. The reason for the difference is that some of the surfaces lie in this slice; in particular, a portion of the structure surrounding the largest void is nearly in the plane of this slice and appears diffuse. Comparison of this slice with Figure 5 continues to support the large-scale coherence of the structures.

It is instructive to compare these new surveys with the original CfA survey (Davis *et al.* 1982, Huchra *et al.* 1983) to a limiting $m_{B(0)} = 14.5$. Figure 7 shows the 29.5° slice to this limit. Here an M* galaxy has a velocity of 6000 km s^{-1}. Because the "effective depth" of this survey is comparable with the scale of the largest structures in Figures 3-5, these structures could not be detected.

The nearby small void centered at 13^h20^m and 3500 km s^{-1} is visible in both Figures 3 and 7. Note that the fainter galaxies fill in the gaps along the perimeter of the void. This comparison and other deep probes through the

right ascension

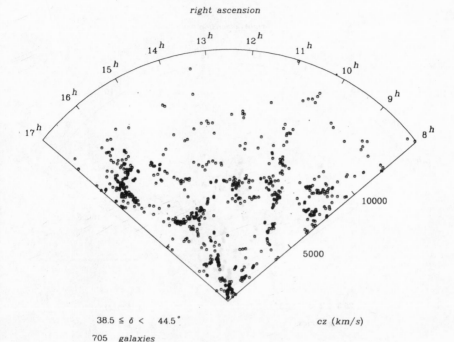

38.5 ≤ δ <   44.5°   cz (km/s)

705   galaxies

FIG. 6. Same as Figures 3 and 4 for slice centered at δ = 41.5°.

29.5° slice (Postman, Huchra, and Geller 1986) indicate that the distribution is insensitive to luminosity for $M_{B(0)} \lesssim - 17.4$. More precisely, the density contrast between the high and low density regions (about a factor of 20) is insensitive to absolute magnitude over this range. Surveys to fainter limiting magnitudes *do* turn up galaxies in the voids, but they turn up proportionately more in the dense structures.

The above conclusion is at odds with the analysis by Giovanelli and Haynes (1988) of structures in the Perseus-Pisces region. The definition of large-scale structure could be a function of properties of individual galaxies other than absolute luminosity. For example, there is a well-known relation between the morphology of a galaxy and local density (Hubble and Humason 1931, Dressler 1980, Dressler 1984, Postman and Geller 1984). It seems very likely that the Giovanelli and Haynes (1988) results are heavily affected by the morphology-density relation.

The change in the fractional coverage of the perimeter of a void as a function of luminosity complicates topological studies like those by Gott, Melott, and Dickinson (1986), and Hamilton, Gott, and Weinberg (1987). The data so far are inadequate to discriminate clearly between "bubble-like" (connected high density regions; isolated low density ones) and "sponge-like" (both high

M.J. GELLER - J.P. HUCHRA

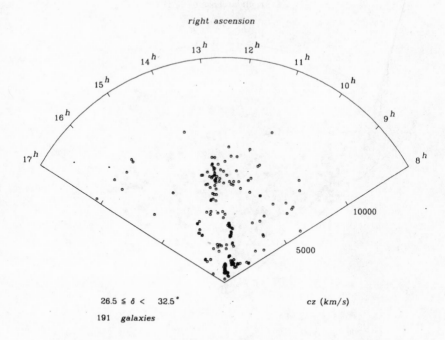

FIG. 7. Observed velocity versus right ascension for the survey strip centered at $\delta = 29.5°$ (Figures 3) but with a magnitude limit $m_{B(0)} = 14.5$.

and low density regions form a connected network) topologies. This distinction is a clue to the initial conditions for large-scale structure formation; a "bubble-like" topology points to non-random phases in the initial perturbation spectrum.

The search for a relation between surface brightness and local density has been prompted by cold dark matter models for the formation of large-scale structure (see e.g. Dekel and Silk 1986). Davis and Djorgovski (1985) published the first analysis of existing data and concluded that a correlation exists in the sense that lower surface brightness objects inhabit regions of lower average density. Later Bothun *et al.* (1986) and Thuan, Gott and Schneider (1987) obtained new data and reached the opposite conclusion: the low and high surface brightness objects trace the same structures as demonstrated by the velocity histograms for galaxies in the direction of the Virgo-Coma void (Figure 8). The Davis and Djorgovski (1985) analysis is incorrect because it was based on a sample with poorly understood selection biases.

The dependence of structure on the properties of individual galaxies can also be examined by comparing surveys based on the IRAS catalog with those based on optical catalogs. At least two complete studies (Smith *et al.* 1987,

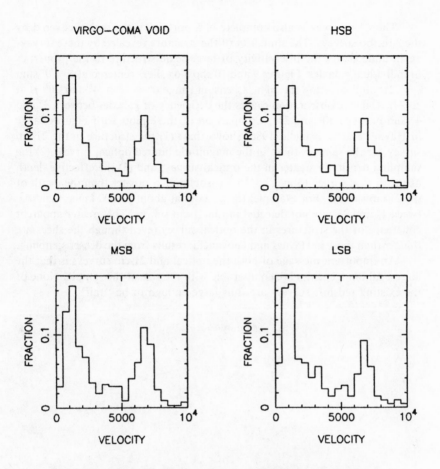

FIG. 8. The velocity distribution of all the Nilson galaxies in the direction of the Virgo-Coma void compared with the velocity distributions of the high, middle and low thirds in surface brightness (from Bothum *et al.* 1986). The underdensity between Virgo and Coma persists at all surface brightness levels.

Strauss and Huchra 1988) indicate that the galaxies in the IRAS catalog have a distribution similar to those in optical catalogs except that the IRAS galaxies are absent from the dense cores of rich clusters like Coma. Smith *et al.* (1987) cover the CfA redshift survey slice in Figure 2 and Strauss and Huchra (1988) survey the void in Boötes. The IRAS survey is not deep enough to see the far edge of the Boötes void, but the underdensity in the velocity range of the void is consistent with the claims based on optical data (~ 20% of the mean). In addition, both the photometric and spectroscopic properties of the IRAS galaxies found in the void are similar to the properties of those galaxies in the boundaries.

The CfA survey is also complete in a portion of the region covered by
the Arecibo surveys. The similarity of the structure revealed by these surveys
again underscores the insensitivity of large-scale structure to the properties
of individual galaxies. Figures 9 and 10 show 6° slices centered at $\delta = 9°$ and
$\delta = 21°$ and covering the right ascension range $0^h \leq \alpha \leq 4^h$ and $20^h \leq \alpha$
$\leq 24^h$. Galactic obscuration causes the deficiency of galaxies between 3^h and
4^h and between 20^h and 21^h. Comparison of these plots with Figures 2a—c
in Haynes and Giovanelli (1986) shows that: (1) the structure in the 21-cm
survey are the same as those in the magnitude limited optical survey; (2) the
sampling density is greater in the optical survey; and (3) the effective depth
of the optical survey to $m_{B(0)} \leq 15.5$ is somewhat greater than the depth of
the 21-cm surveys. For example, the galaxies at about 12,000 km s^{-1} and between 1^h and 23^h are not detected in the 21-cm survey. The greater apparent
coherence of the structures in the optical survey (even though the slices are
thinner than those in Haynes and Giovanelli) results from the denser sampling.

An important message of both the optical and 21-cm surveys is that the
largest inhomogeneities are comparable with the size of the sample. None of
the existing redshift surveys are thus large enough to be "fair".

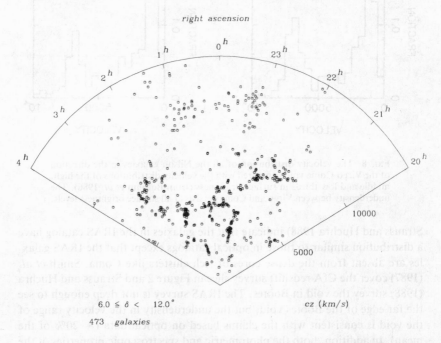

FIG. 9. Observed velocity versus right ascension for the survey strip in the
southern galactic cap centered at $\delta = 9°$. The strip is 6° in declination. Only
the galaxies with velocities $\leq 15,000$ km s^{-1} are shown.

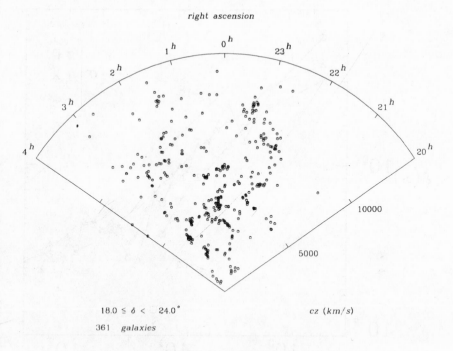

right ascension

18.0 ≤ δ <   24.0°   *cz (km/s)*

361   *galaxies*

FIG. 10. Same as Figure 9 for a slice centered at δ = 21°.

The largest inhomogeneities we detect are the largest we *could* detect within the limits set by the extent of the survey — we have few, if any, reliable direct limits on larger structures in the distribution of light-emitting matter. The size of the inhomogeneities relative to the volume of the surveys may underlie unexplained variations in traditional statistics of the galaxy distribution like the luminosity function (Schechter 1976, KOSS 1983, Bean *et al.* 1983, Davis and Huchra 1982) and the two-point correlation function at large scale (Groth and Peebles 1977, Davis and Peebles 1983, Kirshner, Oemler, and Schechter 1979, Shanks *et al.* 1983). When the inhomogeneities are large compared with the sample volume, mean quantities are not well-defined.

The domination of the sample by large-scale coherent structures and the related ~ 25% uncertainty in the mean density (de Lapparent, Geller, and Huchra 1988) imply that the two-point correlation function is more poorly constrained than previously thought. Figure 11 shows the two-point correlation function $\xi(s)$ where

$$s = \frac{(V_i^2 + V_j^2 - 2V_iV_j\cos\theta_{ij})^{1/2}}{H_o} \tag{1}$$

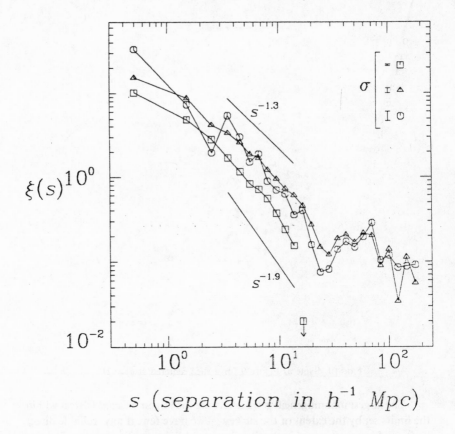

FIG. 11. The two-point correlation function $\xi(s)$ for the sample in Figure 5. The symbols denote ξ_{11} ($\square$), $\xi_{\phi\phi}$ ($\triangle$), and $\xi_{1\phi}$ ($\bigcirc$). Note that the amplitude varies by a factor of two among these estimators.

for the 12° slice with declination between 26.5° and 38.5° (Figure 5). Here V_i and V_j are the velocities of two galaxies separated by θ_{ij} on the sky and H_o is the Hubble constant. We make no correction for the r.m.s. pairwise peculiar velocities of ≤ 350 km s^{-1} (Davis and Peebles 1983, de Lapparent, Geller, and Huchra 1988). The calculation of this correlation function is not seriously affected by the presence of the Coma cluster or of other more poorly sampled clusters in the sample.

Because of the large-scale coherent structures in the sample, the weighting scheme for the calculation of the correlation function *does* affect the result substantially. The calculation of the correlation function and the weighting schemes follow prescriptions in Davis and Peebles (1983). It is probable that none of these schemes are unbiased.

One estimator gives equal weight to both galaxies in the pair. Another is to give each galaxy a weight $\phi(V)^{-1}$, where $\phi(V)$ is the selection function for a magnitude limited sample. The selection function $\phi(V)$ is the probability that a galaxy with velocity V will be included in the sample. This weighting corrects for the variations in sampling as a function of redshift; it weights all volumes of space equally. This weighting increases the noise in the determination of $\xi(s)$. An intermediate procedure weights only one galaxy in the pair by $\phi(V)^{-1}$. The resulting correlation functions are ξ_{11}, $\xi_{\phi\phi}$, and $\xi_{1\phi}$, respectively. For these calculations the parameters of the luminosity function are $M^* = -19.15$, $\alpha = -1.2$, and $\phi^* \simeq 0.025 \ h^3 \ Mpc^{-3} \ mag^{-1}$ (de Lapparent, Geller, and Huchra 1988).

The differences among the estimators in Figure 11 are symptoms of the lack of a fair sample. In the absence of biases introduced by large-scale inhomogeneities in the sample, all the estimators should yield the same result to within the statistical noise shown by the error bars at the left of the Figure. The elongated structure at $\sim 10,000 \ km \ s^{-1}$ increases the amplitude of the correlation function for estimates weighted inversely with $\phi(V)$ relative to the unweighted estimate. This coherent structure contains about a thousand galaxies, nearly half of sample. The selection function $\phi(10,000 \ km \ s^{-1}) = 0.01$ $\phi(1,000 \ km \ s^{-1})$. The distribution appears more highly correlated when this structure is more heavily weighted. (Note that in the CfA survey to $m_{B(0)} = 14.5$, the opposite effect occurs because the most dense structure, the local Supercluster is nearby in that sample (see de Lapparent, Geller, and Huchra 1988).

For least squares fits of the data to the standard power law form,

$$\xi(s) = \left(\frac{s_o}{s}\right)^\gamma, \tag{2}$$

in the range $3.5\text{-}9.5h^{-1}$ Mpc we find that the ranges in the slope and amplitude are, respectively, $1.3\text{-}1.9$ and $5\text{-}12h^{-1}$ Mpc. On scales larger than $20h^{-1}$ Mpc the correlation function is indeterminate because the amplitude of the correlation function is comparable with the uncertainty in the mean density. For an average of the estimators in Figure 11, we obtain

$$\gamma = 1.5 \tag{3}$$

and

$$s_o = 7.5h^{-1} \ Mpc. \tag{4}$$

These values are in agreement with the ones which can be measured from Figure 1 of Davis and Peebles (1983) which shows $\xi(s)$ for the 14.5 CfA sample. The frequently quoted smaller scale and steeper slope are derived from $\omega(r_p)$, the correlation function as a function of projected separation. N-body models have generally been normalized to this smaller $r_{go} = 5.4 \pm 0.3h^{-1}$ Mpc (Davis and Peebles 1983, Davis *et al.* 1985). A direct match of $\xi(s)$ to the models is probably a better procedure. This function is free of assumptions about the peculiar velocities which should be the same in the data and in the models.

Calculation of the correlation function $\xi(r_\varrho, \pi)$ (Davis and Peebles 1983), where

$$r_\varrho = \frac{V_i + V_j}{H_o} \tan \frac{\theta_{ij}}{2} \tag{5}$$

and

$$\pi = V_i - V_j, \tag{6}$$

is a method of examining the distortions in redshift space caused by peculiar velocities. The r.m.s. pairwise dispersion for the CfA survey slices is consistent with previous determinations, 400 ± 100 km s^{-1} on a scale of $2h^{-1}$ Mpc. Because the amplitude of the correlation function is small at large scale, this measure is insensitive to large-scale peculiar velocities.

Because of the limited amount of data and the difficulty of the measurements, the relationship between large-scale coherent flows and the structure in redshift surveys is a wide open question. However, coherent flows should, for example, be associated with large low density regions. If the matter density inside a void is low compared with the average surroundings (i.e. if the galaxy density contrast is a measure of the matter density contrast), the voids expand relative to the average cosmological flow and the structures should appear elongated in redshift space. The exact amplitude of the flow depends upon the underlying physics for the formation of the structure (Schwarz, Ostriker, and Yahil 1975, Ikeuchi, Tomisaka and Ostriker 1983, Peebles 1982, Fillmore and Goldreich 1984, Hoffman, Salpeter, and Wasserman 1983, Bertschinger 1985). For an isolated self-similar void the outward peculiar velocity $v_{pec} \simeq 0.3 \, v_H$ where v_H is the radius of the void in redshift space. The effect of interaction between adjacent shells on peculiar velocities has not been calculated. In a large enough sample containing many voids the intrinsic spatial geometry of the voids averages out and any net elongation could be interpreted as a residual expansion.

The measurement of distances to galaxies in the structures offers a direct probe for large-scale flows associated with voids. These flows are a possible discriminant among theoretical models. In the biased cold dark matter models where the matter density contrast is much smaller than the galaxy density contrast, the outflow velocities should be small. The galaxies on the edges of the voids did not move across the low density regions to their current positions; they merely lit up there. Because many spirals lie in the extended sheets, the infra-red Tully-Fisher technique (Aaronson, Huchra, and Mould 1979) can be used to obtain limits at the few hundred kilometers per second level on scales of fifty megaparsecs, within the theoretically predicted range (Geller *et al.* 1988a). The ~ 300 km s^{-1} limit on peculiar velocities on the large shell in Figure 3 implies $v_{pec}/v_H \lesssim 0.2$, in the murky range for interpretation.

The net elongation of voids is one of several statistics which may be useful for comparing the data with models. In the absence of "fair" samples, one can still examine the properties of individual structures. In discussing these structures a "void" is a region where the density is less than the global average and the contrast is below some well-defined threshold; analogously, the "sheets" are regions above a threshold. Both the voids and the sheets can be characterized quantitatively.

The frequent mention of the "size" of the Boötes void is a demonstration of the power of a measure of the scale of the "largest" observed structure. The spectrum of void sizes is an important test of models; the small-scale end is a constraint on hot dark matter models (Zeldovich 1987, Doroshkevich *et al.* 1980, Centrella and Melott 1983, Centrella *et al.* 1988) and the large-scale end is most demanding for cold dark matter models (Davis *et al.* 1985, White *et al.* 1987) and for the explosive models (Ostriker and Cowie 1981, Ikeuchi 1981, Saarinen, Dekel, and Carr 1986). Determination of the distribution of sizes of voids requires samples much larger than those currently available.

Both cold (White *et al.* 1987) and hot (Centrella *et al.* 1988) dark matter models produce large voids and some thin coherent structures. White *et al.* (1987) argue that a standard cold dark matter model with biased galaxy formation produces a distribution of galaxies which is hard to distinguish from the data in Figures 3 - 6 and Figures 9 and 10 (see their Figure 10). Sampling according to the procedure followed by KOSS, White *et al.* (1987) find that 3 of 25 simulations contain a void as large or larger than Boötes (~ 5000 km s^{-1}).

It is not clear whether the simulations can meet the challenge posed by the increasing number of surveys with more dense sampling than the Boötes survey. There are now five surveys large enough to contain ~ 5000 km s^{-1} voids — all of them do. The sharpness of the structures in Figures 3 - 6 and 9 and 10, and the possibility that they imply non-random phase initial conditions motivate consideration of alternatives to the standard gravitational models for large-scale structure formation (Ostriker and Cowie 1981, Ikeuchi 1981, Ostriker, Thompson, and Witten 1986).

The thickness, coherence, and filling factor of the sheets provide further constraints. The FWHM of the sheets is $\lesssim 500$ km s^{-1} (de Lapparent, Geller, and Huchra 1988). The thickness as a function of orientation with respect to the line-of-sight restricts physical models. If the sheets were collapsing pancakes, we would expect them to be thinner when they are perpendicular to the line-of-sight than when they are parallel to it. On the other hand, any internal velocity dispersion will make the sheets appear thicker when they are perpendicular to the line-of-sight.

The fraction of the survey volume filled by the coherent structures in the distribution of individual galaxies can be calculated by appropriately binning and smoothing the data. The galaxies fill $\leq 20\%$ of the volume and the typical separation of galaxies in the sheets is $3h^{-1}$ Mpc at the survey depth D to which M^* galaxies are included. Remarkably this density is comparable with the surface density of the structures in deep probes (Koo, Kron, and Szalay 1986).

The "uniformity" of the sheets may provide a constraint on Ω (see Peebles 1986). If $\Omega = 1$ and the distribution of galaxies marks the distribution of matter, it is unlikely that a smooth shell can persist for a Hubble time; gravity causes the galaxies to clump up and "fingers" in redshift space should be apparent. If the actual matter density contrast in the sheets is small and the voids are full of nearly uniformly distributed dark matter (with Ω close to 1), the structures could still be in the linear regime. If, on the other hand, $\Omega = 0.2$ or less (as indicated by the dynamical estimates and by analysis of the abundance of the light elements), the structure could set in early on and then just stretch with the universal expansion.

Small groups of galaxies are embedded in the sheets apparent in the slice centered at 29.5° (Ramella, Geller, and Huchra 1988). The properties of these groups are similar to the properties of groups in the 14.5 survey; the median line-of-sight velocity dispersion is $\sigma \simeq 220$ km s^{-1}, the typical scale of the systems is 500 kpc, and the median mass-to-light ratio is 420 $M_\odot/L_\odot$. The centers of these groups trace out the coherent structures in the survey.

In contrast, the richest clusters of galaxies appear to occupy interstitial regions between adjacent shells. The Coma cluster (Figure 3) is an obvious example. The survey provides a warning that the objects in cluster catalogs may not always have the properties expected for real physical systems; in other words fingers in redshift space are not always present at the appropriate location. In Figure 4, for example, the finger at 16.5^h is a cluster identified by Zwicky but not by Abell. Abell includes a cluster between 9^h and 10^h at $\sim$ 7500 km s^{-1}, but its location does not correspond to the true finger in redshift space. When the CfA survey is complete, it will include tens of Abell clusters and should provide a first *direct* measure of the relationship between individual galaxies and clusters of galaxies as tracers of the large-scale matter distribution.

3. THE DISTRIBUTION OF RICH CLUSTERS

Once it is known that galaxies cluster, it is natural to ask whether there is higher order structure. Do clusters of galaxies cluster? If so, are clusters of galaxies and individual galaxies equivalent tracers of the large-scale matter distribution in the universe? One important approach to this issue is the extraction of general statistics from cluster catalogs.

The language of correlation functions provides one way of expressing the difference between galaxies and clusters of galaxies as tracers. Bahcall and Soneira (1983) calculated the two-point correlation function for the Hoessel, Gunn and Thuan (1980; HGT hereafter) survey of 104 nearby Abell clusters (distance class $D \leq 4$, richness class $R \geq 1$). Postman, Geller, and Huchra (1986) calculated the cluster correlation function for a variety of other samples drawn from both the Abell (1958) and Zwicky catalogs and Shectman (1985) calculated the cluster correlation function for a sample derived from the Shane-Wirtanen (1967) counts. For all of these samples, the cluster correlation function is consistent with a power law

$$\xi_c(r) \simeq (r_{co}/r)^{1.8}. \tag{7}$$

Both the power law and the amplitude r_{co} are uncertain. The exponents generally appear to be consistent with the value for the galaxy correlation function $\gamma = 1.8$. With the slope constrained to 1.8, the uncertainty in the amplitude r_{co} for a sample of ~ 150 clusters is $\sim 50\%$. For the samples analyzed so far

$$14h^{-1} \text{ Mpc} \lesssim r_{co} \lesssim 24h^{-1} \text{ Mpc}. \tag{8}$$

The largest value of r_{co} was obtained in the Bahcall-Soneira (1983) analysis. Values at the low end of the range were obtained by Sutherland (1988), by Shectman (1985), and by Huchra et al. (1988).

In general the lower values of r_{co} are for samples which contain less rich clusters, but the value is also affected by the details of the analysis (see Postman, Geller, and Huchra 1986 and Sutherland 1988). Bahcall and Soneira (1983) claim a dependence of the amplitude of the correlation function on the "richness" of systems in the catalog. This dependence seems to be confirmed by Shectman (1985) and by Postman, Geller, and Huchra (1986).

Huchra et al. (1988) analyzed a deep sample of 145 clusters with $R \geq 0$ (103 with $R \geq 1$). The survey covers a 561 square degree region at high galactic latitude,

$$58° \leq \delta \leq 78°,$$

and

$$10^h \leq \alpha \leq 15^h.$$

Figure 12 shows the distribution of the clusters on the sky.

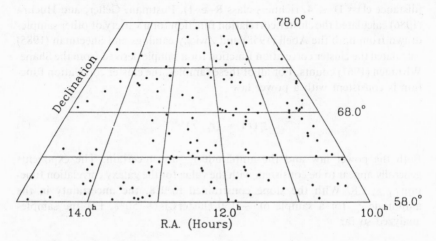

FIG. 12. The distribution on the sky of the 145 Abell clusters in the deep sample of Huchra *et al.* (1988).

The deep sample has little overlap with the HGT sample analyzed by Bahcall and Soneira; 137 of the clusters in the sample have distance class $D \geq 5$. The effective volumes of the two surveys are comparable. The area of the HGT sample is 25 times larger, but the deep survey extends ~ 3 times farther in redshift. The median redshift for the deep survey is 0.17; for the Bahcall-Soneira sample, it is 0.07.

Figure 13 is the cone diagram for the survey. Because clusters are sparse tracers of large-scale structure the apparent voids in the distribution have low significance (see Otto *et al.* 1986).

The amplitude of the correlation function for the $R > 0$ subset of the deep sample is $r_{co} = 17.8h^{-1}$ Mpc (Figure 14). In other words the amplitude of the cluster correlation function for this sample is a factor of ~ 2 less than for the HGT sample; the uncertainty in the amplitude is ~ 40%. The apparent variations in the amplitude of $\xi_c(r)$ are probably not significant. For samples

right ascension

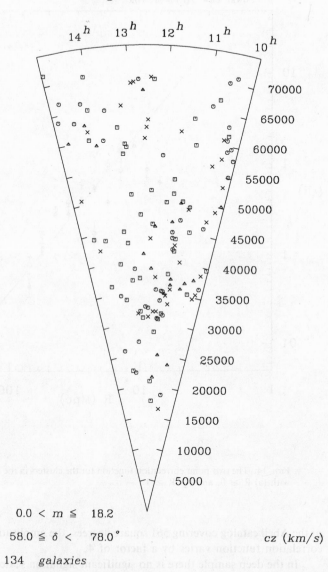

FIG. 13. Observed velocity versus right ascension for the cluster survey strip. Clusters in different declination ranges are plotted as different symbols — box = $58° \leq \delta \leq 63°$, cross = $63° \leq \delta \leq 68°$, circle = $68° \leq \delta \leq 73°$, and triangle = $73° \leq \delta \leq 78°$.

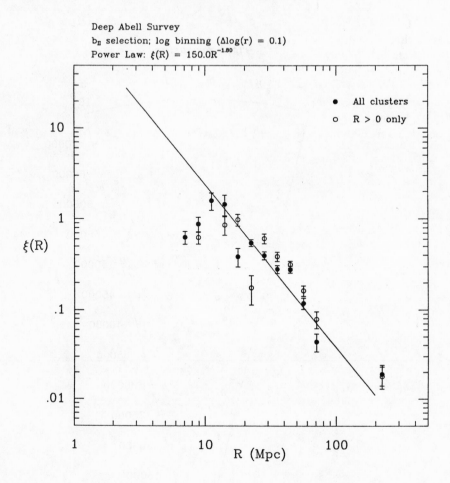

FIG. 14. The two-point correlation function for the clusters in the deep survey
with (a) R $\geq$ 0, and (b) R $\geq$ 1.

of the Abell catalog covering 561 square degrees, the amplitude of the angular
correlation function varies by a factor of 4.

In the deep sample there is no significant signal on scales $\gtrsim 55h^{-1}$ Mpc
(see Figure 14). This limit agrees with Sutherland's (1988) reanalysis of the
Bahcall-Soneira samples and with the analyses by Postman, Geller, and Huchra
(1986). It is inconsistent with claims of significant signal on scales as large
as 150 h^{-1} Mpc (Bahcall and Soneira 1983, Batuski et $al.$ 1988).

Large-scale peculiar velocities are a further issue posed by the surveys of rich clusters. Bahcall and Soneira (1983) examined the distribution of redshift separations for cluster pairs in various ranges of angular separation. At small angular separations, they claim a broadening of the redshift distribution caused by relative pairwise peculiar motions of ~ 2000 km s^{-1} (Bahcall, Soneira, and Burgett 1986). This broadening corresponds to a net elongation of structures along the redshift direction. In the sample of 104 clusters about 30% of the power in the correlation function is contributed by the Corona Borealis supercluster which is elongated along the line-of-sight. It is possible that the elongation reflects the intrinsic geometry of this particular system rather than large-scale peculiar motions. The correlation function approach is only valid when the sample is large enough for the statistics to represent an average over many systems.

Large-scale peculiar motions are not supported by the analysis of the Huchra *et al.* (1988) deep sample. For this sample, relative pairwise peculiar motions are ≤ 1000 km s^{-1}. There are several possible explanations for the different results. One is the effect of a single system like Corona Borealis on the average statistics. Another is the greater difficulty of separating close pairs of clusters in the deeper sample; in other words, if Corona Borealis were present in the deeper sample, a few of the members might not be counted as separate clusters. Undercounting of close pairs could lead to underestimation of r_{co}.

Several caveats are important in consideration of this and other cluster surveys. The redshift measurements for the Huchra *et al.* (1988) sample demonstrate that a large fraction of the clusters are complex systems; many consist of several lesser agglomerations projected along the line-of-sight. Of the 90 clusters in the sample with multiple redshifts, 40% (36 clusters) show multiple redshift systems or significant (≥ 5000 km s^{-1}) extent in redshift space. In these cases, the redshifts in Figure 13 are a best guess of the Abell "cluster" redshift. The problems of foreground and background contamination are significant for all richness and distance classes in the sample. Indeed, in a few cases, it is impossible to identify any cluster at all.

4. CONCLUSIONS

Analyses of galaxy and cluster redshift surveys raise a number of important issues. In both cases, the samples are probably not large enough to be representative. In the case of galaxy surveys, the largest well-defined structures are comparable with the extent of the survey. In the case of cluster surveys single systems could dominate the average statistics.

Some have argued (see e.g. da Costa *et al.* 1988) that the large-scale galaxy distribution is heterogeneous. In other words, the distribution could contain filaments, sheets, voids and diffuse structures. Until the selection criteria for different surveys can be carefully compared, it will not be clear whether this conclusion is a function of sampling or of genuinely variable structure.

A general description of large-scale structure also depends upon having surveys which are large enough to be representative. Deeper surveys extending over large angular scales are important for enabling the identification of structures larger than those which have already been observed; the largest structures provide one of the tightest constraints on models. At least two surveys are underway to meet this goal: one in the North (Geller *et al.* 1988b) and one in the South (KOSS 1988). Both of these surveys reach to a limiting apparent magnitude $m_{B(0)} = 17.5$ and span $\sim 100°$ across the sky.

The largest cluster redshift surveys are limited to ~ 150 objects. The sampling fluctuations alone lead to variations of a factor of ~ 4 in the amplitude of the correlation function. Perhaps even more serious, cluster catalogs are subject to a large number of poorly understood selection biases. Not the least of these is the problem of superpositions along the line-of-sight.

Further progress in mapping the large-scale structure of the universe requires reliable photometric catalogs. Even at the depth of the Zwicky catalog, there is no uniform survey of the sky. Systematic variations in the magnitudes from one region to another in a single catalog (not to mention variations from one catalog to another) almost surely compromise detailed analyses of the properties of the structures. The advent of large format CCD's has made fundamentally important digital surveys possible. An intermediate depth (m $\leq 20^{th}$ magnitude), multicolor survey should be the next major step in the study of large-scale structures.

ACKNOWLEDGEMENTS

We would like to thank Pat Henry for data in advance of publication. We also thank Valérie de Lapparent and Marc Postman for several of the plots displayed here. This work is supported by NASA Grant NAGW-201 and by the Smithsonian Institution. Results reported here are based partly on observations obtained with the Multiple Mirror Telescope, a joint facility of the Smithsonian Institution and the University of Arizona.

REFERENCES

Aaronson, M., Huchra, J.P., and Mould, J. 1979. *Ap J.* **229**, 1.

Abell, G.O. 1958. *Ap J Suppl.* **3**, 211.

Bahcall, N.A. and Soneira, R.M. 1983. *Ap J.* **270**, 20.

Bahcall, N.A., Soneira, R.M., and Burgett, W.S. 1986. *Ap J.* **311**, 15.

Batuski, D.J. *et al.* 1988. preprint.

Bean, A.J., Efstathiou, G., Ellis, R.S., Peterson, B.A., and Shanks, T. 1983. *MN.* **205**, 605.

Bertschinger, E. 1985. *Ap J.* **295**, 1.

Bothun, G.D., Beers, T.C., Mould, J., and Huchra, J.P. 1986. *Ap J.* **308**, 510.

Centrella, J.M. and Melott, A.S. 1983. *Nature.* **305**, 196.

Centrella, J.M., Gallagher, J.S., Melott, A.S., and Bushouse, H.A. 1988. preprint.

da Costa, L.N., Pellegrini, P.S., Sargent, W.L.W., Tonry, J., Davis, M., Meiksin, A., and Latham, D.W. 1988. *Ap J.* **327**, 544.

Davis, M. and Djorgovski, S. 1985. *Ap J Letters.* **299**, L15.

Davis, M. Efstathiou, G., Frenk, C. and White, S.D.M. 1985. *Ap J.* **292**, 371.

Davis, M. and Huchra, J.P. 1982. *Ap J.* **254**, 437.

Davis, M. and Peebles, P.J.E. 1983. *Ap J.* **267**, 465.

Davis, M., Huchra, J.P., Latham, D.W., and Tonry, J. 1982. *Ap J.* **253**, 423.

Dekel, A. and Silk, J. 1986. *Ap J.* **303**, 39.

de Lapparent, V., Geller, M.J., and Huchra, J.P. 1986. *Ap J Letters.* **202**, L1. (dLGH)

_____. 1988. *Ap J.* **332**, 44.

Doroshkevich, A.G., Kotok, E.V., Novikov, I.D., Polyudiv, A.N., Shandarin, S.F., and Sigov, Yu.S. 1980. *MN.* **192**, 321.

Dressler, A. 1980. *Ap J.* **236**, 351.

_____. 1984. *Ann Rev Astron Astrophys.* **22**, 185.

Einasto, J., Jôeveer, M., and Saar, E. 1980. *MN.* **193**, 353.

Fillmore, J.A. and Goldreich, P. 1984. *Ap J.* **281**, 9.

Geller, M.J. *et al.* 1988a. in preparation.

Geller, M.J. *et al.* 1988b. in preparation.

Giovanelli, R. and Haynes, M.P. 1988. in *Large Scale Structures of the Universe.* eds. J. Audouze, M.-C. Pelletan, and A. Szalay, p. 113. Dordrecht: Kluwer Academic Publishers.

Giovanelli, R., Haynes, M.P., and Chincarini, G. 1986. *Ap J.* **300**, 77.

Gott, J.R., Melott, A. and Dickinson, M. 1986. *Ap J.* **306**, 341.

Groth, E.J. and Peebles, P.J.E. 1977. *Ap J.* **217**, 385.

Hamilton, A.J.S., Gott, R., and Weinberg, D. 1987. *Ap J.* **309**, 1.

Haynes, M.P. and Giovanelli, R. 1986. *Ap J Letters.* **306**, L55.

Hoessel, J.G., Gunn, J.E., and Thuan, T.X. 1980. *Ap J.* **241**, 486. (HGT)

Hoffman, G.L., Salpeter, E.E., and Wasserman, I. 1983. *Ap J.* **268**, 527.

Hubble, E. and Humason, M.L. 1931. *Ap J.* **74**, 43.

Huchra, J.P. and Geller, M.J. 1988. in preparation.

Huchra, J.P., Davis, M., Latham, D.W., and Tonry, J. 1983. *Ap J Suppl.* **52**, 89.

Huchra, J.P., Henry, P., Postman, M., and Geller, M.J. 1988. in preparation.

Ikeuchi, S. 1981. *PASJ.* **33**, 211.

Ikeuchi, S., Tomisaka, K., and Ostriker, J.P. 1983. *Ap J.* **265**, 538.

Jôeveer, M., Einasto, J., and Tago, E. 1978. *MN.* **185**, 357.

Kaiser, N. 1987. *MN.* **227**, 1.

Kirshner, R.P., Oemler, A., and Schechter, P. 1979. *AJ.* **84**, 951.

Kirshner, R.P., Oemler, A., Schechter, P.L., and Shectman, S.A. 1981. *Ap J Letters.* **248**, L57. (KOSS 1981)

Kirshner, R.P., Oemler, A. Schechter, P.L., and Shectman, S.A. 1983. *AJ.* **88**, 1285. (KOSS 1983)

Kirshner, R.P., Oemler, A., Schechter, P.L., and Shectman, S.A. 1987. *Ap J.* **314**, 493. (KOSS 1987)

Kirshner, R.P., Oemler, A., Schechter, P.L., and Shectman, S.A. 1988, private communication. (KOSS 1988)

Koo, D. Kron, R. and Szalay, A. 1986. in *13ᵗʰ Texas Symposium on Relativistic Astrophysics.* ed. M. Ulmer. Singapore: World Scientific.

Nilson, P. 1973. *Uppsala General Catalogue of Galaxies, Uppsala Astr. Obs. Ann.* **6**.

Ostriker, J.P. and Cowie, L.L. 1981. *Ap J Letters.* **243**, L127.

Ostriker, J.P., Thompson, C., and Witten, E. 1986. *Phys Lett B.* **180**, 231.

Otto, S., Politzer, D., Preskill, J., and Wise, M. 1986. *Ap J.* **304**, 62.

Peebles, P.J.E. 1982. *Ap J.* **257**, 438.

_____. 1980. *Large-Scale Structure in the Universe.* Princeton: Princeton University Press.

Peterson, B.A., Ellis, R.S., Efstathiou, G., Shanks, T., Bean, A.J., Fong, R. and Zen-Long, Z. 1986. *MN.* **221**, 233.

Postman, M. and Geller, M.J. 1984. *Ap J.* **291**, 85.

Postman, M., Geller, M.J., and Huchra, J.P. 1986. *AJ.* **91**, 1267.

Postman, M., Huchra, J.P., and Geller, M.J. 1986. *AJ.* **92**, 1238.

Ramella, M., Geller, M.J., and Huchra, J.P. 1988, in preparation.

Saarinen, S. Dekel, A., and Carr, B.J. 1986. preprint.

Schechter, P.L. 1976. *Ap J.* **203**, 297.

Schwarz, J., Ostriker, J.P., and Yahil, A. 1985. *Società Italiana di Fisica.* **1**, 157.

Shane, C.D. and Wirtanen, C.A. 1967. *Publ Lick Obs.* **Vol. XXII**, Part 1.

Shanks, T., Bean, A.J., Efstathiou, G., Ellis, R.S., Fong, R., and Peterson, B.A. 1983. *Ap J.* **274**, 529.

Shectman, S.A. 1985. *Ap J Suppl.* **57,** 77.

Smith, B., Kleinmann, S., Huchra, J. and Low, F. 1987. *Ap J.* **318,** 161.

Strauss, M. and Huchra, J. 1988. *AJ.* **95,** 1602.

Sutherland, W. 1988. preprint.

Thuan, T.X., Gott, J.R., and Schneider, S.E. 1987. *Ap J Letters.* **315,** L93.

White, S.D.M., Frenk, C.S., Davis, M., and Efstathiou, G. 1987. *Ap J.* **313,** 505.

Zeldovich, Ya. B. 1970. *AA.* **5,** 84.

Zwicky, F., Herzog, W., Wild, P., Karpowicz, M. and Kowal, C. 1961-1968. *Catalog of Galaxies and of Clusters of Galaxies.* Pasadena: California Institute of Technology.

LARGE-SCALE STRUCTURE IN THE LOCAL UNIVERSE: THE PISCES-PERSEUS SUPERCLUSTER

MARTHA P. HAYNES

National Astronomy and Ionosphere Center and Department of Astronomy, Cornell University

and

RICCARDO GIOVANELLI

National Astronomy and Ionosphere Center

1. INTRODUCTION

At the time of the Study Week on *Astrophysical Cosmology* in 1981 the inhomogeneity of the large-scale structure and the lack of some obvious preferred scale were already well known. Oort (1982) presented the results providing evidence at that time that coherent features with dimensions on the order of $70h^{-1}$ Mpc exist and that a rich variety of structures characterized the galaxy distribution: cells, filaments, voids. In the intervening years, further terminology has come into common usage in the taxonomy of the galaxy distribution: fractals, sheets, strings, bubbles, swiss cheese, meatballs and sponges, in addition to those already mentioned. Some topologies are the direct result of model predictions while others cannot be uniquely represented.

From the observational perspective of the question of large-scale structure, the 1980's has been the decade of an explosion in the number of extragalactic redshift measurements. Technological advances made primarily in detectors and spectrometers available both to optical and radio astronomers have increased dramatically the efficiency of measuring accurate redshifts for galaxies of moderate brightness. The acquisition of redshift information for a large number of galaxies now allows us to investigate the three dimensional nature of the local universe. In some regions of the sky, redshift surveys are sufficiently complete that we can begin to study coherent structures as individual units. Questions of characteristic scales and density contrasts, of environmental alteration and initial conditions and of Hubble flow and peculiar motions can be addressed to specific locations within the galaxy distribution.

In this review, we shall discuss what is currently known about one of the major large-scale features seen in the northern sky: the Pisces-Perseus (P-P) supercluster. Located at a characteristic distance corresponding to cz = +5000 km s^{-1}, the main ridge of P-P appears as a linear structure at least 45h^{-1} Mpc in length and includes the rich clusters A 262, A 347, and A 426, as well as lower density enhancements in Pegasus and around NGC 383 and NGC 507. At the present time, redshifts have been measured for some 5000 galaxies in the region occupied by the supercluster. A great deal of study remains to be made of the supercluster, but we have begun to use it as a laboratory for understanding the dynamics and structure of the galaxy distribution on large scales. Because of its relative proximity, observational data can be obtained for large numbers of supercluster members at a variety of wavelengths. Surely, P-P may not be a representative environment within the local universe. Rather it allows us to probe in depth the range of densities characteristic of local volumes.

2. THE PISCES-PERSEUS REGION

The P-P supercluster is evident clearly in maps of the brightest galaxies such as those catalogued by Shapley and Ames (1932). Even though most galaxies of similar apparent brightness are significantly closer, the main ridge of the supercluster stands out among the distribution of NGC galaxies and thus Bernheimer (1932), among others, was able to recognize its "metagalactic" nature. Figure 1a shows the distribution of the 235 galaxies brighter than B$_T$ = 13.2 mag included in the *Revised Shapley-Ames Catalog* (Sandage and Tammann 1981; RSA) on a equal area projection of the portion of sky that we shall refer to as the "P-P region", that is, the area bounded by 22^h < R.A. < 04^h, 0° < Dec. < +50°. In the lower panel Figure 1b, only those 88 with heliocentric recessional velocities V$_0$ in the range +4500 < V$_0$ < +6000 km s^{-1}, characteristic of the P-P supercluster, are included. The main ridge of the supercluster includes the diagonal string of galaxies betweem 0^h 30^m and 2^h 30^m R.A. and +20° to +40° Dec.

With the acquisition of small numbers of redshifts, the reality of the connectivity of the structure became apparent. Gregory *et al.* (1981) noticed that most of the clusters in the region were located at the same distance and hence belonged to a single supercluster of approximately the same depth as width. Einasto *et al.* (1980) published maps suggesting connective structures in the region deduced from an examination of the distribution of galaxies and clusters in the *Catalog of Galaxies and Clusters of Galaxies* (Zwicky *et al.* 1961-8; CGCG). Although relatively few redshifts were available for galaxies in the

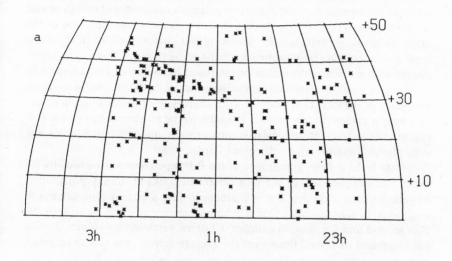

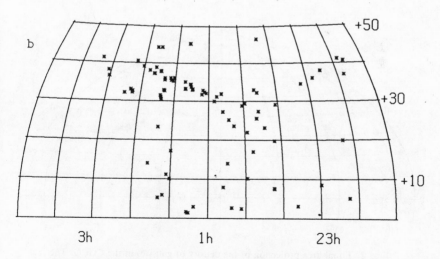

FIG. 1. (a) Equal area projection of the distribution of 273 galaxies brighter than $B_T = 13.2$ included in the RSA in the region of the P-P supercluster; (b) Distribution of the 88 galaxies shown in (a) with redshifts in the range $+4500 < V_o < +6000$ km s^{-1}.

supercluster at that time, the latter authors recognized much of the complexity of the large-scale structure.

Figure 2, reproduced from Giovanelli and Haynes (1982), illustrates the surface density distribution of galaxies in one quarter of the sky on an Aitoff equal area projection. The shade intensity (on an eight step scale) is proportional to the logarithm of the density of galaxies contained within cells of one square degree. The P-P ridge is clearly visible in the western portion of the map; by invoking extragalactic zoomorphism, we sometimes refer to the supercluster structure revealed by such maps as the "lizard". Via this representation, one sees clearly the limitations imposed by the zone of avoidance (the blank region running diagonally across the center of the map), and the southern boundary of the CGCG at Dec. = —2.5 degrees. Heavy extinction in excess of 1 mag is evident at low galactic latitudes to the north and east of the P-P region; the effects of the galactic obscuration on our estimates of the supercluster's extent will be discussed later.

How is the distinct appearance of the P-P supercluster influenced by the depth and completeness of the CGCG? If one looks at the sky distribution of galaxies from a catalog that is characterized by a certain completeness in magnitude or size, one may expect inhomogeneities to stand out against the background and foreground galaxies. Clusters, representing volume density enhancements of several times over the average density, are easy to see, even in maps of deep catalogs. Shallower density enhancements will appear more

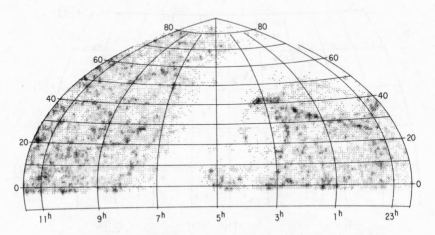

FIG. 2. Equal area projection of the density of galaxies in the CGCG. The shade intensity is proportional to the number of galaxies in each 1° by 1° cell. The blank region cutting across the map is the zone of avoidance. The catalog has a southern boundary at about Dec. = − 3°. (Giovanelli, Haynes, and Chincarini 1986).

or less conspicuous in a surface density map, depending not only on the volume density contrast of the enhancement and its extent along the line of sight, but also on the relative size of its angular extent vis-a-vis that of the sampled region and, especially, on its distance from the observer and the depth of the catalog in question. In a volume — or absolute magnitude — limited catalog, i.e. one that contains all galaxies brighter than a given luminosity limit, out to a given distance r_{lim}, a density enhancement of $1 + \Delta\varrho/\langle\varrho\rangle$ will yield the same surface density contrast $1 + (\Delta\varrho/\langle\varrho\rangle)(\Delta r/r_{lim})$ where Δr is the line of sight extent of the enhancement, independent of its distance from the observer. In an absolute magnitude-limited sample, on the other hand, a homogeneously distributed population of galaxies at that redshift yields a peaked redshift distribution, i.e., a subsample of galaxies at that redshift yelds the highest surface density counts. In this case, the peak of the distribution depends on the catalog limit and on the characteristics of the luminosity function. A density enhancement will therefore be more conspicious in a sky map if it is located at a redshift near the peak. For example, a Gaussian density enhancement of a full width at half maximum of 400 km s^{-1} and a volume overdensity of 90, centered at $cz = +5000$ km s^{-1} will produce a surface density enhancement of about 10 in the counts of CGCG galaxies. On the other hand, a similar overdensity would be practically washed out in the Lick galaxy counts. In order to produce a surface density enhancement comparable to that produced in the CGCG, a given volume density enhancement should be three times deeper and three times further away (Giovanelli *et al.* 1986).

For any magnitude-limited sample we can estimate the redshift distribution one would have expected from a homogeneously distributed, well-behaved population of galaxies (i.e. one that adheres to a universal luminosity function). In general, we can write the number of galaxies with apparent magnitude between m and m + dm, located in the distance interval (r, r + dr), within a solid angle $\Delta\Omega$ as

$$n(m,r) \, dr \, dm\Delta\Omega = D(r) \, \Phi \, [m - 5 \log(r/10)] \, \Delta\Omega r^2 \, dr \, dm \qquad (1)$$

where $D(r)$ is the space density of galaxies and we adopt a Schechter (1976) luminosity function with the parameters $M^* = -20.6$ and $\alpha = -1.25$ (appropriate for blue magnitudes and $h = 1/2$). The number counts integrated for all r for any complete magnitude limited sample are given by

$$n(m) \, dm = n_0 \int_0^\infty \Phi \, [m - 5\log(r/10)] \, D(r) \, r^2 \, dr \, dm. \qquad (2)$$

Substituting $r = v/H$ to obtain a radial velocity distribution, where v is the radial velocity, the expected distribution for each magnitude interval is

$$N(m,v) \, dv \, dm \, \Delta\Omega = D(v/H) \, \Phi \, [(m - 5\log(v/10H)] \, \Delta\Omega \, (v/H)^2 d \, (v/H) \, dm. \qquad (3)$$

The expected velocity distribution is then obtained simply by integrating

$$N(v) \ dv \ = \ dv \int N(v,m) \ dm. \tag{4}$$

One can similarly show that the distribution of a sample complete to a magnitude limit m_{lim} peaks at a distance, corresponding to a velocity v^*, that is obtained from the relation

$$v^* \ = \ 5.248 \ dex \ (0.2 \ m_{lim}). \tag{5}$$

Thus, for a limiting magnitude of $m_{lim} = +14.9$, the redshift distribution is expected to peak at $v^* = +5000$ km s^{-1}. The CGCG is supposed to be roughly complete to a magnitude limit of $+15.7$. At the faintest magnitudes, errors in the magnitude measurements increase, and some systematic errors in the magnitude scale become difficult to correct for galaxies fainter than about $m = +15.4$ (Giovanelli and Haynes 1984). However, if a universal luminosity function applies to all galaxies, then the CGCG best traces volume density enhancements that have characteristic sizes on the order of 5 to 20 Mpc if they are located at a redshift of about 5000 to 7000 km s^{-1}. The distinctiveness of such features is not an artifact of the CGCG, but rather the CGCG will best illustrate such overdensities if they exist.

Other features are also apparent in similar displays of the other quarter of the sky also contained in the CGCG (Fontanelli 1984; de Lapparent *et al.* 1986; Bicay 1987). The Coma-A 1367 and Hercules superclusters are also apparent, although with differing clarity because they both lie at larger distances. In some locations, the observed increase in the surface density is an overstimate because of the chance projection of overlapping structures at different distances. For many purposes of identifying structures, the CGCG has been a useful indicator of obvious large-scale structure in the local universe, and the P-P supercluster is perhaps the most obvious large-scale structure outside the Local Supercluster visible in the CGCG maps.

3. The HI Line Survey

Redshift surveys have been approached from a number of perspectives, dependent partly on the scientific objectives and partly on the availability of telescopes and telescope time. Over the last decade, we and our co-workers (Giovanelli and Haynes 1984; Giovanelli, Haynes and Chincarini 1986; Giovanelli *et al.* 1986; Haynes *et al.* 1988a) have been conducting a survey of spiral galaxies in the P-P region primarily using the 21 cm line of HI as the carrier of redshift information. Twenty years ago 21 cm line emission had been measured in about 140 galaxies (Roberts 1969). Today, HI spectra exist for some 6000 objects. From the initial observations conducted as part of our P-P survey a decade ago to today, our efficiency of measuring redshifts for galaxies has increased by about a factor of five, so that although we are today observing galaxies that are smaller and fainter than the targets of the 1970's, our detection rate remains essentially unchanged for the same observing time. Of historical interest, one should keep in mind that prior to 1975, Arecibo did not operate at 21 cm. In the intervening years, the resurfacing of the antenna surface, reduction of the noise temperature of the receiver, remote tuning of the feed and construction of a wide-bandwidth 1.6 bit correlator have produced a system whereby redshifts are measured for some 1500-2000 galaxies per year even though the telescope is used only 15% of the time for such observations.

The survey we are currently undertaking aims to measure redshifts for a sample of galaxies throughout the P-P region that is complete in magnitude to m = +15.7, the CGCG limit, and in angular size to a = 1.0 arcmin, the limit of the *Uppsala General Catalog* (Nilson 1973; UGC). Three fourths of the catalogued galaxies in the region are spirals and as such are good targets for 21 cm emission searches. Objects at declinations south of Dec. = +38° have been observed with the Arecibo 305 m telescope, while those to the north have been surveyed with the 91 m antenna of the National Radio Astronomy Observatory at Green Bank. The total redshift sample includes almost 5000 measurements of which about 3300 were obtained from 21 cm spectra. The completion percentages vary somewhat over the P-P region and are lower in the northern — and southern — most portions as a result of lowered sensitivity at high zenith angles with the current feed system available at Arecibo and the lower sensitivity and limited tracking at Green Bank. At Arecibo the 21 cm observations are normally conducted over a frequency range corresponding to $v < +15000$ kms^{-1}; searches at higher velocities are presently marred by man-made interference, a circumstance that planned telescope upgrading should alleviate. At Green Bank, the typical search is not quite so deep: $v < +13000$ km s^{-1}. It is likely that most 21 cm non-detections in fact lie at redshifts beyond the search range. In addition, the 21 cm line observations are restricted to HI rich objects, galaxies with morphologies of S0a and later, so that the extension of the survey to earlier types depends on optical redshift measurements. It should be noted that we have obtained morphological classifications and have measured angular sizes of all CGCG galaxies in the P-P region that are not included in the UGC. Such measurements were performed on glass reproductions of the Palomar Sky Survey blue plates. Fortunately, the morphological segregation that is responsible for the more pronounced clustering of early-type objects and their characteristically higher optical surface brightness make E's and S0's favorable targets of optical observers. The combination of optical and HI samples is thus complementary and yields an evenly-represented sample.

To illustrate the current limitations that are characteristic of the survey, Figure 3 illustrates the current completeness statistics for the survey. Figure 3a shows the completeness percentages for galaxies contained in the CGCG across the P-P region with $+21°30' <$ Dec. $< +33°30'$. Of the 1460 redshift measurements available in this strip, 70% were obtained from 21 cm emission spectra. Completion here is defined for each bin of 0.1 mag as the ratio of the number of galaxies in the strip that are listed in the CGCG, and is plotted separately for all types (triangles) and for spirals (open circles). When all types are included, completion is better than 50% for all bins except the faintest m = +15.7. The current survey covering the entire P-P region achieves that level of completion for magnitudes of m = +15.5 and brighter. Figure 3b shows a similar diagram with declination being the tracked variable; separate indications are given for samples complete to the indicated apparent magnitudes. Note that the completeness is high for declinations between about $+3° <$ Dec. $< +33°$, the region of high sensitivity for the Arecibo telescope.

The population ratios E:S0:Sp of the redshift sample included in Figure 3 are 6:11:83, marginally more spiral-rich than those found for the RSA sample of bright field galaxies. Complementary optical observations and efforts by other groups are helping to reach completion of the early-type populations.

For any well-designed redshift program, one should be able to estimate a "completeness function" and then employ that correction function to estimate the redshift distribution that one would have expected from a random population of similar objects. In the case of magnitude limited samples, we can follow the derivation of the expected redshift distribution given for a homogeneous distribution of galaxies characterized by a Schechter luminosity function $\Phi(M)$. The completeness function c(m) then is simply the ratio of the observed apparent magnitude distribution for the sample $n_s(m)$ dm and that of a complete sample n(m) dm. Under the assumption that, within each bin of apparent magnitude, the galaxies for which redshifts have not been measured are not distinguished by any other characteristic, then we can estimate the expected redshift distribution for the incomplete observation sample by computing.

$$N(v) \; dv = dv \int_0^\infty N(v,m) \; c(m) \; dm. \tag{6}$$

For the simplest models, the counts of galaxies are expected to grow with apparent magnitude as dex (0.6 mag). In fact, simulations using a clustered distribution over a large area indicate that the irregularities are smoothed out, and the dex (0.6 mag) law is preserved (Giovanelli et al. 1986). A serious problem in the northern and eastern portions of the P-P region, however, arises because of the encroaching galactic extinction. The effect of Δm magnitudes of galactic obscuration on the observed number counts for a catalog complete

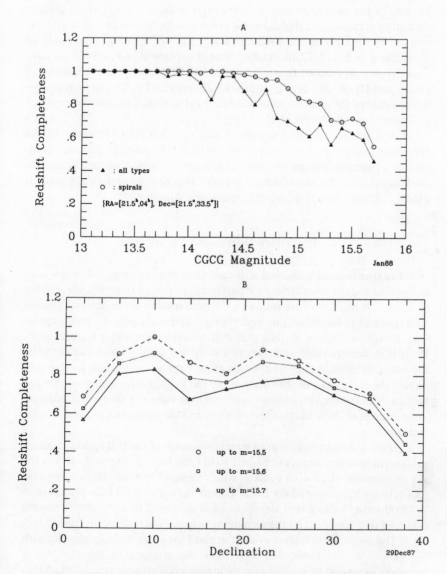

FIG. 3. Panel (a) shows the completeness fraction in 0.1 magnitude bins of the redshift sample within a declination strip across the P-P region with +21°30′ < Dec. < +33°30′. Shown separately is the completeness for all types (triangles) and spirals only (open hexagons). Panel (b) shows the completeness as a function of declination, throughout the P-P region, with separate symbols for galaxy samples complete to the indicated limiting magnitude.

to a limiting magnitude m_{lim} is to dim the galaxies in each interval of apparent magnitude (m_{lim}- m, m_{lim}) beyond the limit of the catalog. In performing analysis of the surface density distribution in selected regions where galactic extinction is important, this dimming effect can be treated to a first approximation by multiplying the counts in a given bin of apparent magnitude by the factor g = dex (0.6 Δm). As discussed in Giovanelli *et al.* (1986), we have used the prescriptions of Burstein and Heiles (1978) and a tape tabulation of galaxy and HI column density counts kindly provided by D. Burstein to construct a map of the galactic extinction in the P-P region. That map is presented as Figure 2 of Giovanelli *et al.* (1986).

For a diameter-limited sample, a size function S(L) analogous to the Schechter luminosity function can be derived. The angular diameter completeness function c(a) can be obtained in a similar fashion to compensate for the limitations in the observational sample. This complementary approach is used to tighten the inferences obtained using equation (6).

4. THE THREE-DIMENSIONAL STRUCTURE

The true three-dimensional structure in such a region as P-P is not easy to describe and the boundaries of separate dynamical units are not always easy to deduce. A number of techniques have been tried to aid in determining the structure and in identifying its underlying preferred scales. A visual impression, though subjective, is quite useful in conveying the major features of P-P: its high density ridge separates relatively low density holes and connects the major dynamical units all the way from Pegasus to Perseus. In trying to picture the complexity of the structures in the P-P supercluster region, we use both two-dimensional projections over restricted ranges of the third coordinate in an attempt to show the thinness of walls and the connectivity of the longest features.

Figure 4 shows a selection of equal area maps of the P-P region in which galaxies in specific ranges of redshift have been plotted. Figure 4a shows the sky distribution of all 4166 galaxies with measured redshifts that correspond to a velocity V_0, corrected for Local Group motion (300 sin l cos b), less than + 12000 km s^{-1}. Note that the impression conveyed by the redshift sample is indeed that conveyed by the galaxies in selected intervals of increasing redshift. The supercluster is most evident in panel (d) which includes galaxies with + 4500 < V_0 < + 6000 km s^{-1}. Note the presence of Perseus cluster members in several panels because their measured velocity typically includes a large component of dynamical distortion of a smooth Hubble flow.

Figure 5 and 6 show a selection of cone diagrams for galaxies located in separate slices in declination (Figure 5) and right ascension (Figure 6). The size of each wedge is 10° in Figure 5 and 30^m in Figure 6. While stretching

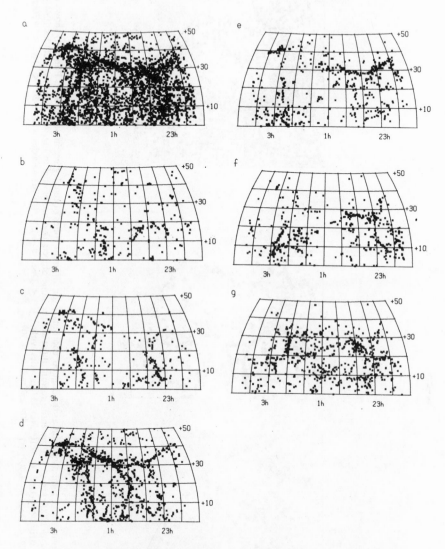

FIG. 4. Sky distribution of galaxies of known redshift on an equal area projection. Panel (a) shows all 4166 galaxies with $V_0 < +12000$ km s^{-1}; (b) 325 galaxies with $V_0 < +3000$ km s^{-1}; (c) 412 galaxies with $+3000 < V_0 < +4500$ km s^{-1}; (d) 1305 galaxies with $+4500 < V_0 < +6000$ km s^{-1}; (e) 732 galaxies with $+6000 < V_0 < +7500$ km s^{-1}; (f) 584 galaxies with $+7500 < V_0 < +9000$ km s^{-1}; and (g) 798 galaxies with $+9000 < V_0 < +12000$ km s^{-1}.

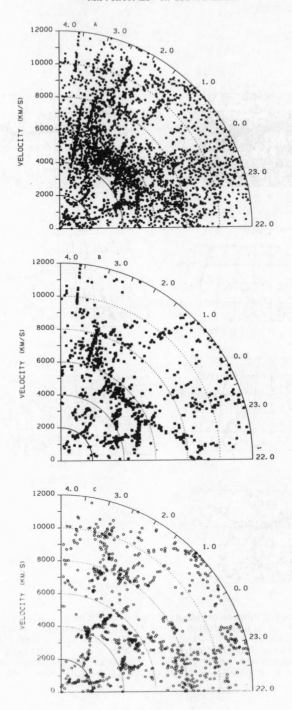

FIG. 5. Cone diagrams with right ascension as the angular coordinate for different zones of ten degrees in declination. Panel (a) shows all 4176 galaxies with $V_0 < +12000$ km s^{-1} in the entire region; these are the same objects that are plotted in Fig. 4a. Panel (b) includes 1044 galaxies in the declination strip $0° <$ Dec. $< +10°$; (c) 937 galaxies with $+10° <$ Dec. $< +20°$; (d) 963 galaxies with $+20° <$ Dec. $< +30°$; (e) 847 galaxies with $+30° <$ Dec. $< +40°$; (f) $+40° <$ Dec. $< +50°$.

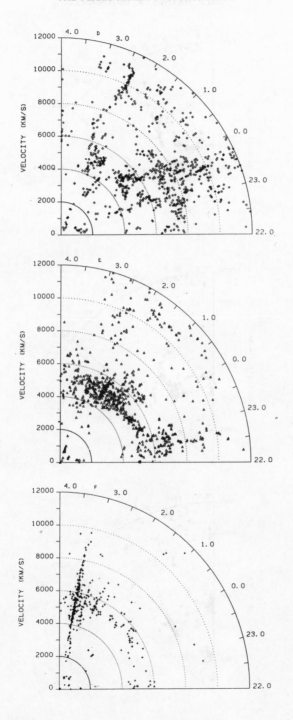

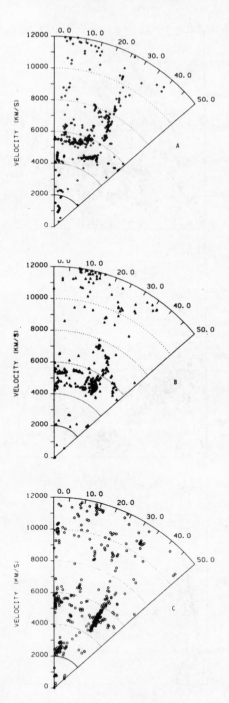

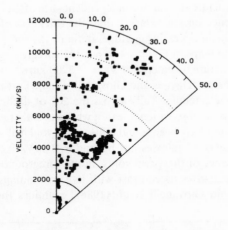

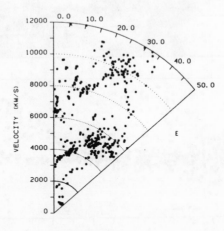

Fig. 6. Cone diagrams with declination as the angular coordinate for different 30 minute slices in right ascension. Panel (a) includes 351 galaxies with 0^h < R.A. < 0^h30^m; (b) 301 galaxies with 0^h30^m < R.A. < 1^h; (c) 384 galaxies with 1^h < R.A. < 1^h30^m; (d) 402 galaxies with 1^h30^m < R.A. < 2^h; (e) 412 galaxies with 2^h < R.A. < 2.5^h.

along the line of sight is seen in the direction of clusters, such as the Perseus cluster in Figure 5f, in many of the other plots "strings" of galaxies that trace what appear to be narrow structures do not seem to arise from cluster dynamics. Moreover, the structure seen in one independent slice often is seen in the next, indicating coherence on scales larger than the thickness of a single wedge.

Besides the main supercluster ridge, perhaps the most prominent other feature seen in the maps in Figures 4 to 6 is the foreground void between us and P-P. This structure is centered at $V_0 = +3200$ km s^{-1} and covers at least 15° in right ascension (Haynes and Giovanelli 1986). Furthermore, it is best described as a tube extending fully across the 50° of declination (Haynes *et al.* 1988a). There is no observational reason whatsoever that galaxies of average luminosity would not have been found to lie in this void had such existed there. Note also the absence of galaxies just beyond P-P at $V_0 = +6000$ to $+7000$ km s^{-1}. The presence of these two voids, in the foreground and background of P-P, further enhances its contrast with its surroundings.

Figure 7, from Giovanelli *et al.* (1986), examines the distribution of

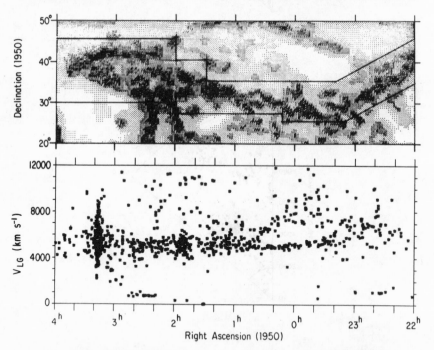

FIG. 7. Redshift distribution along main supercluster ridge. The upper panel presents a shade diagram where the shade intensity indicates the local density distribution along the ridge. In the bottom panel, individual redshift measurements are shown as a function of right ascension for galaxies within the highest density enhancement outlined in the upper panel (from Giovanelli *et al.* 1986).

measured velocities for galaxies contained within the highest density ridge of the P-P supercluster, the lizard's tail, back and head. The upper panel is a shade display in which the shade intensity is proportional to a quantity that measures the local density of galaxies. The sampled high density regime is outlined. In the lower panel the measured velocity is shown at the location in R.A. of each galaxy in the outlined region in the upper panel. It is clear that the main ridge is characterized both by a small width on the sky and depth in the radial direction. The velocity confinement into narrow lanes connecting the larger velocity dispersion clusters is striking. Because systematic velocity gradients across the ridge have not been removed in constructing Figure 7, the observed histogram at any narrow interval of right ascension would appear markedly sharper. Velocity broadening is also contributed by the large velocity dispersions found in clusters. Figure 8 attempts to show the contrast between the two portions of the supercluster, the northern, high density ridge, and the southern, more dispersed component. On a single cone diagram, with right ascension as the angular coordinate, galaxies in two adjacent declination strips are indicated by separate symbols. Galaxies found in the higher den-

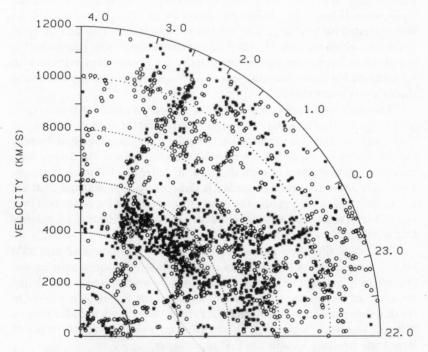

FIG. 8. Cone diagram with right ascension as the angular coordinate for galaxies in the P-P region with $+25°$ < Dec. < $+35°$ (asterisks) and $+15°$ < Dec. < $+25°$ (open circles).

sity portions to the north with $+25° <$ Dec. $< +35°$ are indicated by asterisks, while open circles show the locations of the more southerly objects in the range $+15° <$ Dec. $< +25°$ Notice the degree of correlation among objects in the independent samples. To varying degrees, the foreground is underdense.

In trying to describe the kinds of structures seen in the P-P supercluster and the distribution of galaxies within it, we have used the redshift sample to obtain an absolute magnitude limited subsample very nearly complete to a distance of $V_0 = +7500$ km s^{-1} and M $= -19.5$ over the P-P region. This subsample includes only about 20% of the total number of galaxies, but with minor corrections for incompleteness, it should well reflect the true three-dimensional distribution and hence allow us to obtain a volume density map of the supercluster as outlined by the bright galaxies. The question of whether this same picture would also be displayed by the lower luminosity objects will be deferred to Section 5.

Slices of this volume density map are presented in Figure 9, from Giovanelli and Haynes (1988). The gridding of the volume is cartesian, centered on the Earth and with cells of 200 km s^{-1} on a side. The x-axis points toward R.A. $= 22^h$, Dec. $= 0°$ the y-axis toward R.A. $= 4^h$, Dec. $= 0°$ and the z-axis, toward Dec. $= 90°$. Before the calculation of n (x, y, z), radial velocities were corrected for local large-scale deviations from Hubble flow and for virial distortions within clusters. The latter correction was applied to galaxies within one Abell radius (or an equivalent quantity for poorer clusters) and with velocities within three times the rms line-of-sight velocity dispersion of the cluster's systemic redshift.

The outer boundaries are the celestial equator, a sphere of radius $V_0 = +7500$ km s^{-1} (the boundary of completeness), and the cone of the Dec. $= +45°$ surface. The two slices shown in Figure 9 are taken perpendicular to the z-axis, that is, parallel to the celestial equator. In each, the dashed lines indicate the intersections of each slice and the latter two boundary surfaces. The height above the celestial equator is respectively $V_z = 600$ and 2400 km s^{-1}, as indicated in each panel. The slices are integrated in the z-direction over one cell length, i.e. 200 km s^{-1}. Contours are of equal density of galaxies brighter than M $= -19.5$.

The nearer slice shown in panel (a) contains a "void" centered near 3000 km s^{-1}, surrounded by slight density enhancements and a more conspicuous, but not fully outlined one beyond 6000 km s^{-1}, which extends outside the boundary of the volume. The coherent nature of the P-P ridge is evident in the slice shown in panel (b); note that the Perseus cluster is not included because it lies at somewhat higher V_z. These two maps well illustrate the types of large-scale structure seen in the P-P supercluster.

Such representations allow one to sample the types of structures that can be identified in this region after removing the "polar bias" that affects Earth-

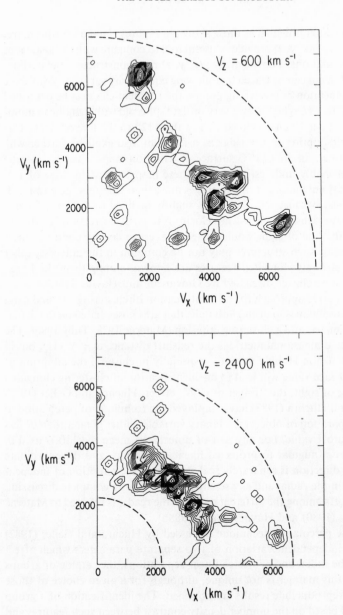

FIG. 9. Volume density contours in two slices across the density array constructed for the absolute magnitude limited subsample. Both are parallel to the the (V_x, V_y) plane (the celestial equator). The outer dashed curve is the intersection of the slice with the sphere of radial velocity $V_0 = +7500$ km s^{-1}; the inner dashed curve is the intersection with the cone of the Dec. = $+45°$ surface.

centered cone diagrams such as those shown in Figures 5 and 6. Furthermore, the density array n(x, y, z) provides a useful tool to compare with the analogous numerical simulations. For the array n(x, y, z) thus constructed, the outlines of the P-P filament can be traced by following equidensity surfaces. With some subjective truncation at branching points, the filament can best be described as a single surface roughly fit by a very prolate spheroid with a major to minor axis ratio of about 10 and inclined by less than 12° to the line of sight. This topological description of the ridge as a filament also explains further why it is so easy to see in the CGCG surface distribution maps. Even so, its true contrast is probably underestimated. In these maps, the main ridge of P-P shows a density enhancement that averages more than ten times the mean density of galaxies. Because the filament is slightly inclined to the line of sight, smearing of the density contrast in velocity results. A correction for the inclination to the line of sight would raise the average density contrast closer to twenty. While such structures may not be common in the universe, other examples of strongly elongated features are known, among them the Lynx-Ursa Major supercluster discussed by Giovanelli and Haynes (1982).

A quantitative approach to the identification of clustering can make use of numerical techniques to distinguish units that are chosen based on the definition of some metric and a clustering criterion (Materne 1978, Tully 1980). The simplest three-dimensional metric is the redshift distance $d = V_o/H_o$, but in the presence of non-Hubble motions, especially in clusters, the adoption of this metric at face value will lead to misidentification of clustering elongated along the line of sight, the "finger of God" effect. Huchra and Geller (1982) and Geller and Huchra (1983) have employed the technique of percolation in which the separation of objects is a binary variable, either greater than or less than some cutoff value. For any pair of objects the mean redshift is used in converting from angular to projected linear separation, and the coordinate in the radial direction is separately scaled through the adoption of independent cutoffs in the radial and sky separations. For more complete discussion of the trade-offs among the different methods the reader is referred to Materne (1978), Tully (1980) and Haynes *et al.* (1988b).

Since the percolation technique discussed by Huchra and Geller (1982) involves the independent variation of two separate parameters which affect the metric, the radial and projected linear separations, the catalog of groups produced in this manner is not unique, although for a given choice of those parameters, reproducibile results are obtained. The identification of a group is established based on the number density contrast between such features and the mean density of the sample under study. As defined by equation (4) of Huchra and Geller (1982), the contrast in number density is a quantitative estimate of the density enhancement relative to the mean and is governed by the choice of cutoff in the projected separation and the assumed luminosity function, since the cutoff is scaled to compensate for the fact that the sampling of the luminosity function varies with increasing redshift (the Malmquist bias). Our choice of parameters for the luminosity function, reference redshifts, and cutoffs are based on fits to the present sample for the luminosity function and trials of several sets of choices (e.g. Haynes *et al.* 1988b).

As a preliminary test of the percolation technique, we have applied the technique of Huchra and Geller (1982) to a section of the P-P region where the survey coverage is more complete and less affected by galactic extinction: $0^h < $ R.A. $< 03^h$, $+ 10° <$ Dec. $< +35°$. In that subregion, the redshift sample contains 1162 galaxies with recessional velocity less than $+ 12000$ km s^{-1}, the value chosen as the cutoff redshift. At present, we impose a magnitude limit for the complete sample of $+ 14.9$ over this range in order to correct properly for the Malmquist bias. This restriction reduces the number of galaxies used in the percolation to 425.

Even with the limitations of this sample, application of the percolation technique easily identifies 51 overdensities with a contrast of 20 or greater and containing at least two members; only 77 of the galaxies are found in regions less dense. At the present stage of our analysis, we are most interested in the properties of groups within the P-P supercluster, that is restricted to the range of recessional velocity $+ 4000$ km $s^{-1} < V_0 < +6500$ km s^{-1}. The ten groups that are likely supercluster members and that contain more than five members as identified by this process are listed in Table 1. Included in the table are the number of galaxies, the coordinates $\langle$R.A.$\rangle$ and $\langle$Dec.$\rangle$ and the mean velocity $\langle V_0 \rangle$ of the group centroid, the line-of-sight velocity dispersion σ, and hl, the maximum projected separation of any pair in Mpc. These ten aggregates contain masses implied by the virial theorem from 10^{13} to 10^{15} M$_\odot$ and have mass-to-light ratios from 90 to 1500.

TABLE 1

GROUP PROPERTIES FOUND BY PERCOLATION

Group	N	$\langle$R.A.$\rangle$ hhmm	$\langle$Dec.$\rangle$ sddmm	$\langle V_0 \rangle$ km s^{-1}	σ km s^{-1}	hl Mpc
1	6	0006	+3232	4557	108	2.9
2	5	0013	+1730	4737	528	4.4
3	11	0017	+2944	5146	1076	6.5
4	12	0020	+2313	5792	1418	6.4
5	31	0044	+3045	5027	736	11.
6	66	0117	+3250	4767	634	12.
7	11	0146	+1131	4928	275	4.1
8	8	0157	+2447	4721	109	3.8
9	41	0216	+3202	4649	437	13.
10	12	0229	+2109	4279	623	6.8

Figure 10 shows the distribution of mean velocities $\langle V_0 \rangle$ for all groups (N > 2) at the P-P redshift, in the northern portion of this restricted region $+20° <$ Dec. $< +35°$. The number of galaxies included in each group is indicated by the size of the circle; the groups with larger numbers of galaxies are included in Table 1. Note again, the narrow range of redshift characteristic of the group found in this region. A relaxation of the contrast level from 20 to 10 will permit almost the entire sample to percolate into a single structure.

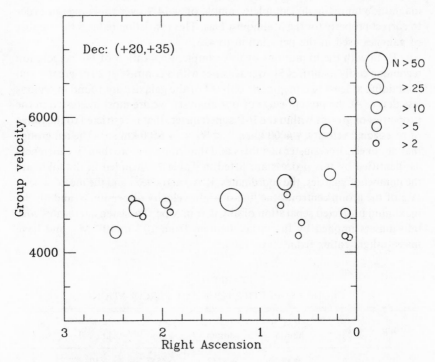

FIG. 10. Distribution of mean velocities for the 18 groups with $+4000 < V_0 < +6500$ km s^{-1} found to lie between $0^h <$ R.A. $< 3^h$, $+20° <$ Dec. $< +35°$. The number of galaxies included in each group is indicated by the size of the circle.

Much additional work in the application of numerical algorithms for the recognition of three-dimensional structure needs to be done. Completion of the sample to a faint limiting magnitude will aid the search for larger-scale structures because identification of the largest connective structures by such numerical techniques requires sampling in the low density regions. Just as the brightest galaxies outline the main ridge in Figure 1b, it takes a sample complete to a fainter magnitude to delineate the true extent of the supercluster from Pegasus to Perseus, because most of the links are of lower contrast than the groups and clusters they connect.

5. Segregation of Galaxy Properties

Segregation of morphological types in clusters of galaxies was first noticed over a half century ago (Hubble and Humason 1931). More recently, numerous authors (Oemler 1974; Melnick and Sargent 1977; Dressler 1980) have found that the relative population of ellipticals, lenticulars, and spirals in clusters and their peripheries is mainly a function of the local galaxian density. While some 80% of field galaxies are spirals, as few as 15% of the galaxies in condensed clusters like Coma show spiral structure. In his quantitative study of 55 rich clusters, Dressler found that the morphology-density relation appeared to be independent of cluster morphology and degree of central concentration.

The monotonic variation in the population fraction with local density found by Dressler extends in a uniform manner from the high-density cluster cores to the low density regimes of typical groups. Nearby loose groups are almost always dominated by spirals. Excluding the Virgo region, only two groups in the list of de Vaucouleurs (1975) contain predominantly bright early-type galaxies, and in both, a bright X-ray galaxy sits at the center of a higher density core surrounded by a more diffuse cloud of spirals. The more quantitative treatment of galaxy groups undertaken by Postman and Geller (1984) has shown that Dressler's morphology-density relation extends smoothly over six orders of magnitude in space density. Only at very low densities where the dynamical timescale is comparable to or greater than the Hubble time does the population fraction not reflect variations in the local galaxy density.

As established by Dressler the variation in population fraction with local density is monotonic but slow. In the highest density regions mechanisms that lead to alteration of a galaxy's morphology, such as galaxy-galaxy and galaxy-intracluster gas interactions, are believed to be relatively efficient. The known segregation of ellipticals to high density regions raises the question of whether the observed differentiation in morphology is inbred during the era of galaxy formation or shortly thereafter, or whether it is a continuing process. The scale over which segregation occurs and the degree of continuity of the variation in population fraction throughout all regimes of density will constrain models predicting the relative formation times of galaxies and their surrounding large-scale structure. Because its members are found in a wide variety of local densities, P-P will serve as a useful laboratory for studying the segregation of morphologies and other galaxian properties.

The most comonly-used statistic employed to quantify the degree and scale of clustering is the two-point spatial correlation function $\xi(r)$ and its angular form w (θ). Both are normally fit by power law forms so that the angular correlation function can be written in terms of an amplitude A and slope β, w (θ) = A θ^β. Using the UGC which contains the largest homogeneous compilation of galaxies including their morphological classification, Davis and Geller (1976) have compared w(θ) computed for galaxies brighter than m = +14.5 separately for different morphologies, and have found that the clustering properties for each type subsample are, in fact, different. Elliptical-elliptical clustering is characterized by a power law with a slope steeper that that appropriate for spiral-spiral clustering. The case for S0-S0 clustering is intermediate. For their sample, the number of galaxies found within 1 Mpc of a random elliptical galaxy is about twice that found within the same distance of a random spiral.

Since we know from Section 3 that the majority of galaxies contained in the P-P region are in fact members of the supercluster, we can gain a visual impression of the degree of morphological segregation simply by plotting separately the sky distribution of UGC galaxies of different morphological classes as was done in Figure 6 of Giovanelli et al. (1986). Here, we make use of the supercluster's distance. Figure 11 shows three panels which illustrate the sky distribution of different morphologies in the current redshift sample restricted to the velocity range = 3500 km s^{-1} < V_0 < +7000 km s^{-1}. The upper panel shows the distribution of all galaxies with available morphological types. The lower two panels display respectively only the elliptical and S0 galaxies (middle) and only the Sbc and Sc galaxies (bottom). Notice that the number of objects in the bottom two panels is nearly identical. Galaxies of progressively earlier morphology are segregated to the higher density regimes; the supercluster ridge is most evident in the distribution of elliptical and S0 galaxies. The visual impression of large-scale structure conveyed by the later-type spirals is quite different from that suggested by the early-type objects.

The angular correlation function w(θ) can also be used on the P-P sample. In dealing with an observational sample such as that compiled for Pisces-Perseus, the calculation of w(θ) must take into account the serious sample biases. Not only does the sample itself have observational constraints and selection effects, but the variability of the galactic extinction over such a large region adds to unevenness of the sampling. In practice, the correlation function must be calculated as the excess probability of finding pairs of galaxies of a given separation in the sample catalog as compared to that found in an identically-selected catalog containing a random distribution of objects. The operational definition of the angular correlation function becomes.

$$w(\theta) = \frac{N_g(\theta)}{N_r(\theta)} - 1 \qquad (7)$$

where $N_g(\theta)$ is the number of pairs in the observed sample with separations in the range ($\theta - d\theta$, $\theta + d\theta$) and $N_r(\theta)$ is the number of pairs with angular

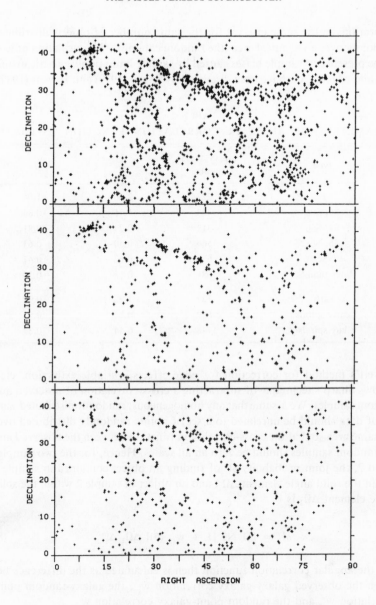

FIG. 11. Sky distribution of galaxies of separate morphological classes in the current redshift catalog restricted to the velocity range $+3500 < V_o < +7000$ km s^{-1}. Classification has been performed for CGCG objects on the Palomar Sky Survey prints following the UGC system. The upper panel shows the distribution of all objects within those redshift limits. Elliptical and S0 galaxies are included in the middle panel while the bottom panel contains Sbc and Sc galaxies.

separations in the same interval for the same number of objects distributed randomly over an identical area (the random catalog). The calculation of $w(\theta)$ in the presence of sample biases caused by sample limitations, galactic extinction and edge effects have been discussed by Sharp (1979) and Hewett (1982).

TABLE 2

PARAMETERS OF ANGULAR CORRELATION FUNCTION ESTIMATES

type	$w(\Theta) = A\Theta^{\beta}$ number	A	β
E	227	2.60	-1.06
S0, S0a	423	1.75	-0.84
Sa, Sab	312	0.29	-0.81
Sb, Sbc	566	0.50	-0.63
Sc	678	0.62	-0.47
later than Sc	681	0.58	-0.30
early	725	1.92	-0.90
early spirals	689	0.74	-0.65
late spirals	1548	0.54	-0.37

Hewett's method for correction for edge effects, variable extinction, etc., follows Sharp's technique of subtracting a cross-correlation of observed and random samples. We assume that any homogeneous, randomly distributed sample of data should be unrelated to a set of points randomly distributed over the same area so that the cross-correlation function between the observed and the random samples should be zero an all scales. Hence, for the two samples 1 and 2, the joint probability δP of finding an object contained in sample 1 within the solid angle element $\Delta\Omega_1$, and an object in sample 2 within the solid angle element $\Delta\Omega_2$ is

$$\delta P = N1 \, N2 \, [1 + w_{12}(\theta)] \, \Delta\Omega_1 \, \Delta\Omega_2 \tag{8}$$

and the angular correlation function then is calculated as the difference between the observed galaxy-galaxy correlation w_{gg}, the galaxy-random point correlation w_{gr} and the random point-galaxy correlation w_{rg}

$$w(\theta) = w_{gg}(\theta) - w_{gr}(\theta) - w_{rg}(\theta). \tag{9}$$

Since random points should always be randomly distributed around any galaxy when summed, the galaxy-random point correlation should always equal zero.

On the other hand, since galaxies are not necessarily randomly distributed with respect to random points, $w_{rg}(\theta)$ is not always zero, and is partially determined by the location of the galaxies relative to the sample boundaries.

The application of Hewett's method to the P-P region was made by Giovanelli *et al.* (1986) whose results for A and β are summarized in Table 2 and Figure 12. Figure 12 shows the calculated correlation function for three subsets of galaxy morphology. In addition to the global variation of clustering characteristics among ellipticals, S0's and spirals noted by Davis and Geller (1976), the analysis of galaxies in the P-P region shows the spiral classes themselves show a smooth variation: spirals of type Sa and Sab tend to cluster on smaller angular scales than do later spirals. The correlation analysis thus quantifies the visual impression of segregation evident in Figure 11.

Similar to the morphology-density relation of Dressler (1980), the population fraction can be calculated over bins of projected local density. As illustrated in Figure 12 of Giovanelli *et al.* (1986) a progressive change is seen in the slope of the variation of population fraction with density, not only from early to late morphologies as reported previously. The trend toward segregation observed in the cores of rich clusters and their peripheries appears to continue monotonically to the regions of lowest population fraction. In fact, depending on the choice of morphological class, a gradient in the population fraction can be observed at nearly every value of local density. Of perhaps equal importance, the progressive variation in the population fraction, as functions of both density and type, is striking not just within the major subgroupings but within the spiral sequence itself. The smooth change in the observed population fraction among the spiral types implies that the conditions that lead to the currently observed morphological segregation arise to a large extent from the matter density at the time of galaxy formation, or at least shortly thereafter. Diffusion of galaxies formed in different density environments is not viable over these scales within a Hubble time.

Morphology may not be the only parameter that shows segregation according to density. The study of the distribution of dwarf galaxies may be an important link to the understanding of morphological segregation and the likelihood of biased galaxy formation. Sharp, Jones and Jones (1978) have examined the distribution of DDO dwarf galaxies relative to all galaxies found in the CGCG. Those authors concluded that dwarfs do in fact tend to avoid clusters, but otherwise are found preferentially in the vicinity of bright galaxies and are intrinsically different from ordinary galaxies of the same apparent magnitude.

An additional approach examines the clustering properties in terms of the galaxies' surface brightness characteristics, after allowing for the morphological dependence on clustering. Both Davis and Djorgovski (1985) and Bothun *et al.* (1986) have considered the clustering of galaxies as a function of surface brightness. On the one hand, Davis and Djorgovski concluded that

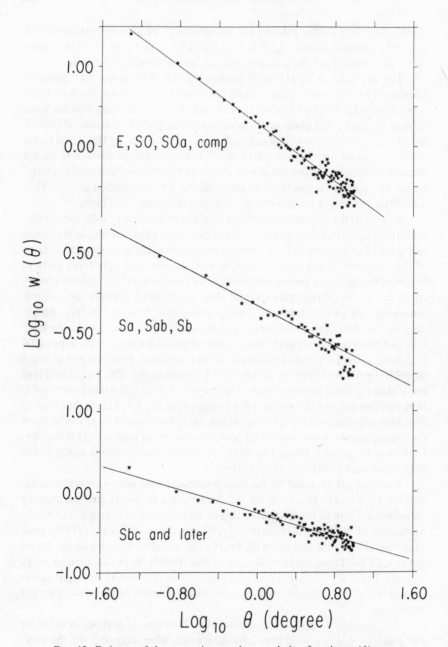

FIG. 12. Estimate of the two-point angular correlation function $w(\theta)$ versus θ, on logarithmic scales, for the subgroupings of morphological types indicated. Parameters of the best-fit power laws are listed in Table 1 (from Giovanelli *et al.* 1986).

the angular correlations are weaker and shallower for the low surface brightness (LSB) objects. At the same time, Bothun *et al.* contend that while the higher surface brightness galaxies have a smaller spatial clustering scale than do LSB's, there is no evidence that the redshift voids are filled by LSB objects.

If the distribution of light in the universe is a good tracer of the mass distribution, then the spatial correlation function $\xi(r)$ should be the same for both giant and dwarf galaxies. Environmental influences may result in a reinforcement of the clustering of ellipticals and S0's with respect to spirals as discussed above. But another critical key to the understanding of galaxy formation is held by the distribution of galaxy luminosities - the luminosity function. The recent study of the luminosity function of galaxies in the Virgo cluster by Binggeli *et al.* (1985) has proved the difference in the shapes of the luminosity function derived for different morphologies. Those authors have found that the luminosity distribution is different for ellipticals, S0's, and spirals and among the Sa/Sb, Sc, and Sd/Sm galaxies as well. The universality of the luminosity function is commonly assumed but is not well established. In fact, there is growing evidence that the galaxy luminosity function varies both with morphological type and local density. Future studies will address the question of whether this variation in the luminosity function with local environment results from different mixing of morphological types or from true luminosity segregation.

From the density array constructed for our volume - and absolute magnitude limited sample, we can estimate the dependence of galaxian properties on the value of the local density n. The value of n associated with each galaxy is that obtained for the occupied cell dxdydz. In order to compare the distribution of objects of different luminosities or in different volumes, we must construct density arrays as before but now complete to somewhat different depths in absolute magnitude. It should be noted that the densities obtained from subsamples are not directly commensurable because they have been constructed over different luminosity depths.

Figure 13 shows the contrast between the actual number of galaxies found within a given redshift shell and the number expected from a uniform density distribution for three subsamples. The upper curve, marked "filament", contains galaxies within the solid angle subtended by the density enhancement represented by the main P-P ridge. As noted previously, the observed density enhancement is a factor of ten over the mean, but the true contrast is probably even greater. A slight underdensity is present in the foreground, related to the obvious foreground void.

Similarly, from the density array, we can outline the solid angle subtended by the foreground void and can compare galaxy counts as observed with those expected for a homogeneous population. The lower two curves in Figure 13 illustrate the results for the void region, plotting separately the "bright" ($M < 19.5$) galaxies and "faint" ($M > 19.5$) galaxies. The latter galaxies

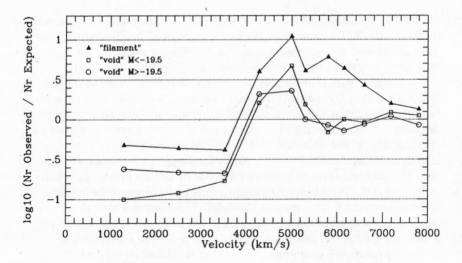

Fɪɢ. 13. Density contrast profiles showing the distribution relative to the mean of galaxies in several subunits of the P-P region. Each point corresponds to the ratio of the observed number of objects to that expected for a random distribution of galaxies within a shell in redshift centered on the point's abscissa. The "filament" line refers to the solid angle occupied by the P-P ridge. The two "void" curves show the results for two different bins of absolute magnitude in the region surrounding the void (from Giovanelli and Haynes 1988).

include all objects to the catalog limit within a given redshift shell. The partial overlap of solid angles subtended by the void and the filament is responsible for the overdensity near +5000 km s^{-1}. Figure 13 shows that the mean underdensity of the void is at least a factor of five. In agreement with Bothun *et al.* (1986) we also see that bright and faint galaxies outline the same general structures, but the density contrast is better outlined by the bright galaxies than by the faint ones.

Based on this analysis Giovanelli and Haynes (1988) conclude that on scales larger than cluster size, density contrasts within the galaxy distribution between high and low density regions are observed to be as large as a factor of 100, and that delineation of large-scale structure and the degree of density enhancement appear to be dependent on the luminosity of the galaxies counted.

The determination of a luminosity dependence on any parameter that may be related to distance is threatened by a Malmquist bias. Because the linear scale of the density inhomogeneities is comparable to that of the volume sampled, high and low density regions are not well-mixed, so that the median distances of cells of low and high density are not the same. As a result, a plot of luminosity versus density that included all of the galaxies in the sample would exhibit a Malmquist bias in the sense that higher luminosities would be obtained for the densities with the higher median distance. In our sampled volume, the higher density regions tend to be found in the P-P supercluster and lie somewhat farther from us than do the low density regions such as the foreground void. In order to avoid the bias introduced by this difference, we have analyzed separately different intervals of redshift, requiring that within each interval the high and low density cells must have similar median redshifts. Figure 14, from Giovanelli and Haynes (1988), shows the results for two redshifts intervals. The

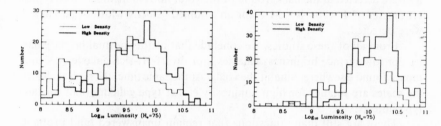

FIG. 14. Luminosity distribution, in solar units, of galaxies in high and low density regions for galaxies in separate regimes of redshift. The left panel shows galaxies with velocities $0 < V_0 < +4000$ km s^{-1}. Space densities were derived from a subsample nearly complete to M $= -18.4$. The right panel includes galaxies with $+4000 < V_0 < +6200$ km s^{-1}, and the subsample is nearly complete to M $= -19.0$ (from Giovanelli and Haynes 1988).

left panel (a) shows histograms of luminosity for galaxies in the redshift range $0 < V_0 < +4000$ km s^{-1}, separately for galaxies in high and low density regions; the right panel (b) shows similar histograms for galaxies with $+4000 < V_0 < +6200$ km s^{-1}. In order to optimize statistics, we have adopted different cutoffs in the absolute magnitude in defining density in the two redshift regimes. In the nearer, density is defined by the subsample of galaxies brighter than -18.4, while in the more distant sample, all galaxies brighter than -19.0 are included. In both instances, there are relatively fewer high luminosity galaxies in the low density regions. Although the present incompleteness demands some caution, the opposite appears to be true for the low luminosity galaxies. A similar analysis using surface brightness leads to the conclusion that galaxies in high density regions, which as shown in Figure 14 tend to be brighter, also have a higher surface magnitude than galaxies in low density volumes.

Some currently popular theories of galaxy formation require that the overall mass distribution must be significantly smoother than that traced by the luminous matter. One way to meet these conditions is to postulate the formation of luminous galaxies in a biased process (e.g. Dekel and Rees 1987). This bias would translate precisely to a spatial segregation of the low luminosity, low surface brightness galaxies from the bright, massive ones so that the low luminosity galaxies would fill the voids which themselves were outlined only by the bright objects. As such, the hole in the foreground of P-P mentioned in Section 3 provides a suitable location to search for the "missing" dwarfs. However, Oemler (1987) reported that a search by Eder, Schombert, Dekel, and Oemler for dwarf galaxies found no galaxies in the void. Furthermore, the most commonly-used total-power (beam switching) mode of extragalactic 21 cm line observing samples a range of redshift in each observation and in each position (ON- and OFF-source). For all of the directions in which galaxies were detected in the background of the void, no void members with typical HI masses of $4 \times 10^7 \, h^{-2} \, M_\odot$ (for an assumed velocity width of 30 km s^{-1}) were detected.

From all of these studies, we conclude that neither normal bright galaxies nor low surface brightness galaxies, nor dwarf-HI-rich galaxies, as commonly found elsewhere, inhabit the voids. At the same time, the typical clustering scales are smaller for high luminosity earlier type galaxies than for later type objects.

There are still many questions that remain unanswered (and probably more, unasked). The relationship of the ridge to the gravitational potential and conditions during the formation epoch remain to be explored, and the possible coherence of structure in Perseus and the surrounding supercluster should be checked. Strom and Strom (1978) showed that Perseus cluster ellipticals are aligned so that their major axes preferentially follow the main cluster chain at a position of 70° - 80°. On larger scales, Gregory *et al.* (1981) found evidence for alignment and perpendicularity within the supercluster ridge. The range of local densities represented by the P-P supercluster implies that the environmental influences on an individual galaxy, both past and present, have been dramatically different from one location to another. P-P will continue to be a prime site for the study of the mechanisms that might produce the observed segregation.

6. CONNECTIONS TO OTHER STRUCTURE

One of the most striking aspects of the sky distribution of galaxies in the P-P region is the location of the most massive dynamical unit, the Perseus cluster, at the northeasternmost end of the main density enhancement, as seen, for example, in Figure 2. To the north and east of Perseus, the galactic obscuration seriously limits tracing the large-scale structure and few galaxies are included in the CGCG or UGC. Kent and Sargent (1983) found an average visual extinction of 0.6 magnitudes in the direction of A426 itself, and the extinction quickly exceeds one magnitude in the region east of R.A. = 3^h. After allowing for the obscuration, Hauschildt (1987) has traced the supercluster as far east as the 3C129 cluster at R.A. = 4^{h}45^m, Dec. = +45°.

After an analysis of the distribution of redshifts of radio galaxies in the region of the zone of avoidance, Burns and Owen (1979) suggested that the supercluster might extend across the obscured region toward A569, located at R.A. = $07^h 05^m$, Dec. = $+48° 43'$, and a redshift $V_o = +5800$ km s^{-1} (Struble and Rood 1987). Giovanelli and Haynes (1982) identified a low-density enhancement in Lynx-Ursa Major at about the same redshift as P-P but containing no high density clusters; they suggested that this low contrast filament could be the eastern extension. Focardi et al. (1984) noted the presence of an excess of galaxies behind the galactic plane detected in the red survey of galaxies within two degrees of the galactic equator compiled by Weinberger (1980). Focardi et al. (1984) concluded that P-P does indeed extend across the zone of avoidance and connects to the cloud around A569 rather than the Lynx-Ursa Major supercluster. However, the survey of the latter authors actually misses much of the Lynx-Ursa Major supercluster and hence their arguments are not applicable. With the data currently available outside the P-P region the redshift structure cannot be well-enough understood. The Perseus cluster itself is characterized by a systemic velocity somewhat inbetween the Lynx-Ursa Major supercluster and A569, which is itself complex, and it is unknown whether P-P continues across the zone of avoidance as a single feature in redshift or branches out somewhere in the region of obscuration (Hauschildt 1987; Haynes et al. 1988a).

The effects of galactic obscuration are a hindrance both to optical and to radio observers; for practical reasons, the latter must know where to point the radio telescope. The existence of galaxies invisible to optical instruments but detectable in the 21 cm line even within a few degrees of the galactic plane has been demonstrated by Kerr and Henning (1987). However, undertaking such a blind search at 21 cm over a wide area suspected to be part of the Perseus supercluster would be extremely time-consuming. The total power mode of observation, typically employed in extragalactic 21 cm line studies, always samples two points on the sky (one "on" source, and another reference "off" source position); typical redshift searches cover frequency ranges corresponding to at least 6000 km s^{-1}. In the survey of Haynes et al. (1988a), fewer than one percent of observations reveal another galaxy at any redshift in the off-source spectrum.

Spiral galaxies in the Perseus supercluster whose optical images are hidden in the zone of avoidance may be among the point sources detected by the Infrared Astronomical Satellite IRAS. Dow et al. (1988) have discussed the feasibility of using far infrared color criteria to find galaxies close to the galactic plane. Haynes et al. (1988a) have obtained 21 cm line spectra for a small sample of galaxies detected by IRAS close to the zone of avoidance, in order to demonstrate the potential for using this method to trace the P-P supercluster to the east beyond Perseus. Future 21 cm observations are planned to search for supercluster members among the point sources detected by IRAS.

The current limitations on the number of available redshifts preclude the compilation of statistically complete samples, and we must confess that it has been frustrating to try to analyze the redshift distribution in this region. However, it seems so unlikely that a supercluster structure as rich as P-P would end so abruptly at the Perseus cluster, that we feel confident that there must be many galaxies and perhaps clusters behind the zone of avoidance. We hope so, because we have already begun the hunt to find them.

Haynes and Giovanelli (1986) have underscored the existence of a well-defined connection between P-P and the Local Supercluster, apparent in Figure 5b in the form of two filamentary-like features which also outline the foreground void discussed earlier. Other authors have employed various clustering techniques to attempt to trace structure on even larger scales, not by tracing the galaxies but by treating the clusters as individual points. Batuski and Burns (1985) have used a percolation analysis to examine the distribution of Abell clusters with $m_{10} < 16.5$ and medium-distant CGCG clusters in a large region of the south galactic cap that overlaps the P-P region. Those authors find evidence that the P-P supercluster is part of a possible filament of galaxies and galaxy clusters that extends over $300h^{-1}$ Mpc. Tully (1986, 1987) has suggested that the relatively nearby P-P supercluster links with the more distant Pisces-Cetus complex. In fact, Tully proposes that P-P is a filamentary enhancement within a planar structure that has coherence on a scale of 0.1c.

7. SEARCH FOR NON-HUBBLE STREAMING

Deviations from the Hubble flow in the local universe have been detected as the motions of the Milky Way with respect to the microwave background, as infall motion into the Virgo cluster (Tully and Shaya 1984), and as the familiar elongated shapes in redshift space seen toward rich clusters. On scales larger than those of clusters, the magnitude of the peculiar velocities is largely unknown, and fast becoming more intriguing as evidenced by the lively discussions during this Study Week. The questions of possible large-scale streaming motions have attracted new attention with the availability of deeper redshift surveys of larger areas of sky. The "CfA slice" (de Lapparent et al. 1986) and the P-P survey have shown both similar and dissimilar structures: the apparent sheets or bubbles seen in both surveys and the high density main ridge unique to the P-P supercluster. Especially interesting is the apparently common appearance of large "voids", regions of low number density of galaxies surrounded by regions of higher number density. The streaming motion around such voids has been modelled as galaxies evacuating the region of the void (Davis et al. 1982) and alternatively as evolving negative density perturbations (Hoffman et al. 1983). The former approach, which is equivalent in result to that of explosive galaxy formation scenarios, predicts streaming motions around a void on the order of 50% of the Hubble expansion; in contrast, the latter favors velocities of only about 10% of the Hubble velocity. Biased galaxy formation pictures may not require any streaming motions at all.

In order to investigate the dynamics in a supercluster environment and to measure possible deviations from the Hubble flow, it is necessary to obtain redshift-independent relative distances to galaxies, i.e. distances up to a scale factor h in the Hubble constant. In the past, there have been essentially two opposite approaches to the task of predicting luminosities for galaxies, and thereby deriving their distances: the first, deriving a luminosity distance from an indicator that is distance independent and has been determined for a large number of galaxies; and the second, the so-called method of "sosies" that is based on the comparison of individually-matched galaxies for which many properties have been measured and are similar. The first approach has been pioneered by Tully and Fisher (1977). An example of the application of the "sosies" method is the measurement of the distance to the Hercules supercluster presented by de Vaucouleurs and Corwin (1986).

The acquisition of large numbers of HI line data in the P-P region makes attractive the future use of the Tully-Fisher relation to predict luminosity distances. At present, the detection of possible streaming motion is hampered by the limited success of deriving redshift-independent distances of galaxies via application of the Tully-Fisher relation. Current limits on the detection of large-scale streaming motions in surveys such as this one are imposed by: (1) the uncertainties in the available blue magnitudes and the necessary corrections; (2) the large intrinsic scatter in the Tully-Fisher relation for blue magnitudes; and (3) the small number of galaxies suitable for application of the relation within regions that can reasonably be assumed to be part of the dynamical structure under study.

The original application of the Tully-Fisher relation using blue or photographic magnitudes relies on an understanding of the numerous corrections for galactic and internal extinction, inclination, and the varying contribution of the disk and bulge (e.g. Tully and Fouqué 1985). Magnitudes have often been measured on different systems and with different apertures. In few instances have data on a complete sample been assembled from a homogeneous survey.

The movement of the luminosity measurement toward longer wavelengths has been driven primarily by the desire to minimize internal extinction effects so critical in the blue and at the same time, so uncertain. A secondary reason for the use of redder magnitudes has been the desire to lessen the influence of recent, and possibly variable, star formation which also would contribute more in the blue. The improvement in the application of the Tully-Fisher relation as a secondary distance indicator using infrared magnitudes has been demonstrated (Aaronson and Mould 1983; Aaronson et al. 1986). Because of the need for CCD images for large numbers of galaxies, use of the R or I bands is more practical. While the sky background is still noisy and the internal extinction corrections are not negligible at I relative to H, the sensitivity of current detectors is sufficient to provide quality I-band images in typically ten minutes or less (Bothun and Mould 1987). We have begun a program to obtain I-band images for galaxies in the P-P supercluster. The imagery will be used to derive magnitudes, diameters, and disk scale lengths in an attempt to reduce the scatter in the Tully-Fisher relation by means of multivariate analysis.

The general goal is to predict the distance d of a galaxy from $i = 1 \ldots n$ observed quantities a_i, given that at least one of them is distance-independent. One needs to determine how many independent parameters are needed to describe the scatter of the sample in parameter space. The use of Principal Component Analysis (PCA) has been discussed by such authors as Brosche (1973) and Whitmore (1984), and offers the promise of reducing the scatter in the Tully-Fisher relation for spirals in much the same way as a three-parameter fit has helped the analogous method for ellipticals. Application of such an analysis to a similar sample in the Hercules supercluster has been presented by Freudling *et al.* (1988).

At the present time, the large dispersion in the distance estimates makes it impossible to derive the non-Hubble velocity for any single object. However, although the luminosity distances derived from the Tully-Fisher relation are currently less accurate than the Hubble distances, they should be independent of peculiar velocity and hence unbiased. By investigating the difference between the Hubble distance and the luminosity distance as a function of the luminosity distance, we can test that the average of those residuals is zero for a sample of galaxies that is assumed to have a random peculiar distribution and is used as a comparison with other samples that might themselves show a streaming motion. An all-sky sample of Sc galaxies is being constructed for this purpose. Such an approach avoids the apparent distance dependent increase of the Hubble constant (Giraud 1985) in a residual versus redshift diagram, the result of the unavoidable Malmquist bias of the sample (Teerikorpi 1984). However, the large dispersion in the luminosity distance introduces the effect that the residuals will, in the end, be averaged over distance bins. Galaxies will frequently be counted in a "wrong" distance bin, that is the distance bin assigned to it by the luminosity distance rather than the true distance. A galaxy with a distance too high for its bin will appear to have a positive peculiar velocity. This usually does not matter, because for each galaxy too distant for its bin, there will be another one too close. However, if the number of galaxies is much higher at a certain distance (for a cluster or because the completeness of the survey depends on distance), then the distance bin before this concentration will contain more galaxies from higher distances than from lower ones. The opposite happens behind a cluster. As a result, an apparent streaming motion will be seen towards any concentration of observed galaxies. This spurious result can be avoided for all bins but the zero velocity bin and the bin with the highest distance (if the bin size is larger than the errors for individual galaxies) by taking a sample of galaxies which is equally distributed in redshift or by weighting the average according to the local density of observed galaxies. The former approach is usually more straightforward, but unfortunately results in a significant reduction in the number of galaxies actually used in the analysis. By drawing samples of similar objects in each redshift bin from an all-sky sample and from the sample within a supercluster or around a void, this method should be able to detect streaming motions. Freudling *et al.* (1988) have set a preliminary upper limit of 400 kms^{-1} on infall motions within the Hercules supercluster. Similar results are obtained for P-P. The future acquisition of homogeneous data samples and good magnitudes will hopefully permit measurement of departures from the Hubble flow within P-P and in the vicinity of the foreground void.

8. SUMMARY AND CONCLUSIONS

Of all the nearby superclusters Pisces-Perseus is perhaps the easiest to recognize in maps of the surface density distribution of CGCG galaxies, partly because of its geometry and richness and partly because of its distance from us relative to the depth of the catalog. The P-P region is not a randomly selected volume of space, but rather has been deliberately chosen for study because of its obvious structure. The main ridge is at least 45h^{-1} Mpc in length and has an axial ratio close to ten. The supercluster also contains a more widely dispersed population that is separated from foreground and background structures by relatively empty volumes. Connective density enhancements surround the foreground void and appear to link the Local Supercluster to P-P. The wealth of structure displayed by P-P reminds us that the large-scale topology is complex and not easily described. The most conspicuous high density enhancements do however exhibit filamentary morphology.

The range of local densities characteristic of the supercluster environment extends from the highest density cluster cores found in A426 and A262 to the outer supercluster periphery. In the latter locale, the interaction timescale between galaxies is much longer than the Hubble time, and objects there must have existed for virtually their entire lifetimes in such low density regions. Hence, they are objects unaffected by neighbors and occupy volumes of both low gas and galaxy density. Gas removal and disruption mechanisms believed to be responsible for the alteration of morphologies in high density environments cannot be expected to be efficient in the supercluster periphery. Comparison of the intrinsic properties of galaxies in different density regimes will allow us to determine the relative properties of galaxies in the varying environments.

Because of its location and extent, P-P is also difficult to trace at optical wavelengths and hence is especially suited to an HI line redshift survey. Its maximum extent and possible connection with Lynx-Ursa Major or A 569 in the east or the Pisces-Cetus complex in the west must still be confirmed. The boundaries of the foregound void need further delineation via a large sample of low luminosity, low surface brightness objects. If biasing mechanisms do govern the morphological segregation at early epochs, the void might still be expected to contain a population of HI-poor, low luminosity galaxies or gas clouds.

While time will provide more observations so that completion of the sample to some fainter apparent magnitude is possible, problems with understanding incompleteness and the definition of a "fair" sample will remain. If the luminosity function is dependent not only on the the morphological type of a galaxy, but separately also on the local density, then samples of unequal morphological mix may not carry the same indication of luminosity distribution. As long as galaxies are not standard candles, we will have to worry about whether the galaxies in the observational sample are properly representative. Spirals and ellipticals do not occupy the same volumes; some allowance for this segretation must be included. The interpretation of large-scale flows from the derivation of luminosity distances currently assumes that the intrinsic properties of galaxies are everywhere constant. If the luminosity function varies, even for a single type, the use of the Tully-Fisher relation to predict distances will have to be carefully applied to galaxies located in different density regimes.

In many respects, fortune provides us with a special view of the Pisces-Perseus supercluster. Is its complex and filamentary structure unique or is such structure just easy to recognize? We can ask whether we would still identify it if its main ridge did not lie nearly in the plane of the sky, if its axial ratio were not so great or if its typical volume density enhancement were not so large. Attempts to trace the extent of Perseus deliver the warning that the size of the largest apparent structure is comparable to the depth of the volume surveyed; will even greater structures be found in deeper surveys? Pisces-Perseus provides a laboratory for the exploration of intergalactic environments and their effects on galaxies and will continue to serve as a testing ground for our ideas of the origin of large-scale structure.

We wish to acknowledge that the National Astronomy and Ionosphere Center is operated by Cornell University under a cooperative agreement with the National Science Foundation.

REFERENCES

Aaronson, M. and Mould, J. 1983. *Ap J.* **265**, 1.

Aaronson, M., Bothun, G., Mould, J., Huchra, J., Schommer, R., and Cornell, M. 1986. *Ap J.* **302**, 536.

Batuski, D.J. and Burns, J.O. 1985. *Ap J.* **299**, 5.

Bernheimer, W.E. 1932. *Nature.* **130**, 132.

Bicay, M.D. 1987. *Ph. D. thesis.* Stanford University.

Binggeli, B., Sandage, A., and Tammann, G.A. 1985. *AJ.* **90**, 1759.

Bothun, G.D. and Mould, J.R. 1987. *Ap J.* **313**, 629.

Bothun, G.D., Beers, T.C., Mould, J.R., and Huchra, J.P. 1986. *Ap J.* **308**, 510.

Brosche, A. 1973. *AA.* **23**, 259.

Burns, J.O. and Owen, F.N. 1979. *AJ.* **84**, 1478.

Burstein, D. and Heiles, C.1978. *Ap J.* **225**, 40.

Davis, M. and Djorgovski, S. 1985. *Ap J.* **299**, 15.

Davis, M. and Geller, M.J. 1976. *Ap J.* **208**, 13.

Davis, M., Huchra, J., Latham, D.W., and Tonry, J. 1982. *Ap J.* **253**, 423.

Dekel, A. and Rees, M.J. 1987. *Nature.* **326**, 455.

de Lapparent, V., Geller, M.J., and Huchra, J.P. 1986. *Ap J Letters.* **302**, L1.

de Vaucouleurs, G. 1975. in *Galaxies and the Universe.* ed. by A. Sandage, M. Sandage, and J. Kristian. p. 557. Chicago: University of Chicago Press.

de Vaucouleurs, G. and Corwin, H.G. 1986. *Ap J.* **308**, 487.

Dow, M.W., Lu., N.Y., Houck, J.R., Salpeter, E.E., and Lewis, B.M. 1988. *Ap J Letters.* **324**, L51.

Dressler, A. 1980. *Ap J.* **236**, 351.

Einasto, J., Jôeveer, M., and Saar, E. 1980. *MN.* **193**, 353.

Focardi, P., Marano, B., and Vettolani, G. 1984. *AA.* **136**, 178.

Fontanelli, P. 1984. *AA.* **138**, 85.

Freudling, W., Haynes, M.P., and Giovanelli, R. 1988. *AJ.* submitted.

Geller, M.J. and Huchra, J.P. 1983. *Ap J Suppl.* **52**, 61.

Giovanelli, R. and Haynes, M.P. 1982. *AJ.* **87**, 1355.

_____. 1984. *AJ.* **89**, 1.

_____. 1988. in *Large-Scale Structures of the Universe.* ed. J. Audouze, M.-C. Pelletan, and A. Szalay, p. 113, Dordrecht: Kluwer Academic Publishers.

Giovanelli, R., Haynes, M.P., and Chincarini, G.L. 1986. *Ap J.* **300**, 77.

Giovanelli, R., Haynes, M.P., Myers, S.T., and Roth, J. 1986. *AJ.* **92**, 250.

Giraud, E. 1985. *AA.* **153**, 125.

Gregory, S.A., Thompson, K.A., and Tifft, W.G. 1981. *Ap J.* **243**, 411.

Hauschilt, M. 1987. *AA.* **184**, 43.

70   M.P. HAYNES - R. GIOVANELLI

Haynes, M.P. and Giovanelli, R. 1986. *Ap J Letters.* **303**, L127.

Haynes, M.P., Giovanelli, R., Starosta, B.M., and Magri, C. 1988a *AJ.* **95**, 607.

Haynes, M.P., Giovanelli, R., and Magri, C. 1988b (in preparation).

Hewett, P.C. 1982. *MN.* **201**, 867.

Hoffman, G.L., Salpeter, E.E., and Wasserman, I. 1983. *Ap J.* **268**, 527.

Hubble, E. and Humason, M.L. 1931. *Ap J.* **74**, 43.

Huchra, J.P. and Geller, M.J. 1982. *Ap J.* **257**, 423.

Kent, S.M. and Sargent, W.L.W. 1983. *AJ.* **88**, 697.

Kerr, F.J. and Henning, P.A. 1987. *Ap J Letters.* **320**, L99.

Materne, J. 1978. *AA.* **63**, 401.

Melnick, J. and Sargent, W.L.W. 1977. *Ap J.* **215**, 401.

Nilson, P. 1973. *Uppsala General Catalog, Uppsala Astr. Obs. Ann.* **6** (UGC).

Oemler, G. 1974. *Ap J.* **194**, 1.

_____. 1987. in *Nearly Normal Galaxies.* ed. S.M. Faber. p. 213. New York: Springer-Verlag.

Oort, J.H. 1982. in *Astrophysical Cosmology.* eds. H.A. Bruck, G.V. Coyne, and M.S. Longair. p. 127. Città del Vaticano: Pont. Acad. Scient.

Postman, M. and Geller, M.J. 1984. *Ap J.* **281**, 95.

Roberts, M. 1969. *AJ.* **74**, 859.

Sandage, A. and Tamman, G. 1981. *Revised Shapley-Ames Catalog.* Washington: Carnegie Institution.

Schechter, P. 1976. *Ap J.* **203**, 297.

Shapley, H. and Ames, A. 1932. *Harvard Obs Ann.* **88**, No. 2.

Sharp, N.A. 1979. *AA.* **74**, 308.

Sharp, N.A., Jones, B.J.T., and Jones, J.E. 1978. *MN.* **185**, 457.

Strom, S. and Strom, K. 1978. *AJ.* **83**, 732.

Struble, M.P. and Rood, H.J. 1987. *Ap J Suppl.* **63**, 543.

Teerikorpi, P. 1984. *AA.* **141**, 407.

Tully, R.B. 1980. *Ap J.* **237**, 390.

_____. 1986. *Ap J.* **303**, 25.

_____. 1987. *Ap J.* **323**, 1.

Tully, R.B. and Fisher, J.R. 1977. *AA.* **54**, 661.

Tully, R.B. and Fouqué, P. 1985. *Ap J Suppl.* **58**, 67.

Tully, R.B. and Shaya, E.J. 1984. *Ap J.* **281**, 31.

Weinberger, R. 1980. *AA. Suppl.* **40**, 123.

Whitmore, B.C. 1984. *Ap J.* **278**, 61.

Zwicky, F., Herzog, E., Karpowicz, M., Kowal, C.T., and Wild, P. 1961-68. *Catalog of Galaxies and Clusters of Galaxies.* Pasadena: California Institute of Technology, six volumes.

MORPHOLOGY OF LARGE-SCALE STRUCTURE

R. BRENT TULLY

Institute for Astronomy, University of Hawaii

ABSTRACT

The opportunity will be taken to provide a brief summary of some characteristics of large-scale structure. In the first section, properties of the distribution of galaxies within 3000 km s^{-1} will be described. Then, in the second, there will be a review of why one should believe in structure on a scale of 30,000 km s^{-1}.

1. STRUCTURE DELINEATED BY GALAXIES WITHIN 3000 KM S^{-1}

The *Nearby Galaxies Atlas* (Tully and Fisher 1987) provides a graphic description of the distribution of galaxies with velocities less than 3000 km s^{-1}. Some general characteristics of the galaxies within this volume are discussed by Tully (1988) and, specifically, the delineation of a clustering hierarchy is presented by Tully (1987a). Some of the important points made in those publications and in Figs. 1 and 2 are as follows:

a. Essentially all nearby galaxies can be assigned to entities called 'clouds'. Roughly 70% of these can be assigned to 'groups' that are probably bound and mostly collapsed, a further 20% lie in 'associations' that are probably *not* bound, and 10% lie at-large in the clouds. There is no convincing example of even a single isolated galaxy within the surveyed volume.

b. The Local Group sits right at the edge of what will be called the 'Local Void'. This void is apparently empty of galaxies across a diameter of 1600 km s^{-1} and only contains 14 known galaxies across a diameter of 2200 km s^{-1} (12 of these lie in two small clouds). Unfortunately, the region is split by the zone of obscuration, which diminishes the significance of the above assertions. However, if voids delineated by luminous galaxies were actually filled by gas-rich dwarfs, such entities would have been seen easily in the Local Void.

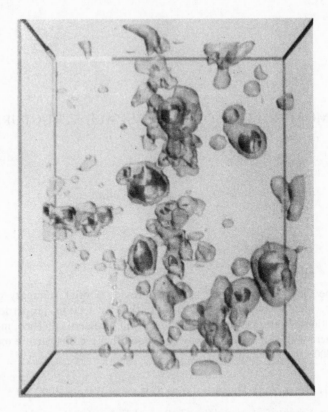

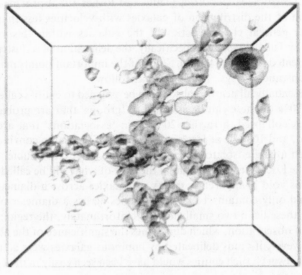

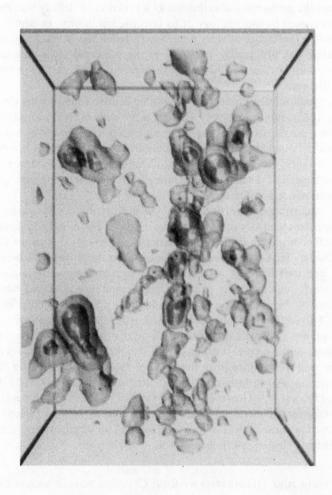

FIG. 1. Three orthogonal views of the structure in the distribution of galaxies in the Local Supercluster. Dimensions of the cube are 3750 km s⁻¹ by 3000 km s⁻¹ by 2625 km s⁻¹. The outer contour represents a surface of density 0.5 galaxies Mpc⁻³ (H_0 = 75 km s⁻¹ Mpc⁻¹). The Virgo Cluster is located at the region of highest density near the center of the volume. The plane of the Local Supercluster is horizontal in the lower-left projection and vertical in the upper-right projection. The 'louvered wall' is vertical in both the lower-left and upper-left projections.

c. A more quantitative assessment of the dependence on environment is possible. The local density in the vicinity of each nearby galaxy was determined, and then the probability distribution as a function of density was found for galaxies classed by morphology or by intrinsic luminosity. In addition to the well-known preference of early systems to inhabit high-density regions, there is a subtle but progressive variation in the preferred environment of late systems, with later or less luminous systems more likely to lie in regions of lower density.

d. Qualitatively, there is a remarkable connectedness to the clouds that creates a filamentary network. There is a problem in trying to quantify this claim because the structures are large compared with the survey volume and are usually bounded by the zone of obscuration or by serious incompletion due to distance. However, the general phenomenon is particularly striking in the south galactic hemisphere because, outside of the few interconnected filaments, space is quite empty.

e. An extraordinary characteristic of the distribution of nearby galaxies is the apparent tendency of galaxies to lie in layers. The most prominent feature is the adherence of many nearby galaxies to the so-called plane of the Local Supercluster. However, almost all clouds, even those well off the principal plane, are stretched out in structures elongated at modest angles to the supergalactic equator. The principal plane extends across and beyond the boundary of the presently surveyed volume, a diameter of 6000 km s^{-1}.

f. There is another apparent characteristic of nearby structure that again defies quantitative evaluation because it occurs on a scale comparable to the dimensions of the survey. It would appear that there is a concentration of galaxies in a 'wall' *perpendicular* to the plane of the Local Supercluster. The wall has dimensions of at least 6000 km s^{-1} (the sample dimension) by 3000 km s^{-1} by 1500 km s^{-1}. However, viewed face-on to the major dimensions, the wall is fragmented into the apparent layers parallel to the supergalactic equator mentioned previously (it is a 'louvered wall'). The Virgo, Antlia, Centaurus, and Hydra I clusters all lie in the wall.

2. Structure Delineated by Rich Clusters within 30,000 km s^{-1}

The claim was made by Tully (1986, 1987b) that rich clusters tend to congregate in 'supercluster complexes'. There are parts of five such complexes in the well-observed region of the sky within 0.1c, each one apparently containing roughly 50 rich clusters and extending across 30,000 km s^{-1}. A supercluster complex would contain roughly 10^{18} $M_{\odot}$ and a million big galaxies. Roughly two-thirds of nearby rich clusters are associated with a supercluster complex.

It is claimed that we reside in the 'Pisces-Cetus Supercluster Complex'. In addition to the general properties described above, this entity has the special

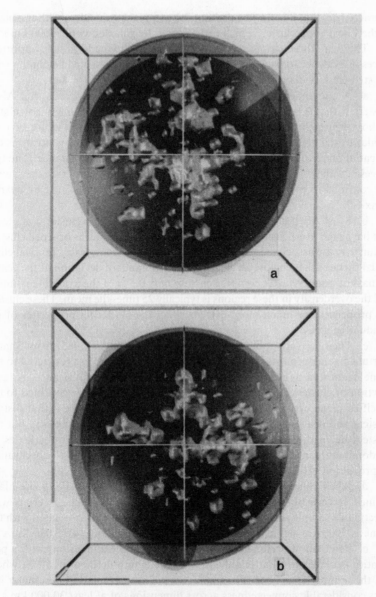

FIG. 2. Two views of the structure in the distribution of rich clusters. The radius of the sphere is 30,000 km s^{-1}. The outer contour represents a surface density of 1.7×10^{-5} clusters Mpc^{-3} ($H_0 = 75$ km s^{-1} Mpc^{-1}). (a) North galactic pole, which contains the Hercules-Corona Borealis (top), Leo (lower right), and Ursa Major (middle left) supercluster complexes. (b) South galactic pole, which contains the Pisces-Cetus (middle right) and Aquarius (top) supercluster complexes.

characteristic that it is flattened about 7:1 to a plane coincident with the plane of the Local Supercluster, which is delineated by galaxies within 3000 km s^{-1}.

There are three strong pieces of evidence and two additional supportive pieces of evidence to back up these claims. These are presented beginning with the strongest evidence and progressing to the weakest:

a. A one-dimensional two-point correlation analysis reveals that Abell clusters within 0.1c in the south galactic hemisphere are strongly correlated at short projected spacings when projections are taken against a vector perpendicular to the plane defined by nearby galaxies. This analysis neutralizes the potential biases due to obscuration and incompletion with distance. The test suggests, with 99% confidence, that a large fraction of the clusters in the sample are confined to one or more strata parallel to the plane defined by nearby galaxies.

b. Most Abell clusters lie in a small number of supercluster complexes that have high overdensities and low filling factors. Two-thirds of Abell clusters within 0.1c are associated with only five supercluster complexes. Triaxial ellipsoidal surfaces that enclose all clusters within a standard deviation of the center of mass of a supercluster complex enclose only 0.5% of the available volume, and the overdensity in these regions is typically 25 times the mean. These clustering properties are much more extreme than the clustering properties of individual galaxies.

c. The relatively poor clusters defined by Shectman (1985) from Shane-Wirtanen galaxy counts in the south galactic hemisphere are restricted to the plane delineated by nearly galaxies to a high degree of significance. A large fraction of clusters at a typical velocity of 15,000 km s^{-1} are confined to the specified plane with FWHM of 2000 km s^{-1}. The coincidence is that distant clusters and nearby galaxies lie on the same great circle on the sky, but the clusters lie in the south galactic hemisphere while the nearby galaxies lie predominantly in the north. Hence, the coincidence is not due to obscuration or projection effects.

d. Abell clusters within 0.1c in the south also tend to lie in the plane defined by nearby galaxies—the supergalactic equator. The overlap with Shectman's sample is small. The FWHM of Abell clusters associated with the plane is 3000 km s^{-1} at a typical distance corresponding to 15,000 km s^{-1}.

e. Percolation occurs across of order the sample dimensions with a percolation scale length that is only 70% of the length defined by $n^{-1/3}$, where n is the number density of the sample. Such a low percolation parameter implies considerable connectedness across dimensions of at least 30,000 km s^{-1}.

In summary, the evidence for structure on a scale approaching 0.1c is surprisingly strong. If true, the clustering of clusters is much more extreme than the clustering of galaxies. For example, it would not have been suspected that galaxies cluster strongly with an essentially all-sky sample of only 300-400 systems. There have been fears that the largest-scale structure is an artifact of sample biases. However, it is unlikely that biases would lead to pronounced structure in three dimensions that is not obvious in two dimensions, very unlikely to produce a feature as thin in three-dimensional projection as the Pisces-Cetus plane, and most unlikely to produce the coincidence in position angle of the Pisces-Cetus and Local Supercluster planes demonstrated by the one-dimensional two-point correlation analysis.

REFERENCES

Shectman, S. A. 1985. *Ap J Suppl.* **57**, 77.
Tully, R.B. 1986. *Ap J.* **303**, 25.
_____ . 1987a. *Ap J.* **321**, 280.
_____ . 1987b. *Ap J.* **323**, 1.
_____ . 1988. *AJ.* **96**, 73.
Tully, R. B. and Fisher, J. R. 1987. *Nearby Galaxies Atlas.* Cambridge: Cambridge University Press.

LARGE-SCALE STRUCTURE AND MOTION
TRACED BY GALAXY CLUSTERS

NETA A. BAHCALL

Space Telescope Science Institute, Baltimore

1. INTRODUCTION

The study of the large-scale structure and motion in the universe is critical to our understanding of the formation and evolution of galaxies and structure. The large-scale structures observed today provide the cosmic fossils of conditions that existed in the early universe. The classic method of investigating structure in the universe is the observation of the spatial distribution of galaxies. Extensive surveys of thousands of galaxy redshifts are needed in order to cover large enough volumes and scales. Such has been carried out by several groups (see review papers by Oort 1983, Chincarini and Vettolani 1987, and Rood 1988; also Gregory and Thompson 1978, Gregory *et al.* 1981, Davis *et al.* 1982, Giovanelli *et al.* 1986, de Lapparent *et al.* 1986, da Costa *et al.* 1988).

A different approach is emphasized in this paper: using the high density peaks of the galaxy distribution, i.e., the rich clusters of galaxies, as tracers of the large-scale structure (see Bahcall 1988a for a more comprehensive review). As the mountain peaks trace mountain-chains on earth, rich clusters, with their low space density and large mean separation, serve as an efficient tracer of the largest-scale structures. While $\sim 10^6$ galaxies cover the high-latitude northern hemisphere to $\sim 18^m$, only ~ 500 rich clusters highlight the structure in the same volume of space. Recent results, summarized in this paper, show that clusters do indeed provide an efficient and effective tracer of the large-scale structure of the universe.

A Hubble constant of $H_o = 100h$ km s^{-1} Mpc^{-1} is used throughout this paper.

2. The Cluster Correlation Function

2.1 - Abell Clusters

The Abell (1958) catalog of rich clusters has been analyzed by many investigators using different techniques in an attempt to determine the spatial distribution of rich clusters. Abell (1958, 1961) found that the surface distribution of the clusters in his statistical sample was highly nonrandom and reported evidence suggesting the existence of superclusters; Bogart and Wagoner (1973), Hauser and Peebles (1973), and Rood (1976) (see also references therein) also found, using nearest-neighbor distributions and/or angular correlation functions, strong evidence for superclustering among the Abell clusters. The studies dealt primarily with the surface distribution of clusters and, in some cases, used approximate estimates for cluster redshifts. More recently, Bahcall and Soneira (1983, 1984) and, independently Klypin and Kopylov (1983), used redshift measurements of complete samples of clusters to determine directly the spatial distribution of rich clusters. I discuss below these results, as well as more recent results obtained by other investigators.

The two point spatial correlation function of clusters, $\xi_{cc}(r)$, was determined by Bahcall and Soneira (1983) (hereafter BS83) using Abell's (1958) statistical sample of rich clusters of galaxies of distance class $D \leq 4$ ($z \lesssim 0.1$), with redshifts for all clusters reported by Hoessel et al. (1980). This sample includes all 104 Abell clusters at $D \leq 4$ that are of richness class $R \geq 1$ and are located at high galactic latitude ($|b| \geq 30°$). A summary of the sample properties and its division into distance and richness classes, as well as into hemispheres, is presented in BS83. Also listed in the above reference are properties of the much larger and deeper $D = 5 + 6$ statistical sample ($z \lesssim 0.2$) that includes 1547 clusters. While only a small fraction of the redshifts are measured for this sample, it was used, because of its much larger number of clusters, in various comparison tests to strenghten and confirm the results obtained from the $D \leq 4$ sample.

The frequency distribution, $F(r)$, for all pairs of clusters with separation r in the sample was determined. In order to minimize the influence of selection effects on the determination of $\xi(r)$, a set of 1000 random catalogs was constructed, each containing 104 clusters randomly distributed within the angular boundaries of the survey region, but with the same selection functions in both redshift, $n(z)$, and latitude, $P(b)$, as the Abell redshift sample. The frequency distribution of cluster pairs was determined in both the real and random catalogs and the results were compared. This procedure ensures that the selection effects and boundary conditions will affect the data and random catalogs in the same manner.

The spatial correlation function was determined from the relation

$$\xi_{cc}(r) = F(r) / F^R(r) - 1, \tag{1}$$

where $F(r)$ is the observed frequency of pairs in the Abell sample and $F^R(r)$ is the corresponding frequency of random pairs (as determined by the ensemble average frequency of the 1000 random catalogs).

The resulting correlation function is presented in Figure 1. Strong spatial correlations are observed at separations $\lesssim 25h^{-1}$ Mpc. Weaker correlations are observed to larger separations of at least $\sim 50h^{-1}$ Mpc, and possibly $\sim 100h^{-1}$ Mpc, where $\xi_{cc} \sim 0.1$; beyond $150h^{-1}$ Mpc no statistically significant correlations are observed in the present sample.

The correlation function of Figure 1 can be well approximated by a single power law relation of the form $\xi_{cc}(r) = 300r^{-1.8}$, for $5 \lesssim r \lesssim 150h^{-1}$ Mpc. The function is smooth, with little scatter at $r \lesssim 50h^{-1}$ Mpc. At $r > 50h^{-1}$ Mpc, the scatter and uncertainties increase, but weak correlations of the order of 0.2 are still detected at these very large separations. When corrected for velocity broadening among clusters, the intrinsic rich ($R \geq 1$) cluster correlation function was determined by Bahcall and Soneira to be:

$$\xi_{cc}(r) = 360r^{-1.8} = (r/26)^{-1.8} \qquad r \lesssim 100h^{-1} \text{ Mpc}. \qquad (2)$$

In comparison, the correlation function of galaxies is given by (Groth and Peebles 1977, Davis and Peebles 1983):

$$\xi_{cc}(r) = 20r^{-1.8} = (r/5)^{-1.8} \qquad r \lesssim 20h^{-1} \text{ Mpc}. \qquad (3)$$

The rich cluster correlation function has the same shape and slope as that of the galaxy correlation function, but is considerably stronger at any given scale, by a factor of ~ 18, than the correlation function of galaxies. The cluster correlations also extend to greater separations than the scales observed in the galaxy correlations. The cluster correlation scale-length, i.e., the scale at which the correlation function is unity, is $r_0 \simeq 26h^{-1}$ Mpc (eq. 2), as compared with $r_0 \simeq 5h^{-1}$ Mpc for galaxies. The extent of the rich cluster correlation function beyond the reported $\simeq 15h^{-1}$ Mpc break in the galaxy correlation function (Groth and Peebles 1977) suggests the existence of large-scale structure in the universe ($\geq 15h^{-1}$ Mpc). While the reason for the strong increase of correlation strength and scale from galaxies to clusters is still a theoretical challenge, some possible explanations are discussed in Section 6 and in Bahcall (1988a). The cluster correlation function determined above places constraints on models for the formation of galaxies and structure.

In order to ensure that the spatial correlation function is not due to some special peculiarities in the nearby $D \leq 4$ sample, Bahcall and Soneira carried out several tests that are discussed below.

First, the angular correlation function of the much larger and deeper $D = 5 + 6$ sample (1547 $R \geq 1$ clusters to $z \lesssim 0.2$) was determined and compared

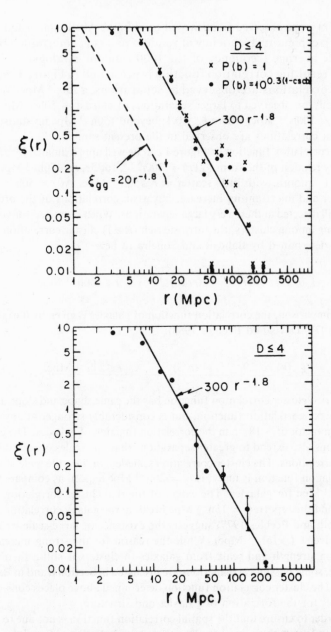

FIG. 1. Top: The spatial correlation function of the $D \le 4$ Abell cluster sample (BS83). Crosses refer to no correction for latitude selection function; dots refer to the full correction of $P(b)$. The solid line is the best fit 1.8 power law to the data. The dashed line is the galaxy-galaxy correlation function of Peebles and co-workers. Bottom: same, but plotted in larger bins at large separations.

with that expected from the spatial correlation function above (2), as well as from the expected scaling-law (Peebles 1980a) of the $D \leq 4$ angular correlation function. The angular correlation functions of the nearby $D \leq 4$ and distant $D = 5 + 6$ samples are determined to be (BS83)

$$w_{D \leq 4}(\theta) \simeq 3\theta^{-1} \qquad 0°5 \leq \theta \leq 25° \qquad (4)$$

$$w_{D = 5+6}(\theta) \simeq 0.8\theta^{-1} \qquad 0°2 \leq \theta \leq 14°. \qquad (5)$$

The angular correlation scale as expected from the scaling law applied to their respective distances. A comparison of the scaled functions is shown in Figure 2. If the correlations were mainly due to patchy obscuration or other omissions by Abell, the (observed) scaling would not be expected. The scaling agreement indicates that any possible projection biases in the catalog (e.g., Sutherland 1988) are rather small and do not significantly affect the correlation results (see also Dekel 1988). The reduced correlation scale suggested by Sutherland may result from overcorrecting the actual correlation power on large scales. A comparison of the $D = 5 + 6$ angular function with that expected from the spatial correlation function of equation (2), when integrated over the relevant redshift distribution, is given by BS83. The agreement between the $D \leq 4$ and $D = 5 + 6$ functions is excellent. This agreement indicates that the $D \leq 4$ redshift sample is a fair sample of the much larger sample, and that the observed correlations represent real correlations of clusters in space. The scaling-law of the angular functions was also studied by Hauser and Peebles (1973) who reached similar conclusions with regard to the reality of the intrinsic correlations.

Second, the angular correlation function was compared with the pure redshift (i.e, line-of-sight) correlations of the clusters. If the correlations were mostly due to patchy obscuration on the sky or other similar biases, no extensive redshift correlations would be expected. It is observed (BS83) that the redshift correlations are indeed positive and extend to large scales similar to those in the projected correlations, further strengthening the reality of the correlations.

Third, the angular correlation function of the $D = 5 + 6$ sample was determined in different regions of the sky, yielding consistent results within the uncertainties (Figure 3).

These tests, and those listed in Section 2.5 below, suggest that the observed cluster correlation function is mostly due to physical clustering of rich clusters of galaxies that extends to large scales.

Since the correlation strength appears to increase from galaxies to clusters, Bahcall and Soneira also investigated whether a similar trend is observed between the correlation function of poor and rich clusters. The angular correla-

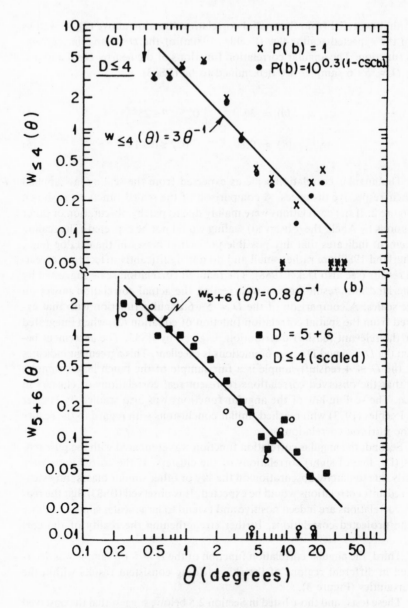

FIG. 2. Top: The angular correlation function of the $D \leq 4$ Abell cluster sample (BS83); Bottom: the angular correlation function of the deep $D = 5 + 6$ sample (squares). Open circles are the angular correlation function of the $D \leq 4$ sample (Figure 2) scaled by the standard scaling law of an intrinsic spatial correlation using the distance ratio of the two samples (Sec. 2). Correlations of ≤ 0.05 are rather uncertain. The position of the mean Abell radius is indicated by the arrow.

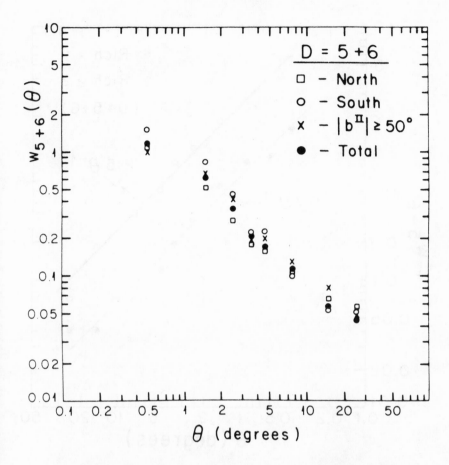

FIG. 3. The angular correlation function of the $D = 5 + 6$ Abell cluster sample for different zones: northern hemisphere (*squares*); southern hemisphere (*circles*); high latitude, $|b| \geq 50°$ (*crosses*); and the total sample (*dots*). A very strong south polar apparent supercluster increases somewhat the southern correlations at small separations. Other zones of low latitude ($|b| = 30° - 50°$) and different longitudes yield similar results (BS83).

tion functions of different richness classes ($R = 1$ and $R \geq 2$) were determined for the large $D = 5 + 6$ sample (1125 $R = 1$ clusters, 422 $R \geq 2$ clusters). The amplitude of the correlation function was found to be strongly dependent on cluster richness, with richer clusters ($R \geq 2$) showing stronger correlations by a factor of ~ 3 as compared with the poorer ($R = 1$) clusters. The results are shown in Figure 4. Both richness classes exhibit the same power-law shape correlation function as observed in the total sample; they satisfy

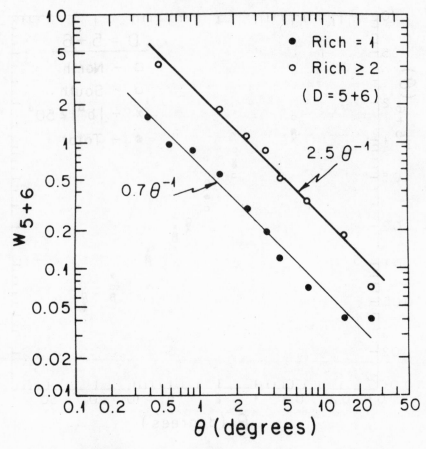

FIG. 4. The angular correlation function of richness 1 and richness ≥ 2 Abell clusters in the $D = 5 + 6$ sample (BS83).

$$w_{5+6}(\theta) \simeq 0.7\theta^{-1} \qquad R = 1 \tag{6}$$

$$w_{5+6}(\theta) \simeq 2.5\theta^{-1} \qquad R \geq 2. \tag{7}$$

The implied spatial correlation can then be represented by:

$$\xi_{cc}(r) \simeq (r/24)^{-1.8} \qquad R = 1 \tag{8}$$

$$\xi_{cc}(r) \simeq (r/48)^{-1.8} \qquad R \geq 2. \tag{9}$$

The amplitude of the total ($R \geq 1$) correlation function is dominated by the lower amplitude of the poorer, but more numerous, $R = 1$ clusters. BS83 suggested, therefore, that the correlation function depends on the richness of the system, increasing in strength from single galaxies to poor and rich clusters (see also Section 4 for a more updated richness dependence). The galaxy-cluster cross-correlation function (Seldner and Peebles 1977; see, however, Efstathiou 1988; also Section 2.4) is consistent with the cluster correlations and the trend observed above. Recent observations of clusters of different types and richnesses (see summary below) yield results that are consistent with the richness trend suggested by BS83 (Section 4).

Klypin and Kopylov (1983) investigated the spatial correlation function of a nearby sample of Abell clusters similar to the one described above, supplementing available redshift data with their own observations. Their sample includes 158 Abell clusters of all richness classes ($R \geq 0$; i.e., including the somewhat incomplete class of $R = 0$ clusters) in distance group $D \leq 4$, and located at $|b| \geq 30°$. Their results are consistent with those of BS83. They find $\xi_{cc}(r) = (r/25)^{-1.6}$ for their observed range of $r \lesssim 50h^{-1}$ Mpc. The approximately 10% difference in slope is within the 1σ uncertainty of the slope determination estimated by BS83.

The earlier work of Hauser and Peebles (1973) used power-spectrum analysis and angular correlations to investigate the distribution of clusters in the Abell catalog. They also find evidence for strong superclustering of clusters, and show that the degree and angular scale of the apparent superclustering varies with distance in the manner expected if the clustering is intrinsic to the spatial distribution rather than a consequence of patchy local obscuration.

Additional recent investigations of the spatial distribution of rich clusters of galaxies in the Abell catalog include Kalinkov et al. (1985), Batuski and Burns (1985), Postman et al. (1986), Shvartsman (1988), Szalay et al. (1988), and Huchra (1988). These studies include investigations of different subsamples of the catalog, to different distances, regions, and/or richnesses, as well as apply different techniques and/or corrections. All the investigations yield consistent results with those described above, as summarized below.

Kalinkov et al. (1985) find a spatial correlation function for rich ($R \geq 1$) Abell clusters, using new redshift estimator calibrations and richness corrections, of $\xi_{cc}(r) = (r/22.4)^{-1.9}$ for $r \lesssim 80h^{-1}$ Mpc.

Batuski and Burns (1985) determined the spatial correlation function for Abell clusters of all richness groups ($R \geq 0$) to $z \simeq 0.085$. The sample includes 226 clusters. (The higher spatial density of this sample as compared with the $R \geq 1$ sample is due to the inclusion of the $R = 0$ clusters). For this sample they find $\xi_{cc}^{R \geq 0}(r) = 65r^{-1.5}$ for $r \lesssim 150h^{-1}$ Mpc. The somewhat shallower slope, while within 2σ of the 1.8 slope, may be partially due to the use of some estimated rather than measured redshifts, which reduces the correlations on small scales and flattens the slope (see BS83). When approximated as a 1.8 power-law slope, the function is $\xi_{cc}^{R \geq 0}(r) \simeq 200(r)^{-1.8} \simeq (r/19)^{-1.8}$. This correlation function is one order of magnitude stronger than the galaxy correlations, and about 50% lower than BS83 correlation function for $R \geq$ 1 clusters. The somewhat reduced correlation strength is consistent with the richness dependence suggested by BS83 and Bahcall and Burgett (1986).

Postman *et al.* (1986) re-analyzed the $D \leq 4$ sample used by Bahcall and Soneira, as well as a sample of 152 Abell clusters to $z \leq 0.1$. Their results are consistent with the BS83 correlation functions.

Shvartsman (1988) and Kopylov *et al.* (1987) used the 6-meter USSR telescope to measure redshifts of all very rich ($R \geq 2$) Abell clusters to $z \leq 0.23$, located at $b > 60°$. They calculated the spatial correlation function of this deep sample of very rich clusters, that includes 50 clusters in the redshift range $0.10 \leq z \leq 0.23$. They find $\xi_{cc}^{R \geq 2}(r) = (r/40)^{-1.5 \pm 0.5}$ for the range $5 \leq r \leq 50h^{-1}$ Mpc, consistent with the BS83 correlations of very rich ($R \geq 2$) clusters, and with the suggested increase of correalation strength (and length) with richness. The correlation scale for the $R \geq 2$ clusters is $\sim 40h^{-1}$ Mpc, while the correlation scale for the $R \geq 1$ clusters is $\sim 25h^{-1}$ Mpc. The above authors also report weak but positive correlations at much larger separations: $\xi(100\text{-}150h^{-1} \text{ Mpc}) = 0.47 \pm 0.14$. A similar result is suggested by Batuski *et al.* (1988). This is comparable to the supercluster correlation results of Bahcall and Burgett (1986) (Sec. 3) where similar marginal (3σ) correlations are detected. Systematic effects, however, which may be important on these scales, are difficult to assess.

Geller and Huchra (1988) used a deep redshift sample ($z \leq 0.2$) of Abell clusters complete over a small region of the northern sky. They find $\xi_{cc}^{R \geq 0}(r)$ $\sim (r/20)^{-1.8}$, consistent with the results discussed above.

The new southern hemisphere catalog of rich clusters (Abell, Corwin, and Olowin 1988) can also be analyzed for structure. Bahcall *et al.* (1988b) analyzed the distribution of clusters in this catalog. Preliminary results suggest that the correlation function of clusters in the southern sky is consistent with the results presented above for northern clusters.

2.2 - Shectman Clusters

Shectman (1985) used the Shane-Wirtanen (1967) counts to identify clusters of galaxies by finding local density maxima above a threshold value, after slightly smoothing the data to reduce the effect of the sampling grid. A total of 646 clusters of galaxies was identified using the specified selection algorithm.

The radial velocity distribution of these clusters is similar to the radial velocity distribution of Abell clusters of distance class $D \leq 4$ as determined by Shectman from comparisons of velocity data for a complete sample of 112 clusters. The space density of the Shectman clusters is therefore ~ 6 times greater than the space density of the 104 $R \geq 1$, $D \leq 4$ Abell cluster sample.

The angular two-point correlation function of the Shectman clusters at $|b| \geq 50°$ (a sample of 488 clusters in total) was determined by Shectman

(1985). The implied spatial correlation function is $\xi_{cc}(r) \simeq 180r^{-1.8} \simeq (r/18)^{-1.8}$. This correlation function is about ten times larger than the galaxy correlation (eq. 3), and is about a factor of two lower than the rich ($R \geq 1$) cluster correlations (eq. 2). Since the space density of the Shectman clusters is ~ 6 times higher than the density of the $R \geq 1$ clusters, and the identifications of the former are therefore with poorer clusters, the results of the Shectman cluster correlations are consistent with those of the Abell clusters, and consistent with the trend suggested by Bahcall and Soneira of increased correlation strength with cluster richness (Sec. 4).

2.3 - Zwicky Clusters

The angular distribution of clusters in the Zwicky catalog (1968) was analyzed by Postman *et al.* (1986). The cluster selection algorithm in the Zwicky catalog differs markedly from the cluster selection definition of Abell (e.g., Bahcall 1977, 1988a). Abell's definition of a cluster relates to the cluster intrinsic properties (i.e., number of galaxies within a given linear scale, and a given absolute magnitude range) and thus is independent of redshift (except for standard selection biases). Zwicky's clusters are defined relative to the mean density of the field, with varying cluster sizes and contours, and consider all galaxies down to the plate limit. Therefore, the cluster selection is by definition strongly dependent on redshift. A direct comparison between the correlation function of Zwicky and Abell clusters is, therefore, not straightforward. However, an uncorrected comparison will test to some extent the universality of the cluster correlation function, with its suggested dependence on richness, as well as further test the sensitivity of the correlation function to the cluster identification procedure.

It is found that in the distance range where Abell and Zwicky identify clusters of comparable overdensity (1173 distant Zwicky clusters with $z \simeq 0.1 - 0.14$), the correlation functions of the Abell and Zwicky clusters are indeed the same in the scale range studied ($r \leq 60h^{-1}$ Mpc). The angular correlation functions of the two nearer samples of the Zwicky clusters (377 Near clusters and 680 Medium-Distant clusters) are observed to be weaker (when scaled to the same depth as the $D \leq 4$ Abell sample) than the rich ($R \geq 1$) Abell clusters. Since these nearer Zwicky clusters are by definition much poorer clusters, with a considerably higher space density than the $R \geq 1$ Abell clusters, they are expected to have a weaker correlation strength (BS83 and Sec. 4).

A comparison of the cluster correlation functions determined by the various investigators discussed above using different catalogs and samples is summarized in Figures 5a-5b. A general agreement is observed among all the results. The consistency of the correlation functions determined from different catalogs, cluster selection criteria, redshift and richness ranges, and by different investigators, strongly supports the reality and universality of the cluster correlations described in this section.

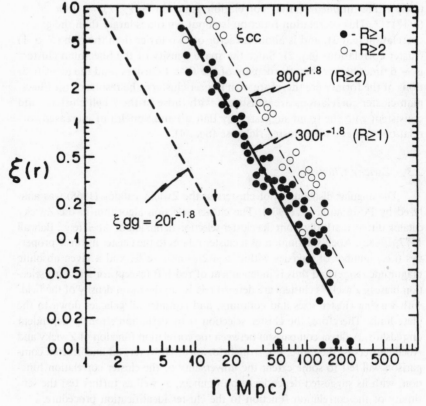

Fig. 5a. A composite of the spatial cluster correlation function determined by different investigators from different cluster samples. Abell clusters are to different depths ($z \simeq 0.08$ through $z \simeq 0.24$), different richnesses, and different regions; (Sec. 2). The BS83 correlation function, $300r^{-1.8}$, is shown (see Figure 1); the results of the different samples are all consistent with this function. The richer $R \geq 2$ clusters exhibit stronger correlations as suggested by BS83 (Sec. 4).

2.4 - Galaxy-Cluster Cross-Correlations

The angular cross-correlation between the galaxy distribution in the Shane-Wirtanen galaxy counts and the positions of rich Abell clusters was studied by Seldner and Peebles (1977), and more recently by Efstathiou (1988). This cross-correlation function, $w_{gc}(\theta)$, measures the excess probability, over random, of finding a galaxy within a given separation from a cluster; i.e., it describes the enhanced density of galaxies around a cluster.

Seldner and Peebles (1977) find that the angular function $w_{gc}(\theta)$ scales with cluster distance-class D as expected from the galaxy luminosity function.

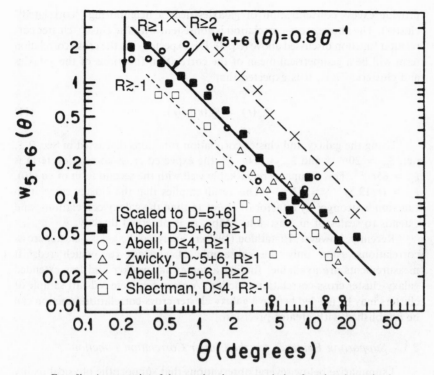

FIG. 5b. A composite of the angular cluster correlation function determined from different catalogs and samples, indicated by the different symbols (Sec. 2). All results are scaled to the $D = 5 + 6$ distance. The BS83 correlation function for the $D = 5 + 6$ clusters (Figure 2) is indicated by the solid line. The consistency among the different samples, as well as the dependence of the correlation strength on richness (Sec. 4), are apparent.

The $w_{gc}(\theta)$ estimates are reasonably well fitted by a two-power-law model for the function, $\xi_{gc}(r)$ (Peebles 1980b):

$$\xi_{gc}(r) = (r/7)^{-2.5} + (r/12.5)^{-1.7} \qquad r \lesssim 40h^{-1}\mathrm{Mpc}. \qquad (10)$$

The enhancement of Lick counts around cluster centers is traced to $r \sim 40h^{-1}$ Mpc before it is lost in the noise.

The first term of the galaxy-cluster cross-correlation (eq. 10) represents the "standard" internal density profile of galaxies in a cluster (which generally has the shape of a bounded isothermal sphere; e.g., Bahcall 1977). The more slowly varying part of the cross-correlation function found at larger scales and represented by the $r^{-1.7}$ part of eq.(10) is produced by the clustering of clusters, as discussed in the previous subsections (i.e., galaxies from one cluster provide excess concentration of galaxies near a neighboring "correlated" cluster). The above cross-correlation is consistent with the cluster-cluster correlation function discussed above (eq. 2). It is expected that the cross-correlation term will be a geometrical mean of the correlation functions of the galaxies and clusters. Thus, it is expected that

$$\xi_{gc}(r) \simeq \xi_{gg}^{\frac{1}{2}}(r)\xi_{cc}^{\frac{1}{2}}(r). \tag{11}$$

Using the galaxy and cluster correlation functions discussed in Sec. 2.1, i.e., $\xi_{gg} \simeq 20r^{-1.8}$ and $\xi_{cc} \simeq 360r^{-1.8}$, the expected cross-correlation term is $\xi_{gc} \simeq 85r^{-1.8}$. This compares remarkably well with the second term of eq.(10), $\xi_{gc} \simeq (r/12.5)^{-1.7} \simeq 73r^{-1.7}$. This result implies that the cluster correlation function is stronger by a factor of about 16 than the galaxy correlations, and extends to scales of at least $40h^{-1}$ Mpc, as is observed directly.

Recently, however, Efstathiou (1988) re-analyzed the galaxy-cluster cross-correlations using only the subsample of clusters for which redshift measurements are available, finding a somewhat weaker and less extended galaxy-cluster cross-correlation function. A more complete redshift sample of clusters may be needed before a galaxy-cluster cross-correlation function can be established with greater precision.

2.5 - Supporting Evidence for the Cluster Correlation Function

I summarize below several observations that support the physical reality of the cluster correlation function discussed above.

a. The angular cluster correlation function scales with depth as expected from spatial correlations, rather than from patchy obscuration or systematic omission (Hauser and Peebles 1973, BS83).

b. The projected and redshift cluster correlation functions yield consistent results, thus indicating the physical reality of the correlations.

c. The cluster correlation function yields consistent results in different large regions of the sky (e.g., north versus south, high versus low latitudes, different longitude ranges; BS83). The estimated scatter, or uncertainty, in the correlation function is approximately $\pm 15\%$ in the correlation scale (i.e., approximately $\pm 25\%$ in the amplitude).

d. The cluster correlation determined from the Abell sample is consistent with more recent results using other samples and catalogs (e.g., Shectman clusters - Shectman 1985; Zwicky clusters - Postman et al. 1986; and subsamples of different regions, redshifts, and richnesses in the Abell catalog - BS83; Batuski and Burns 1985; Postman et al. 1986; Shvartsman 1988; Szalay et al. 1988; see Figure 5.). These comparisons provide a strong test of the sensitivity of the correlation function to the cluster identification procedure. The scatter in the correlation scale-length is $\lesssim \pm 15\%$, as discussed above.

e. The cluster correlation function is consistent with the galaxy-cluster cross-correlation function determined by Seldner and Peebles (1977). It is less consistent, however, with the cross-correlation results of Efstathiou (1988).

f. A preliminary estimate of the completeness limit of the *nearby* Abell sample obtained by comparisons with X-ray data of galaxy clusters yields a reasonably high completeness level for the sample (work in progress by Bahcall, Maccacaro, Gioia, *et al.*). Of the 25 nearest clusters in the sample, all but one are detected as extended X-ray sources with luminosities appropriate to rich clusters (the 25th cluster has an upper limit consistent with the expected luminosity). In addition, preliminary results of the Einstein Medium Deep X-ray Survey show that no extended X-ray cluster is found that should have been a rich cluster in Abell's nearby sample, but was missed (out of approximately four real clusters expected within the survey area; i.e., completeness of better than $\sim 75\%$).

The suggestion of a possible "projection" effect in the Abell catalog, in which the existence of one cluster close to a neighboring cluster enhances the richness of the latter (Sutherland 1988) has only a minimal effect on the resulting correlation function (e.g., Dekel 1988). At a separation of $\sim 10h^{-1}$ Mpc from the cluster center, typically less than ~ 1-3 galaxies from the tail of the foreground cluster will contribute to the neighboring cluster. A negligible effect is expected on scales much larger than the above. A significant projection effect would also be inconsistent with the observed scaling-law between the nearer and distant samples. The consistent results obtained for the correlation functions of clusters from different catalogs and samples also indicate that any such selection effects are likely to have only a minimal impact on the resulting correlations.

The evidence listed above supports the reality of the cluster correlation function and suggests that it is unlikely that the correlations are mainly a result of catalog biases or omissions. A determination of the cluster correlation function from catalogs with automated selection procedures will improve the accuracy of the intrinsic cluster correlations, especially at large separations where the correlations are rather weak.

3. SUPERCLUSTER CORRELATIONS

Bahcall and Burgett (1986) carried the study of rich galaxy clusters one step further by studying the spatial distribution of superclusters. The sample used was the Bahcall-Soneira (1984) complete catalog of superclusters to $z \leq 0.08$, where superclusters are defined as groups of rich clusters and identified by a spatial density enhancement of clusters. All volumes of space with a spatial density of clusters f times larger than the mean cluster density are identified in the above catalog as superclusters for a specified value of f. The supercluster selection process was repeated for various overdensity values f, from

$f = 10$ to $f = 400$, yielding specific supercluster catalogs for each f value. A total of 16 superclusters are cataloged for $R \geq 1$ and $f = 20$, and 26 superclusters for $R \geq 0$ and $f = 20$.

The spatial correlation among the superclusters was determined by Bahcall and Burgett (1986) for samples of different richness and overdensity. Because of the large size of the superclusters themselves, no meaningful correlations are expected at small separations (≤ 50 h^{-1} Mpc). In addition, no detectable correlations are expected at very large separations (> 200 h^{-1} Mpc), since this scale is comparable to the limits of the sample. Any observable correlations are therefore expected only in a separation "window" around ~ 100 h^{-1} Mpc.

The results, presented in Figure 6, reveal correlations among superclusters on a very large scale: ~ 100-150 h^{-1} Mpc. Because of the small size of the supercluster sample, the statistical uncertainty is appreciable; the observed effect is at the 3σ level (as determined by comparisons with numerical simulations of random catalogs). In addition, all the samples with different overdensities and cluster richnesses show a similar effect at a similar scale length. The results imply the existence of very large-scale structures with scales of ~ 100-150 h^{-1} Mpc.

Similar results have been recently obtained by Kopylov *et al.* (1987) by studying correlations of very rich clusters to $z \leq 0.2$ (Sec. 2.1). They report $\xi_{cc}(100$-150h^{-1} Mpc$) = 0.47 \pm 0.14$. Tully's (1988, 1987) observations of very large-scale structures in the cluster distribution, up to ~ 300h^{-1} Mpc, may also reflect the above observed tendency of superclusters to cluster.

Figure 6 shows that the supercluster correlation strength is stronger than that of the rich cluster correlations by a factor of approximately 4. It is approximately two orders of magnitude stronger than the galaxy correlation amplitude. While this enhancement is observed in the ~ 100-150h^{-1} Mpc range, it is possible that the supercluster correlation function also follows an $r^{-1.8}$ law. If the correlations follow an $r^{-1.8}$ law, then the function would satisfy the relation

$$\xi_{sc,sc}(r) \simeq 1500r^{-1.8} \simeq (r/60)^{-1.8}. \qquad (12)$$

The implied correlation scale of superclusters would be 60h^{-1} Mpc, as compared with 5h^{-1} Mpc for the correlation scale of galaxies (Groth and Peebles 1977) and 25h^{-1} Mpc for rich ($R \geq 1$) clusters (BS83). This apparent increase in correlation strength is consistent with the earlier prediction of BS83 of increased correlations with richness (luminosity) of the system. The supercluster correlation amplitude fits well the predicted trend (Sec. 4).

4. RICHNESS DEPENDENCE OF THE CORRELATIONS

As discussed above, the cluster correlation function appears to depend strongly on cluster richness (BS83), with richer clusters showing stronger cor-

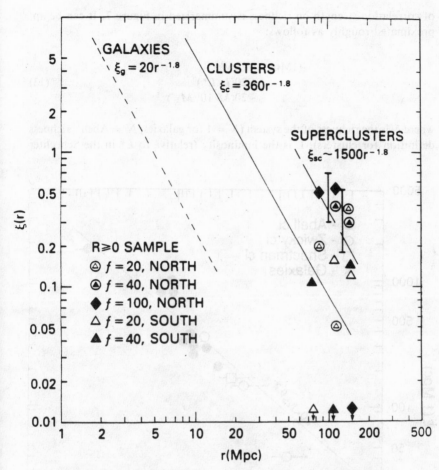

FIG. 6. The spatial correlation of superclusters for the $R \geq 0$ sample (Bahcall and Burgett 1986). Different sub-samples are indicated by different symbols. No meaningful correlations are expected below $\sim 50h^{-1}$ Mpc.

relations than poorer clusters. This result, combined with the lower correlation amplitude of individual galaxies, led Bahcall and Soneira to the conclusion that progressively stronger correlations exist, at a given separation, for richer or more luminous galaxy systems (Sec. 2.1). Several recent studies of the correlations of other types and richnesses of clusters, reviewed above (Batuski and Burns 1985; Shectman 1985, and Postman *et al.* 1986 for poorer clusters; Kopylov *et al.* 1987 for richer clusters; Bahcall and Burgett 1986 for superclusters) appear to be consistent with the trend suggested by Bahcall and Soneira and later expanded by Bahcall and Burgett (1986). This dependence of correlation strength on richness is summarized in Figure 7. It can be approximated roughly as follows:

$$\begin{aligned}
\xi(1\mathrm{Mpc}) \ &\sim 20N^{0.7} \\
&\sim 20(L/L^*)^{0.7} \\
&\sim 20(M/10^{12}M_\odot)^{0.5}
\end{aligned} \tag{13}$$

where N is the richness of the system ($N = 1$ for galaxies; $N =$ Abell's richness definition for clusters), L is the luminosity (relative to L^* in the Schechter

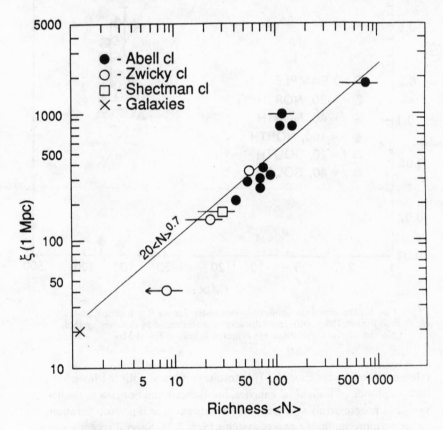

FIG. 7. The dependence of the correlation function strength on the mean richness ($\propto$ luminosity) of the system (Bahcall and Burgett 1986). The results are for clusters from different catalogs (Abell, Zwicky, and Shectman, as indicated by the symbols), determined by different investigators for samples of different richnesses, depths, and regions (Sec. 2). The correlation strength for galaxies and superclusters are also included. The solid line indicates the approximate dependence on richness.

luminosity function), and M is the mass of the system. This relation suggests an average trend in the data and should not be regarded as an exact formula. (Obviously, the relation between N, L, and M is not unique; for a given N, different L's and M's may apply and vice versa. The difference between the M versus the L slope is due to the higher observed M/L ratios for clusters than for galaxies).

The correlation-richness dependence suggests that rich clusters populate the large-scale structures, or superclusters, more abundantly than galaxies do relative to their mean space densities. It also implies that rich clusters are indeed an efficient tracer of large-scale structure in the universe.

Several galaxy formation models such as biased cold dark matter, hybrid scenarios, and cosmic strings can reproduce a trend of increasing correlation strength from galaxies to clusters.

5. A Universal Correlation Function

The increase of correlation strength with richness implies that rich, luminous systems are more strongly clustered, at a given separation, than poorer systems. The power-law of the correlation functions is also observed to be identical in the various systems studied. Either initial conditions, or subsequent evolution, may be responsible for the observed phenomena. Since the observed correlation functions follow the same power law ($r^{-1.8}$), the effect of increased correlation strength with richness (at a given separation) can also be expressed as a scale shift in the correlation functions (Szalay and Schramm 1985). In Figure 8 I plot the amplitude of the correlation functions of the various systems (galaxies, poor and rich clusters, superclusters) as a function of the mean separation of objects in the sample, d (see Bahcall and Burgett 1986; Bahcall 1987). The mean separation is related to the mean spatial density of objects in the sample, n, through $d = n^{-1/3}$. For example the mean separation of galaxies is about 5 Mpc, while the mean separations of $R \geq 1$ and $R \geq 2$ clusters are, respectively, about 50 Mpc and 70 Mpc.

It is apparent from Figure 8 that the correlation strength increases with the sample's mean separation. Moreover, a dimensionless correlation function normalized to the sample's mean separation, d, appears to yield a constant, universal function for nearly all the systems studied (some enhancement is required for galaxies, as described below). This universal dimensionless correlation function has the form

$$\xi_i(r) \simeq 0.3(r/d_i)^{-1.8} \simeq (r/0.5d_i)^{-1.8}, \tag{14}$$

where the index i refers to the system being considered, and d_i is its mean separation. Relation (14) implies a universal dimensionless correlation

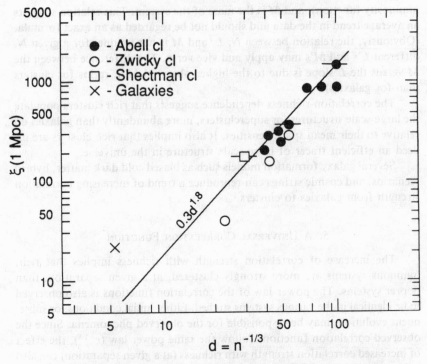

FIG. 8. The dependence of the correlation function on the mean separation of objects in the system. The results are for clusters from different catalogs (Abell, Zwicky, and Shectman, as indicated by the symbols), determined by different investigators for samples of different mean densities (i.e., mean separations). The correlation strength for galaxies and superclusters are also included. The solid line represents a $d^{1.8}$ dependence (e.g., Szalay and Schramm 1985, Bahcall and Burgett 1986).

amplitude of ~ 0.3, and, equivalently, a universal correlation scale of $r_o \simeq 0.5d_i$. The correlation function of galaxies is stronger than expressed by relation (14) by a factor of about four (Figure 8). The universality of the correlation function suggests a scale-invariant clustering process (Szalay and Schramm 1985). The stronger dimensionless galaxy correlations may imply gravitational enhancement on smaller scales. If a non-linear process, other than gravity, participates in galaxy formation, and this process is scale-invariant, the created structure will have a single power law correlation function, the slope of which (α) is related to the geometry of the structure, i.e., its fractal dimension (β). The latter is related to the correlation function slope via $\alpha = \beta - 3$ (see, e.g., Mandelbrot 1982). The fractal dimension of the universal structure implied by the above data is therefore $\beta \simeq 1.2$. Small-scale gravitational clustering may break the scale invariance and increase the dimensionless correlation amplitude for galaxies.

We do not know yet what physical process can create a scale invariant structure with $\beta \simeq 1.2$. An innovative suggestion involves cosmic strings as the primary agent in the formation of galaxies and clusters; this model appears to create such a scale-invariant infrastructure (Turok 1985). The model yields a scale-invariant correlation function similar to that observed, with a power-law of -2 (as implied by one-dimensional "string" structures with fractal dimension of unity). More detailed calculations with cosmic strings models are currently being carried out by several investigators (Bouchet and Bennet 1988, Turok 1988).

6. Phenomenological Clustering Models

6.1 - Long Tails to Galaxy Clusters

The galaxy correlations depend, at least partially, on the rich cluster correlations since clusters contain galaxies. If all galaxies were members of rich clusters, the two correlation functions should be approximately the same on large scales. The fraction of galaxies in clusters is clearly less than unity. The fraction of galaxies, f, that are associated with rich clusters, represents the probability that a randomly chosen galaxy is correlated with a rich cluster. These associations may include large structures (tens of Mpc), comparable to the separations observed in the cluster correlation function (and well above the standard Abell radius of $1.5h^{-1}$ Mpc).

The galaxy correlation function contains contributions from three terms (Bahcall 1986): galaxy pairs from the fraction f of galaxies that are cluster members; pairs from the fraction $1-f$ of galaxies that are non-cluster members ("field"); and cross-term pairs. Inserting the analytic expressions for each of these terms into the expression for the overall galaxy correlation function yields:

$$\left(\frac{\xi_{cc}}{\xi_{gg}}\right)^{\frac{1}{2}} = \frac{1 - (1 - f)\,(\xi_{gg}^f/\xi_{gg})^{\frac{1}{2}}}{f}. \tag{15}$$

The above ratio of the cluster to galaxy correlation strength depends on two parameters: the fraction of galaxies in clusters, f, and the ratio of the "field" galaxy correlation strength, ξ_{gg}^f (i.e., the correlation of the $1-f$ fraction of galaxies outside the clusters) to the overall galaxy correlation ξ_{gg}. If all galaxies were associated with rich clusters, $i.e., f = 1$, then the galaxy and cluster correlations are identical on large scales, as expected. However, for any fraction $f < 1$, the galaxy correlations will be weaker than parent cluster correlations due to the reducing effect of the less clustered "field" galaxies. Figure 9 represents graphically relation (15). The curves are the expected $\xi_{cc}/\xi_{gg}(f)$ relations for selected values of the parameter $\chi \equiv \xi_{gg}^f/\xi_{gg}$. The observed cor-

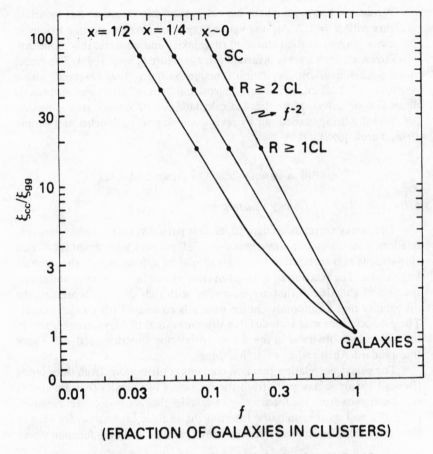

FIG. 9. The ratio of the cluster to galaxy correlation functions predicted from Bahcall's (1986) model (eq. 15) is plotted as a function of the fraction f of galaxies associated with the clusters. The different curves represent different values of the "field" (non-cluster) correlation strength, ξ_{gg}^f, in terms of the ratio parameter $x = \xi_{gg}^f / \xi_{gg}$. The observed correlation strengths of $R \geq 1$ clusters, $R \geq 2$ clusters, and superclusters are indicated by the points.

relation strengths are represented by the data points. The observed ratio $\xi_{cc} / \xi_{gg} \simeq 18$ for $R \geq 1$ clusters yields a fraction of galaxies in clusters that ranges from $f \simeq 25\%$ for $\xi_{gg}^f / \xi_{gg} \simeq 0$ to $f \simeq 15\%$ for $\xi_{gg}^f / \xi_{gg} \simeq \frac{1}{4}$. Therefore, if approximately 20% of all galaxies are associated with rich ($R \geq 1$) clusters, the galaxy correlation function will be, as observed, ~ 18 times weaker than the cluster correlations.

The model suggests (Bahcall 1986) that the fraction of galaxies associated with rich clusters is considerably larger than previously believed; most of these galaxies would be distributed in the outer tails of the clusters, which may extend to at least $\sim 30h^{-1}$ Mpc. Most clusters are therefore predicted to be embedded within much larger structures.

The model makes testable predictions that can be studied with complete redshift surveys. Redshift surveys of galaxies (e.g., Gregory and Thompson 1978; Gregory *et al.* 1981; Chincarini *et al.* 1981; Haynes and Giovanelli 1986; de Lapparent *et al.* 1986) indeed suggest that clusters are generally embedded in large elongated structures that contain a considerable fraction of galaxies. This picture is qualitatively consistent with the phenomenological model described above.

The long-tail model may also explain the negligible correlations observed among Ly-α clouds in QSO spectra (Sargent *et al.* 1980). If the clouds can only exist in the field (non-cluster environment) because clusters have too large an ambient pressure, they would not be expected to have significant correlations.

6.2 - The Shell Model

Galaxies may be distributed on surfaces of shells (or cells), with rich clusters located at shell intersections. This picture is suggested by redshift surveys of galaxies (Gregory *et al.* 1981, Giovanelli *et al.* 1986, de Lapparent *et al.* 1986). In order to test this "shell" model and its agreement with the observed galaxy and cluster correlations, Bahcall, Henriksen, and Smith (1988) (see also Bahcall 1988b) made simulations in which they placed galaxies on surfaces of randomly distributed shells, and formed clusters at the shell intersections. They found that the model cluster correlations are consistent with the observed cluster correlations, including the large increase in correlation strength from galaxies to clusters. The results are not very sensitive to the exact parameters used. The model galaxy correlations appear to be consistent with observations on small scales, but exhibit a tail of weak positive correlations at larger separations that are not seen in the data.

An example of results from a typical model is shown in Figure 10; more details are given in Bahcall *et al.* (1988a). The results suggest that the observed cluster correlations may be simply due to the geometry of clusters positioned on randomly placed shells or similar structures; the typical structure size is best fit with a radius of approximately $20h^{-1}$ Mpc.

7. PECULIAR MOTION OF CLUSTERS

The discussion in the previous sections summarizes evidence for the existence of structures on the scale of $\sim 10\text{-}150h^{-1}$ Mpc. A question of critical importance is what are the velocity fields in these structures. Peculiar velocities of clusters on these scales may indicate the existence of large amounts of (dark)

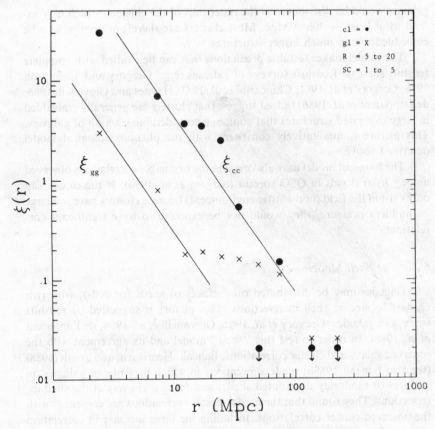

Fig. 10. Shell-model correlation functions (Bahcall *et al.* 1988) for clusters (dots) and galaxies (crosses), and their comparisons with observations (ξ_{gg}, and ξ_{cc} lines). The plotted model represents shell radii distribution in the range 5 to 20 h^{-1} Mpc, and, depending on size, 1 to 5 clusters forming at shell intersections (Sec. 6.2).

matter and are of fundamental importance for models of galaxy and structure formation. Early discussions of possible peculiar velocities among clusters in superclusters are presented by Abell (1961) and Noonan (1977). Noonan observed a tendency of clusters with neighboring Abell clusters to have a greater scatter on the Hubble diagram, which was interpreted as a gravitational perturbation on the cluster redshifts due to the neighboring clusters. More recently, Bahcall *et al.* (1986) used the complete redshift sample of $D \leq 4$ rich Abell clusters to study the possible existence of peculiar motion and/or structural anisotropy on large scales. They find strong broadening in the redshift distribution that corresponds to a cluster velocity of ~ 10^3 km s^{-1}. These findings are summarized below.

7.1 - Redshift Elongation: The "Finger-of-God" Effect

The distribution in space of the $D \leq 4$ redshift sample of Abell clusters was studied by Bahcall *et al.* (1986) by separating the three-dimensional distribution into its components along the line-of-sight (redshift) axis and the perpendicular axes projected on the sky. All clusters were assumed to be located at their Hubble distances as indicated by their redshifts, and their pair separations in Mpc were determined in the three components. A scatter-diagram of the cluster pair separations in the redshift (z) direction (R_z) versus their separations in α or δ $(R_\alpha$ or $R_\delta)$ was then determined.

If all clusters were located at their Hubble distances with negligible peculiar motion, and if the sample was not dominated by elongated structures in a given direction, a symmetric scatter-diagram should be observed. If a large peculiar velocity exists among clusters, it would manifest itself as an elongated distribution along the z-direction in the R_z-R_α and R_z-R_δ diagrams. This elongation, i.e., the so-called "Finger-of-God" effect, is normally interpreted as peculiar motion. However, the effect may also be caused by geometrically elongated structures, if they dominate the sample (with elongation in the z-direction; see below).

The results are presented in Figures 11 to 13. The scatter diagrams are plotted in Figure 11 for both the $R \geq 0$ and $R \geq 1$ samples. Frequency distributions representing these diagrams are presented in Figure 13. A strong and systematic elongation in the z-direction exists in all the real samples studied. Scatter-diagrams for sets of random catalogs do not exhibit any conspicuous elongation (Figure 12), as expected; a symmetric distribution in all directions is observed. As an additional test, Bahcall *et al.* (1986) also determined the scatter-diagrams in the projected plane, R_α - R_δ, of the cluster sample (Figure 12). Again, as expected, a symmetric distribution is observed in this plane. These tests strengthen the conclusion that the observed elongation is real. The effect of elongation is strong; statistically it corresponds to approximately 8σ in a single sample (assuming, for illustrative simplicity, Gaussian statistics). It is therefore unlikely that the observed redshift elongation is a chance fluctuation. The effect becomes more apparent in the larger $R \geq 0$ sample; this increase is expected if the effect is real.

A similar effect was observed by BS83 in their comparison between the cluster correlation function in the redshift and spatial directions. A broadening in the redshift direction was observed in that study, similar to the present finding. The elongation is unlikely to be caused by background/foreground contamination of galaxies and clusters (e.g. Sutherland 1988), since this would yield an excess of pairs at any R_z separation, as well as any R_α or R_δ, rather than the excess (i.e., broadening) observed specifically at small separations $(\Delta z \leq 0.015)$. The effect is also much larger than either the uncertainties in the redshift measurement or the uncertainties caused by the internal velocity dispersion within the clusters (see below).

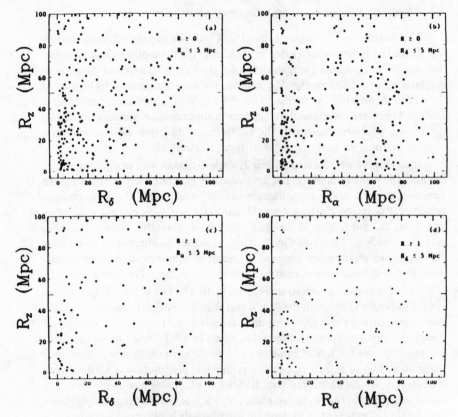

FIG. 11. Scatter-diagrams of Abell cluster pair separations in Mpc in the R_z-R_α and R_z-R_δ planes (Bahcall *et al.* 1986). (The pair separations along the third axis, perpendicular to each plane, are limited to ≤ 5 Mpc). All cluster pairs with a total spatial separation ≤ 100 Mpc are included. Figures a, b and c, d represent, respectively, the $R \geq 0$ and $R \geq 1$ richness samples. The elongation in the redshift direction is apparent in all cases.

To determine what velocity could cause the observed effect, the authors convolved the frequency distribution observed along the projected axis, which is unperturbed by peculiar motion, with a Gaussian velocity distribution. A Gaussian form is assumed for convenience in estimating the velocity broadening. This convolved distribution should match the broadened distribution observed in the redshift direction. The best fit is obtained for a velocity width of $\sqrt{2}\sigma \simeq 2000$ km s^{-1}. The estimated uncertainty on this mean velocity is approximately $+1000/ -500$ km s^{-1}. The above result is consistent with the result of BS83, using the redshift broadening observed in the cluster correlation function.

The above value for the velocity width includes all contributions to the

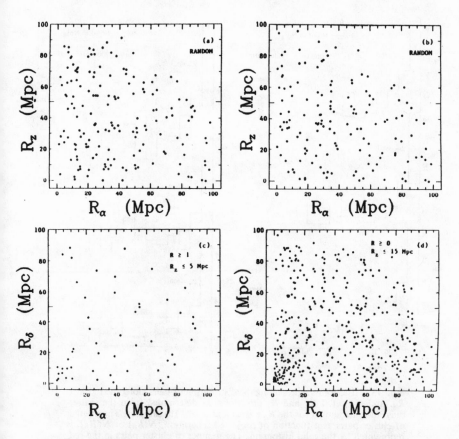

FIG. 12. Same as Figure 11 but for typical random distribution of clusters (Fig. a, b), and for the *projected* distribution (i.e. R_δ-R_α plane) of the actual cluster samples (Fig. c, d). No elongation is expected in either case and none is observed. The clustering of clusters is apparent in the data sample of Fig. c, d.

broadening effect, such as redshift measuring uncertainty and possible deviations from the true cluster redshift due to individual galaxy velocities in the clusters. Redshift measuring uncertainties are negligible compared to the 2000 km s^{-1} observed. The effect of peculiar motion within the clusters (for those clusters that have only a small number of measured galaxy redshifts) was estimated by comparing cluster redshifts from the current sample with those obtained using a larger number of measured galaxy redshifts, when available. For the latter study, the redshift catalogs of Sarazin, Rood, and Struble (1982), and Fetisova (1981) were used. A root mean square deviation for these cluster redshifts of approximately 300 km s^{-1} is observed due to the above effect. This value is reasonable considering that the full velocity dispersion in clusters

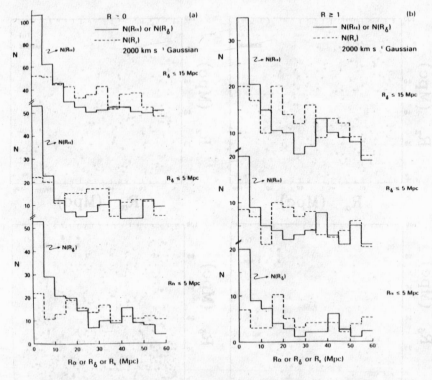

FIG. 13. Histograms representing the distribution of pairs along the redshift, R_z, and projected R_α and R_δ directions, as determined from the scatter-diagrams, are shown for the $R \geq 0$ (a) and $R \geq 1$ (b) samples. The number of cluster pairs as a function of projected separation, $N(R_\alpha)$ or $N(R_\delta)$, is represented by the solid histogram. The number of cluster pairs in the red-shift direction, $N(R_z)$, is represented by the dashed histogram. The dotted curve represents a convolution of the projected distribution, $N(R_\alpha)$ or $N(R_\delta)$, with a Gaussian of 2000 km s^{-1} width. This convolved profile is in general agreement with the broadened distribution observed in the redshift direction, $N(R_z)$.

is typically $\sim$ 1000 km s^{-1}, and that the redshifts measured are for the brightest centrally located galaxies; these galaxies are generally close to the central velocity of the cluster. Subtracting quadratically a possible deviation of $\sqrt{2}$ 300 km s^{-1} from the observed 2000 km s^{-1} yields 1950 km s^{-1}, i.e., a negligible change. Even if we assume, conservatively, $\sim$ 700 km s^{-1} for the internal broadening, the net cluster pair velocity is still 1740 km s^{-1}. Thus, a considerable elongation effect of approximately 10^3 km s^{-1} per cluster remains after correction for internal motion.

The observed elongation may be caused by either peculiar motion of clusters or a true geometrical elongation of superclusters. These are briefly discussed below.

7.2 - Explanations of the Redshift Elongation

If the observed elongation is caused primarily by peculiar motion of clusters in superclusters, the net cluster pair motion in the line-of-sight is approximately 1700 km s^{-1}, or, equivalently, about 1200 km s^{-1} for single cluster motion. Most of this effect arises in the central parts of the rich superclusters. A large peculiar velocity could be caused by the gravitational potential of the superclusters or by non-gravitational effects such as explosions.

To estimate a supercluster mass which may support this velocity, a typical supercluster size of ~ 25h^{-1} Mpc (= cluster correlation scale-length) is used and the virial relation $M \propto v^2 r$ is assumed. This yields a typical supercluster mass of

$$M_{sc} \simeq 2 \times 10^{16} M_\odot, \qquad r \lesssim 25h^{-1}\text{Mpc}. \qquad (16)$$

This mass is comparable to the mass of ~ 20 rich clusters, while typically only ~ 3-5 rich clusters are members of a supercluster. Even when the luminous tails of clusters are accounted for, the results may still suggest an excess of dark matter in superclusters as compared with clusters. Using an observed luminosity and/or density profile of r^{-3} or $r^{-2.5}$ around a rich cluster, we estimate an M/L for superclusters that is typically twice that of rich clusters, i.e., M/L ~ 500.

Redshift observations of two individual higher-redshift superclusters (Ciardullo et al. 1983) appear to indicate a much lower velocity for the superclusters than suggested even by a free expansion. This suggests, for these two systems, either a flat face-on geometry of the superclusters or a slow-down of the initial expansion due to the supercluster mass. In either case, it is likely that individual superclusters are at different stages of their evolution as well as at different observed orientations. The Corona Borealis supercluster (Bahcall et al. 1986) appears to show a redshift elongation in the distribution of both its clusters and the galaxies.

The elongation observed in the scatter diagrams may also be caused, at least partially, by a geometrical elongation of superclusters. If the most prominent superclusters are elongated in the line-of-sight direction, an apparent elongation in the distribution of pair separation along this axis may result. I discuss below an observational test to distinguish between peculiar velocity and geometrical elongation of large-scale structures.

If the observed redshift elongation is caused by geometrical elongation, cluster redshifts should be correlated with the magnitude of their standard galaxies, following Hubbles's law. No such magnitude-redshift correlation should be present if the effect is entirely due to peculiar velocity. More generally, an independent distance-indicator (such as the magnitude of the brightest cluster galaxy or Tully-Fisher type relations) could be used to determine the actual distances to the clusters, and thus to interpret the origin of the observed

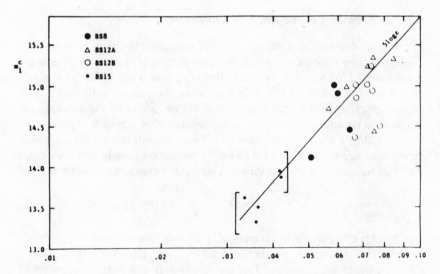

FIG. 14. The magnitude-redshift relation for the brightest cluster galaxies for clusters in the richest Bahcall-Soneira superclusters (Sec. 7.2).

redshift broadening (by comparing the actual distances with the observed redshifts).

The dependence of galaxy magnitudes on redshifts in the close cluster pairs was studied by Bahcall *et al.* (1986). The magnitude of the brightest galaxy in each cluster, m_1^c, corrected for the cluster morphological type and richness as given by Hoessel *et al.* (1980), was used as a distance indicator. If the observed redshift elongation is caused by geometrical anisotropy, a proper (Hubble) correlation of m_1^c with z is expected within individual superclusters. This correlation should not exist if peculiar velocity is the cause of the observed elongation. The expected magnitude difference for a cluster pair with a redshift separation of about 0.01-0.015, assuming Hubble distances, is ~ 0.3 to 0.5 mag (depending on z). This difference is large enough to be measured with accurate observations of standard galaxy magnitudes. A marginal m_1^c dependence was found (Bahcall *et al.* 1986) for some individual superclusters (Figure 14), suggesting that at least some of the redshift broadening observed may be due to geometrical elongation of the large structures. Increased accuracy and greater statistics for galaxy magnitudes may clarify the significance of the results. It is possible that both geometrical elongation and peculiar velocity of clusters contribute to the observed redshift broadening. Other distance indicators, such as Tully-Fisher or Faber-Jackson relations, should also be applied to the problem in order to help distinguish between peculiar motion and geometry. Recently, comparable velocities of ~ 10^3 km s^{-1} between some cluster pairs were also suggested by Mould (1988) and Burstein (1988) using actual distance indicators of galaxies.

REFERENCES

Abell, G. O. 1958. *Ap J Suppl.* **3**, 211.

———. *AJ.* **66**, 607.

Abell, G. O., Corwin, H. G., and Olowin, R. 1988. *Ap J Suppl.*, to be published.

Bahcall, N. A. 1977. *Ann Rev Astron Astrophys.* **15**, 505.

———. 1986. *Ap J Letters.* **302**, L41.

———. 1987. *Comm Astrophys.* **11**, 283.

———. 1988a. *Ann Rev Astron Astrophys.* **26**.

———. 1988b. in *Large-Scale Structures of the Universe*, eds. J. Audouze, M.-C. Pelletan, and A. Szalay. Dordrecht: Kluwer Academic Publishers, p. 229.

Bahcall, N. A. and Burgett W. S. 1986. *Ap J Letters.* **300**, L35.

Bahcall, N. A. and Soneira, R. M. 1983. *Ap J.* **270**, 20 (BS83).

———. 1984. *AP J.* **277**, 27 (BS84).

Bahcall, N. A., Soneria, R. M. and Burgett, W. S. 1986. *Ap J.* **311**, 15.

Bahcall, N. A., Herinksen, M. J. and Smith, T. E. 1988. *Ap. J.*, to be submitted.

Bahcall, N. A., Batuski, D. J., and Olowin, R. 1988b. *Ap J. Letters.* **333**, L13.

Batuski, D. J. and Burns, J. O. 1985. *Ap J.* **299**, 5.

Batuski,D. J., Burns, J. O., Laubscher, B. E., and Elston, R. J. 1988. *Ap J.*, to be published.

Bogart, R. S. and Wagoner, R. V. 1973. *Ap J.* **181**, 609.

Bouchet, F. R. and Bennet, D. P. 1988, in *Large-Scale Structures of the Universe*, eds. J. Audouze, M.-C. Pelletan, and A. Szalay. Dordrecht: Kluwer Academic Publishers, p. 289.

Burstein, D. 1988. this volume.

Chincarini, G. and Vettolani, G. 1987. in *Observational Cosmology,* eds. A. Hewitt *et al.* Reidel, p. 275.

Chincarini, G, Rood, H. J. and Thompson, L. A. 1981. *Ap J Letters.* **249**, L47.

Ciardullo, R., Ford, H., Bastko, F., and Harms, R. 1983. *Ap J.* **273**, 24.

da Costa, L.N., Pellegrini, P.S., Sargent, W.L.W., Tonry, J., Davis, M., Meiksin, A., and Latham, D. 1988. *Ap J.* **327**, 544.

Davis, M. and Peebles, P.J.E. 1983. *Ap J.* **267**, 465.

Davis, M., Huchra, J., Latham, D.W., and Tonry, J. 1982. *Ap J.* **253**, 423.

Dekel, A. 1988, this volume.

de Lapparent, V., Geller, M., and Huchra, J. 1986. *Ap J Letters.* **302**, L1.

Efstathiou, G. 1988, this volume.

Fetisova, T.S. 1981. *Astr Zh.* **58**, 1137.

Geller, M.J. and Huchra, J. 1988, this volume.

Giovanelli, R., Haynes, M.P., and Chincarini, G. 1986. *Ap J.* **300**, 77.

Gregory, S.A. and Thompson, L.A. 1978. *Ap J.* **222**, 784.

Gregory, S.A., Thompson, L.A., and Tifft, W.G. 1981. *Ap J.* **243**, 411.

Groth, E. and Peebles, P.J.E. 1977. *Ap J.* **217**, 385.

Hauser, M.G. and Peebles, P.J.E. 1973. *Ap J.* **185**, 757.

Haynes, M.P. and Giovanelli, R. 1986. *Ap J Letters.* **306**, L55.

Hoessel, J.G., Gunn, J.E., and Thuan, T.X. 1980. *Ap J.* **241**, 486.

Kalinkov, M., Stavrev, K., and Kuneva, I. 1985. *AN.* **306**, 283.

Klypin, A.A. and Kopylov, A.I. 1983. *Sov Astron Letters.* **9**, 41.

Kopylov, A.I., Kuznetsov, D.Yu., Fetisova, T.S., Shvartsman, V.F. 1987. *The Large Scale Structure of the Universe, Seminar Proceedings.* Special Astrophysical Observatory, September 1986.

Mandelbrot, B.B. 1982. *The Fractal Geometry of Nature.* San Francisco: Freeman.

Mould, J. 1988, this volume.

Noonan, T. 1977. *AA.* **54**, 57.

Oort, J. 1983. *Ann Rev Astron Astrophys.* **21**, 373.

Peebles, P.J.E. 1980a. *The Large Scale Structure of the Universe.* Princeton: Princeton University Press.

_____ . 1980b. *Physical Cosmology, Les Houches, Session XXXII.* eds. R. Balian *et al.*

Postman, M., Geller, M., and Huchra, J. 1986. *AJ.* **91**, 1267.

Rood, H.J. 1976. *Ap J.* **207**, , 16.

_____ . 1988. *Ann Rev Astron Astrophys.* **26**.

Sarazin, C.L., Rood, H.J., and Struble, M.F. 1982. *AA Letters.* **108**, L7.

Sargent, W.L.W., Young, P.J., Boksenberg, A., and Tytler, D. 1980. *Ap J Suppl.* **42**, 41.

Seldner, M. and Peebles, P.J.E. 1977. *Ap J.* **215**, 703.

Shane, C.D. and Wirtanen, C.A. 1967. *Pub Lick Obs.* **22**, 1.

Shectman, S. 1985. *Ap J Suppl.* **57**, 77.

Shvartsman, V.F. 1988. in *Large-Scale Structures of the Universe*, eds. J. Audouze, M.-C. Pelletan, and A. Szalay. Dordrecht: Kluwer Academic Publishers, p. 129.

Sutherland, W. 1988. *MN.* to be published.

Szalay, A.S. and Schramm, D.N. 1985. *Nature.* **314**, 718.

Szalay, A.S., Hollosi, J., and Toth, G. 1988. preprint.

Tully, R.B. 1986. *Ap J.* **303**, 25.

_____ . R.B. 1987. *Ap J.* **323**, 1.

Turok, N. 1985. *Phys Rev Letters.* **55**, 1801.

_____ . 1988. in *Large-Scale Structures of the Universe*, eds. J. Audouze, M.-C. Pelletan, and A. Szalay. Dordrecht: Kluwer Academic Publishers, p. 281.

Zwicky, F., Herzog, E., Wild, P., Karpowicz, M., and Kowal, C.T. 1961-1968. *Catalog of Galaxies and Clusters of Galaxies, 6 Volumes.* Pasadena: Calif. Inst. of Technology.

SUMMARY OF SESSION ONE

G. EFSTATHIOU

Institute of Astronomy, Cambridge

We have heard and seen a lot of interesting things in this session. I have been sitting here wondering how it would appear to an extraterrestrial being. For the sake of argument, I will call him Pisces-Cetus, or PC for short. Let me tell you about PC. He has no eyes! We might consider this to be a great disability, but PC isn't bothered because he is blessed with an extraordinary capacity for assimilating and understanding numerical data. Correlation functions, likelihood functions, and all the rest, pose no problems for PC; he digests them with lightning speed. What would PC have made of this session?

Brent Tully has shown us pictures of structures that extend almost right across the observable universe. PC can't assess this, so he looks at the statistic. But wait a minute, wasn't the statistic motivated by the visual appearance of the map? PC is confused.

Neta Bahcall and John Huchra presented important statistical results on the clustering of Abell clusters. They are clustered much more strongly than galaxies. PC is impressed, but is a bit unsure about the Abell catalogue. Wasn't it constructed by eye? Are these humans really as objective as they claim? Nick Kaiser discussed work by Will Sutherland which suggests that spurious clustering in two-dimensions enhances the amplitude of the three-dimensional function measured in redshift space. The implication is that Abell found an enhanced number of clusters nearby other clusters on the sky. PC is worried and asks if the amplitudes of the angular correlation functions scale with distance class in the expected way; but there is no quantitative answer. Avishai Dekel discussed one mechanism that could lead to a bias, and concluded that it probably couldn't produce enough in the way of spurious clustering. Are there really problems with the Abell catalogue?

Marc Davis discussed results from the new Southern Sky redshift survey. The amplitude of the galaxy correlation function and the peculiar velocities between pairs of galaxies agrees well with the results from the CfA survey. PC thinks this is very good. The results have stabilised with sample size, so our applications of the cosmic virial theorem and our determinations of Ω look secure. But Margaret Geller argued that the 15th magnitude CfA redshift survey indicates that we had not yet achieved a fair sample of the universe. For example, she found that the amplitude of the galaxy correlation function was sensitive to her choice of weighting scheme. Furthermore, she pointed out inhomogeneities on the scale of the new survey. Jim Peebles remarked that the definition of a fair sample depends on the statistic that you want to measure. PC nodded in agreement.

Martha Haynes showed beautiful results on the distribution of galaxies in the Pisces-Perseus Supercluster. In particular, the luminous galaxies seem to be more clustered than the faint ones. Is this evidence in support of the idea that galaxies are biased tracers of the mass distribution? Does this, as Marc Davis contended, confirm the predictions of the "biased" cold dark matter model?

Well, PC has gone home to report on all this to his friends. I hope he is understanding! Our lack of precision in discussing these issues stems from two sources. Firstly, surveys of large-scale structure are difficult and very demanding of telescope time. So we don't have enough data. We must continue the effort to get more. Secondly, we have eyes! So we tend to get excited about interesting shapes and patterns instead of focussing our efforts on well designed statistical tests. Wouldn't it be better to first decide on a sensible statistic and then to design an optimal observing program to measure it? Before PC left, he whispered in my ear that if you smoothed the galaxy distribution on scales of 2000 km/s, you would be left with Gaussian random noise. Perhaps he was just being malicious.

II

LARGE-SCALE VELOCITY FIELDS

MOTIONS OF GALAXIES IN THE NEIGHBORHOOD
OF THE LOCAL GROUP

S. M. FABER

Lick Observatory, University of California, Santa Cruz

and

DAVID BURSTEIN

Department of Physics, Arizona State University

ABSTRACT

The velocity field of galaxies relative to the cosmic microwave background is investigated to a distance of 3000 km s^{-1} using two samples of spiral galaxies as well as elliptical galaxies. Velocity-field models are optimized that include motions due to a spherically symmetric Great Attractor, a Virgocentric flow, and a Local Anomaly, of which the Local Group is a part. The predictions of these models are compared graphically in a number of ways. We find that the spiral samples agree well in a formal sense with the Great-Attractor-Virgo model recently proposed to explain the motions of elliptical galaxies. However, new observations indicate that the Great Attractor is not spherically symmetric in its inner regions, which could require future modifications to this model. The Local Group shows an anomaly of 360 km s^{-1} with respect to the above model, which is shared by the cloud of galaxies around it out to 700 km s^{-1}. The amplitude and dimensions of this Local Anomaly seem to be typical of other deviant patches in the velocity field. It is likely that the Local Anomaly is the result of the irregular gravitational attraction of nearby, visible galaxies. Virgocentric infall models are heavily modified by the inclusion of the Great Attractor flow. The local Virgocentric infall velocity at the Local Group is poorly determined, but is quite a bit smaller (85-133 km s^{-1}) than conventional values. The gross properties of the velocity field that are defined by the present data are sketched in terms of a minimal, cylindrical, geometric model.

1. Introduction

We are part of a group currently investigating non-uniformities in the Hubble expansion based on the motions of nearby elliptical galaxies (Dressler *et al.* 1987, Lynden-Bell *et al.* 1988 [hereafter LFBDDTW]). We have discovered evidence for a large-scale flow of galaxies towards the constellation Centaurus in the southern hemisphere. This flow reaches a velocity of up to 1000 km s^{-1} in places and encompasses the entire Local Supercluster, including Virgo, Fornax, Eridanus, Ursa Major, and Leo (see Tully and Fisher 1987, hereafter TF87), as well as the region that has come to be known as the Hydra-Centaurus Supercluster (e.g., Chincarini and Rood 1979). Virgocentric infall exists as one sub-flow within it; in this paper we will present evidence for the existence of other (but smaller) such flows.

Guided by the familiar Virgo flow models, we have modeled the larger flow using a spherically symmetric infall centered on a point 4350 km s^{-1} distant (LFBDDTW). This infall center is located in the same direction as the Centaurus clusters but lies well beyond them. Because the model implies a major mass concentration quite distant from us, it has garnered the nickname "Great Attractor" (attributed to Alan Dressler).

A major difference between Virgo infall and the Great Attractor (hereafter GA) is the fact that the present data penetrate barely to the center of mass, far enough to show us infall towards the GA, but not infall from the back side. While this is a weakness, we will show that the data do strongly prefer a convergent flow model over simple bulk motion. The signature for this is a quadrupole term in the velocity field, which is equivalent to a tide in the spherical infall model. This quadrupole shows itself as a compression of the Hubble flow perpendicular to the direction of the GA and a stretching of the flow towards it. The interpretation of a quadrupole as convergent infall was first suggested by Lilje, Yahil, and Jones (1986, hereafter LYJ). In addition to noting the quadrupole in the elliptical galaxy data, LFBDDTW also reviewed evidence for a peak in the density of galaxies at the right distance and direction to be the GA. This evidence consisted of a map of galaxies on the sky in Centaurus, compiled by Ofer Lahav, plus published radial velocity surveys in the region.

Since that time, we have been testing the GA model against other data sets. There are four data sets of interest: the catalog of Tully-Fisher distances to nearby spiral galaxies compiled by Aaronson *et al.* (1982a) and supplemented by Bothun *et al.* (1984, the combination of both data sets will be referred to as Aaronson *et al.*); a similar catalog compiled by de Vaucouleurs and Peters (1984, hereafter DVP), a catalog of Tully-Fisher distances to nearby rich clusters by Aaronson *et al.* (1986, hereafter ABM86), and Sc I spiral distances by Rubin *et al.* (1976a, b). The present paper is limited to motions of nearby galaxies and concentrates on the first two samples, which have their major weight within 2500 km s^{-1}. The ABM86 cluster sample is used briefly to set the far-field zero-point for the Hubble expansion. This is equivalent to determining the Hubble constant in most studies, but since our distances are always given in units of km s^{-1}, the analogous scale factor for us is a dimensionless constant (see LFBDDTW).

In the present study, we investigate the following questions:

a. How well do both spiral and elliptical galaxies with distances less than 3000 km s^{-1} fit the adopted velocity field model, the dominant component of which is motion induced by the GA?

b. How well do the galaxies (almost all spirals) closest to the Local Group fit the model? LFBDDTW noted that the Local Group itself fits poorly, with a residual velocity of about 360 km s^{-1}. If there is a Local Anomaly, how far does it extend, and how does it merge with the flow pattern at larger distances?

c. Existing studies of Virgo infall (e.g., Aaronson *et al.* 1982b, hereafter AHMST) have usually modeled Virgo infall in the rest frame of the sample galaxies. Virgocentric infall and bulk motion have thus been removed, but the tidal field of the GA is not modeled. Since the local velocity of the GA flow (~ 550 km s^{-1}) is greater than previous estimates of Virgo infall (250 km s^{-1}), the tidal distortion of the former has potentially influenced previous derivations of the amplitude of Virgo infall. We therefore examine what is left of Virgo infall after the GA-induced motion has been removed.

d. What is the typical coherence length for peculiar motions relative to the GA-induced flow? Is there evidence for localized infalling regions other than the Virgo cluster? How typical is the residual motion of the Local Anomaly?

e. For completeness, we also review the most recent data on the size and structure of the GA.

Most of the quantitative results of the present paper are based on a series of three-component, maximum-likelihood fits to the spiral galaxy velocity data alone. As such, these complement the models of LFBDDTW, which use only the elliptical galaxy data. Since it is difficult to represent three-dimensional motions on a two-dimensional page, we show graphical representations of the data before and after fitting to various models. By studying these graphs in conjunction with maps of the galaxy distribution provided here (Figs. 1 and 2) and the more informative maps of TF87, it should be possible for the reader to construct a mental image of the local velocity field.

An interesting by-product of this work is the discovery that the spiral data sets are of varying accuracy and are also internally inhomogeneous, including even the Aaronson *et al.* field data. The DVP data are so heterogeneous, in fact, that we have used only those galaxies with Tully-Fisher distances. The best Aaronson *et al.* data seem to have an error that is appreciably better than that usually quoted for the Tully-Fisher method.

We also discuss briefly the pros and cons of various methods of maximum-likelihood fits, but details will appear in a later version of this work.

2. THE LOCAL TOPOGRAPHY

Tully and Fisher have made a major contribution to our understanding of the nearby space distribution of galaxies with their recent *Nearby Galaxies Atlas* (TF87). It is unfortunate that we cannot reproduce their beautiful maps here, but a combination of the cartoon in Fig. 1 and the maps of Fig. 2 will have to suffice. Fig. 1 shows a cube of the local volume 10,000 km s^{-1} on a side in supergalactic coordinates. The major structures — Virgo, Ursa Major, Centaurus, Perseus-Pisces, and Pavo-Indus-Telescopium — lie either

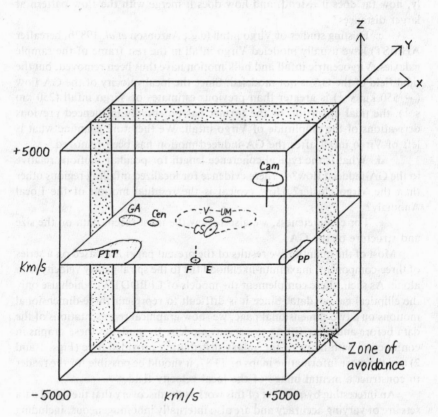

FIG. 1. A cartoon of the local volume of space that is sampled by the data sets used in this paper. The positions of the main structures are noted: (GA = Great Attractor; C = Centaurus; V = Virgo cluster; CS = Coma-Sculptor Cloud; UM = Ursa Major cluster; FE = Fornax-Eridanus; Cam = Camelopardalis; PP = Perseus-Pisces; PIT = Pavo-Indus-Telescopium.

directly in the supergalactic plane or close to it. Fornax and its neighbor, Eridanus, lie somewhat below the plane in the south, and the Leo clouds (not shown) lie below the plane under Virgo-Ursa Major. TF87 identify an important new grouping, the highly flattened Coma-Sculptor cloud located at the center of the cube, of which more is said below. The Local Group is a resident of this cloud, which in turn appears to be an appendage to the Virgo-Ursa Major complex and also part of the supergalactic plane.

The position of the GA as inferred from the elliptical galaxy data is shown by the large dot in the upper left-hand corner of the SG plane. The GA lies close to the plane, but the mass distribution in its vicinity is not well understood. The two Centaurus clusters, plus several smaller neighbors, lie in the foreground of the GA and are falling into it (away from us) with velocities approaching 1000 km s^{-1}.

Figs. 2a, b, and c show three orthogonal supergalactic projections of all galaxies with measured distances, both spirals and ellipticals, with X, Y and Z components defined in the usual manner. The data from the four samples are plotted with different symbols, as given in the caption. Several groupings from Fig. 1 are labelled, as well as the direction of the Perseus-Pisces complex. The SG plane is most visible in the Y-Z projection, which also shows the highly flattened Coma-Sculptor Cloud (center). Fig. 2 agrees with the analogous diagrams in TF87 very well, despite the fact that the diagrams in TF87 are based on radial velocity distances (a point which is discussed further below). Qualitatively the galaxies with measured distances here appear to be a fair tracer of the general galaxy population, with about 20% of the galaxies in TF87 having measured distances.

A major feature of Fig. 2 is the galactic zone of avoidance. Fortuitously, this is almost exactly perpendicular to the SG plane and also to the Y axis. The wedges empty of galaxies are especially visible in the X-Y and Y-Z projections.

3. Assumed Flow Models

The flow model we have been investigating is a three-component flow that is an embellishment of the model in LFBDDTW. It has the following components:

a. GA flow: This flow is assumed to be spherically symmetric about a point located at galactic coordinates $l_A = 309°$, $b_A = +18°$ at a distance 4200 km s^{-1} from the Local Group (slightly revised from LFBDDTW). From recent work, we have realized that the velocities of ellipticals in that direction peak at 3000 km s^{-1} and fall beyond (see Fig. 10k), indicating that the flow model requires a core radius. The velocity model adopted is

$$u_A = v_A [r_A/d_A] [(d_A^2 + c_A^2)/(r_A^2 + c_A^2)]^{(n_A + 1)/2} \qquad (1)$$

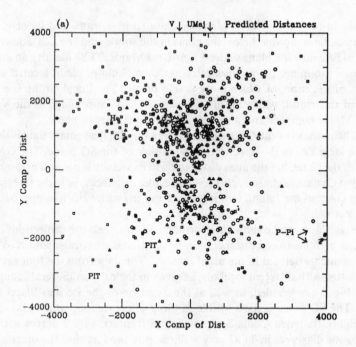

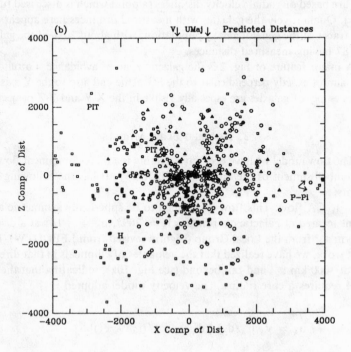

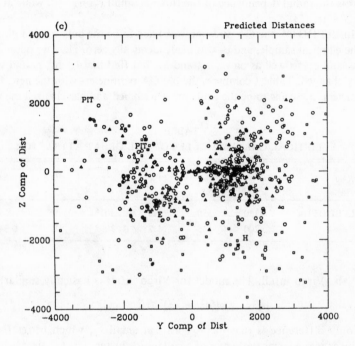

Fig. 2. Projected positions of galaxies in supergalactic coordinates, according to the predicted distance of each galaxy (Malmquist-bias corrected as discussed in the text). The four data samples are coded by different symbols: triangles denote Aaronson *et al.* "Good" data; squares denote elliptical galaxy data from LFBDDTW; hexagons denote Aaronson *et al.* "Fair" data; and circles denote de Vaucouleurs and Peters Tully-Fisher data. The general positions of clusters are either marked or pointed to by arrows: V = Virgo cluster; UMaj = Ursa Major; F = Fornax; E = Eridanus; P-Pi = Perseus-Pisces region (mostly off the maps); PIT = the Pavo-Indus-Telescopium region, which is marked in two places. The X, Y and Z coordinates are as defined in the RC2. Fig. 2a, the X-Y plane; Fig. 2b, the X-Z plane; Fig. 2c, the Y-Z plane. These maps compare well with those of TF87 (who used primarily radial velocities) with respect to both the correspondence of samples and to the relative positions of groups and clusters.

where u_A is total space velocity radially directed toward the infall center. The quantity r_A is the distance of a galaxy from the infall center, d_A is the distance of the infall center from the Local Group, v_A is the flow velocity at the radius of the Local Group, and c_A is the core radius of the flow. The quantity n_A controls the radial dependence of the flow: at small r_A, $u_A \sim 0$, while at large r_A, $u_A \sim (r_A)^{-n_A}$.

In LFBDDTW the GA direction was fixed to be the bulk motion direction for the elliptical sample, and c_A was made identically zero. Here we optimize on l_A, b_A, and c_A, as well as on r_A, v_A, and n_A, but find that l_A, b_A, r_A and v_A are hardly changed. Table 1 compares the old GA parameters with the new. Of all parameters, n_A seems to be the least well determined and has changed the most.

TABLE 1

ELLIPTICAL GALAXY MODELS OF THE GREAT ATTRACTOR

Model	d_A (km s^{-1})	v_A	n_A	l_A	b_A	c_A
Old (LFBDDTW)	4350	570	1.0	307°	9°	—
New	4200	535	1.7	309°	18°	$0.34d_A$

b. Virgo infall: The model for Virgo infall is basically similar:

$$u_V = v_V \, [(d_V^2 + c_V^2)/(r_V^2 + c_V^2)]^{n_V/2} \tag{2}$$

The only difference is that $u_V \sim$ const. at small r_V, which better fits the nonlinear regime near the center of a collapsed cluster. The far-field behavior is again $u_V \sim (r_V)^{-n_V}$.

c. Local Anomaly: We show below that the velocity anomaly associated with the local Group extends to a region that is at least 700 km s^{-1} in radius around it. We therefore have incorporated a "Local Anomaly Switch", which turns on a bulk-velocity correction vector of 360 km s^{-1} for this region, making it blend smoothly with the surroundings.

d. Errors and Hubble constant: Maximum likelihood requires an error estimate for each point. As in LFBDDTW, this is represented by a measurement error Δ per point (given in magnitudes), plus a "random noise" term, σ_f, for Hubble flow noise (in km s^{-1}). The two errors are adjusted separately to maximize the likelihood. There is also a scale factor analogous to the Hubble constant that relates distance to velocity. How this is determined is explained below.

In treating the spirals we do not group them into clusters as we did with the ellipticals. Each is a separate point of equal weight within each data category.

4. Methods of Fit

We have broadened our methods of maximum-likelihood fitting in the same spirit as AHMST, who also tried multiple methods. The goal is to get from the data to the "best" velocity-field model. So far we have tried two approaches:

Method 1: Get the distance of each galaxy using the distance estimator (e.g., Tully-Fisher for spirals). Use these distances to calculate the velocity residual of every object relative to a smooth Hubble flow, and thus make a "picture" of the velocity-field residuals. Devise a mathematical model that incorporates the basic features of this velocity field, and find the free parameters by minimizing the scatter (maximizing the likelihood) in the velocity-field residuals. This was the method employed in LFBDDTW.

Method 2: Get the distance of each galaxy using its observed velocity, corrected by an *a priori* velocity-field model. With these distances, make a "picture" of the basic distance-indicator relation (for Tully-Fisher, this would be a graph of rotation velocity vs. absolute magnitude). Vary the parameters in the velocity-field model to minimize the scatter (maximize the likelihood) in the distance-indicator relation.

Each fitting method has its pros and cons, and each one is at some stage indispensable. Method 1 must be used initially to the point of making a picture of the observed velocity field. This is required to decide what terms to include in the velocity-field model. Method 1 also has residuals in km s^{-1}, and hence Hubble-flow noise yields Gaussian errors for all galaxies, near and far. This is important for nearby galaxies, for which Hubble-flow noise dominates. Without Gaussian errors, maximum-likelihood becomes much more complicated.

However, the raw distance estimates in Method 1 are biased too small due to Malmquist-bias effects and must be corrected (see LFBDDTW). For a uniform space distribution, the distance correction is a multiplicative constant given by:

$$r = r_{raw} (1 + 0.74 \Delta^2) \tag{3}$$

where Δ is the observational error in magnitudes.

The least accurate spiral samples have distance errors of $\pm$ 0.5 mag and Malmquist corrections of 19%, which are large. Worse, in clumpy regions, the correction depends on the density gradient along the line-of-sight and can even be negative. Using Virgo as an example, we show below how this effect can introduce spurious features into the velocity field in clumpy regions.

Perhaps the worst feature of Method 1 is that the galaxy distance estimator, being inherently logarithmic, does not yield symmetric, Gaussian errors in km s^{-1}. Distance errors of even $\pm$ 0.4 mag produce a seriously skewed error distribution in velocity space. This in turn introduces a bias into certain parameters, notably the scale factor (Hubble constant), and the distance error, Δ. Errors in the latter quantity are serious because they fold back into the Malmquist-bias correction.

Method 2 is the opposite of Method 1 in almost every way. It yields a picture of the distance-indicator relation rather than the velocity field. This picture can be used to fit the mathematical form of the relation from *field* galaxies. However, this approach can introduce undesirable cross-talk between the shape of the relation and features in the velocity field (LFBDDTW). It is preferable if possible to obtain the shape of the relation from independent data, e.g., from calibrating clusters. Unfortunately the AHM86 cluster galaxies do not span quite enough range in magnitude for this, and some appeal must be made to field data after all. (Yes, Jim Peebles, we have sinned, but, like all sinners we claim we can't help it.).

A great advantage of Method 2 is the fact that scatter about the Tully-Fisher relation is naturally Gaussian and constant with distance. The Gaussian maximum-likelihood method is therefore well suited to all but the nearest galaxies, and most of the flow parameters are therefore unbiased. This is crucial for galaxies beyond 1000 km s^{-1}, for which distance errors generally dominate over Hubble-flow noise. Method 2 minimizes naturally in log v rather than v.

Although the Malmquist-bias problem is smaller for Method 2, Malmquist bias still afflicts the distances — caused in this case by distance errors due to Hubble-flow noise. However, the typical distance error is usually smaller — say 10% rather than 20-25% with Tully-Fisher — and the resulting Malmquist bias is only a few percent if velocity-distance is near linear.

On the debit side, Method 2 gives no picture of the velocity-field: for that we must resort to Method 1. A second drawback is that Hubble-flow noise yields non-Gaussian residuals for nearby galaxies, for which the method thus gives biased results. (This situation is just the opposite of Method 1, which had trouble with distant galaxies, for a similar reason). The derived Hubble-flow noise parameter and the Hubble constant for nearby galaxies are most affected. LFBDDTW also show that the likelihood in Method 2 is not as "pure" as in Method 1 and maximizes in a slightly different location in parameter space. The effect on the derived velocity-field parameters is not yet clear.

As long as one is committed to Gaussian maximum-likelihood, there is clearly no single method that rigorously treats both near and far galaxies and deals well with Malmquist bias. Instead one must use an interplay of both methods, depending on need. AHMST also explored both methods.

A final drawback to Method 2 is the fact it cannot yield unique distances to galaxies in the so-called triple-valued regions around infall centers. In the local volume, there are two at least two such regions, the familiar Virgo region and a larger one around the GA. As a temporary way of dealing with this we are using a hybrid version of Method 2 that employs radial velocity as the basic distance indicator but Tully-Fisher distance to calculate the velocity flow correction to the model. In effect, this method has the Malmquist-bias properties of Method 1 but the other properties of Method 2. We will examine the "pure" Method 2 in a later paper.

In all models we have limited the sample to a certain subset of the Aaronson *et al.* spirals with exceptionally small errors (see below). In view of the systematic biases that can creep in due to errors, we feel that it is better to derive quantitative results from the best data only and then to compare the other samples graphically. Such comparisons are shown in Section 6 below.

The last issue is the scale factor, or Hubble constant. This is equivalent to fixing the absolute magnitude zero-point of the Tully-Fisher relation. From the discussion above, it should be clear that Method 2 yields a good zero-point when applied to sufficiently distant galaxies. Accordingly, we have used a standard GA-Virgocentric flow model to derive velocity-field distances to the ABM86 cluster spirals. Since these clusters are 3500 to 11,000 km s^{-1} distant and are in the opposite hemisphere from the GA, their corrected velocity-field distances should be fairly accurate. With these clusters determining the zero-point, the peculiar velocity of Coma with respect to cosmic rest turns out to be -200 ± 300 km s^{-1}, which agrees well with the value of -220 km s^{-1} from the ellipticals in LFBDDTW.

5. THE LOCAL ANOMALY

The first step in model fitting is to discover how far the Local Anomaly extends. To find this, we carried out bulk-motion solutions on the Aaronson *et al.* field spirals inside successively larger shells around the Local Group (Table 2). (Since these involved nearby galaxies, Method 1 was used.) The resultant velocity vector agrees well with the motion of the Local Group within the errors (50-100 km s^{-1}) out to a radius of 700-800 km s^{-1}, at which point it begins to swing toward the GA at (309, 18). This shift in the motion of nearby galaxies was first pointed out by de Vaucouleurs and Peters (1968) and later explored by de Vaucouleurs and Peters (1984) using a variety of distance-indicator methods. The shared motion of the Local Group and nearby galaxies was also emphasized recently by Peebles (1987) using the same Aaronson *et al.* spirals as here. This locally coherent motion is the reason why Hubble-flow noise has been estimated as low as 100 km s^{-1} (e.g., Sandage 1972).

The homogeneous motion of the local patch is illustrated in Fig. 3, which plots radial velocity residuals relative to cosmic rest versus the cosine of the angle to the direction of the Local Group microwave vector (269, 28). Galax-

TABLE 2

BULK MOTIONS OF NEARBY GALAXIES

Shell Limits (km s^{-1})	v_{BLK} (km s^{-1})	l_{BLK}	b_{BLK}	No. of Galaxies	Errors[a] (km s^{-1})
0 - 500	623 ± 89	275° ± 12°	27° ± 4°	18	138, 43, 31
500 - 700	491 ± 94	275° ± 15°	29° ± 7°	14	132, 76, 46
700 - 900	645 ± 78	295° ± 9°	22° ± 5°	14	94, 78, 57
900 - 1100	497 ± 79	309° ± 10°	28° ± 4°	22	97, 76, 48
1100 - 1500	659 ± 85	309° ± 8°	1° ± 3°	53	94, 77, 30
Loc. Gr.	614	269°	28° with respect to the CMB		
Local GA flow vector	535 ± 70	309° ± 10°	18° ± 10°		

[a] The three axes of the error ellipse. Because of the flattening of the local cloud, the largest error tends to point roughly perpendicular to the SG plane (47,6 in galactic coordinates). The smallest error is perpendicular to the galactic plane, and the middle error is perpendicular to the other two.

ies within a sphere of radius 850 km s^{-1} are plotted, and galaxies in specific regions (i.e., towards the Virgo cluster, away from the Virgo cluster, or in the Leo Spur) are plotted with different symbols. Galaxies streaming along with the Local Group should lie on the straight line, which has total amplitude ± 614 km s^{-1}, the CMB velocity of the Local Group. A maximum-likelihood fit for the velocity dispersion inside 700 km s^{-1} is only 80-90 km s^{-1}, indeed confirming that the local Hubble flow inside the patch is quite quiet (Sandage 1972).

The local patch is the only one for which we can unambiguously determine the full three-dimensional space motion and compare it to the geometry of the local galaxy distribution. Figs. 4a, b, and c show supergalactic projections inside the local volume analogous to Fig. 2. The highly flattened cloud at the center is the Coma-Sculptor Cloud of TF87. If this cloud has the same motion as the Local Group, its total bulk velocity with respect to the CMB is X, Y, Z = (−417, 249, −376) in SG coordinates. A more interesting quantity is the difference between this motion and that of the surrounding galaxies, which agree well with the GA-Virgo model. This difference is about 360

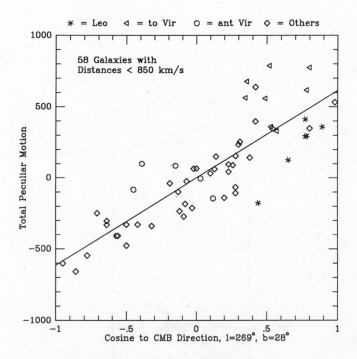

Fig. 3. Peculiar velocities relative to the CMB rest frame of all galaxies with predicted distances less than 850 km s^{-1}, plotted versus the cosine of the angle to each galaxy with respect to the direction of the Local Group's CMB motion (l = 269°, b = 28°). Galaxies moving with the Local Group should lie on the line. Galaxies located near the Virgo cluster are denoted by triangles; galaxies located the furthest from Virgo are denoted by hexagons; galaxies located in the Leo 'Spur' (TF87) are denoted by stars, and all remaining galaxies are denoted by diamonds. The most deviant galaxies (non-diamonds) tend to lie at the edges of the volume (see Fig. 4).

km s^{-1} towards (199,0), or (169, −30, −317) in SG coordinates. Since this component is locally generated in the gravitational instability picture (see Lynden-Bell and Lahav 1988), there should be a correlated feature in the nearby galaxy distribution. There is no conspicuous peak towards (199,0), but there is a striking dearth of galaxies in the opposite direction, which TF87 call the Local Void. This is visible in Fig. 2 as the empty volume towards minus X and positive Y (note that the Local Void is partially obscured by the Galactic plane). Lynden-Bell and Lahav (1988) call attention to the impact of the Local Void in swinging the motion of the Local Group away from the GA. The strong negative Z velocity of −317 km s^{-1} is probably due to the strong excess of galaxies below the SG Plane (see TF87 and also Fig. 2).

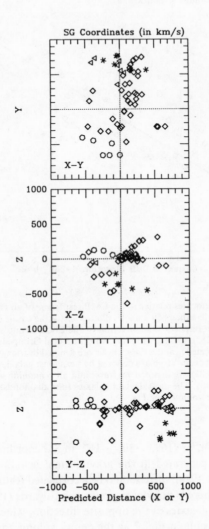

FIG. 4. The distribution in supergalactic coordinates of the galaxies in Fig. 3 (same symbols). The highly flattened structure defined by the diamonds is the Coma-Sculptor cloud of TF87. Non-diamonds are galaxies that deviate the most in Fig. 3. The peculiar motions of the six galaxies at high Z agree well with the overall cloud motion.

In short, there is a strong probability that the local peculiar motion of the Coma-Sculptor cloud (and the Local Group) can be accounted for via the gravitational attraction of visible nearby matter. The flattening of the Cloud is therefore also probably due to gravity-induced motions. This fact sets important upper limits on the role of non-gravitational forces, such as cosmic explosions, in determining the distribution and kinematics of local matter (Peebles 1987).

Peebles has also searched for internal collapse motions within the Cloud and has set stringent upper limits on flow velocities perpendicular to the plane of flattening. Once the surrounding galaxy density is mapped (e.g., using IRAS-selected or optically-selected galaxies) such motions can be used to estimate Ω in a manner analogous to Virgo infall. The only points that deviate significantly in Fig. 3 are six galaxies with large Y distance near Virgo at the edge of the volume (see Fig. 4), which are perturbed by Virgo infall and the GA tidal field, and four galaxies in the spur connected to the Leo Cloud. In particular, the six Coma-Sculptor galaxies with significant Z distances (and not in the Leo Spur) have infall motions toward the Supergalactic plane that appear to be small.

To summarize, the local Coma-Sculptor cloud is a highly flattened structure with low internal velocity dispersion but large bulk motion both perpendicular to, and parallel to, the flattening plane. It is striking that the boundary of coherent motion coincides closely with the boundary of the Cloud. The peculiar motion of the Coma-Sculptor cloud probably originates in the asymmetric distribution of nearby matter, notably in the paucity of galaxies in the Local Void and the excess of galaxies on one side of the SG plane compared to the other. The peculiar velocity relative to a large-scale flow model is about 360 km s^{-1}, which converts to 200 km s^{-1} line-of-sight. This amplitude and the dimensions of the Cloud — some 1500 km s^{-1} in diameter — are qualitatively typical of the sizes and motions of patches of galaxies elsewhere in the local volume (see Fig. 10).

6. Picturing the Velocity Field

6.1 - Distance Indicator Relationships

The hybrid form of Method 2 is used in Fig. 5 to construct four versions of the distance indicator relationship (Tully-Fisher for spirals, D_n - central velocity dispersion for ellipticals) for galaxies with distances less than 4600 km s^{-1}. Four different velocity-field models are used: (Panel 1) observed radial velocity with respct to the CMB; (Panel 2) radial velocity in CMB coordinates corrected for a Virgocentric infall of 250 km s^{-1}; (Panel 3) velocity in the rest-frame of the Aaronson *et al.* sample (i.e., with the bulk motion of the galaxies removed), corrected for a Virgocentric infall of 200 km s^{-1};

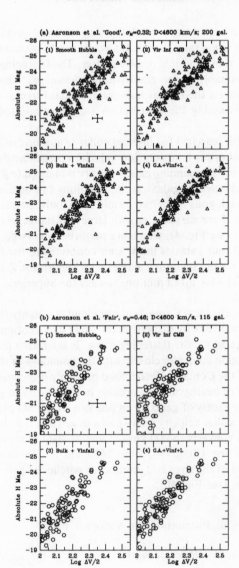

FIG. 5. These figures show the basic distance-indicator relations for the four data samples under four different assumption about the velocity field: 1) a smooth Hubble flow relative to the CMB rest frame, with no other motions; 2) a smooth Hubble flow combined with only a Virgocentric infall model, with $v_V = 250$ km s^{-1}, $n_V = 1.0$, $c_V = 0.0$; 3) a bulk motion for the whole volume, combined with the Virgocentric flow of panel 2, but with $v_V = 200$ km s^{-1}; 4) the Great Attractor + Virgocentric infall + Local Anomaly model having $v_A = 530$ km s^{-1}, $r_A = 4350$ km s^{-1}, $n_A = 1.62$, $c_A = 0.34$, $l_A = 309°$, $b_A = 18°$, $v_V = 150$ km s^{-1}, $n_V = 1.0$ and $c_V = 0.0$. Only galaxies with

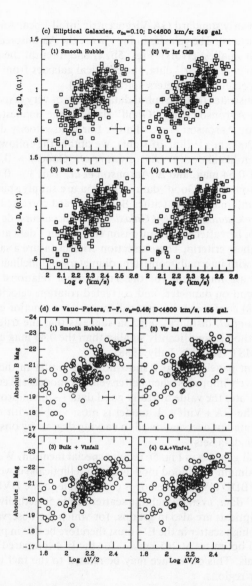

(c) Elliptical Galaxies, $\sigma_{D_n}=0.10$; D<4600 km/s; 249 gal.

(1) Smooth Hubble   (2) Vir Inf CMB

(3) Bulk + Vinfall   (4) G.A.+Vinf+L

Log D_n (0.1')

Log σ (km/s)

(d) de Vauc-Peters, T-F, $\sigma_M=0.46$; D<4600 km/s, 155 gal.

(1) Smooth Hubble   (2) Vir Inf CMB

(3) Bulk + Vinfall   (4) G.A.+Vinf+L

Absolute B Mag

Log $\Delta V/2$

predicted distances less than 4600 km s^{-1} are plotted. Fig. 5a gives Aaronson et al. 'Good' sample, 200 galaxies; Fig. 5b gives Aaronson *et al.* "Fair" sample, 115 galaxies; Fig. 5c gives elliptical galaxies, 249 galaxies; and Fig. 5d gives de Vaucouleurs and Peters galaxies that do not overlap with the Aaronson *et al.* data, 155 galaxies. Note the improvement of the relationship for each data sample with each successive velocity field model; the improvement is most marked for that sample with the smallest observational error, the Aaronson *et al.* "Good" sample in Fig. 5a.

and (Panel 4) a nearly standard GA-Virgocentric infall-Local Anomaly model (see caption for details). The galaxies in Fig. 5a are the Aaronson *et al.* "Good" spirals (data with smallest errors). Figs. 5b, c, and d repeat the same process for the Aaronson *et al.* "Fair" data, the elliptical galaxies from LFBDDTW, and the de Vaucouleurs and Peters spirals (DVP).

Fig. 5 demonstrates that the spiral data sets are of very disparate quality. The errors in the Aaronson *et al.* "Good" data are the smallest, followed by DVP, and then by Aaronson *et al.* "Fair". For field velocity dispersion, σ_f, in the range 80-200 km s^{-1} (line-of-sight), we obtain the following values of Δ from maximum-likelihood: Δ (Aaronson *et al.* 'Good') = 0.34-0.40 mag, Δ (DVP) = 0.42-0.54 mag, and Δ (Aaronson *et al.* 'Fair') = 0.49-0.52 mag.

The Aaronson *et al.* 'Good' data in Fig. 5a are simply spirals that have both diameters and magnitudes (either B_T or Harvard-corrected) given in the *Second Reference Catalog of Bright Galaxies* (de Vaucouleurs, de Vaucouleurs, and Corwin 1976; hereafter RC2). Aaronson *et al.* "Fair" data are those galaxies not meeting these criteria. These selection criteria ensure a sample of well-studied galaxies with internally-consistent diameters and inclinations. Both of these quantities are important for the Tully-Fisher relationship, as the H magnitudes depend on diameter, and corrected rotation velocity on inclination. Tully (1988) confirmed that poor diameters are a major cause of error in the Aaronson *et al.* distances. Fig. 5a suggests that the true error of the Tully-Fisher method is significantly smaller than the 0.45 mag conservatively quoted by AHMST (see also Tully 1988).

The different panels in Fig. 5 show how the scatter in the Tully-Fisher relation successively improves when better velocity models are employed. The figure also illustrates the value of really good data — the improvement in the last step using the GA + Vinf + L model is most visible with the Aaronson *et al.* "Good" data but overshadowed to varying degress by observational errors in the other data sets.

The elliptical galaxies in Fig. 5c require special mention. We believe from clusters and maximum-likelihood models that Δ (ellipticals) is well-determined at 0.45 mag (LFBDDTW). This error is comparable to the DVP and Aaronson *et al.* "Fair" data, even though the scatter in Fig 5c actually looks worse. Many of the ellipticals are also in groups, for which the observational errors are smaller. The high scatter in the E's must therefore be due in part to a higher σ_f, which indeed optimizes at 245 km s^{-1} for the E's compared to 80-200 km s^{-1} for the spirals. This difference may be related to the fact that ellipticals populate denser regions.

6.2 - Velocity Field Maps

Method 1 can be used to construct pictures of the observed velocity-field residuals relative to the cosmic rest frame and to various models. The following diagrams compare velocity field maps for the four data sets used in Fig. 5.

Figs. 6a, b, c and d present plots similar those given in LFBDDTW. These figures show observed peculiar velocities relative to the CMB for galaxies within ± 22.5° of four planes with poles given in each panel. Together, these four planes cover the sky. The second panel in each figure coincides with the supergalactic plane, and the axes correspond closely to SG coordinates. The fourth panel is perpendicular to the SG plane but nearly parallel to the galactic plane. The axis about which the planes are rotated is coincident with the X-axis of Figs. 2 and 4.

In these and the following pictures of the velocity field, we have applied the Malmquist-bias correction of Eq. 3, which moves the galaxies away by 8-18% depending on the sample. This correction is not valid for clumpy regions such as Virgo, but is required for more uniform regions elsewhere. Thus, without correcting for the space density of the sample precisely, there is no way of making velocity-field plots that are rigorously correct everywhere. Model velocity fields that are fit to these residuals suffer from a similar bias. This is demonstrated below by quantitative model-fitting to Virgocentric infall. The effect of the Malmquist-bias correction on the velocity field around Virgo is illustrated explicitly in Fig. 11.

Comparing the four data sets in Figs. 6a, b, c and d one sees the same features in all of them: Virgocentric infall, a strong flow toward negative X, which is the main GA flow, and a general compression inwards along the Y-axis, which we interpret as the tidal signature of the GA. It is this compression that indicates radial convergence towards a point several thousand km s $^{-1}$ distant in Centaurus. In the spiral samples, which are basically local, this tidal field is well modeled by a quadrupole with long axis towards the convergent point and short axes perpendicular to it. Pioneering this method on the Aaronson *et al.* data, LYJ found a convergent point some 3000 km s $^{-1}$ distant in the direction of the GA. We find the same distance using a full infall model, but we argue below that this distance to the GA is not as accurate as found by other methods.

Figs. 6e, f, g and h show the residual velocities of the same galaxies after fitting to the standard GA + Vinf + L model used in Fig. 5 (fourth panel there). The improvement in every data set is dramatic. Galaxies from different samples clearly tend to have similar residual motions in the same regions of space. However, some of this is probably due to correlated Malmquist-bias errors in clumpy regions, which show up the same way in all samples.

Finally, Figs. 7a and b summarize our current picture of the overall velocity field by plotting all four data sets together. The agreement among the samples is striking indeed.

Figs. 6 and 7 also illustrate important differences in the spatial coverage of the various samples. The spirals blanket the local volume more densely than the ellipticals, many of which lie outside the volume shown here. However, the spiral samples lack coverage in the southern hemisphere and have relatively

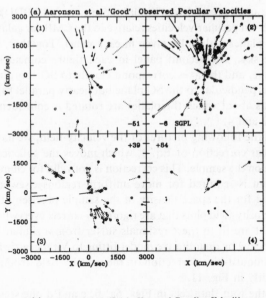

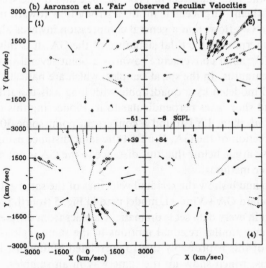

FIG. 6. Velocity field maps like those in LFBDDTW. All galaxies within 22.5°
of the plane perpendicular to l = 227°, b = t°, where t is indicated in each
panel. The four planes so defined cover the whole sky; the second panel (t = -6°)
is the supergalactic plane and the fourth panel (t = +84°) corresponds closely
to the galactic plane. (Note that the X-axes on these diagrams are the mirror-
image of those published by LFBDDTW.) Motions away from the Local Group
are denoted by solid circles and solid lines; motions towards the Local Group
are denoted by open circles and dotted lines. The observed peculiar velocities

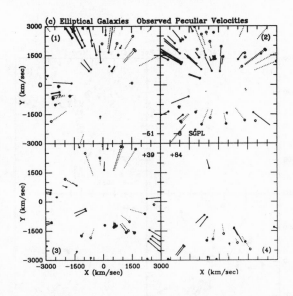

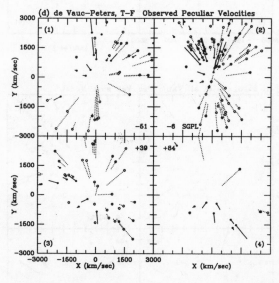

for galaxies in the CMB rest frame relative to a smooth Hubble flow are given in Figs. (a) for the Aaronson *et al.* "Good" sample; (b) for the Aaronson *et al.* "Fair" sample; (c) for the elliptical galaxies analyzed by LFBDDTW; and (d) for the de Vaucouleurs and Peters Tully-Fisher galaxies that do not overlap the Aaronson *et al.* sample. Analogous plots for peculiar velocities with respect to the standard GA + Vinf + L model used in Fig. 5 are given for each data set in Figures 6(e) — (h).

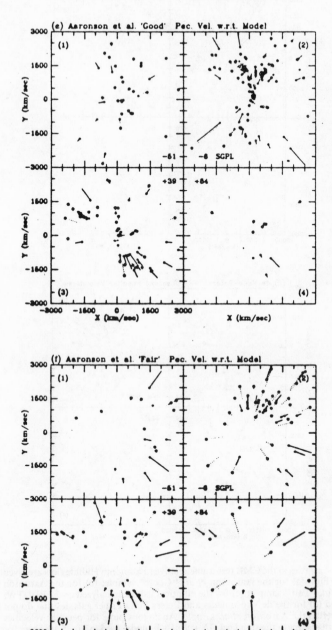

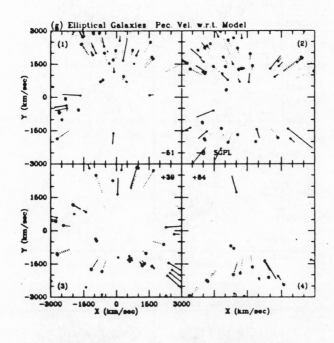

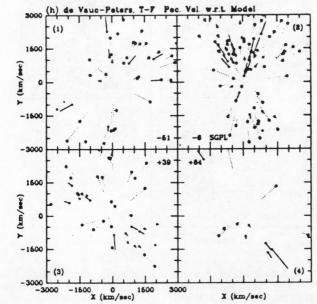

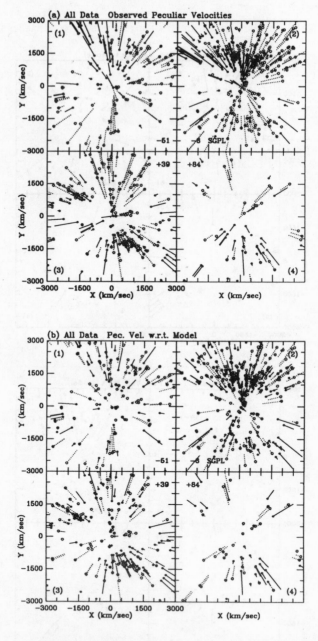

FIG. 7. Combined data for all samples in Fig. 6. Fig. 7a gives observed peculiar velocities with respect to the CMB and Fig. 7b gives peculiar velocities with respect to the GA + Vinf + L model.

few objects close to the GA (negative X axis). The information they provide
about the GA is largely local, through the quadrupole term, and they can say
relatively little about how the GA flow increases as one approaches its center
of mass.

A major result of this Study Week was a better correspondence of the
velocity-field maps with the density maps from the IRAS survey (see Yahil
1988 and Strauss and Davis 1988). Fig. 8 presents an enlarged version of Fig.
7a (second panel) giving the observed peculiar motions for galaxies within
± 22.5° of the SG plane. This map is designed to match a similar one by Yahil
(1988) of the velocity field predicted from IRAS in the SG plane. As presented
at the Study Week, the two showed encouraging agreement, and Strauss and
Davis (1988) present additional comparisons that suggest the two methods are
beginning to converge.

Our impression from comparing the two maps at the conference is that
the local flow due to Virgo is well matched but that the IRAS map relative

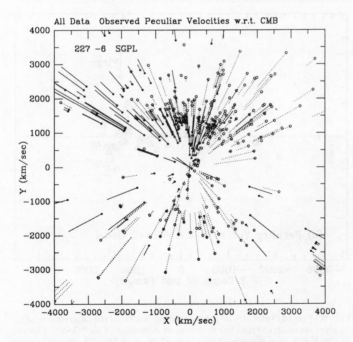

FIG. 8. An enlarged version of Fig. 7a, panel 2, showing the observed peculiar
velocities of all galaxies within ± 22.5° of the supergalactic plane and over
a larger region. This figure should be compared to a similar figure published
by Yahil (1988), which predicts peculiar velocities based on density distributions derived from IRAS galaxies.

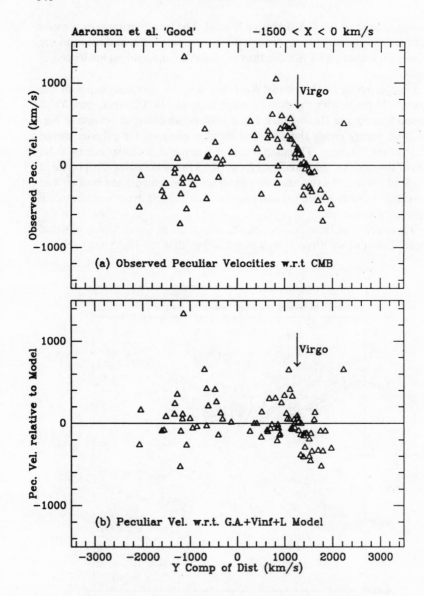

FIG. 9. Peculiar velocities of galaxies within 'slices' of the local volume. Fig. 9a gives the observed peculiar velocities of Aaronson *et al.* "Good" galaxies in the Virgo slice bounded by $-1500 < X < 0$ km s^{-1}, plotted versus Y distance. The Y position of the Virgo cluster is marked. Fig. 9b gives the peculiar velocities of the same galaxies as in Fig. 9a, but this time velocities are with respect to the standard GA + Vinf + L model of Fig. 5. Figs. 9c and d give the same Virgo 'slice' as Figs. 9a and b, but show the Aaronson *et al.* "Fair" sample (hexagons), elliptical galaxy sample (squares) and

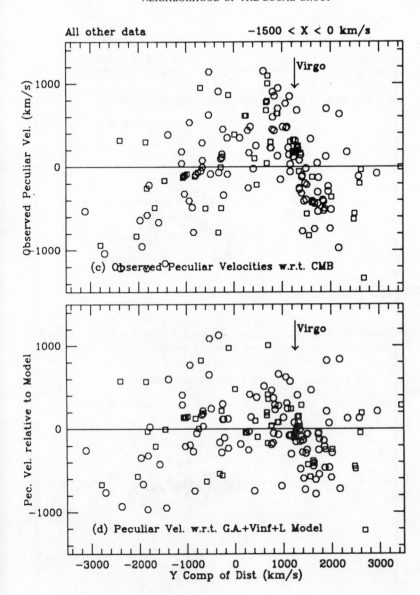

All other data   −1500 < X < 0 km/s

(c) Observed Peculiar Velocities w.r.t. CMB

(d) Peculiar Vel. w.r.t. G.A.+Vinf+L Model

de Vaucouleurs-Peters sample (circles). Figs. 9e and f give peculiar velocities of Aaronson *et al.* "Good" galaxies in the Ursa Major slice bounded by 1500 < X < 0 km s⁻¹, plotted versus Y distance. The Y position of the Ursa Major cluster is marked. Fig. 9e gives observed peculiar motion with respect to the CMB, Fig. 9f shows the residual motion after fitting to the standard model. Figs. 9g and h are the same as Figs. 9e and f but show the other three samples. Details are given in the text.

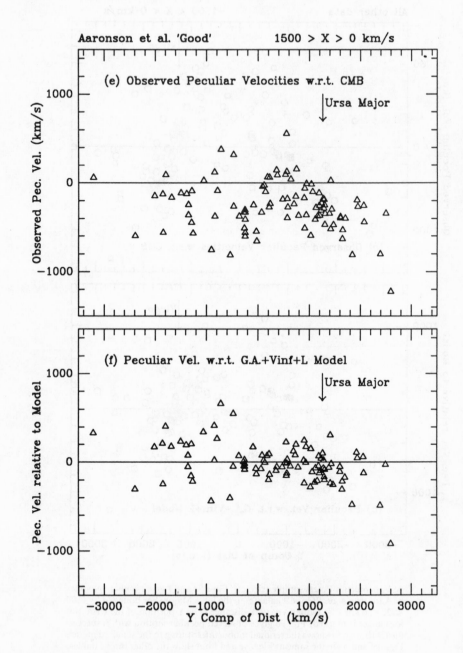

Aaronson et al. 'Good'   1500 > X > 0 km/s

(e) Observed Peculiar Velocities w.r.t. CMB

Ursa Major

(f) Peculiar Vel. w.r.t. G.A.+Vinf+L Model

Ursa Major

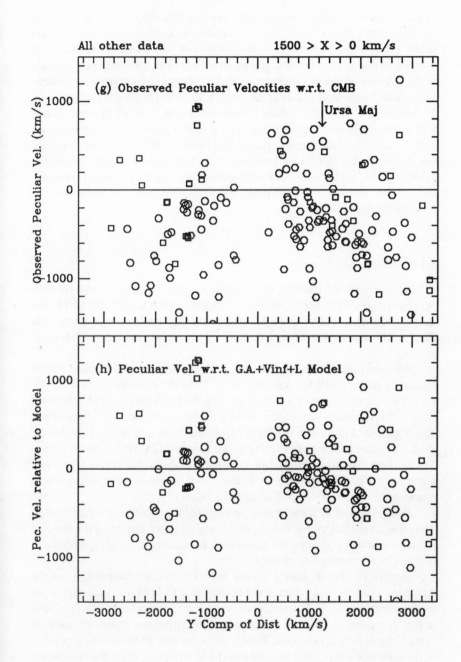

All other data   1500 > X > 0 km/s

(g) Observed Peculiar Velocities w.r.t. CMB

Ursa Maj

(h) Peculiar Vel. w.r.t. G.A.+Vinf+L Model

to ours shows too little contribution from the GA. This shows not only as too little direct flow towards the GA but also as too little tidal compression in the Y direction; an example is the net inward flow in Ursa Major, which is strong in our map (in the first quadrant) but not in IRAS. This underestimate of the GA may be due to a preliminary IRAS treatment of density in the galactic plane, which now simply fills in the plane with uniform density. Any portion of the GA hidden by the Galaxy is thus minimized. Lynden-Bell and Lahav (1988) report a similar deficit using optical counts and the same simple treatment of galactic plane density. However, Dressler (1988) has shown evidence (see below) that a significant fraction of the GA may be hidden behind the galactic plane, perhaps as much as one-half. Beefing up the IRAS contribution by that amount would improve the agreement between prediction and observation considerably.

6.3 - X-Slices and Regions

Another way of showing the peculiar velocity fields is to take slices in SG coordinates. The slice between X = 0 and -1500 km s^{-1}, parallel to the Y axis and perpendicular to the SG plane runs through the Virgo cluster (see Fig. 1). The slice next door with $0 < X < 1500$ km s^{-1} runs through the Ursa Major cluster.

We term these 'X-slices', and in Fig. 9a we plot observed peculiar velocity relative to the CMB for Aaronson *et al.* "Good" galaxies in the Virgo X-slice. This plot is basically a picture of Virgocentric infall and GA tidal field close to the SG plane. Virgocentric infall is the strong wiggle on the right, which shows galaxies falling into the cluster from the front and the rear. Figure 9b shows the same galaxies after fitting to the standard GA-Vinf-L model of Fig. 5. The model removes the overall systematic motions quite well. The small apparent residual Virgo infall may be due to real deviations or to errors, which mimic infall in this kind of diagram (see below).

Figures 9c and 9d show analogous plots for the other three data samples. The same trends are visible but the points scatter much more and the residual infall around Virgo after model fitting is higher. Both effects are symptomatic of the higher errors of these data.

Figures 9e and 9f show a similar before-and-after comparison for the Aaronson *et al.* "Good" data in the neighboring Ursa Major slice. Again the model does a good job. The systematic negative velocities in the observed data in Fig. 9e are due to GA tidal compression, not Virgo, and cannot be removed using Virgocentric infall alone. Finally, Figs. 9g and 9h show the same comparison for the other three data sets in the Ursa Major slice. The trends are similar but the scatter again is larger. The elliptical galaxies with strong positive residuals belong to the N1600-N1700 group, which is very badly fit by the model (see LFBDDTW).

Complementary to the slices are figures that plot peculiar velocity versus distance for different regions of the sky. In choosing such regions we were guided by the spatial limits of groups and clusters identified by TF87. In Fig. 10 we present "region" plots for 16 selected regions around the sky. Regions are defined for convenience either in galactic coordinates (Figs. 10a-1) or supergalactic coordinates (Figs. 10m-p). These plots cover all of the principal groups and clusters within the sampled volume. Two plots are given for each figure: the left-hand (LH) side shows observed peculiar velocity relative to the CMB versus predicted distance; the right-hand (RH) side shows residual velocity relative to the standard GA + Vinf + L model of Fig. 5, and also versus predicted distance. The line drawn in the RH plot corresponds to a distance error of ± 250 km s^{-1}.

The negative slope of the distance error line presents a fundamental ambiguity in interpreting this kind of diagram. The positive-negative pattern of distance errors looks like true infall when centered on a cluster. Indeed, if infall velocity varies as r from the center, infall precisely mimics observational errors. With *a priori* knowledge of the observational errors it is possible to assess the reality of infall in each case. With this as prologue, we give short comments about the motions in each region, against which the reader can compare his/her own interpretations:

N5846 (Fig. 10a): This region includes the well-known cluster whose name it bears. Nearly all of the galaxies along this line-of-sight seem to be associated with the cluster, which has a net motion of about 200 km s^{-1} relative to the CMB (LH side) and a motion of $\sim$ 100 km s^{-1} relative to the model (RH side). A distance error of 17% for the Aaronson *et al.* "Good" data implies a scatter of about ± 300 km s^{-1}, which is slightly less than actually observed. The evidence for infall is suggestive but marginal.

Leo Cloud (Fig. 10b): Upon correction for the model, what looks like one large collapsing cloud, with a range in peculiar velocity of over 1000 km s^{-1}, appears to break up into two separate structures with the present data. The further cloud has a residual of ~ -200 km s^{-1} with respect to the model, and each cloud extends $\sim$ 1000 km s^{-1} in line-of-sight distance. The apparent infall is consistent with errors.

N1549 (Fig. 10c): Nearly all the galaxies along this line-of-sight are associated with the N1549 group. As with N5846, the evidence for infall is suggestive but marginal. The center of mass has no significant left-over motion with respect to the model.

Fornax and Eridanus (Figs. 10d, e, and f): The galaxies in Fornax are localized to the cluster center and have velocities that look virialized (i.e., no correlation with predicted distance). The galaxies in Eridanus are more spread out along the line-of-sight and are less virialized. Both clusters are somewhat

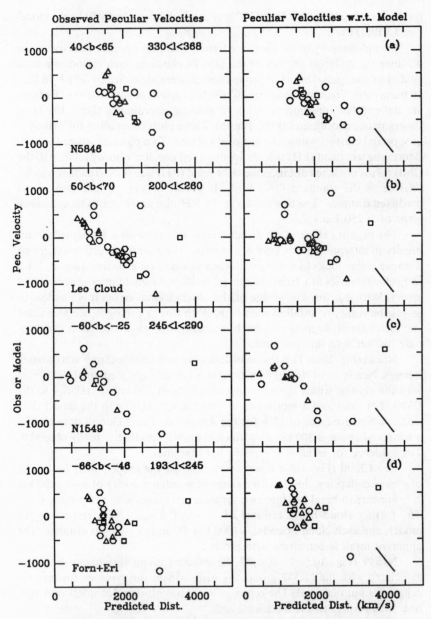

FIG. 10. Peculiar velocity versus predicted distance for galaxies in different 'regions' on the sky. The name and coordinate limits for each region are given in the left-hand figure; Figs. 10a to 10l in galactic coordinates; Figs. 10m to 10p are in supergalactic coordinates. The data for galaxies from all four samples are used; symbols have the same meaning as in Fig. 2. The left-hand side of

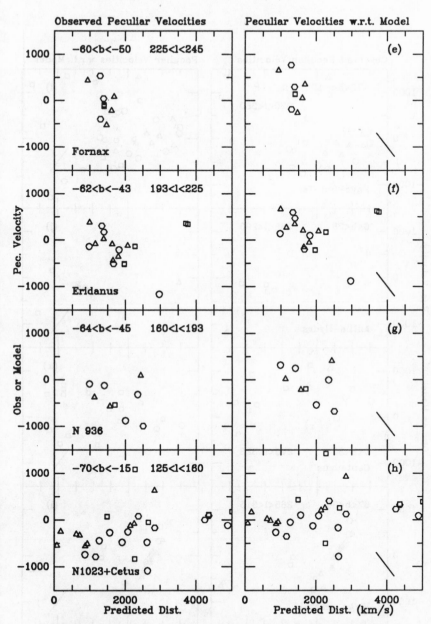

each figure plots observed peculiar velocity in the CMB rest frame versus predicted distance; the right-hand side plots the peculiar velocity with respect to the standard model. The line shown on the right-hand side of each figure represents a "distance-error" slope of ± 250 km s^{-1}. Details are discussed in the text.

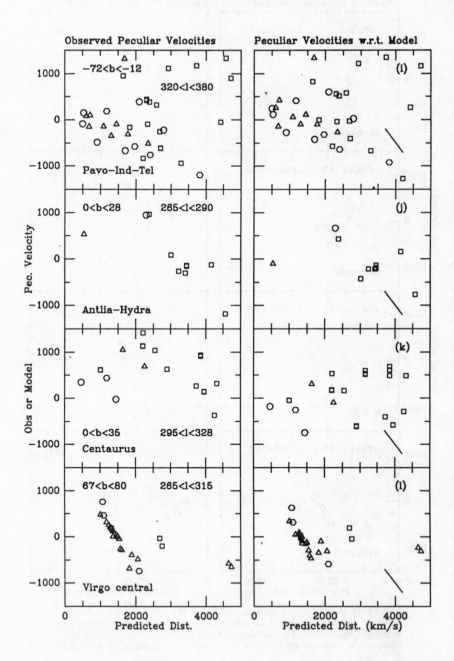

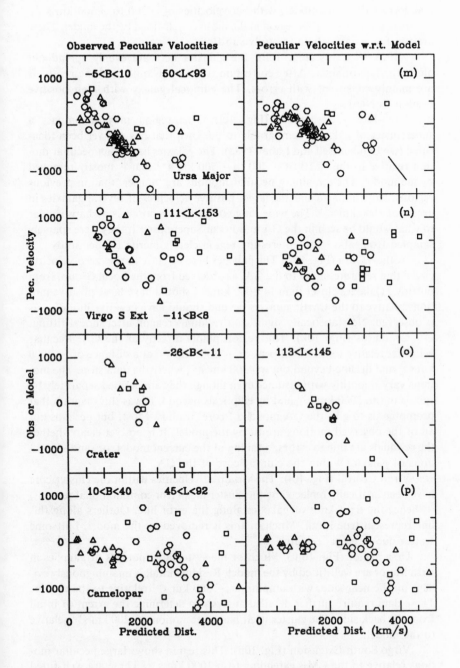

overcorrected by the model, with net velocities of $+200$ to $+300$ km s^{-1}.

N936 (Fig. 10g): This region in the mean is well-fitted by the model. There are too few galaxies here to study sub-flows.

N1023 + Cetus (Fig. 10h): the galaxies here are evenly spread out along the line-of-sight. After correction, the flow is quiet and the residuals are mainly consistent with errors. The elliptical galaxy with high positive residual is N661.

Pavo-Indus-Telescopium (Fig. 10i): This region may encompass a supercluster of galaxies comparable to our own that has not yet been identified (see Lynden-Bell and Lahav 1988). The galaxies have a net peculiar motion relative to the CMB of -400 to -500 km s^{-1} that is mostly removed by the model. The remaining peculiar motions are 'noisier' than in previous regions. Note in particular the large, positive velocities of several galaxies in excess of 1000 km s^{-1}. The wide range of peculiar motions is perhaps similar to what would be seen in the GA and Local Supercluster if they were sparsely sampled from afar. This interesting region clearly merits further study.

Antlia-Hydra (Fig. 10j): The galaxies here lie close to the 'zero-velocity-circle' that separates infall to the GA, as observed from the Local Group, from outflow. Galaxies closer than $\sim$ 3000 km s^{-1} should have large positive motions relative to the CMB; galaxies beyond should have negative motions, as is observed. No statistically significant residuals remain after model fitting.

Centaurus (Fig. 10k): This region points directly at the GA. Peculiar velocities relative to the CMB peak at $\sim$ 1000 km s^{-1} at a distance of $\sim$ 3000 km s^{-1} and decline beyond out to 4300 km s^{-1}, where the data stop. The motions vary smoothly with distance even though they are defined separately by spirals out to 2000 km s^{-1} and by ellipticals beyond. It was this diagram that prompted us to give the GA model a "core" radius. Most, but perhaps not all of the observed motions are fit by the model. It is not yet clear whether the residuals are due to errors, a failing of the current model, or random motions of galaxies within the clumpy core of the GA.

Virgo Central (Fig. 10l): The virialized velocities within the cluster core have been artifically replaced by the cluster mean (for computational reasons) — hence the tight knot of galaxies along the error line. Outliers along this line represent true infall. Much of this is removed by the model, but some scatter due to errors remains.

Ursa Major (Fig. 10m): Much of the systematic motions of galaxies in Ursa Major are well-fitted by the model. Residual infall is unambiguously evident on the near side, with amplitude $\sim$ 300 km s^{-1} relative to the cluster. These data require further detailed analysis to determine the extent of infall from the back side. The cluster itself has a net motion of 200 km s^{-1} relative to the model.

Virgo South Extension (Fig. 10n): This region shows large peculiar motions relative to the CMB extending to $+1000$ km s^{-1}. These are well-fitted by the model on average, but large scatter remains. The 'noisy' residuals resemble those in Pavo-Indus-Telescopium.

Crater (Fig. 10o): As with the N1549 and N5846, the galaxies here appear to be associated with one group. The group as a whole has a peculiar motion of ~ 300 km s^{-1} relative to the CMB that is moderately well-fitted by the model, leaving a smaller net velocity of ~ 100 km s^{-1}. The evidence for true infall into this cluster is somewhat stronger than for the other two clusters (the Aaronson et al. "Good" data show a wider scatter than expected), but is still marginal.

Camelopardalis (Fig. 10p): Within 1700 km s^{-1}, the motion is quiet and coherent. More distant galaxies show 'noisy' residuals, but much of this may be due to errors.

Three general features of the peculiar velocity field are evident from the 'X-slice' and 'region' diagrams. First, the dominant motion in this volume is due to the GA, as shown both by direct acceleration and by tidal compression. This large-scale coherence is why the distribution of galaxies in distance-space (Fig. 2) appears very similar to the distribution of galaxies in redshift-space (TF87).

Second, the motions of galaxies that remain *relative* to the GA flow are of three kinds: a) "Bulk" motions that have a low internal velocity dispersion. Examples include the Local Anomaly, the Leo Cloud, N1023/Cetus and the near region of Camelopardalis; b) Localized clusters with possible infall (Virgo, Ursa Major, N5846, N1549, Crater, Fornax and Eridanus). The evidence for infall is convincing for Virgo and Ursa Major but is less so for the other five. Virgo is certainly the most massive of these nearby clusters; c) Regions with internally 'noisy' motions. These include Pavo-Indus-Telescopium, Virgo South Extension, Centaurus and possibly the more distant galaxies in Camelopardalis.

Finally, systematic deviations from the standard GA model occur with amplitudes of up to 300 km s^{-1}. Deviant regions include groups (Crater, Fornax, Ursa Major) and also larger regions 1000 km s^{-1} or more across (near-side Ursa Major, Leo, Eridanus and perhaps the near region in Camelopardalis). Compared to these, the local Coma-Sculptor cloud seems fairly typical in terms of its scale size (1400 km s^{-1}) and residual motion (360 km s^{-1} total space velocity, 200 km s^{-1} line-of-sight).

7. QUANTITATIVE MODELS

This section reports on quantitative model fits using Eqs. 1 and 2. The results are preliminary since we are still in the stage of revising both the data and the fitting methods (hence the models here sometimes differ slightly from those in the graphs). The Aaronson et al. "Good" data are used exclusively, and the "Local switch" is always turned on so that the Local Anomaly out to 700 km s^{-1} (see Table 2) is smoothed out. The errors Δ and σ_f have been optimized separately in each case to give the best likelihood, but we do not take their relative values too seriously (see below).

7.1 - Malmquist Effects in Virgo: Method 1 vs. Method 2

The first two models (Table 3) show the sensitivity of Virgo infall to the Malmquist-bias correction in Method 1. In Model 1 we have applied a bias correction from Eq. 3 that corresponds to uniform space density and a Δ of 0.37 mag. Model 2 is the same without the bias correction. Both models assume the "new" GA flow from Table 1.

The resultant Virgo infall parameters are given in the last columns. The Malmquist-bias correction in Model 1 clearly pumps up the Virgo flow. The quantity v_V is larger, and the fall-off away from Virgo is more gradual (smaller n_V). An intuitive appreciation can be had from Fig. 11, which shows the observed peculiar velocities relative to the CMB around Virgo in both fits. Figure 11a shows the residuals with Malmquist-bias included, Fig. 11b without. Since the correction multiplies all distances by a constant factor, the distances of all galaxies in Fig. 11a are shifted outward relative to Fig. 11b. This has a systematic effect on the velocity vectors behind Virgo and Ursa Major, which are lengthened. The model responds by choosing a slower fall-off of Virgocentric infall velocity away from the cluster center.

This Malmquist-bias correction is clearly invalid in a clumpy region like Virgo-Ursa Major. The example shows how an invalid Malmquist-bias correction can create an illusory velocity field. However, many of the Aaronson et al. galaxies are more uniformly distributed in space, and for them the uniform-density correction may not be a bad approximation. The situation serves to show how hard it is to to get an unbiased picture of the velocity field. It also illustrates how important it is, when making pictures of the velocity field, to have Δ as small as possible. Since the Malmquist correction goes as Δ^2, the differences between Figs. 11a and 11b would be significantly larger if the Aaronson et al. "Fair" or DVP data sets were used.

7.2 - Virgo Infall and the Great Attractor

The rest of the models use the hybrid version of Method 2, which effectively minimizes in log v but otherwise has the same Malmquist-bias properties as Method 1. Because we are primarily interested in Virgo we have turned off the uniform-density Malmquist correction based on the previous section, which showed that the correction biases the Virgo infall solution unphysically. Model 3 is a repeat of Model 2 with this new method. The differences, though slight, illustrate the effects of minimizing in log v rather than in v.

TABLE 3

VIRGO INFALL MODELS USING AARONSON ET AL. "GOOD DATA" ONLY

| Model | GA Parameters | | | | | | σ_f^a (km s^{-1}) | Δ^a mag | Virgo Parameters | | |
	d_A (km s^{-1})	v_A	n_A	c_A	l_A (°)	b_A			v_V (km s^{-1})	n_V	c_V
1. Method 1 Malm. on GA solution from ellipticals	4200	535	1.7	0.34	308	18	145	0.37	260	0.5	0.00
2. Method 1 Malm. off	4200	535	1.7	0.34	308	18	77	0.45	133	1.0	0.10
3. Method 2	4200	535	1.7	0.34	308	18	90	0.41	100	1.5	0.25
4. Method 2 GA solution from spirals	3000	555	1.1	0.20	305	18	79	0.41	85	1.9	0.30
5. Method 2 AHMST clone	None, use bulk flow only						80	0.41	300	0.6	0.40

All models listed here use Aaronson et al. "Good" data only. GA parameters are taken either from Table 1 (Nos. 1, 2 or 3) or optimized here (No. 4).

a The relative values of σ_f and Δ are highly uncertain — see text.

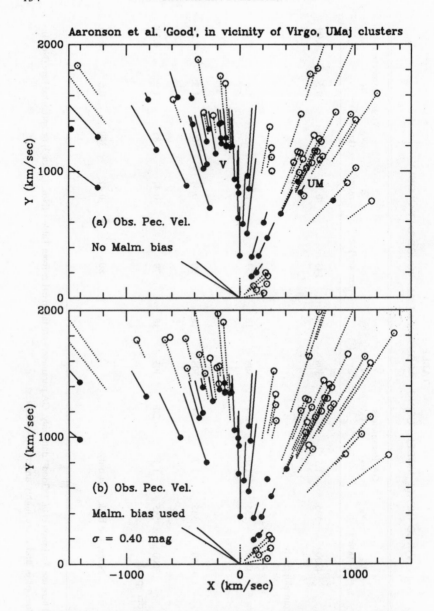

FIG. 11. The observed peculiar velocity field in the CMB rest frame, in the vicinity of the Virgo cluster, as defined by the galaxies in the Aaronson *et al.* "Good" sample, under two different assumptions: (a) Peculiar velocity predicted with no Malmquist-bias correction. (b) Peculiar velocity with a Malmquist-bias correction as in Eq. 3 and an observational error of 0.40 mag. Note the apparent increase of infall velocity on the back side of the Virgo cluster in (b) relative to (a).

Both models have a steep fall-off (high n_V) and low v_V. With its steep profile and low infall velocity at the Local Group, this picture of Virgo infall differs considerably from the conventional one, but agrees with the visual picture presented in the last section. More on this presently.

Whereas Model 3 takes its GA parameters from the elliptical galaxies, Model 4 optimizes them from the Aaronson *et al.* "Good" data themselves. The two solutions agree well with regard to flow velocity, v_A, at the Local Group and the GA direction on the sky. The fall-off away from the GA is shallower in Model 4, but this is not strongly constrained. On the other hand, Model 4 puts the GA at only 3000 km s^{-1}, compared to 4200 km s^{-1} from the ellipticals, a difference which is several times the formal error. However, the spirals determine the distance to the GA mostly through the quadrupole term, which is easily perturbed by the small-scale systematic flows that we know exist, but are not contained in the model (see above). If the recently-measured spirals in Centaurus were included (see Mould 1988) their outward velocities would push the GA further away, as do the ellipticals (see LFBDDTW). With these differences taken into account, the agreement between the local spirals and the GA model is excellent.

In computing Model 4, we were struck by its small value of local Virgo infall (85 km s^{-1}). Aside from Model 1, of dubious validity, our models never return infall velocities in the 250-300 km s^{-1} range, despite numerous attempts to perturb them. When we finally tried to recreate a model exactly equivalent to AHMST, the situation became clear. In this model (Model 5), there is no GA flow, and all radial velocities are taken with respect to the rest frame of the Aaronson *et al.* "Good" sample. The fit prefers a high v_V of 300 km s^{-1} and a very shallow fall-off with $n_V = 0.6$.

The crucial difference in this model is the lack of a GA, specifically the GA tidal compression along the Y axis. Because it produces a similar compression, Virgo infall increases to take up the slack. This is illustrated in Figs. 12a, b and c, which repeat the X-slice through Virgo shown in Figs. 9a-d. Fig. 12a shows observed peculiar velocities relative to the CMB, as in Fig. 9a. Fig. 12b shows the Virgocentric infall signal that is left after a standard GA model is removed. This is comparable to the Virgocentric infall signal in Models 2-4, where the GA is fit simultaneously with Virgo. One sees how well the GA removes all the large-scale trends, leaving only a small-scale deviation in the immediate neighborhood of Virgo. Fig. 12c shows the radial velocities after transformation to the sample rest frame, which corresponds to the Virgo signal in Model 5 and in AHMST. Here the signal continues through the position of the Local Group to the left, where negative velocities look like Virgocentric inflow. This is the primary source of cross-talk between Virgocentric flow models and the GA flow.

Figs. 12d-f show the same three cases for the neighboring Ursa Major X-slice. Removing the GA flow again has a major effect. The GA accounts

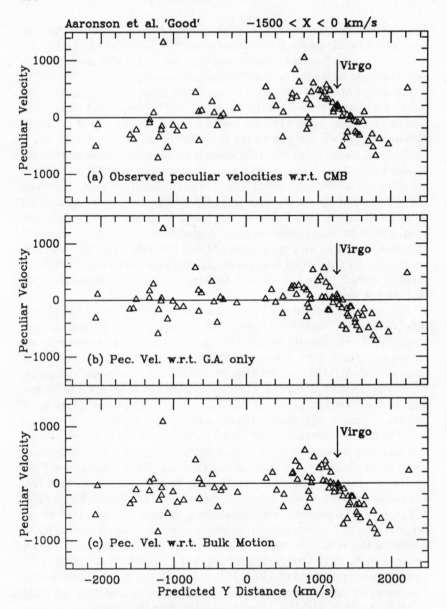

FIG. 12. 'X-Slices' illustrating the cross-talk between the peculiar velocity fields
of the GA and Virgo. Only data from the Aaronson *et al.* "Good" sample
are used. Figs. 12a-c plot the peculiar velocities of galaxies versus Y distance
in a slice bounded by $-1500 < X < 0$ km s^{-1} that highlight motions relative
to the Virgo cluster. Figs. 12d-f highlight motions of galaxies relative to the
Ursa Major cluster in a slice bounded by $1500 < X < 0$ km s^{-1}. These plots
differ from those in Fig. 8 in that only galaxies within a distance of 2500 km

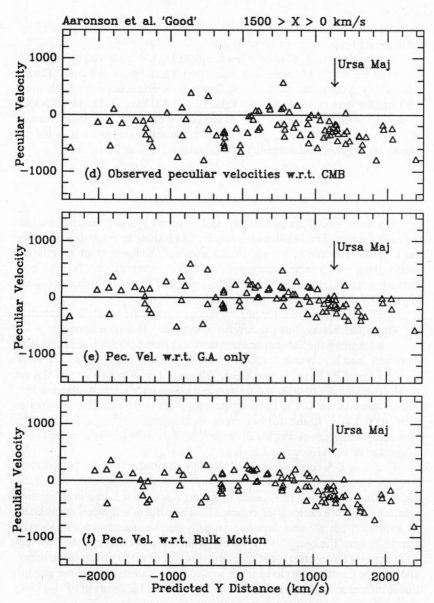

Aaronson et al. 'Good'   **1500 > X > 0 km/s**

(d) Observed peculiar velocities w.r.t. CMB

(e) Pec. Vel. w.r.t. G.A. only

(f) Pec. Vel. w.r.t. Bulk Motion

Ursa Maj

s^{-1} are used. Figs. 12a and d show observed peculiar velocities relative to the CMB. Figs. 12b and e plot peculiar velocities relative to a model that subtracts out only the velocity field due to the GA. Figs. 12c and f plot peculiar velocities relative to the bulk motion of all galaxies (in all four data samples) with distances less than 2500 km s^{-1}. With the GA subtracted, the remaining Virgo infall is highly localized to the cluster. Subtracting just the bulk motion leaves behind negative velocities at negative Y, which mimic Virgo infall.

for the greater part of all the systematic motion in this slice, leaving a localized infall around the center of Ursa Major.

We conclude that, beyond a few hundred km s^{-1} from Virgo, Virgocentric infall is a fragile phenomenon that is dominated by the GA flow. The infall velocity at the radius of the Local Group is uncertain but is probably quite a lot smaller than the conventional value of 250-300 km s^{-1}. The infall velocity at a distance of 400 km s^{-1} from the center of Virgo is well-determined at approximately 500 km s^{-1} (see Fig. 12b). However, this point is inside the non-linear regime, and its utility for calculating Ω is doubtful.

7.3 - Errors and Error Correlations

The above models are preliminary, and we do not yet have a full understanding of the errors. The likelihood maximization is done by brute-force searching in parameter space, so we do not produce the usual error correlation matrix. However, running many solutions gives us an intuitive feel for how the various parameters interrelate. We give some of the main results here, leaving a fuller discussion for a future paper.

a. The Great Attractor parameters as a group are highly insensitive to the Virgo parameters, but the reverse is not true, as shown above.

b. Among the GA parameters the local flow velocity is insensitive to all others, and this is true for both the spirals and the ellipticals. The error on v_A is about 70 km s^{-1} from either data set. The direction of the GA on the sky is insensitive to other parameters and has an uncertainty of about $\pm 10°$. The distance to the GA is fixed by the ellipticals to be at a distance greater than 4000 km s^{-1}; the formal error is estimated to be ± 300 km s^{-1}. However, the notion of an "infall center" is likely to be highly idealized since the center of the GA may be clumpy.

c. The GA core radius, c_A, and slope parameter, n_A, are only slightly coupled. They are determined at present only through the elliptical galaxy data. The nominal error in c_A is $\sim \pm 0.5$ and in n_A it is $\sim \pm 0.3$, and both increase together. The core radius is determined by a handful of ellipticals at distances of 3000-4000 km s^{-1} and depends critically on the assumption of spherical symmetry near the core. New evidence from Alan Dressler (see below) suggests that, on the contrary, the mass distribution near the core is quite lumpy and that the Centaurus region itself is a major sub-condensation whose gravitational influence on us is considerable. The extreme idealization of the inner regions in the present model may thus prove untenable.

d. If a fixed GA flow is assumed, then the two most closely coupled parameters for Virgo are v_V and n_V. Because the infall amplitude is firmly anchored at a few hundred km s^{-1} from the cluster (cf. Fig. 12b), a steeper slope must be compensated by lower infall farther out. With a fixed GA, the nominal uncertainty in v_V is ± 30 km s^{-1}, but the weakness of the signal far out and the high sensitivity to the large-scale velocity field of the GA imply larger systematic errors.

e. The error estimates for Δ and σ_f are strongly anti-correlated. Maximum-likelihood distinguishes between them by the way they fall off with distance: in magnitude units, measurement error is constant with distance, whereas field-dispersion error declines. We have seen that σ_f is well-determined locally to be 80-90 km s^{-1} out to 700 km s^{-1}. The best-fitting model over the whole volume adjusts by choosing a relatively low σ_f of 90-120 km s^{-1} and a high compensating Δ of 0.39-0.41 mag (Aaronson *et al.* "Good").

Maximum-likelihood assumes that σ_f is constant everywhere, which is risky. Patches like the Local Anomaly may be quiet internally (80-90 km s^{-1}) but deviate from the flow model by significantly larger amounts (see Sec. 6). The local value of σ_f may thus be too low to typify the whole volume. As a guide, we can use the Local Anomaly, whose magnitude of 360 km s^{-1} converts to 200 km s^{-1} line-of-sight. Taking this as an upper limit to σ_f gives = 0.34 mag for the Aaronson *et al.* "Good" data. This probably brackets the range. A better value could come from good diameters and inclinations for the spirals in clusters (see discussion by Tully 1988), which can give Δ independent of σ_f.

7.4 - Parameter Significance

The significance of each parameter is given by the change in likelihood that results from its addition to the model. Adding an extra parameter should decrease the likelihood by 0.5 units just by random chance. Likelihoods for Method 2 are given in Table 4 for four models. In each case, Δ and σ_f have been optimized to give the best likelihood. The first model has no motions whatsoever and provides a zeroth-order comparison. The second model is the best bulk model with no internal flows (i.e., local switch "off"). The third model turns on the local switch, which adds three parameters, and reoptimizes the bulk motion. The fourth model adds Virgo infall, which adds another three parameters, and again reoptimizes the bulk motion. Finally the fifth model adds the Great Attractor but takes away the bulk flow, which is no longer needed. This adds six parameters but takes away three, for a net gain of three. The number of free parameters in each model is given in the table.

The table shows that the addition of extra parameters is strongly merited at every stage, giving a likelihood change that at minimum is at least four times larger than expected by chance. The improvement due to the GA comes in two stages: from Model 1 to Model 2, which adds the bulk flow, and from Model 3 to Model 4, which adds the tidal term. Thus, the GA is somewhat more important than Virgo in maximizing the total likelihood.

TABLE 4
LIKELIHOOD CHANGES

No.	Model	No. Param.	Likelihood (Method 2)	Δ (Likelihood)
1	No motions	1[a]	−201.41	——
2	Bulk motion, no internal flows	4[b]	−164.82	+36.6
3	Turn on local anomaly	7[c]	−158.62	+42.8
4	Turn on Virgo[e] infall	10[d]	−121.86	+79.6
5	Turn on GA[g] turn off bulk	13[f]	−112.96	+88.5

[a] No motions permitted. Free parameter is the Hubble constant.
[b] Hubble constant plus three bulk-motion components.
[c] Add three Local Anomaly components.
[d] Add three Virgo parameters: v_v, n_v, c_v.
[e] Virgo parameters: $v_v = 225$ km s^{-1}, $n_v = 0.8$, $c_v = 0.3$ d_v. These differ from Model 5, Table 3 owing to slightly different σ_f and Δ.
[f] Add six GA parameters: d_A, v_A, n_A, c_A, l_A, b_A. Take away three bulk-motion components.
[g] GA parameters are "new" from Table 1. Virgo parameters: $v_v = 100$ km s^{-1}, $n_v = 1.7$, $c_v = 0.25$ d_v.

8. THE NATURE OF THE GREAT ATTRACTOR

In LFBDDTW we reviewed the scant information available at that time on the structure of the Great Attractor. With the help of Ofer Lahav we constructed a map of optical galaxies that showed a marked concentration of galaxies in that direction covering a large solid angle on the sky. A crude comparison to Virgo indicated 15 to 30 times as many galaxies within the region of strong overdensity, with perhaps as many more obscured by the galactic plane. This ratio was consistent with the mass excess inferred from scaling the GA flow to Virgo infall. The mass excess in absolute units was estimated to be about 5×10^{16} h^{-2} M$_\odot$.

We also referenced a radial velocity survey by da Costa et al. (1986) that showed a prominent peak at 4500 km s^{-1} in the same direction. If this were the center of the GA, we reasoned, it would not be expected to move very fast relative to the CMB, and hence its observed radial velocity would be in good agreement with the distance of 4350 km s^{-1} predicted by the elliptical galaxies.

A major unanswered question, though, was our failure to detect any ellipticals falling into the GA from the backside. The sample in the north penetrated far enough: why were the comparable galaxies missing in the GA? This lack was especially worrisome since the the claimed over-density in the Great Attractor ought to enhance the population of early-type galaxies.

The answer to this question has since been at least partially clarified. The GA is indeed rich in early-type galaxies, as shown by Hubble types in the ESO catalog. However, there is also a tendency in the same catalog to classify small E's as S0's, so they were omitted from our original target sample. Furthermore, our target sample coverage is slightly less complete in the south than in the north. Togther these two effects mean that we penetrate about 30% less deep in the south, and this seems to be enough to make the difference. Dressler and Faber are currently beginning a deeper survey of the region around the Great Attractor in hopes of detecting ellipticals on the far side.

The major advance since LFBDDTW is a large radial velocity survey of the GA region by Dressler (1988), who has targeted 1400 galaxies in the large region shown in Fig. 13. Galaxies were selected in the same way, i.e., by apparent diameter, as the Southern Sky Redshift Survey (SSRS) of da Costa et $al.$ (1987), which provides a comparison sample. The present results are based on a first sample of 900 galaxies chosen randomly from the whole. Fig. 14 shows a velocity histogram of these galaxies compared to a properly scaled histogram from the SSRS and one also from the northern Harvard Redshift Survey (Geller and Huchra 1988), which is also comparable. Fig. 14 shows two strong peaks centered at 3000 km s^{-1} and 4500 km s^{-1}. The nearer peak is spatially confined on the sky to a region around the Centaurus clusters, which are visible in Fig. 13 near (298, 22).

The velocity of this peak is identical to the velocity of the closer of the two Centaurus clusters. Both the ellipticals and the new spiral survey by Aaronson et $al.$ (Mould 1988) show that the nearer Centaurus cluster has a distance of around 2400 km s^{-1}, and that it has a large positive peculiar motion away from us. If this conclusion applies to all the galaxies in the 3000 km s^{-1} peak, then as a group they lie closer to us than their radial velocities would suggest.

The situation with the 4500 km s^{-1} peak is even more complex. The velocity of this peak coincides with that of the more distant Centaurus cluster, but the galaxies that comprise it are spread over virtually the entire area of Fig. 13 (see also Lucey et $al.$ 1986b). Furthermore, both LFBDDTW and Aaronson et $al.$ show that the more distant Centaurus cluster, like the nearer cluster, is moving rapidly away from us towards the GA. A possible interpretation is that a part of the 4500 km s^{-1} peak is closer to us but that most of it — the diffuse background — is the GA and is at rest with respect to the CMB. This interpretation is consistent with the GA model, but it is clearly not unique. Independent distance estimates to galaxies in the region are badly needed.

If the above interpretation is correct, it is possible to estimate the local overdensity in the Centaurus-GA region by comparing to the other two surveys,

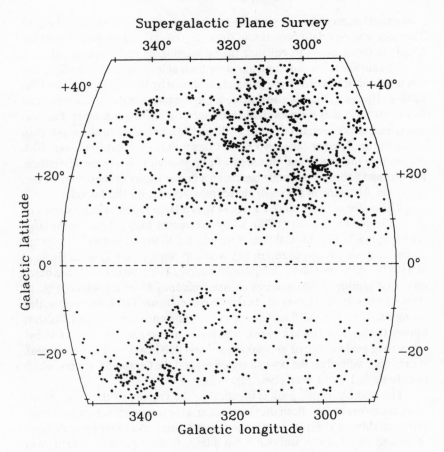

FIG. 13. The locations of galaxies targeted in Dressler's radial velocity survey of the GA region (Dressler 1988).

and then to integrate this average value to find the overdensity within a sphere centered on the GA inside the Local Group. Dressler finds enough galaxies to infer a peak $\Delta\varrho/\varrho$ of about 3 and an integrated $\Delta\varrho/\varrho$ of 1.5. This implies $\Omega = 0.1$-0.2 using the usual infall formula, so the visible mass is adequate to induce the local flow velocity of ~ 550 km s^{-1} attributed to the GA. More interesting are the relative masses of the 3000 km s^{-1} and 4500 km s^{-1} peaks. Allowing for the shallower sampling and larger volume of the 4500 km s^{-1} peak, Dressler concludes that its mass is actually four times larger than the nearer peak, unlike the impression given by Fig. 14. This bolsters the interpretation of the far peak as the GA.

However, this mass ratio also predicts that the two regions are comparable in their gravitational effects on the Local Group. This says that sub-clumping

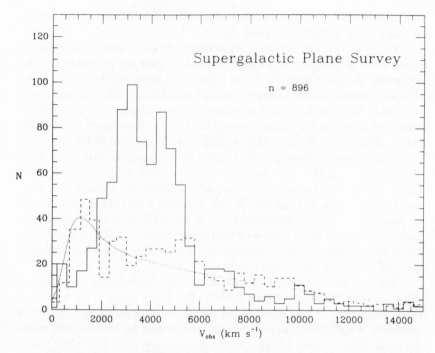

FIG. 14. Velocity histrogram for 896 galaxies measured by Dressler (solid line). These are compared to the histogram of 1657 galaxies measured by SSRS (observed data = dashed line, smoothed data = dotted line), normalized to Dressler's area of 0.85 steradian.

(including, perhaps, the Centaurus clusters) within the GA is gravitationally important. At present our homogenized, spherically symmetric model ignores such effects. Sub-clumping has two important implications: 1) We may expect to detect strong departures from radial symmetry when motions of other galaxies in the GA region are mapped. The local spirals may already be showing this in their quadrupole term; 2) It may be highly misleading to construct a spherically symmetric infall model of the GA based on just the volume currently sampled. The mass in the Centaurus region that is foreground to the GA could be as near as 2400 km s^{-1}, and hence one of the major contributors to the local acceleration field. If a mass concentration like Centaurus is not present on the far side of the GA, then the hypothetical infall velocity there at our distance from the GA could be as low as 300 km s^{-1}, all other things being equal. The spherically-averaged infall velocity at a radius of ~ 4000 km s^{-1} from the center of infall thus probably lies somewhere between 400 km s^{-1} and 550 km s^{-1}. This is important in certain kinds of cosmological estimates, such as the spherical approximation for infall velocity versus $\Delta M/M$

(see LFBDDTW). Comparisons with theory should be limited to the volume of space so far actually surveyed.

To aid in these comparisons, we attempt here (with some trepidation) to sketch out a simple geometric model of what we think the present data show minimally. A sphere does not fit the region with good data very well, so we choose a cylinder centered roughly on the Local Group with long axis aimed at the GA. The flow is parallel to the axis of this cylinder, with a velocity of about 500 km s^{-1} at the far end, some 1500 km s^{-1} distant from us and 5700 km s^{-1} distant from the GA. This distance is set by the limit of the good data in that direction, as shown in Fig. 15. The flow velocity rises to a maximum of 1000 km s^{-1} at the other end near the GA, which is located about 3000 km s^{-1} away from us, at the distance of the Centaurus clusters. The total length of the region with substantial flow velocities is thus about 4500 km s^{-1}. Its diameter is not well determined, but is probably of order 3500 km s^{-1}, based on the extent of the cross-wise tidal compression. This sketch is obviously highly schematic and neglects, for example, that the flow is actually convergent and not parallel to the cylinder walls.

Since we measure the radial velocity component only, Peebles asked at this conference whether the flow might in fact break down in the middle of the cylinder near us and whether the two flows at opposite ends might be unrelated. We do not believe this to be the case because, looking sideways where the break should occur, we see a strong compression from both sides that is most simply ascribed to tidal compression by a coherent, large-scale infall. A multi-flow picture offers no explanation for this. It will be possible to answer the question definitively soon when the IRAS density maps (or comparable data) allow us to trace the three-dimensional flow pattern through the region of the Local Group. The preliminary IRAS maps at this conference already suggest that the flow pattern is indeed continuous and coherent through our location, as the cylinder model implies.

As a last point we stress that a large fraction of the GA may be hidden by the galactic plane. Two studies at this conference actually suggest this. Lynden-Bell and Lahav (1988) report that the optical peculiar acceleration converges too rapidly on nearby galaxies, whereas they expected a larger contribution from the GA — perhaps as much as a factor of two more. Strauss and Davis (1988) and Yahil (1988) likewise show diagrams that suggest that the velocity field of the GA is about a factor of two too weak. Extra mass of this order could be hidden by the galactic plane (see Fig. 13). Not knowing how much mass is hidden could be a real handicap to future studies of the region.

9. SUMMARY

Our current major conclusions on the velocity field in the neighborhood of the Local Group are as follows:

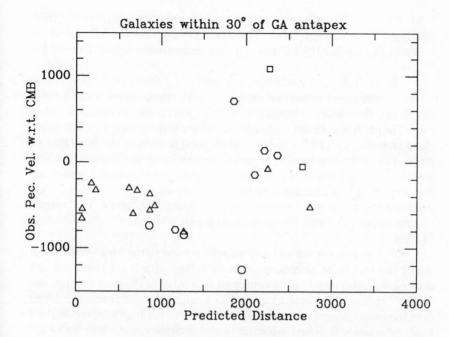

FIG. 15. The observed peculiar velocities with respect to the CMB for galaxies within 30° of the antapex of the GA (l = 129°, b = −18°), plotted versus
predicted distance. Galaxies from all four data sets are plotted as in Fig. 2.
The flow to the GA appears to persist to at least 1500 km s⁻¹.

1. The local spirals within 2500 km s^{-1} generally corroborate the
velocity flow pattern seen in ellipticals. When fit to a Virgo infall— GA model,
they yield the same direction for the GA and v_A, but put the GA closer to
us, at 3000 km s^{-1} rather than 4200 km s^{-1}. However, if new spiral
measurements in Centaurus were added, the distance would be pushed farther away, closer to the elliptical value.

2. From a radial velocity survey by Dressler, evidence is mounting for
a substantial overdensity of galaxies in the GA direction at about the right
distance and amplitude to cause the observed flow pattern. The exact spatial
distribution of galaxies there is unknown but appears to be highly clumped.
The Centaurus association, with the same velocity as the closer Centaurus
cluster, is a major foreground subcondensation that contributes a substantial
share of the gravitational acceleration at the Local Group, in addition to the
GA.

3. Virgocentric infall models are heavily modified by inclusion of the
GA flow. Much of the apparent infall behind Virgo is due to the GA tidal

field. With the GA removed, the remaining Virgo infall signature far from the cluster is weak. Virgo infall at the Local Group is poorly determined but is likely to be smaller (85-133 km s^{-1}) than conventional values (250-300 km s^{-1}).

4. The sphere of galaxies out to 700 km s^{-1} around the Local Group shows an anomalous motion of 360 km s^{-1} with respect to the Virgo infall— GA model. This motion is identical within the errors to that of the Local Group itself. The patch of coherent motion is coextensive with the local Coma-Sculptor cloud identified by TF87, which is also the local portion of the Supergalactic plane. The anomalous motion is mostly perpendicular to this plane but also has a component parallel to it. It seems likely that the motion can be ascribed to irregularities in the local gravity field and is caused partly by the Local Void and partly by an excess of nearby galaxies below the SG plane. The internal velocity dispersion within the local Cloud is only 80-90 km s^{-1} along the line-of-sight.

5. The peculiar velocities of galaxies relative to the Virgo infall— GA model fall into three categories: quiet bulk flow with low internal velocity dispersion; group membership combined with infall to the group center; and "noisy" regions containing galaxies with a range of peculiar motions larger than the observational errors. Almost all galaxies are found in one of these kinds of regions. Residual velocities in one dimension can range up to 200 km s^{-1} on scales 1000 km s^{-1} and larger. The size and motion of the Local Anomaly are typical in comparison to other nearby regions.

6. Although a spherically symmetric GA model fits the present data well for both spirals and ellipticals, it is dangerous to use it for cosmology. On account of significant sub-clumping, the flow pattern around the GA is likely to be quite anisotropic. We have sketched a minimal, interim model that might be useful for cosmological calculations. The model involves a coherent flow pattern along a cylinder aimed at the GA that is roughly 4500 km s^{-1} long and 3500 km s^{-1} wide. The flow velocity varies from 500 km s^{-1} at the far end of the cylinder up to 1000 km s^{-1} near the Great Attractor, and then declines closer to the center of mass. The quadrupole term in the local galaxies suggests that the flows at opposite ends of this cylinder are, in fact, smoothly connected into one coherent flow through the location of the Local Group.

ACKNOWLEDGEMENTS

We would like to thank our colleagues Roger Davies, Alan Dressler, Donald Lynden-Bell, Roberto Terlevich, and Gary Wegner, who could not be co-authors on this preliminary report for lack of time. They have contributed much to our understanding of this subject in recent years, and we expect that they will be able to join us on the final version. We also thank Ofer Lahav for his efforts on our behalf.

REFERENCES

Aaronson, M., Huchra, J., Mould, J., Schechter, P.L., and Tully, R.B. 1982b. *Ap J.* **258**, 64. (AHMST)

Aaronson, M., Bothun, G.D., Mould, J.R., Huchra, J., Schommer, R., and Cornell, M. 1986. *Ap J.* **302**, 536. (ABM86)

Aaronson, M., Huchra, J., Mould, J.R., Tully, R.B., Fisher, J.R., van Woerden, H., Goss, W.M., Chamaraux, P., Mebold, U., Siegman, B., Berriman, G., and Persson, S.E. 1982a. *Ap J Suppl.* **50**, 241.

Bothun, G.D., Aaronson, M., Schommer, R., Huchra, J., and Mould, J. 1984. *Ap J.* **278**, 475.

Chincarini, G. and Rood, H.J. 1979. *Ap J.* **230**, 648.

da Costa, L.N., Nunes, M.A., Pellegrini, P.S., and Willmer, C. 1986. *AJ.* **91**, 6.

da Costa, L.N., Pellegrini, P.S., Sargent, W.L.W., Tonry, J., Davis, M., and Latham, D.W. 1987. preprint.

de Vaucouleurs, G. and Peters, W.L. 1968. *Nature.* **220**, 868.

———. 1984. *Ap J.* **287**, 1. (DVP)

de Vaucouleurs, G., de Vaucouleurs, A., and Corwin, H.G. 1976. *Second Reference Catalog of Bright Galaxies.* Austin: University of Texas Press. (RC2)

Dressler, A. 1988. *Ap J.* in press.

Dressler, A., Faber, S.M., Burstein, D., Davies, R.L., Lynden-Bell, D., Terlevich, R.J., and Wegner, G.W. 1987. *Ap J Letters.* **313**, L37.

Geller, M.J. and Huchra, J.P. 1988. this volume.

Lilje, P., Yahil, A., and Jones, B.J.T. 1986. *Ap J.* **307**, 91. (LYJ)

Lucey, J.R., Currie, M.J., and Dickens, R.J. 1986a. *MN.* **221**, 453.

———. 1986b. *MN.* **222**, 417.

Lynden-Bell, D. and Lahav, O. 1988. this volume.

Lynden-Bell, D., Faber, S.M., Burstein, D., Davies, R.L., Dressler, A., Terlevich, R.J., and Wegner, G.W. 1988. *Ap J.* **326**, 19. (LFBDDTW).

Mould, J. 1988. this volume.

Peebles, P.J.E. 1987, preprint.

Rubin, V.C., Thonnard, N., Ford, W.K., and Roberts, M.S. 1976a. *AJ.* **81**, 719.

Rubin, V.C., Thonnard, N., Ford, W.K., Roberts, M.S. and Graham J.A. 1976b. *AJ.* **81**, 687.

Sandage, A. 1972. *Ap. J.* **178**, 1.

Strauss, M.A. and Davis, M. 1988. this volume.

Tully, R.B. 1988. this volume.

Tully, R.B. and Fisher, J.R. 1987. *Nearby Galaxies Atlas.* New York: Cambridge University Press. (TF87)

Yahil, A. 1988. this volume.

DISTANCES TO GALAXIES IN THE FIELD

R. BRENT TULLY

Institute for Astronomy, University of Hawaii

ABSTRACT

Three topics will be discussed that are related to the determination of distances to individual galaxies and the Hubble Constant. First, there will be a brief description of recent work on the calibration of luminosity—line width relations now that photometric information based on CCD observations is available. Second, the problem of Malmquist bias will be reviewed, and it will be argued that there is a straightforward way to avoid bias in our situation. Third, the preliminary results from a program to determine the distances to individual galaxies will be summarized. It will be argued that there is a 'local velocity anomaly' and that it is this happenstance that is a principal reason for the perpetuation of a controversy over the value of H_0.

1. CALIBRATION OF LUMINOSITY—LINE WIDTH RELATIONS

New data are leading to improvements in the correlations between luminosity and H I profile line width (Tully and Fisher 1977), and there are positive implications regarding the use of these relations for the estimation of distances. Pierce and Tully (1988) discuss the calibration of B, R, and I-band relationships based on CCD imaging data and the recalibration of the H-band relationship for samples drawn from the Virgo and Ursa Major clusters. Figure 1 is taken from this reference.

The results of this new work are summarized as follows: (1) CCD photometry of galaxies in the Virgo and Ursa Major clusters was accomplished with fields-of-view that are large compared with the dimensions of the galaxies. Images were acquired in B, R, and I passbands. Resultant measured fluxes should be accurate to ± 0.03 mag. (2) In the Ursa Major Cluster, an almost complete sample of spiral galaxies brighter than $B_T = 13.3$ mag was observed.

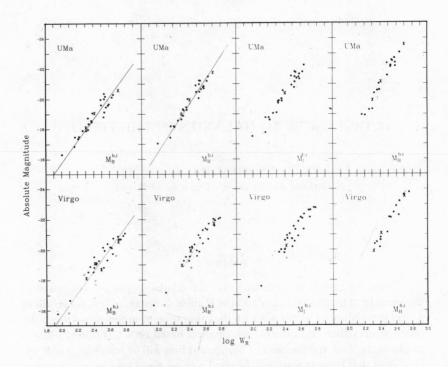

FIG. 1. Luminosity-line width relations. Top panels: Ursa Major Cluster; bottom panels: Virgo Cluster. Panels from left to right: B, R, I, and H passbands. Data have been shifted to fit to the three local calibrators (M31, M33, NGC 2403) indicated by crosses. The Virgo Cluster fits do not include the open circles, which correspond to galaxies thought to be viewed in projection falling into the cluster.

In the Virgo Cluster completion is only to 12.0 mag. Galaxies as faint as 15.2 mag were observed in these clusters. (3) There is now relatively little problem with the quality of the H I profile information for systems in these two clusters. In particular, galaxies in the Virgo Cluster can be observed with the Arecibo telescope (e.g., Helou, Hoffman, and Salpeter 1984). (4) A significant improvement has come about because the CCD images can be used to give relatively reliable inclinations, so line width deprojections are more accurate. Uncertainties are now ± 3° rms, down from ± 7° in earlier work. This information has been used to tighten the $H_{-0.5}$ band relationship (Aaronson, Huchra, and Mould 1979). (5) Scatter is lower in the Ursa Major relationships than in the Virgo relationships and lowest in the R, I, and H passbands. The observed rms scatter in Ursa Major is 0.30 mag and, given the inferred depth of the cluster, the scatter for a single galaxy must be about 0.25 mag. (6) It is argued that the greater scatter in the case of the Virgo Cluster is probably due to inclusion of systems slightly to the foreground and background of the cluster that are falling in. (7) There is only weak evidence of separation by morphological type. A 2σ effect is seen in the B-band samples, and only a 1σ effect is seen at R, I, and H bands. (8) The slope of the relationships steepen toward longer wavelengths. However, slopes are shallower with total luminosities than with luminosities within the restricted aperture used with the H-band work. The new calibration suggests that in the infrared $L_T \sim V^{3.2}$. (9) The three nearby systems M31, M33, and NGC 2403 are used to calibrate the zero point. The two Local Group galaxies were observed in B and R with a "1-inch" telescope that provided a field of view of $5°$. All three calibrators have H-band measurements so this band can be given a consistent calibration. In the I band NGC 2403 provides the sole calibration. (10) The two clusters are subsequently found to be at essentially the same distances (Virgo: 15.6 ± 1.5 Mpc; U Ma: 15.5 ± 1.2 Mpc). (11) Given observed velocities and a model of Virgocentric retardation that influences our motion by 300 km s^{-1}, a value of 85 km s^{-1} Mpc^{-1} is found for the Hubble Constant.

2. MALMQUIST BIAS

A lot has been said recently about Malmquist bias in the specific context of the luminosity—line width relations. Sandage (1988) says it is a serious problem. Bottinelli *et al.* (1986) discuss biased and unbiased domains. Giraud (1986) agrees there is a bias but feels these authors overestimate the effect. Lynden-Bell *et al.* (1988) describe a way of compensating for the problem in the case of the related $D_n - \sigma$ relationship. Schechter (1980) and Aaronson *et al.* (1982) prescribe a recipe that should largely negate the bias.

It will be argued here that the latter set of authors were on the right track but that a simpler procedure can be used that is still bias-free (see also Kraan-Korteweg *et al.* 1986). It is essential, however, that line widths be measured in an unbiased fashion. Also, one must have access to a complete sample that is representative of the sample to be studied, both in intrinsic properties and in the quality of the data.

The standard procedure in the past has been to fit a straight line to data in a plot of magnitude versus log W^i (W^i is the deprojected line width), where either the line is the regression that involves minimization of residuals in magnitudes or it represents the combination of two orthogonal regressions. In either case, if one assumes that a field galaxy with observed W^i has the absolute magnitude of the mean relationship, then the distance modulus $(m - M)$ that is determined will tend to be too low because an apparent magnitude cutoff to the sample will select in favor of brighter galaxies at a given distance. In other words, galaxies fainter than the mean will be rejected in greater numbers than galaxies brighter than the mean. The result is Malmquist bias.

However, if the line formed from the regression that involves minimization of residuals in line width is taken, and the rest of the operation is performed as before, there is no bias. As long as a straight line adequately describes the relationship and there is a normal distribution about the mean, there should be as many objects of given absolute magnitude drawn from the sample with W^i larger than the mean as smaller than the mean. This is the key point. To rephrase it, the regression on line width bisects the sample in the unbiased dimension. Every galaxy of given absolute magnitude and distance in the sample under investigation is drawn from a distribution in which there is equal *a priori* probability of a complementary galaxy of identical absolute magnitude and distance but with a line width that differs from the mean for that magnitude by the opposite sign. Neither of these galaxies has preferred entry into the sample over the other. Hence, equal numbers of objects in the sample will be drawn from the left of the regression as from the right and the assumption that the object has the mean luminosity for a given observed W^i is as likely to be too high as too low.

This intuitive concept was tested with simulated data that had the general characteristics of the actual data available. The details of the simulation will be described elsewhere (Tully 1988b), but the result was confirmation of the proposal made here. In practice, it is the combined Virgo and Ursa Major Cluster samples that provide the necessary calibration of the regression on W^i. The assumption is required that the cluster galaxies are representative of systems in other environments. At present, the Ursa Major sample is essentially complete to $M_B = -17.7$, providing a volume-limited sample complete to that limit.

3. THE LOCAL VELOCITY ANOMALY AND THE HUBBLE CONSTANT

In response to the challenge of this Study Week, two of us seem to have independently discovered manifestations of the same phenomenon and used the same terminology to describe it. Burstein, Faber, and Lynden-Bell have reported on work with their collaborators that leads them as well to deduce the existence of the 'local velocity anomaly'. It will be argued that this anomaly conspired to generate the controversy that has arisen over the value of the Hubble Constant.

The H-band sample discussed by Aaronson *et al.* (1982) will be used for the ensuing discussion. The calibration is in the system described in Section 1, and distances are determined in a way that should be statistically free of bias, as described in Section 2. The other information that is used is the specification of cloud membership for each object in the sample provided by the *Nearby Galaxies Catalog* (Tully 1988a).

Now, the ratio of the estimated distance to a predicted distance from a kinematic model can be determined for each galaxy. Two specific kinematic models will be considered: (I) pure Hubble expansion and (II) the Virgocentric retardation model discussed by Tully and Shaya (1984). Means and dispersions in the ratio of estimated to predicted distance can now be found for all members of a common cloud. If the kinematic model has been based on the proper value of the Hubble Constant, then on the average, the mean ratio for all clouds should be unity. Deviations from unity would arise if some clouds have motions that are poorly described by the specific kinematic model.

The results for all clouds with at least five independent measurements are shown in Figure 2a (for model I) and in Figure 2b (for model II). In the case of model I, a value of $H_0 = 80$ km s^{-1} Mpc^{-1} was required to force to unity the mean ratio: measured distance/kinematic distance. However, two of the three nearest clouds are seen to lie high in Figure 2a. We reside in the nearest one: the Coma-Sculptor Cloud. The other, very high point is associated with the Leo Spur, which is an appendage to our Coma-Sculptor Cloud. These two nearby entities contain many of the galaxies that historically have played a major role in the determination of H_0. If only galaxies within these two clouds are considered, then H_0 would have a value in the range 60-65.

In the case of model II, the two nearby clouds stand out much more prominently as special cases. After 3σ rejection of three anomalous points, a value of $H_0 = 95$ km s^{-1} Mpc^{-1} drives the mean ratio of measured to kinematic distances to unity. Still, if only data from the two nearby anomalous clouds are considered, one would conclude $H_0 \simeq 63$ km s^{-1} Mpc^{-1}.

These results suggest very strongly the resolution to the decade-old controversy over the value of the Hubble Constant. One hint is the interconnectedness of H_0 and the Virgocentric mass model as reflected in the mean ratio of the observed to predicted distances in Figures 2a and 2b. Another hint is the way the two nearest clouds particularly stand out in this ratio in the Virgocentric mass model case. Combine these hints with a statement that must be true in general: an observer *within* a region of retarded expansion due to a mass concentration and looking only at systems within the same region will tend to measure a value of H_0 that is *too low*.

This latter point is surely the explanation for why $H_0 = 80$ gives the best fit for model I (no mass concentration) and $H_0 = 95$ gives the best fit for model II (roughly $10^{15} M_\odot$ in Virgo). This is the same problem that must be confronted if the Virgo Cluster is used alone to estimate H_0, where observed velocity must be corrected for 'infall'. If one accepts that the Virgo Cluster has a significant gravitational effect on us, then it follows that direct measurements of H_0 from field galaxies in the local vicinity will probably give results that are *biased low*.

Provisionally accept that model II provides a better description of the Local Supercluster than the pure Hubble expansion description of model I. However, this model manifestly fails to explain the motions observed in our cloud and the Leo Spur, since the ratios of observed to expected distances are

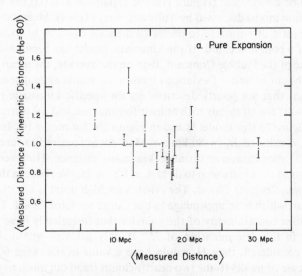

FIG. 2a. Ratio of measured to expected distances: pure expansion model. Each data
point corresponds to the mean ratio associated with all galaxies with measured distances
within a single cloud. Only clouds with at least five measurements are included. Model
assumes $H_0 = 80$ km s^{-1} Mpc^{-1}.

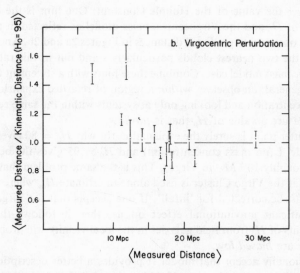

FIG. 2b. Ratio of measured to expected distances: Virgocentric retardation model.
Data points correspond to the same clouds as in Fig. 2a. Model assumes
$H_0 = 95$ km s^{-1} Mpc^{-1}.

much greater than unity in these two cases. It is submitted that a second effect, to be referred to as the 'local velocity anomaly', is being brought into focus.

An evident possibility is that there is sufficient mass within the Coma-Sculptor Cloud and associated Leo Spur to cause a local retardation to expansion. Within the Coma-Sculptor Cloud, typical deviant velocities are -200 km s^{-1} at observed velocities of 450 km s^{-1} and distances of 7 Mpc. An effect of this amplitude could be caused by $1 \times 10^{14} M_{\odot}$. The cloud has an integrated blue luminosity of $5 \times 10^{11} L_{\odot}$. A mass-to-light ratio of $200 - 300$ $M_{\odot}/L_{\odot}$ is implied.

In actual fact, model II may well *overestimate* the influence of the Virgo Cluster. That model was developed to describe the pattern of motions in close proximity to the cluster and only to a lesser extent was influenced by motions across the full domain of the Local Supercluster. With model II, points associated with clouds in the south galactic hemisphere (the opposite hemisphere from Virgo) tend to lie *higher than the mean* in Figure 2b, whereas with model I, the equivalent points lie *lower than the mean*. This situation suggests that a correct description is *intermediate* between models I and II.

In addition, it is now becoming rather convincing that we are under the influence of a mass concentration on a scale larger than Virgocentric: that of the Great Attractor (Lynden-Bell *et al.* 1988). There is evidence in support of such a proposition in Figure 2b; points associated with clouds nearest the Great Attractor lie low, suggesting there are motions away from us relative to the motions anticipated by the Virgocentric mass model.

4. SUMMARY

We are at a very early stage in our efforts to model the motions of nearby galaxies. Evidence is accumulating for the Great Attractor mass concentration of $5 \times 10^{16} M_{\odot}$ at 4300 km s^{-1}, there is a general consensus for 10^{15} $M_{\odot}$ centered on the Virgo Cluster at 1000 km s^{-1}, and now it is suggested there is a local velocity anomaly caused by $10^{14} M_{\odot}$ at 450 km s^{-1}. Judging by the clumpiness in the distribution of galaxies, an accurate description would have to include even more components.

This situation has a severe consequence because of the systematic underestimation of H_0 that occurs when the observer and the sample are both within an overdense region. In the specific case that confronts us, the nearest mass concentrations probably have the largest effects (local anomaly: 200 km s^{-1} at 450 km s^{-1}; Virgo: 300 km s^{-1} at 1000 km s^{-1}; Great Attractor: 600 km s^{-1} at 4300 km s^{-1}). The result is an increase in the apparent value of the Hubble Constant with distance.

Such a situation could also be a consequence of Malmquist bias. That has been the point-of-view of Bottinelli *et al.* (1986), Tammann (1987), and

Sandage (1988). It was argued in Section 2, though, that the present analysis avoids Malmquist bias. This claim is sure to be met with scepticism, but note that Figures 2a and 2b do not display the signature of Malmquist bias, which would be a steady or accelerating decrease in the ratio of measured to expected distance with increasing distance. Instead, this ratio is apparently constant with distance if one disregards the local cloud and its spur. (Kraan-Korteweg, Cameron, and Tammann (1986) also discovered a manifestation of the same phenomenon, but they dismissed it as a probable observational artifact.)

Detailed modeling of the velocity field in the Local Supercluster is in progress. The case is substantial, though, for the existence of a 'local velocity anomaly' that, abetted by the Virgo Cluster velocity anomaly, gives rise to deceptively low values of the Hubble parameter in samples dominated by nearby galaxies and creates the artifact that H_0 increases with distance, a phenomenon misinterpreted as Malmquist bias. It is argued that this perversity of nature is at the origin of the controversy over the value of H_0 and, incidentally, may have provided some sustenance to the claim that the Hubble law has a quadratic form (Segal 1976). However, this perversity would recur commonly throughout the universe since most cosmic observers probably live near mass concentrations.

With the zero-point calibrations used in this analysis, a value of $H_0 \simeq 63$ km s^{-1} Mpc^{-1} would have been determined from galaxies within the local cloud and its spur, if the effect being discussed is ignored. On this zero-point, the true value is almost surely greater than the 80 km s^{-1} Mpc^{-1} suggested by the pure expansion model I, but probably not as high as the value of 95 km s^{-1} Mpc^{-1} required by model II. The most recent Aaronson et al. (1986) result based on clusters within 10,000 km s^{-1}, with a minor adjustment to the zero-point used in this paper, is $H_0 = 88$ km s^{-1} Mpc^{-1}. This obviously compatible value is derived over a domain that is larger than the presently documented mass concentrations and may represent a measurement of H_0 unaffected by peculiar motions.

REFERENCES

Aaronson, M., Huchra, J. P., and Mould, J. R. 1979. *Ap J.* **229**, 1.

Aaronson, M., Huchra, J. P., Mould, J. R., Schechter, P. L., and Tully, R. B. 1982. *Ap J.* **258**, 64.

Aaronson, M., Bothun, G., Mould, J. R., Huchra, J. P., Schommer, R. A., and Cornell, M. E. 1986. *Ap J.* **302**, 536.

Bottinelli, L., Gouguenheim, L., Paturel, G., and Teerikorpi, P. 1986. *AA.* **156**, 157.

Giraud, E. 1986. *AA.* **174**, 23.

Helou, G., Hoffman, G. L., and Salpeter, E. E. 1984. *Ap J Suppl.* **55**, 433.

Kraan-Korteweg, R. C., Cameron, L., and Tammann, G. A. 1986. in *Galaxy Distances and Deviations from Universal Expansion.* ed. B. F. Madore and R. B. Tully. p. 65. Dordrecht: Reidel.

Lynden-Bell, D., Burstein, D., Davies, R. L., Dressler, A., Faber, S. M., Terlevich, R., and Wegner, G. 1988. *Ap J.* **326**, 19.

Pierce, M. J. and Tully, R. B. 1988. *Ap J.* **330**, 579.

Sandage, A. 1988. *Ap J.* **331**, 583.

Schechter, P. L. 1980. *AJ.* **85**, 801.

Segal, I. E. 1976. *Mathematical Cosmology and Extragalactic Astronomy.* New York: Academic Press.

Tammann, G. A. 1987. in *IAU Symp. 124. Observational Cosmology.* ed. A. Hewitt, G. Burbidge, and L. Fang. p. 151. Dordrecht: Reidel.

Tully, R. B. 1988a. *Nearby Galaxies Catalog.* Cambridge: Cambridge University Press.

———. 1988b. *Nature,* **334**, 209.

Tully, R. B. and Fisher, J. R. 1977. *AA.* **54**, 661.

Tully, R. B. and Shaya, E. J. 1984. *Ap J.* **281**, 31.

LARGE PECULIAR VELOCITIES IN THE HYDRA-CENTAURUS SUPERCLUSTER

JEREMY MOULD

Palomar Observatory, California Institute of Technology

ABSTRACT

Six clusters forming part of the Hydra-Cen Supercluster and its extension on the opposite side of the galactic plane have been reported to show large peculiar velocities relative to the Hubble flow. Additional observations required to verify the infrared Tully-Fisher distances on which these peculiar velocities are based include improved measurements of the isophotal diameters of the galaxy sample, and investigation of the Tully-Fisher and Faber-Jackson relations with environment. Examination of the local volume bounded by 8000 km/sec redshift suggests clusters in the Hydra-Cen complex which may provide the gravitational acceleration of the measured clusters. These are the Hydra and IC 4329 clusters on the north side of the galactic plane, and the Indus cluster and NGC 6769 group on the south side. The Cen 45 subcluster does not appear to be one of the attractors, however; it shows a very large positive peculiar velocity. It is still curious, moreover, that large peculiar velocities have not been observed in other similar mass concentrations, namely the Coma supercluster and the Perseus-Pisces supercluster. A detailed program of mapping these velocity fields with Tully-Fisher distances remains ahead.

1. INTRODUCTION

At the recent Balatonfured meeting Aaronson *et al.* (1988) presented preliminary evidence that five clusters forming part of the Hydra-Centaurus Supercluster have positive peculiar velocities ranging from zero to 1000 km/sec. A sixth cluster in Pavo showed a similar large outflow velocity. These velocities are measured with respect to a frame comoving with the expansion and in which an observer would see no microwave dipole anisotropy. The peculiar veocities are inferred from IR Tully-Fisher distances to the six clusters, based on 21 cm data from A.N.R.A.O. at Parkes and infrared photometry from Las Campanas Observatory. In the present discussion I want to slightly update the observational data and suggest some reasons for keeping an open mind about these peculiar velocities, until a number of additional critical observations are available.

2. Isophotal Diameters

CCD photometry of Hydra-Centaurus galaxies has been obtained at Cerro Tololo Interamerican Observatory by Mark Cornell of the University of Arizona early in 1987. $H_{-0.5}$ magnitudes, from which infrared Tully-Fisher distances are inferred, are defined in terms of the light contained in one third (-0.5 dex) the radius at which the galaxy's surface brightness (corrected for inclination) has fallen to 25 B mag arcsec^{-2}. Accurate isophotal diameter measurements (Burstein 1988) can reduce the random scatter in the IR Tully-Fisher relation relative to eye-estimated diameters (Nilson 1973), which we are sometimes forced to rely on, and remove (Cornell et al. 1987) systematic errors which would otherwise compromise our results.

Aaronson et al. (1987) expressed concern about the quality of the optical diameters for the Hydra-Cen and Pavo galaxies. Only 10 of the 64 galaxies used to estimate the 6 cluster distances have diameters from the Second Reference Catalog (RC2: de Vaucouleurs, de Vaucouleurs, and Corwin 1976); the remaining galaxy diameters are eye-estimates (Lauberts 1982) transformed to the RC2 system (Mould and Ziebell 1982). Recently, Mark Cornell has sent me accurate isophotal diameters for 6 of the sample galaxies (only one of which was in the RC2), showing an average (systematic) difference from the transformed diameters of 0.03 ± 0.01 in log D_{25}. In addition, a new study of the transformation equations between ESO and the RC2 has appeared (Paturel et al. 1987). The similarity between these equations limits systematic errors in the adopted diameters to 0.03 to -0.01 in log D_{25} (the range resulting from a difference in scale factor in the transformations). From the average slope of the infrared growth curve for spiral galaxies ($\Delta H/\Delta log A \approx 2.3$) one can calculate a limit on the implied error in the deduced peculiar velocities of 100 km/sec. Although this is reassuring, it in no way decreases the importance of measuring accurate isophotal diameters for *all* the sample galaxies.

3. Anomalous Surface Brightnesses

Aaronson et al. (1986) pointed out that, even after accurate isophotal diameters had been substituted for the eye-estimates in the Arecibo cluster sample, different clusters populated systematically different relations between mean surface brightness and 21 cm velocity width. These are distance independent quantities and (whatever the physical basis for such a relation) would be expected to define a consistent relation, if cluster galaxies are all drawn from the same population. If cluster galaxies are not all drawn from the same population, it is dangerous to deduce distances from the same observables deployed in a distance dependent manner (Kraan-Korteweg 1983).

Now, surface brightness is much more dependent on accurate isophotal diameters than $H_{-0.5}$ is. The 0.03 dex systematic error I mentioned earlier leads to a 0.15 mag error in Σ_H. However, the inter-cluster anomalies in surface brightness found by Aaronson et al. (1986) range up to 0.5 mag in Σ_H (see their figure 1), and even larger anomalies are seen in some of the Hydra-Cen and Pavo clusters with their current, mainly eye-estimated, galaxy diameters. This problem will be examined carefully when we have measured accurate isophotal diameters. Also under examination is the question of using surface brightness as a second parameter in the Tully-Fisher relation.

In the meantime, a further, related cause for concern is the indication of similar surface brightness anomalies, possibly affecting the Faber-Jackson relation for elliptical galaxies. The data plotted in Figure 1 show all the galaxies assigned the redshifts of the Virgo, Coma and Centaurus clusters by Burstein et al. (1987) and Davies et al. (1987). The effective surface brightness defined in the first paper is plotted against the velocity dispersion adopted in the second. As with the spiral galaxies, the question is, are galaxies in different clusters drawn from a common surface brightness population?

That this matter is not a red herring for either the Tully-Fisher relation or the Faber-Jackson relation can be seen from the following thought experiment. Imagine two galaxies of identical distance and mass distribution. Now take the first galaxy and turn up the luminous output of all the stars in the galaxy by, say, a factor of two. The surface brightness of the galaxy increases everywhere by a factor of two. The dynamical parameters, σ or ΔV, of the two galaxies remain the same. The D_n parameter (the diameter within which the integrated surface brightness is $\Sigma = 20.75$ mag arcsec^{-2}) becomes larger by at least a factor of $\sqrt{2}$, and the $H_{-0.5}$ magnitude becomes brighter by at least 0.75 mag. Application of the (D_n, σ) relation or the IR Tully-Fisher relation then places the first galaxy at a larger distance than the second. The counter-argument is that the first galaxy would not now fit the plane populated by elliptical galaxies in (R_e, σ, I_e) space (Faber et al. 1987), or, as I was told at the meeting, "galaxies like that just don't exist." The surface brightness dependence of the Tully-Fisher relation, on the other hand, remains to be fully investigated (but see Bothun and Mould 1986 for a start in that direction.)

Furthermore, we know that in the outer parts of galaxies there are gradients in M/L. On larger scales the notion of biasing, used to reconcile the observed density of the universe with $\Omega = 1$, suggests that clusters contain galaxies with atypically large dark halos. Can we still be confident that on galactic scales luminous material is linked in a fixed proportion to cold dark

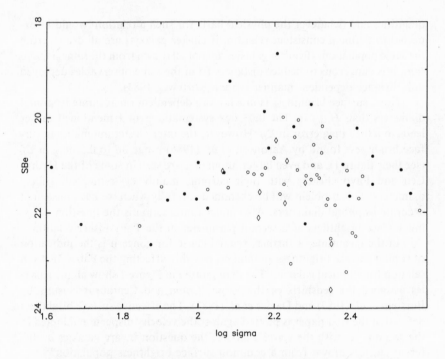

FIG. 1. Effective surface brightness versus the logarithm of the velocity dispersion for elliptical galaxies in Virgo (solid symbols), Coma (open circles), and Centaurus (diamonds). Data from Burstein et al. (1987) and Davies et al. (1987).

matter, a condition which we have known for a long time to be a prerequisite for invariant Tully-Fisher and Faber-Jackson relations (Freeman 1979, Faber 1982)? A vital observational test is to discover whether these relations are dependent on local galaxy density.

4. The Local Region Out to 8000 km/sec

Aaronson et al. (1986) presented distances to 10 clusters generally with statistical errors of 5% or less. At 8000 km/sec this accuracy permits a 2σ detection of an 800 km/sec peculiar velocity. So, if we are going to consider the frequency or uniqueness of large peculiar velocities in Hydra-Cen, we should probably restrict ourselves to such a volume. Figure 2 shows the distribution of clusters within that volume, projected on to the supergalactic plane. This projection is the optimum one; the apparent relative positions of the clusters closely approximate the true one. With the exclusion of the Cancer cluster at supergalactic latitude -48°, the dispersion of the clusters about the supergalactic plane is 18°. Filled symbols indicate clusters from the combined Arecibo

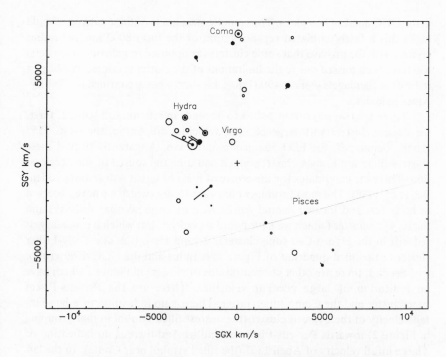

FIG. 2. Peculiar velocities for clusters and groups (denoted as solid symbols) are plotted in a spatial projection on the supergalactic plane. The Local Group is at the origin, and the arrows point radially, since that is the only measureable component. Other clusters without measured peculiar velocities and within the local 8000 km/sec are denoted by open symbols. The circle size indicates the richness of the cluster. The four major superclusters in this volume are labelled.

and Parkes samples to which IR Tully-Fisher distances have been measured, and they are plotted at their inferred velocity in the Hubble flow. A line segment joins this location to the observed velocity of the cluster, indicating the size and sense of the radial component of the peculiar velocity. The resultant peculiar velocities do not depend on H_0 or on the calibration of the Tully-Fisher relation. A different calibration would yield its own value of H_0 but not change the location of the clusters in Figure 2. Open symbols indicate the location of clusters in the samples of Huchra and Geller (1982) from the (northern) CfA survey and of Sandage (1975) from his southern redshift survey. These clusters are plotted at their observed velocities after correction for the motion of the Local Group towards the apparent microwave background apex. The size of the cluster is indicated by the radius of the circle via the following scheme. Galaxies in Virgo larger than 6′ in the UGC were counted in a 15° box. Galaxies were counted in other clusters with scaled angular dimensions in the UGC and ESO catalogs. The completeness limits of these catalogs should make this a fairly unbiased representation of the local 8000 km/sec radius region, with the proviso that some clusters (as opposed to galaxies in clusters) will have been missed due to the limitations of the cluster catalogs. The largest redshift in Sandage's work is 4900 km/s, his survey being confined to Shapley-Ames galaxies.

There are two important points to be made in relation to Figure 2. First, the Parkes clusters with large peculiar velocities are Antlia, the NGC 3557 group, Centaurus, the ESO 508 group and Pavo. Apparently behind these clusters there are known clusters which could be the source of the acceleration. These are candidates for the centre of mass of Great Attractor (Lynden-Bell et al. 1988). The most prominent ones are Hydra (which we have observed to be at rest and located behind Antlia with its large peculiar velocity) and the IC 4329 cluster (which we have not observed yet, but which has the largest redshift in the Hydra-Cen supercluster). Behind Pavo (the one Tully-Fisher cluster in the third quadrant of Figure 2) is Indus and the NGC 6769 group.

Second, there are other concentrations of clusters in Figure 2 which have no indication of large peculiar velocities. These are the Perseus-Pisces supercluster and the Coma supercluster. There is no indication of a large infall velocity of the Pegasus cluster (the nearest filled symbol in that grouping in Figure 2) towards Perseus-Pisces behind it. And there is no indication of a large infall velocity of Abell 1367 (the filled symbol near Coma). In the latter case the motion could be real but transverse.

ACKNOWLEDGEMENTS

In closing, I want to pay tribute to my late colleague and friend, Marc Aaronson, whose leadership and uncompromising zeal for accuracy in measurement has furnished us with so much of the data that make this subject of the large-scale motions in the universe a viable one. The efforts of our many collaborators referenced immediately below are gratefully acknowledged, as is funding by the National Science Foundation of the United States for this project.

REFERENCES

Aaronson, M., Bothun, G., Mould, J., Huchra, J., Schommer, R., and Cornell, M. 1986. *Ap J.* **302**, 536.

Aaronson, M., Bothun, G., Budge, K., Dawe, J., Dickens, R., Hall, P., Lacey, J., Mould, J., Murray, J., Schommer, R., and Wright, A. 1987. in *Large-Scale Structures of the Universe*, eds. J. Audouze, M.-C. Pelletan, and A. Szalay. Dordrecht: Kluwer Academic Publishers, p. 185.

Bothun, G. and Mould, J. 1986. *Ap J.* **313**, 629.

Burstein, D. 1988. this volume.

Burstein, D., Davies, R., Dressler, A., Faber, S. Stone, R., Lynden-Bell, D., Terlevich, R., and Wegner, G. 1987. *Ap J Suppl.* **64**, 601.

Cornell, M., Aaronson, M., Bothun, G., Mould, J., Huchra, J., and Schommer, R. 1987. *Ap J Suppl.* **64**, 507.

da Costa, L., Nunes, M., Pellegrini, P., and Willmer, C. 1986. *Ap J.* **91**, 6.

Davies, R., Burstein, D., Dressler, A., Faber, S., Lynden-Bell, D., Terlevich, R. and Wegner, G. 1987. *Ap J Suppl.* **64**, 581.

de Vaucouleurs, G., de Vaucouleurs, A., and Corwin, H. 1976. *Second Reference Catalogue of Bright Galaxies*. Austin: University of Texas Press. (RC2)

Dressler, A. 1987. *preprint*.

Faber, S.M. 1982. in *Astrophysical Cosmology*. eds. H.A. Bruck, G.V. Coyne, and M.S. Longair, p. 191. Vatican City State: Pontifical Academy.

Faber, S., Dressler, A., Davies, R., Burstein, D., Lynden-Bell, D., Terlevich, R., and Wegner, G. 1987. in *Nearly Normal Galaxies*. ed. S.M. Faber. p. 175. New York: Springer Verlag.

Freeman, K. 1979. in *Photometry, Kinematics and Dynamics of Galaxies*. ed. D. Evans. p. 85. Austin: University of Texas Press.

Huchra, J., and Geller, M. 1982. *Ap J Suppl.* **52**, 61.

Kraan-Korteweg, R. 1983. *AA.* **125**, 109.

Lauberts, A. 1982. *The ESO / Uppsala Survey of the ESO (B) Atlas*. Munich: European Southern Observatory.

Lucey, J., Currie, M., and Dickens, R. 1986. *MN.* **221**, 453.

Lynden-Bell, D., Faber, S., Burstein, D., Davies, R., Dressler, A., Terlevich, R., and Wegner, G. 1988. *Ap J.* **326**, 19.

Mould, J. and Ziebell, D. 1982. *PASP.* **94**, 221.

Nilson, P. 1973. *Uppsala General Catalogue of Galaxies*. Upsala Astr. Obs. Ann. Vol. 6. (UGC)

Paturel, G., Fouqué, P., Lauberts, A., Valentijn, E., Corwin, H., and de Vaucouleurs, G. 1987. *AA.* **184**, 86.

Sandage, A., 1975. *Ap J.* **202**, 563.

LARGE-SCALE MOTIONS FROM A NEW SAMPLE
OF SPIRAL GALAXIES: FIELD AND CLUSTER

VERA C. RUBIN

Department of Terrestrial Magnetism,
Carnegie Institution of Washington

1. INTRODUCTION

At present, extensive observational programs are required in order to obtain sufficient data to permit examination of large-scale deviations from a smooth Hubble flow. Unfortunately, significant telescope time is necessary to enlarge the body of suitable samples. Meaningful answers can only come from high quality data; such data and answers accrue only slowly. Recently I decided to investigate the properties of one set of spirals with observations already at hand: those spirals which my colleagues and I had studied spectroscopically for individual rotation properties. My hope was that these galaxies could produce results concerning large-scale bulk motions without requiring additional observations. Note that I am not here discussing the Sc I spiral sample which Kent Ford and I and our colleagues studied 12 years ago (Rubin *et al.* 1976). For that early study we observed only central velocities of spirals (many of them nearly face-on), and determined distances by assigning to each galaxy the absolute magnitude appropriate to an Sc I galaxy. For the present analysis I use the rotational properties of a sample of field Sa, Sb, and Sc spirals of proper inclination.

2. ROTATION CURVES AS A DIAGNOSTIC OF GALAXY LUMINOSITY

Rotation curves are a good diagnostic of galaxy luminosity, as is well known. Rotation velocities increase rapidly with radius for high luminosity galaxies; rotational velocities rise to higher values for high luminosity galaxies than for low luminosity galaxies. These properties lead to the Tully-Fisher relation. These properties also make it possible to construct synthetic rotation curves (Rubin *et al.* 1985), template rotation curves which can be employed to estimate the absolute magnitude for any spiral which has a known rotation curve and Hubble type.

2.1 - Field Galaxy Sample

We have recently published rotation velocities for 53 field spiral galaxies (Rubin *et al.* 1985; Rubin and Graham 1987) with systematic velocities smaller than 5100 km/s. Although a few of these galaxies lie in environments of moderate density (Whitmore 1984) and one, NGC 4321, is a member of the Virgo cluster, the results which I describe below are not altered when these few galaxies are omitted. The galaxies are well distributed in velocity from nearby to about V = 3800 km/s, with only 6 galaxies between 3800 and 5100 km/s. The median velocity is 1985 km/s. A lower velocity cut-off of 3800 km/s (used in Rubin 1988) does not alter the results presented below. I retain the higher velocity limit here, in order to compare the results for field galaxies with those of clusters at distances near 4000 km/s.

The distribution of the sample on the sky is shown in Figure 1; galaxies are identified in Table 1. The spirals are reasonably well distributed on the sky for declinations above −10°, but there are almost no galaxies below δ = −30°. Sky coverage is just acceptable. No consideration as to sky coverage

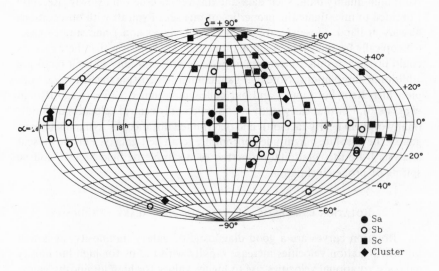

FIG.1. An all-sky plot (right ascension, declination) showing the distribution of spirals used in the analysis. The Milky Way crosses the Equator at 6ʰ and 18ʰ, with a corresponding lack of observed galaxies in those regions.

TABLE 1

RESIDUAL VELOCITIES FOR FIELD AND CLUSTER GALAXIES

Galaxy	V_{LG} km/s	m(i,b)	M(dy)	cos(Apex)	V_{LG}-$V_{(Hubble)}$ km/s	V_{MWB}-$V_{(Hubble)}$ km/s
Field Galaxies						
N701	1825	12.34	−20.7	−0.3826	−203.	−520.
N753	5077	12.55	−22.6	−0.7949	−281.	−727.
N1035	1277	12.27	−20.0	−0.2816	−145.	−378.
N1087	1523	11.14	−20.4	−0.3567	507.	250.
N1325	1499	11.52	−20.5	0.0150	231.	168.
N1353	1422	11.59	−21.6	0.0273	−751.	−802.
N1357	1967	12.24	−21.2	−0.0587	−471.	−557.
N1417	4070	12.28	−22.7	−0.1426	−884.	−996.
N1421	1985	10.86	−21.7	−0.0334	360.	293.
N1515	955	10.69	−20.4	0.4746	129.	299.
N1620	3418	12.70	−21.8	−0.0253	−554.	−557.
U3691	2076	12.15	−20.1	0.2351	667.	920.
I467	2213	11.90	−20.6	−0.5846	632.	433.
N2590	4787	13.00	−22.1	0.6075	−449.	15.
N2608	2059	12.35	−19.9	0.2194	650.	938.
N2639	3238	12.22	−22.2	−0.1179	−590.	−487.
N2715	1487	11.18	−20.1	−0.5277	585.	427.
N2742	1363	11.66	−21.3	−0.2629	−591.	−574.
N2775	1185	10.92	−21.7	0.5829	−486.	−11.
N2815	2270	11.25	−22.5	0.8946	−542.	40.
N2844	1491	13.10	−19.1	0.0850	114.	339.
N2998	4781	12.11	−22.4	0.0374	791.	991.
N3054	2152	11.54	−21.8	0.9395	−176.	429.
N3067	1411	11.93	−19.8	0.2421	302.	617.
N3145	3435	11.60	−22.4	0.8540	280.	873.
N3198	691	10.27	−20.2	0.0243	70.	263.
N3200	3266	11.20	−22.5	0.9025	518.	1124.
N3223	2617	10.96	−22.3	0.9856	373.	980.
N3281	3115	11.83	−20.5	0.9893	1653.	2259.
N3495	925	10.63	−21.1	0.6866	−184.	347.
N3593	520	11.00	−19.1	0.5579	−4.	468.
N3672	1655	11.08	−21.8	0.8270	−229.	350.
N3898	1264	11.30	−21.0	−0.1714	−178.	−108.
N4062	742	11.39	−20.9	0.2263	−693.	−399.

Galaxy	V_{LG} km/s	m(i,b)	M(dy)	cos(Apex)	V_{LG}-$V_{(Hubble)}$ km/s	V_{MWB}-$V_{(Hubble)}$ km/s
N4321	1478	9.98	−22.2	0.4564	114.	517.
N4378	2431	12.22	−21.8	0.6031	−753.	−285.
N4448	650	11.00	−21.1	0.2533	−665.	−364.
N4594	927	8.51	−22.6	0.7666	93.	614.
N4605	288	10.26	−18.4	−0.2930	18.	8.
N4682	2152	12.34	−21.5	0.7403	−779.	−272.
N4698	932	10.93	−20.5	0.5224	−34.	387.
N4800	874	11.91	−20.5	−0.0679	−643.	−524.
N4845	998	11.20	−19.8	0.5940	206.	651.
N5676	2268	11.04	−23.0	−0.2569	−945.	−963.
N6643	1735	11.02	−22.1	−0.7341	−369.	−691.
N7083	2979	11.51	−21.6	0.3031	885.	893.
U11810	4913	13.80	−20.8	−0.7295	754.	208.
N7171	2865	12.40	−21.5	−0.5442	−148.	−616.
N7217	1236	10.56	−22.5	−0.9717	−810.	−1417.
N7537	2863	12.68	−19.6	−0.7780	1434.	870.
N7541	2873	11.50	−22.9	−0.7783	−920.	−1484.
N7606	2371	10.71	−22.3	−0.6166	371.	−126.
N7664	3709	12.66	−22.2	−0.9413	−979.	−1582.

Cluster Galaxies

Galaxy	V_{LG} km/s	m(i,b)	M(dy)	cos(Apex)	V_{LG}-$V_{(Hubble)}$ km/s	V_{MWB}-$V_{(Hubble)}$ km/s
U4329	4556	13.61	−20.0	0.3039	1920.	2245.
N2558	4556	12.97	−21.8	0.3140	59.	389.
U4386	4556	12.49	−21.6	0.3145	1268.	1600.
DC18-42	4353	13.76	−20.4	0.4543	957.	1074.
DC18-2	4353	13.97	−20.1	0.4474	1095.	1208.
DC18-66	4353	13.93	−20.1	0.4329	1154.	1258.
DC18-8	4353	13.71	−20.1	0.4425	1463.	1572.
DC18-10	4353	15.12	−20.2	0.4401	−1441.	−1333.
DC18-24	4353	14.54	−19.8	0.4384	663.	770.
U12417	4062	14.05	−21.1	−0.7925	−1296.	−1866.
N7591	4062	12.89	−21.2	−0.7985	774.	203.
U12498	4062	14.04	−19.2	−0.8135	1839.	1263.
N7608	4062	13.90	−19.4	−0.8159	1777.	1200.
N7631	4062	12.93	−20.9	−0.8135	1145.	570.

entered the original observation program, so it is remarkable that the sky
distribution is even this good.

For each program spiral an absolute magnitude is estimated from its rotation curve, as shown in Figure 2. This procedure, while little different from
the conventional use of the Tully-Fisher relation to estimate an absolute
magnitude from a value of V_{max}, permits the use of a rotation curve which
does not extend as far as V_{max}, and a rotation curve which exhibits velocity
peculiarities. From the absolute magnitude, coupled with a corrected apparent
magnitude, a distance is obtained which is, of course, independent of observed
central velocity. For a value of $H_0 = 50$ km sec^{-1} Mpc^{-1} (the value used
to calculate the absolute magnitude scale for the template rotation curves),
I then calculate ΔVel, the difference between the observed velocity and the
velocity predicted at its distance for a smooth Hubble flow. To search for a
bulk motion of the set of galaxies, I look for a dipole on the sky which will
minimize the residuals of ΔVel. For each galaxy, the velocity in the rest frame
of the Local Group V_{LG}, the corrected apparent magnitude m(i, b), and the
absolute magnitude estimated from the rotation curve M(dy), are listed in col-

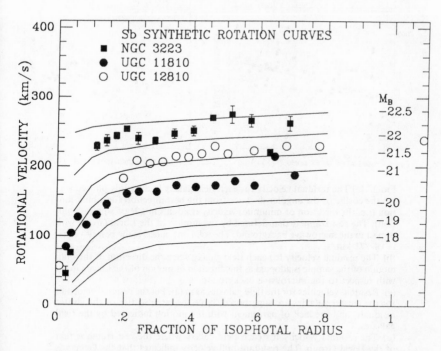

FIG.2. Optical rotation curves for three Sb galaxies, superposed upon synthetic rotation curves for Sb galaxies. For each galaxy, a value of M_B is
estimated from the fit.

umns 2,3, and 4 of Table 1. For one solution, shown below, the rest frame is defined by the Local Group, and the value $\Delta\text{Vel} = V_{LG} - V_{Hubble}$ is listed in Column 6. For all remaining solutions the rest frame is that of the cosmic microwave background (MWB) radiation; $\Delta\text{Vel} = V_{MWB} - V_{Hubble}$ is in Column 7. The inclusion of a Virgo infall component does not alter the solution.

Results from several least squares solutions are shown in Figure 3, where I plot ΔVel for each galaxy versus the cosine of its angular distance from the

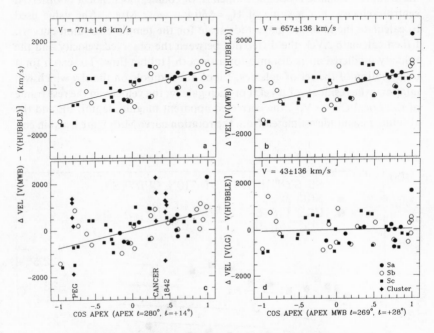

FIG.3. (a) The residual velocity for each field galaxy in the study plotted versus the cosine of the angular distance from the best direction for a bulk motion (i.e. the direction of minimum velocity residuals in a least squares solution). The rest frame is that defined by the motion of the Local Group with respect to the microwave background. The solid line indicates the best fit velocity of 771 km/s.

(b) The residual velocity for each field galaxy when the direction of the bulk motion of the sample is adopted as the direction of motion of the Local Group with respect to the microwave background.

(c) Residual velocities for the field galaxies as in (a), but the residual velocities for the cluster galaxies are superposed. Note the large scatter of the cluster residuals, and the lack of agreement with the motion indicated by the field galaxies.

(d) The residual velocity for each field galaxy where the rest frame is that of the Local Group. The resultant null velocity indicates that the Galaxy is at rest with respect to this sample, and that the entire sample is moving in concert. Note that this figure is essentially just that in (b), rotated so that V is approximately zero.

direction of the bulk motion determined by the least squares procedure. A bulk motion is indicated by a diagonal line on the plot, with the amplitude of the motion equal to the value of ΔVel at cos(Apex) = 1.

The best solution (i.e., the minimum scatter in ΔVel, Figure 3a, and column 5, Table 1) indicates a bulk motion of V = 771 ± 146 km/s toward l = 280°, b = + 14°. This is close in direction and in amplitude to the motion attributed to the Great Attractor (Faber and Burstein 1988, Burstein *et al.* 1986), and is close to the motion of the Local Group with respect to the MWB, for which V = 614 km/s toward l = 269°, b = + 28° (Lubin and Villela 1986 and references therein). In fact, a solution for a bulk motion, in which the apex of the MWB dipole is adopted as the apex, (Figure 3b) differs little from the unconstrained solution. This can be illustrated another way. If we chose as the rest frame not the MWB but the Local Group, then a solution for a bulk motion in the direction of the MWB gives a null result (Figure 3d), as does a solution which is totally unconstrained. Hence, the Local Group is virtually at rest with respect to the spirals which constitute this sample. Because we attribute to the Local Group that motion which produces the MWB dipole, this motion thus becomes a bulk motion for the entire sample as well.

The large-scale motion indicated by this spiral sample agrees well with recent determination from other samples (Lynden-Bell 1988); its concordance with the individual velocity vectors predicted by the IRAS observations is described by Davis (1988). The direction of the motion is also close to the 1976 Rubin and Ford result, after that solution is transformed to the MWB rest frame. Part of the agreement of all of the apices arises because each is the vector sum of a motion of the LG with respect to the sample, plus the larger motion of the LG with respect to the rest frame defined by the MWB.

However, even this good agreement does not put to rest some concerns about the procedures and the results. Are the properties of galaxies, as a function of direction, of distance, and of environment, so systematic that zero-points and slopes of correlations among parameters in one region are reproduced exactly in all other directions? Or are there differences among galaxies in different regions of the sky, which we erroneously interpret as a velocity? At a distance corresponding to V = 2500 km/s, a difference in absolute magnitude by about 0.4 magnitude will mimic a velocity of 500 km/s. Because it seems unlikely that a variation across the sky of twice this amount will have gone unnoticed, a coherent motion for the nearby galaxies seems well established.

2.2 - Cluster Galaxy Sample

More recently, we have initiated observational programs to investigate the properties of spiral galaxies in various environments. We have obtained observations (Rubin *et al.* 1988) of rotation properties for a small sample of spirals in the clusters of Cancer, Pegasus I, and DC 1842-63 (Dressler 1980). I can analyze the dynamics of these cluster galaxies as I have analyzed the field spirals, and ask if their dynamics indicates the same bulk motion as do the field spirals. Because each cluster galaxy is placed at the mean cluster distance, considerable smoothing and reduced scatter is expected on a cosine plot. Surprisingly, this is not the case, as is apparent from Figure 3c, where I superpose the residual cluster velocities on the solution based on field spirals, Figure 3a. Residual velocities for the cluster galaxies show a scatter within a single cluster that is larger than the scatter due to the field galaxies, and do not indicate the bulk motion found for the spirals.

One of the clusters, DC 1842-63, is located only 39° from the direction to the Great Attractor (but 90° from the apex defined by the MWB), and its mean velocity, V = 4353 km/s, is just that accepted for the (distance of the) Great Attractor. Residual velocities for galaxies in this cluster are not too discrepant (Figure 3c). However, residual velocities for galaxies in the Peg I cluster are predominantly positive, in a direction in which all bulk motion models predict negative residuals.

One explanation for the field/cluster discrepancy is that with small number statistics we have managed to observe deviant cluster spirals. However, our selection procedure favored galaxies of relatively normal morphology. More likely, a single bulk motion is too simple an explanation for the large motions observed. Such a conclusion would certainly follow from the flow diagrams produced by Yahil. Still an alternative possibility is that rotation properties of spirals in clusters do not match those of spirals in the field, so that the synthetic rotation curves formed from field spirals are not the proper templates for cluster galaxies. Although we discuss such possibilities elsewhere in this volume, only more observations of extended rotation curves of spirals in clusters will settle this question. No such difference for field/cluster dynamical properties is observed for elliptical galaxies according to Faber, and the cluster spirals observed at HI by Aaronson *et al.* (1986) show no deviant characteristics.

3. Unanswered Questions

Major questions remain. Are the slightly different motions returned by different samples significant, or do the differences merely reflect the different mix of objects investigated? How well do different samples agree in regions of overlap in direction and distance? Initial answers to these questions are given in the contributions here of Faber and Burstein (1988), and of Strauss and Davis (1988). Is the dipole observed in the MWB gravitationally induced, and how do the deviations from a smooth Hubble flow which we have discussed this week relate to this MWB dipole?

An important question arises for studies which mix observations of field galaxies, of cluster galaxies, observations of neutral atomic gas, and ionized gas: are the characteristics of spirals in regions of high galactic density so similar to those isolated in the field that the Tully-Fisher and Faber-Jackson relations have the same slopes and zero points independent of environment? Are these relations so general that they can be correctly determined using 21-cm observations of galaxies whose neutral hydrogen gas disks are known to be asymmetrical and tidally truncated near the cores of rich clusters? And what of the ionized gas disks? Do the asymmetries and truncations observed in the neutral hydrogen disks exist also in the ionized gas disks, or are optical and 21-cm observations looking at features with such different evolutionary histories that their correlations with other galaxy parameters are not likely to be similar? Have bulges acquired disks, have disks acquired bulges, have early type galaxies merged, all-the-while retaining similar correlations among the physical properties we measure? If, as it appears at present, the evolutionary effects on galaxies in clusters have been greatest for the early type galaxies, then perhaps we are not erring in combining 21-cm observations of principally late-type spirals in clusters and in the field with optical observations of principally field spirals. A few of these questions are addressed further in my contribution later in this volume. These are some of the questions which brought us together, and they are questions which will concern many astronomers in the coming years.

ACKNOWLEDGEMENTS

Spectra of the field and cluster galaxies were obtained at Kitt Peak, Cerro Tololo Inter American, and Las Campanas Observatories; I thank the Directors for observing time. I also thank my colleagues for continued support, and Dr. John Graham for comments on an early draft of this paper.

196 V.C. RUBIN

REFERENCES

Aaronson, M., Bothun, G., Mould, J., Huchra, J., Schommer, R., and Cornell, M. E. 1986. *Ap J.* **302**, 536.

Burstein, D., Davies, R. L., Dressler, A., Faber, S. M., Lynden-Bell, D., Terlevich, R. J., and Wegner, G. A. 1986. in *Galaxy Distances and Deviations from Universal Expansion*. eds. B. Madore and R. B. Tully. p. 255. Boston: Reidel.

Dressler, A. 1980. *Ap J Suppl.* **42**, 569.

Faber, S. M., and Burstein, D. 1988, this volume.

Lubin, P. M., and Villela, T. 1986. in *Galaxy Distances and Deviations from Universal Expansion*. eds. B. Madore and R. B. Tully. p. 169. Boston: Reidel.

Lynden-Bell, D. 1988, this volume.

Rubin, V. C. 1988. in *Large-Scale Structures of the Universe*. eds. J. Audouze, M.-C. Pelletan, and A. Szalay. Dordrecht: Kluwer Academic Publishers, p. 181.

Rubin, V. C., and Graham, J. A. 1987. *Ap J Letters.* **316**, 67.

Rubin, V. C., Thonnard, N., Ford, W. K., and Roberts, M. S. 1976. *AJ.* **81**, 719.

Rubin, V. C., Burstein, D., Ford, W. K., and Thonnard, N. 1985. *Ap J.* **289**, 81.

Rubin, V. C., Whitmore, B. C. and Ford, W. K. 1988. *Ap J.* **333**, 522.

Strauss, M. A., and Davis, M. 1988, this volume.

Whitmore, B. C. 1984. *Ap J.* **278**, 61.

III

MOTION OF THE LOCAL GROUP

WHENCE ARISES THE LOCAL FLOW OF GALAXIES?

D. LYNDEN-BELL and O. LAHAV

Mount Stromlo Observatory, Australia
and Institute of Astronomy, Cambridge

SUMMARY

If the apparent large-scale coherence of the velocity field of galaxies out to 3500 km/s is not a chance superposition, the gravity that causes it must arise from distant sources such as the Great Attractor. If light traces mass, the origin of the local gravity field can be found by tracing the origin of the locally observed net flux of extragalactic light. If possible problems with the $-2.5° > \delta > -17.5°$ strip, the zone of avoidance and convergence of the dipole at faint fluxes are ignored, then half of the local extragalactic gravity field arises from galaxies with diameters greater than $4'$. The Great Attractor model would predict only $2'$ corresponding to galaxies at twice the distance. This optical result agrees with a similar IRAS result. Thus the gravity field of the Great Attractor model is not confirmed, although a combination of obscuration in the Milky Way and insuffent depth in the galaxy catalogues could be denying us the true picture.

A mapping of the supergalactic plane in depth provides further evidence that Virgo is at one side of that structure whose centre lies in Centaurus or even further south in the zone of avoidance. Even the presently catalogued distribution of extragalactic light shows that Virgo is relatively weak causing 108 km/s infall and that even here the infall towards Centaurus is twice as strong.

The Local Void gives a significant push across the supergalactic plane which slews the Local Group's motion from $l = 290°$ $b = 28°$ to the observed $l = 261°$ $b = 29°$.

1. MOTION OF THE LOCAL GROUP

We first review the determination of the Local Group's motion relative to the Cosmic Microwave Background (CMB), concentrating on uncertainties in the deduction of the Sun's motion within the Local Group and the peculiar motion of the Local Group relative to the mean flow of galaxies in its neighbourhood.

We then turn to the study of the origin of that motion from the gravity field of external galaxies. Since both gravity and light fall off as distance-squared, the assumption that light traces mass allows us to determine the gravity field from the locally measured flux of extragalactic light. For galaxies greater than 1 arcminute in diameter, existing galaxy catalogues (Nilson 1973, Lauberts 1982) are sufficient to do this provided they are properly calibrated. Exceptions occur in the zone of avoidance and in the band $-2.5° > \delta > -17.5°$ where the existing galaxy catalogue (Vorontsov Velyaminov and Arkipova, 1963-68) is too incomplete and inhomogeneous. To get a first result these zones have been filled in with the mean distibution of galaxies observed at high galactic latitudes $|b| > 40°$. Somewhat similar problems occur with the distribution of IRAS galaxies which has small missing strips at high latitudes and a large uncertain region near the galactic plane caused by galactic cirrus and confusion with galactic sources.

In both cases the assumption that light traces mass can not be precisely true since most of the mass presumably resides in dark halos around galaxies or clusters rather than in light sources. For visible light the assumption may be sufficient to give a good account of the gravity field since the M/L ratios of galaxies are not very widely distributed. Faber et al. (1987) deduce a factor of ~ 2 in ellipticals. For IRAS fluxes there is a far wider spread in mass to infra-red ratio with ellipticals undetected and some active galaxies found with luminosities similar to quasars. Thus IRAS flux at best only follows the mass density statistically and underestimates the high density regions where early type galaxies predominate. With this broad luminosity function it is no surprise that the distribution of IRAS sources on the sky looks like a washed out version of the distribution of optical sources (see e.g., Meurs and Harmon 1988).

The estimation of the gravity field due to the sources of greatest apparent brightness can not be done from IRAS fluxes since they are dominated by 'shot noise' of individual nearby sources which happen to be strong infra-red emitters. In practice this problem has been reduced either by binning in flux and number weighting within each flux bin (Yahil et al. 1986) or by ignoring flux altogether and calculating the direction of the number weighted dipole (Meiksin and Davis 1986). While the latter is useful to get a more stable direction, it loses much of the original motivation since the inverse-distance-weighting is removed. Indeed, in a Poisson universe in which galaxies are placed at random the direction of the number weighted dipole would not converge, but would perform a random walk across the sky as the limiting flux of the sources counted was reduced. Even with the inverse-distance-squared weighting the strength of the dipole in such a universe only converges as distance to the minus one half power.

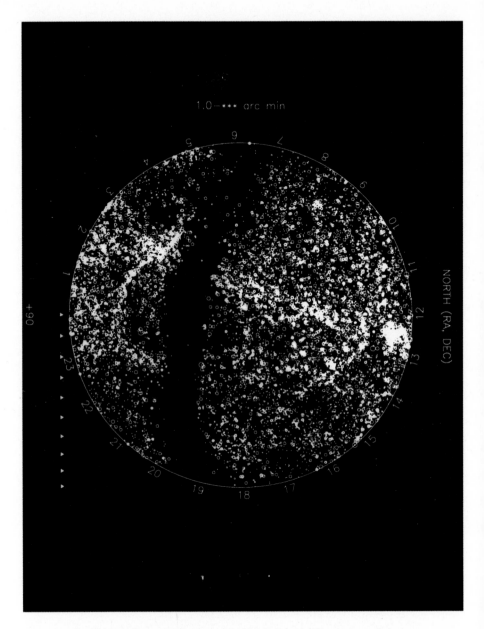

Fig. 1. Equal Area Projection of the North Celestial Hemisphere. Galaxies in Nilson's Uppsala General Catalogue coded by diameter.

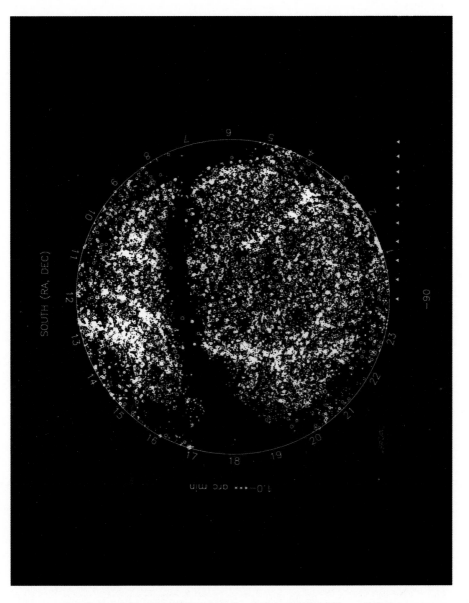

FIG. 2. Equal Area Projection of the South Celestial Hemisphere. Galaxies in Lauberts's ESO/Uppsala Catalogue coded by diameter for δ < 17.5°. For 0° > δ > −2.5° from UGC as Figure 1. For −2.5° > δ > −17.5° the galaxies from the Morphological Catalogue are plotted but this is readily seen to be less complete.

In the optical it is more credible to argue that with suitable modifications for galaxy type, light might trace mass; in which case the gravity field's 'shot noise' due to the apparently brightest half dozen galaxies may really cause an extra peculiar motion of the Local Group with respect to the mean flow of galaxies near here. It would be of interest to see if there is some correlation between the directions of these two vectors.

The distribution of extra-galactic light across the sky is pictured in equal area projection in figures 1-5. Figures 1 and 2 show the northern and southern celestial hemispheres as depicted from the UGC and ESO catalogues. In Figure 2 the missing strip $-2.5° > \delta > -17.5°$ has been filled in from the Morphological Catalogue, but the obvious inhomogeneity of this catalogue not only in declination but apparently also in R.A. is seen by the variable step in galaxy density where it joints the ESO catalogue at $\delta = -17.5°$. The remaining figures show the distribution of galaxies in different diameter ranges in galactic coordinates. Study of these allows one to get a good idea of the structure of the local galaxy distribution not only over the sky but also in depth. As we shall see presently a galaxy with characteristic diameter D_* has a diameter of $1'$ when placed at a distance corresponding to 7000 km/s.

Section 2 discusses the deduction of the optical dipole from these maps and performs a number of experiments to discover where it comes from. The two major components are a pull towards $l = 290°$, $b = 28°$ (between Centaurus and Hydra) and a push from roughly $l = 74°$, $b = -35°$ which arises from a void in the local galaxy distribution at distances < 2000 km/s.

Section 3 gives the directions of observed galaxy streaming and discusses their relationship to the CMB, optical and IRAS dipoles. Section 4 shows how to deduce the selection present in Huchra's radial velocity catalogue by comparing it with diameter limited catalogues. This enables us to get a crude map of galaxy density as a function of velocity within 15° of the supergalactic plane.

2. THE VELOCITY OF THE LOCAL GROUP

The microwave observations (Lubin and Villela 1986) show that the variation in $\Delta T/T$ around the sky is well fitted by a dipole whose direction and magnitude are independent of observing frequency. This dipole is well explained by a motion of the Sun relative to the CMB of 360 ± 27 km/s towards $l = 265° \pm 2°$, $b = 50° \pm 2°$.

Although the line joining us to Andromeda lies in the plane defined by non-local-group bright galaxies, the fact that Andromeda is both much nearer and approaching us has led to the belief that we are both gravitationally bound into the Local Group. Once this concept is adopted the Local Group's motion becomes of greater interest than the Sun's. Unfortunately the Sun's velocity relative to the barycentre of the Local Group is imperfectly known. Table 1 lists determinations starting with the first one by Humason and Wahlquist

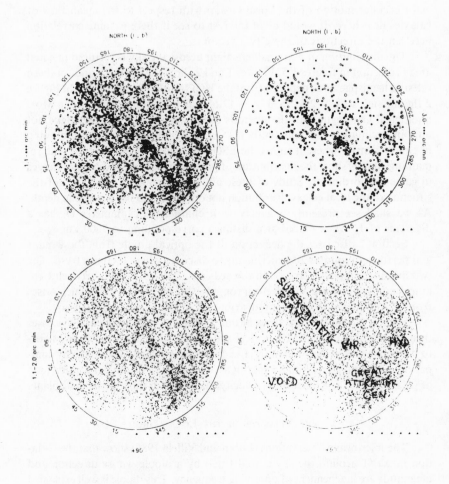

Fig. 3abcd. Equal Area Projection of the North Galactic Hemisphere from the galaxies contained in Figures 1 and 2: (a) all galaxies; (b) bright galaxies $\Theta > 3$'; (c) faint galaxies $2.0' > \Theta > 1.1'$; (d) naming chart.

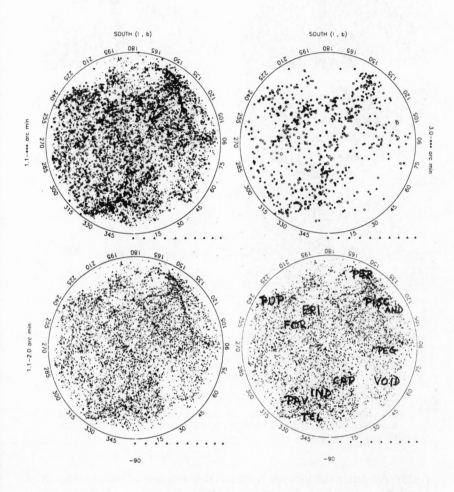

FIG. 4abcd. As Figure 3 but for the South Galactic Hemisphere.

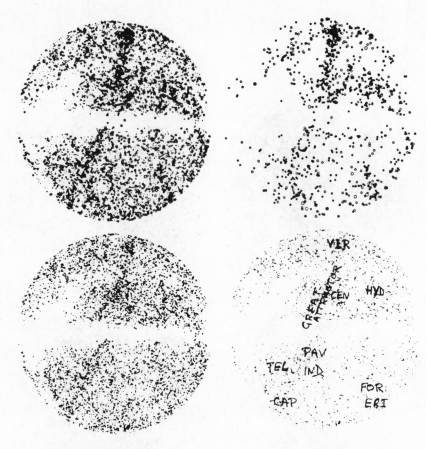

Fig. 5abcd. As Figure 3 but for the Great Attractor region showing the hemisphere centred on the the $l = 317°$, $b = 0°$ intersection of the supergalactic and galactic planes.

(1955). All determinations are from suitable averages of radial velocities of Local Group members. They differ in their treatment of whether members of the M31 subgroup and the Galaxy's subgroup are averaged together or treated as though they were independently moving. However, the main differences stem from whether the solution is an unconstrained kinematic one or is subjected to a dynamical constraint such as no net linear momentum in the barycentre's frame. The most constrained solutions assume that the two major subgroups move on a rectilinear orbit. Since none of the solutions differ from these by more than their probable errors the most constrained solutions should probably be preferred. While the differences are never large, they can be important when considering small motions such as the Local Group's

TABLE 1

THE SUN's MOTION RELATIVE TO THE LOCAL GROUP

Authors	$v \cos l \cos b$	$v \sin l \cos b$	$v \sin b$	Number
Humason and Wahlquist (1955)	-85	277	-35	1
Yahil, Sandage, and Tammann (1977)	-79	295	-38	2
Richter, Tammann, and Huchtmeier (1987)	-50	307	-19	3
Lynden-Bell and Lin (1977) (constrained momentum	-25	313	-9	4
Sandage (1986) (constrained to line)	-37	291	-28	5
This article (constrained to line)	-30	297	-27	6
de Vaucouleurs et al. (1976) standard	0	300	0	7

Sun's motion relative to CMB

$V_{\odot} - V_{CB} = 360 \pm 27 \, (265° \pm 2°, 50° \pm 2°) = (-20, -231, 276)$

    (Lubin and Villela 1986)

Local Group's motion relative to CMB by subtraction of 7.

$V_{LG} - V_{CB} = (-20, -531, 276) = 600 \pm 27 \, (l = 268°, b = 27°)$

peculiar velocity relative to nearby galaxies. In Table 1 the solutions constrained to a line are compounded of the motion of the Sun relative to the LSR, the motion of the LSR around the Galaxy and the motion of the Galaxy relative to the centre of mass of it and Andromeda. Thus writing $\mu = M_A/(M_A + M_G)$ for the fraction of the Local Group's mass which resides in the M31 subgroup, we have

$$V_{\odot} - V_{LG} = (V_{\odot} - V_{LSR}) + (V_{LSR} - V_G) + \mu \, (V_G - V_A) \, \hat{r}_A$$

where $\hat{r}_A$ is the unit vector to Andromeda. Evaluating each term with IAU recommendations gives

$$V_{\odot} - V_{LG} = (9, 12, 7) + (0, 220, 0) + 2/3 \, (123)\hat{r}_A = (-30, 297, -27) \text{ km/s}.$$

This differs little from de Vaucouleurs' standard $(0, 300, 0)$. We therefore use de Vaucouleurs' standard (de Vaucouleurs et al. 1976), since for most applications standardisation is important and it is within the errors of our preferred solution.

    The peculiar motion of the Local Group relative to the mean flow of nearby galaxies is not easily detectable. Richter, Tammann, and Huchtmeier (1987) first find an apparently significant result and then reject it on the ground that the transformation to Local Group axes is equally uncertain. With de

Vaucouleurs' standard reduction their result is a motion $(-28, -9, -94) \pm$ 17 km/s, while with our preferred solution (Table 1, no. 6) it becomes $(2, -6, -66)$. The new results of Faber and Burstein (1988) from the best Fisher-Tully determinations is smaller than its probable error, 66 ± 105 towards $(170° \pm 97°, 6° \pm 31°) = (-64, 11, 7)$ km/s in de Vaucouleurs' frame, our preferred frame $(-34, 14, 34)$. These may be contrasted with the larger values found by Aaronson *et al.* (1982) of $(-172, -39, 24)$ in de Vaucouleurs' frame. The low values are probably to be preferred and the Local Group's peculiar velocity has not yet been detected.

3. THE OPTICAL DIPOLE

Lahav (Lahav *et al.* 1988) has determined the diameter function of both UGC and ESO galaxies using the northern and southern complete redshift surveys. The cumulative diameter function fits a Yahil form well. Let D_* be the characteristic diameter and $t = (D/D_*)^2$. Then Yahil's form for the cumulative function is

$$\Phi (>t) = \Phi_* t^{-\mu} (1 + t/\nu)^{-\nu}.$$

The UGC and ESO diameter distributions can both be fitted with one functional form with $\mu = 0.2$ $\nu = 3.4$ but with different D_*. For UGC, $D_* = 6186 \pm 269$ arcminutes km/s, with the strange diameter unit following because we use km/s as a measurement of distance (6875 arcminutes km/s $= 20$ h^{-1} kpc, for there are 3437.7 arcminutes in a radian). For ESO, $D_* = 6973 \pm 303$ arcminutes km/s. Assuming that in reality the true diameter functions are the same then ESO measurements of diameters are $1.13 \pm .05$ times larger than the UGC measurements. This is no surprise as the ESO plates are deeper and they give greater galaxy counts. This measurement difference is confirmed by the numbers of objects in the faintest bins in which the numbers counted in both UGC and ESO approximate those expected of a uniform universe. To get agreement in numbers a UGC galaxy of 1.00 arcminute diameter has to correspond to an ESO galaxy of 1.10 arcminutes diameter. This is in good agreement with the $1.13 \pm .05$ factor deduced above from the diameter functions. After putting all diameters onto the ESO system the common value of Φ_* is 0.011 galaxies per $(100$ km/s$)^3$. This joint calibration now replaces that used by Lahav (1987) previously which was the preliminary photoelectric calibration of Fouqué and Paturel (1985, which they have since revised, Paturel *et al.* 1987).

Because the ESO catalogue has no magnitude calibration we use the square of the angular diameter in its place. Notice that this automatically introduces the desired inverse square distance weighting. To get the mass weighting we have to assume that the square of the diameter of a galaxy is proportional to its mass. This would true if the product of the M/L ratio and the surface brightness were independent of mass. Since M/L ratios increase like $L^{1/4}$ (Dressler *et al.* 1987, Faber *et al.* 1987) while for bright ellipticals the surface brightnesses decrease as luminosities increase, the product can not be far from constant. It could be that using diameter squared is as good as, if not better than, using luminous flux.

Our calculation of the observed optical dipole now proceeds as follows: (1) for $|b| < 15°$ so few galaxies are in the catalogues that we have replaced this whole band with mean sky. Notice that this gives no contribution to the net dipole. We do the same in the missing band, $-2.5° > \delta > -17.5°$, but due to its offset from a great circle the mean sky replacement does give a contribution to the dipole; (2) above $|b| = 15°$ we allow for exinction as follows: (i) we increase the measured angular diameters by the factor $10^{A_B/5}$. Diameter squared then still behaves like luminosity; (ii) we increase the limiting diameter down to which the catalogue is deemed complete by the same factor. Thus Θ_c now changes over the sky; (iii) whenever the catalogue's absorption-corrected completion limit, Θ_c, is larger than the minimum true diameter, Θ_{min} down to which we are calculating a dipole, we increase the contribution of each galaxy by a factor that accounts for this incompletion. This factor is

$$1 + \frac{M(\Theta_{min}) - M(\Theta_c)}{N(\Theta_c) \; \Theta^2},$$

where $M(>\Theta)$ is the cumulative $\Sigma\Theta^2$ of all galaxies at the pole with angular diameter greater than Θ and $N(>\Theta)$ is their cumulative number.

M and N have been calculated from the $b < -40°$ region of the ESO catalogue to avoid any distortions caused by the proximity of Virgo and Coma in the North. Their normalization is irrelevant to the above formula, which extrapolates the part of the cumulative dipole contribution that can still be seen despite absorption, with a continuous curve of the shape seen in unobscured parts of the sky. The formula used for absorption was that of Fisher and Tully (for details see Lahav *et al.* 1988). Thus the dipole actually calculated is

$$P(>\Theta_{min}) = 3/(4\pi) \; \Sigma \left[\Theta^2 + \frac{M(\Theta_{min}) - M(\Theta_c)}{N(\Theta_c)} \right] \hat{r} \qquad (3\text{-}1)$$

where $\hat{r}$ is the unit vector to the galaxy counted, Θ is its absorption-corrected angular diameter and Θ_c is the similarly corrected limit to the catalogue in that direction. The sum is over all galaxies with $\Theta > \Theta_{min}$ outside the excluded zones described above, and to that sum the contribution from mean sky in the uncounted zones is added. All diameters are put on the ESO scale at the beginning of the process using the factor 1.13 for the UGC diameters. To our delight the direction of the optical dipole so determined lies at $l = 261°$, $b = 29°$, very close to the direction of the Local Group's motion relative to the CMB (268,27). The monopole is defined as the true $\Sigma\ \Theta^2$ per unit solid angle, so it is given by the expression on the rhs of formula 3-1 omitting both the 3 and the $\hat{r}$. The magnitude of the dipole is 4.4×10^3 (arcmin)^2 which is 24% of the monopole of 18.3×10^3 (arcmin)^2. [Meiksin and Davis 1986 define their percentage dipoles without a 3 in the dipole term. To put their % dipoles on the scale used here, multiply their values by 3. Our definition has the property that with a truly dipolar sky with a p% dipole the flux from a small region of sky in the dipole's direction is $1 + p$ times the mean.] However, much of this dipole arises from relatively large galaxies. 3.8×10^3 (arcmin)^2 comes from galaxies of diameter $> 2'$, 3.0×10^3 (arcmin)^2 from galaxies of diameter $> 4'$ and 1.7×10^3 (arcmin)^2 from galaxies of diameter $> 8'$. Thus over a third arises from galaxies of distance ≤ 1000 km/sec and almost two thirds from galaxies of distance ≤ 2000 km/s. Only 1/7 of the dipole arises from galaxies ≥ 4000 km/sec. These numbers are in stark contrast to those predicted by the Great Attractor model for which half the gravity is generated from distances beyond the centre of the attractor at ~ 4300 km/sec distance.

While much of the optical dipole arises nearby, the contribution of the Virgo Cluster is relatively small. The excess contribution from the circle of 12° diameter enclosing the cluster, over that expected from mean sky is only 17% of the whole dipole. Increase of the circle's diameter to 20° only raises that to 18%. The Great Attractor area of sky in Centaurus is more significant. A cone of 30° total angle gives an excess which is 41% of the whole dipole. If we presume that our total dipole of 4437 (arcmin)^2 should not be increased by further contributions from fainter galaxies, nor from the excluded zones, then the gravity corresponding to the whole dipole generates the Local Group's infall of 600 km/s. On such a scale the galaxies in the 20° circle around Virgo would generate an infall of only 108 km/s here and the Centaurus region 246 km/s. Although that number sounds more hopeful for the Great Attractor interpretation a number of the galaxies included are closer and a significant fraction of the signal comes from them. We now recalibrate the dipole in terms of the net optical flux passing through unit area at the Local Group. The total optical dipole of 4437 (arcmin)^2 corresponds to a net flux from the direction of the dipole $4\pi/3$ times larger. To convert this to an approximate optical flux we plotted total B magnitudes against log Θ(max) for numerous galaxies. We find the correlation between Θ and B mag, with Θ in arc minutes, is given by $B = 15.00 - 2.5$ log Θ^2. Inserting 4437 $(4\pi/3)$ for Θ^2 gives the dipole's flux as $B_D = 4^m.33$. Assuming solar colours this corresponds to $m_{BOL} = 3.6$ and the optical flux, F, past the Local Group is thus approximately $F = 9 \times 10^{-7}$ ergs cm^{-2} s^{-1}. This generates an infall of 600 km/s at the Local Group. On this basis each extra $\Theta = 10'$ or $B = 10^m$ galaxy would contribute an infall velocity of 3.2 km/s.

To discover more precisely which areas of sky are responsible for the optical dipole we oriented a cone towards the total dipole and calculated the contribution from within the cone after subtracting the contribution that would have arisen within the cone from a sky that was totally uniform. In Figure 6 we show how this excess dipole within the cone varies as a function of the fraction of the whole sky within the cone. There is a strong rise up to 70% of the total as the cone's semi-angle rises to 45°, but there is also a notable

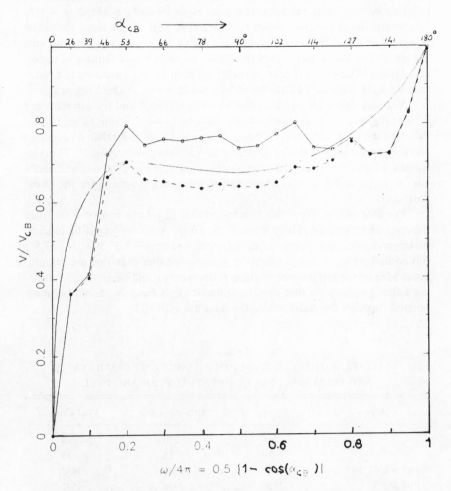

FIG. 6. Fraction of the total optical dipole as a function of the fraction of the sky within the cone centred on $l = 268°$, $b = 27°$ with different opening angles. Full line total dipole in cone; dashed line projection of total dipole onto $l = 268°$, $b = 27°$; light line prediction of $\varrho \propto |r - r_m|^{-2}$ Great Attractor model, $1/2\ [(\pi - \alpha) \sin \alpha + 1 - \cos \alpha - (\pi^2/8) \sin^2 \alpha)]$.

push from the decrement at the back of the cone. At first we thought this push from behind would be missing in the Great Attractor model but the actual prediction is not so bad. To make a better fit the Attractor itself must be more spread out and the void more concentrated but the relative magnitudes of the contributions are about right. To see if the directions of the push and the pull are really colinear we asked for the direction of that 90° (45° semi angle) cone that enclosed the greatest excess dipole. This lies not in the direction of the final dipole, but is directed toward $l = 290°$, $b = 28°$. Similarly the backward pointing 90° cone with the most push has its axis toward $l = 74°$, $b = -35°$. The compound of a push from this direction and a pull from the other represents most of the total dipole. Thus the Local Void is not colinear with the attraction and its push slews the dipole away from the Centaurus region into Hydra. These results may in reality be even stronger because the zones that we have painted with the mean sky are involved in these regions.

We may again get an idea how far away the void and the attractor are by looking at the excess dipoles from these 90° cones. From Table 2 more than half the void's contribution comes from galaxies of more than 8′ diameter and the contribution is almost complete for galaxies greater than 4′. The contribution from the attractive cone comes from galaxies of about half those sizes but that is still at only about half the distance proposed for the Great Attractor.

Possible escapes from this conclusion are: (i) a large contribution from the zone of avoidance, where most of the Great Attractor might lie hidden; (ii) larger dipole contributions from the uncounted strip $-2.5° > \delta > -17.5°$; (iii) contributions to the dipole from galaxies smaller than one arc minute. Some idea of the importance of these refinements could be achieved following Faber's suggestion that the extragalactic sky's features should be interpolated through the zones where the data are missing.

TABLE 2

CONTRIBUTIONS TO THE OPTICAL DIPOLE BY DIAMETER
AND FROM 90° CONES OF GREATEST PUSH AND PULL

Area	All sky	Attractor Cone	Void Cone
l	261	290	74
b	29	28	-35
$P(\Theta > 1.03′)$	4437	3562	1657
$P(\Theta > 2′)$	3820	2496	1541
$P(\Theta > 4′)$	3058	1794	1338
$P(\Theta > 8′)$	1708	1118	906

Note: All values of P are in (arcmin)^2 on the ESO diameter measures.

4. COMPARISON OF DIPOLES AND DIRECTIONS OF INTEREST

Figure 7 shows the direction of the motion of the Local Group relative to the CMB and the motions of different galaxy samples also in the CMB frame. These are not strictly comparable since Virgo infall has been modelled and removed from the galaxy sample but not from the Local Group's motion. If Virgo infalls of 100, 200 or 300 km/s were removed the remaining motion of the Local group would be in the directions labelled 100, 200 or 300, respec-

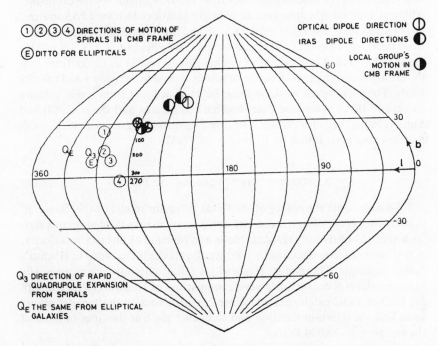

FIG. 7. Interesting directions in galactic coordinates. All mean motions are in the CMB frame.

1	Mean motion of Rubin *et al.* (1976) spirals: 862 km/s
2	Mean motion of de Vaucouleurs and Peters (1985) spirals: 460 km/s
3	Mean motion of AHM spirals according to Lilje, Yahil, and Jones (1986): 502 km/s
E	Mean motion of ellipticals with Re < 8000 km/s: 521 ± 89 km/s (Lynden-Bell *et al.* 1988, Lynden-Bell 1987)
◖	Motion of Local Group: 600 ± 27 km/s
L	Revised optical dipole of Lahav, Rowan-Robinson and Lynden-Bell (1988)
H	Revised IRAS dipole of Harmon, Lahav and Meurs (1987)
100 200 300	Directions of Local Group's motion relative to CMB after removal of a 100, 200, or 300 km/s Virgo infall component.

tively. The new optical dipole described here (Lahav *et al.* 1988) has shifted some 35° from the old direction (Lahav 1987). While some 10° of this shift is due to better allowance for absorption, most of the shift is caused by the new diameter calibration, through the redshift survey in place of the Fouqué and Paturel calibration. The directions of the IRAS dipoles of Yahil *et al.* (1986) and the number weighted IRAS dipole of of Meiksin and Davis (1986) are also plotted. Meurs and Harmon (1988) have made a new study of selection procedures for isolating galaxies from the IRAS point source catalogue. Although their criteria miss some large bright galaxies which are IRAS sources, nevertheless they have more galaxies in all and the flux weighted dipole found from them by Harmon, Lahav, and Meurs (1987) lies at $l = 273°$, $b = 31°$, much closer to the Local Group's motion. This agreement again confirms that the extra-galactic dipole vectors are intimately related to the gravity and velocity fields. The most recent work on these fields as deduced from radial velocity surveys of IRAS galaxies is described here by Strauss and Davis (1988) and Yahil (1988). They also deduce a nearby origin for the gravity field, a result that has also been found by Clowes *et al.* (1987).

5. MAPPING THE SUPERGALACTIC PLANE

To investigate the reality of the Great Attractor model of the 'Samurai' (Lynden-Bell *et al.* 1988, Lynden-Bell 1987) we try to map the density of galaxies within 15° of the supergalactic plane as a function of distance or velocity. A first orientation is obtained by plotting all galaxies in that zone in Huchra's (1981) catalogue. This picture (Figure 8) is full of selection effects especially in its celestial N-S contrast. However, comparison of Huchra's catalogue with the ESO or UGC catalogues which are complete to a given diameter allows us to find the selection function and so deduce the true densities in most of the region $v < 10,000$ km/s.

Consider galaxies in a solid angle of sky Ω centred on the direction $\hat{r}$. Let N_H $(v, \Theta, \hat{r})$ be the number of galaxies in Huchra's catalogue per unit range of radial velocity v, angular diameter Θ and within the solid angle $\Omega(\hat{r})$. We correct all the velocities to the Local Group frame using de Vaucouleurs' standard prescription. To map the supergalactic plane in the space of radial velocities and angles in the sky we use v as though it were the distance. The true diameter of the galaxy whose angular diameter is Θ is then $D = v\Theta$. Let $\Phi(D)$ dD be the differential diameter function giving the fraction of all galaxies whose diameters lie in the range D to D + dD. Further we suppose that Huchra's catalogue has severe selections which are functions of both direction in the sky, $\hat{r}$, and angular diameter of the galaxy Θ. Then

$$N_H (v, \Theta, \hat{r}) = \Omega(\hat{r}) \, n(v) \, v^3 \, \Phi(v\Theta) \, S(\Theta, \hat{r}) \qquad (5\text{-}1)$$

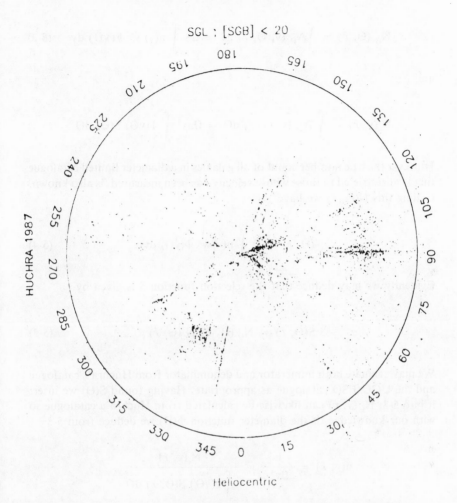

FIG. 8. Chart of all galaxies in Huchra's (1987) catalogue lying within 20° of the supergalactic plane as a function of supergalactic longitude and corrected radial velocity. The region out to 10,000 km/s is shown.

where S is the selection function giving the fraction of the objects that have measured velocities and diameters in Huchra's catalogue and $n(v, \hat{r})$ is the true number density of galaxies at radial velocity, v.

There are too few galaxies in the catalogue to determine N_H as a function of so many arguments. However integrals over N_H can be obtained from the data, in particular the integrals over radial velocity or over angular diameter. We therefore define

$$N_\Theta\,(\Theta,\,\hat{r}) \;=\; \int N_H\,(v,\,\Theta,\,\hat{r})\,dv \;=\; \Omega S \int n(v)\,v^3\,\Phi(v\Theta)\,dv \qquad (5\text{-}2)$$

and

$$N_v(v,\,\hat{r}) \;=\; \int N_H\,(v,\,\Theta,\,\hat{r})\,d\Theta \;=\; \Omega n v^3 \int \Phi(v\Theta)\,S(\Theta)\,d\Theta. \qquad (5\text{-}3)$$

However the true number count of all galaxies in a diameter limited catalogue that is unrestricted to those whose velocity has been measured, is also known. Calling this $N(\Theta,\,\hat{r})$ we have

$$N(\Theta,\,\hat{r}) \;=\; \Omega \int n(v)\,v^3\,\Phi(v\Theta)\,dv. \qquad (5\text{-}4)$$

Evidently we may deduce that the selection function S is given by

$$S(\Theta,\,\hat{r}) \;=\; N_\Theta\,(\Theta,\,\hat{r})/N\,(\Theta,\,\hat{r}) \qquad (5\text{-}5)$$

We may evaluate both numerator and denominator from Huchra's catalogue and the UGC/ESO catalogue as appropriate. Having found $S(\Theta)$ we insert it into 5-3. $N_v\,(v,\,\hat{r})$ can likewise be calculated from Huchra'a catalogue so with our knowledge of the diameter function $\Phi(D)$ we deduce from 5-3

$$n(v,\,\hat{r}) \;=\; \frac{1}{\Omega v^3} \cdot \frac{N_v(v,\,\hat{r})}{\displaystyle\int \Phi(v\Theta)\,S(\Theta,\,\hat{r})\,d\Theta}.$$

Figure 9 shows a plot of n $(v,\,\hat{r})$ with radial velocity plotted radially and supergalactic longitude plotted as azimuth. To get sufficient numbers to calculate S the solid angles chosen are cones of $15°$ semi-angle centred close to the supergalactic plane tracing the maxima of the bright galaxy distribution.

It is already clear from this crude map that Virgo is at one side of the density enhancement in which it lies, the centre of the distribution being displaced towards Centaurus or Pavo.

The need for more reliable cartography of the galaxy distribution is clear. A great start on the nearby galaxies has been made by Tully and Fishers' Atlas (1987). The redshift surveys are rapidly extending this but difficulties will remain in the zone of avoidance.

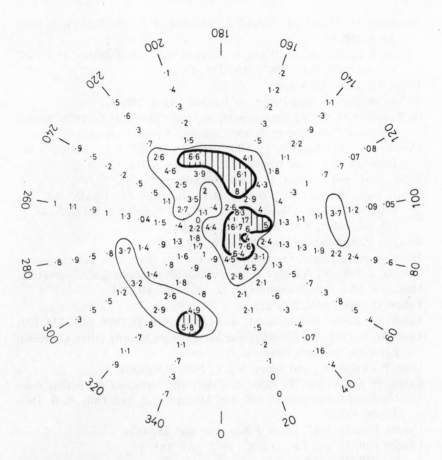

FIG. 9. Map of the number density of galaxies near the supergalactic plane as a function of supergalactic longitude and radial velocity corrected to the LG. The region out to 10,000 km/s is shown with contours at 2.5 and 5 times mean density of the universe. The Perseus-Pisces chain, the Virgo-Centaurus-Pavo complex, and the Coma region all stand out. Densities are averaged over 30° cones.

REFERENCES

Aaronson, M., Huchra, J., Mould, J., Schechter, P.L., and Tully, R.B. 1982. *Ap J.* **258**, 64.

Clowes, R.G., Savage, A., Wang, G., Leggett, S.K., MacGillivray, H.T., and Wolstencroft, R.D. 1987. *MN.* **229**, 27p.

Davis, M. 1988. this volume.

de Vaucouleurs, G. and Peters, W.L. 1985. *Ap J.* **297**, 27.

de Vaucouleurs, G., de Vaucouleurs, A., and Corwin, H.G. 1976. *Second Reference Catalogue of Bright Galaxies.* Austin: University of Texas.

Dressler, A., Lynden-Bell, D., Burstein, D., Davies, R.L., Faber, S.M., Terlevich, R.J., and Wegner, G. 1987. *Ap J.* **313**, 42.

Faber, S.M. and Burstein, D. 1988. this volume.

Faber, S.M., Dressler, A., Davies, R.L., Burstein, D., Lynden-Bell, D., Terlevich, R.J. and Wegner, G.A. 1987. in *Nearly Normal Galaxies.* p. 175. ed. Faber, S.M. New York: Springer Verlag.

Fouqué, P. and Paturel, G. 1985. *AA.* **150**, 192.

Harmon, R.L., Lahav, O., and Meurs, E.J.A., 1987. *MN.* **228**, 5p.

Huchra, J. 1987, ZCAT. tape from Smithsonian Astrophysical Observatory.

Humason, M.L. and Wahlquist, H.D. 1955. *AJ.* **60**, 254.

Lahav, O. 1987. *MN.* **225**, 213.

Lahav, O., Rowan-Robinson, M., and Lynden-Bell, D. 1988. *MN.* **234**, 677.

Lauberts, A. 1982. *The ESO/Uppsala Survey of the ESO(B).* Atlas. Garching: European Southern Observatory.

Lilje, P., Yahil, A., and Jones, B.J.T. 1986. *Ap J.* **307**, 91.

Lubin, P. and Villela, T. 1986. in *Galaxy Distances and Deviations from Universal Expansion.* p. 169. eds. Madore, B.F. and Tully, R.B. Dordrecht: Reidel.

Lynden-Bell, D. 1987. *Quart J Roy Astr Soc.* **28**, 187.

Lynden-Bell, D. and Lin, D.N.C. 1977. *MN.* **181**, 37.

Lynden-Bell, D., Faber, S.M., Burstein, D. Davies, R.L., Dressler, A., Terlevich, R.J., and Wegner, G. 1988. *Ap J.* 326, 19.

Meiksin, A. and Davis, M. 1986. *AJ.* **91**, 191.

Meurs, E.J.A. and Harmon, R.L. 1988. in preparation.

Nilson, P. 1973. *Uppsala General Catalogue of Galaxies.* Upsala Astr. Obs. Ann 6.

Paturel, G., Fouqué, P., Lauberts, A., Valentijn, E.A., Corwin, H.G., and de Vaucouleurs, G. 1987. *AA.* **184**, 86.

Richter, O.G., Tammann, G.A., and Huchtmeier, W.K. 1987. *AA.* **171**, 33.

Rubin, V.C., Ford, W.K., Thonnard, N., Roberts, M.S., and Graham, J.A. 1976. *AJ.* **81**, 687.

Rubin, V.C., Ford, W.K., Thonnard, N., and Roberts, M.S. 1976. *AJ*. **81**, 719.

Sandage, A. 1986. *Ap J*. **307**, 1.

_____ . 1987. *Ap J*. **317**, 557.

Strauss, M.A. and Davis, M. 1988. this volume.

Tully, R.B. and Fisher, J.R. 1987. *Nearby Galaxies Atlas*. Cambridge University Press.

Vorontsov Velyaminov , B.A. and Arkipova, A.A. 1963-68. *Morphological Catalogue of Galaxies*. Moscow: Moscow State University.

Yahil, A. 1988. this volume.

Yahil, A., Sandage, A.R., and Tammann, G.A. 1977. *Ap J*. **217**, 903.

Yahil, A., Walker, D., and Rowan-Robinson, M. 1986. *Ap J*. **301**, L1.

THE STRUCTURE OF THE UNIVERSE TO 10,000 KM S^{-1}
AS DETERMINED BY IRAS GALAXIES

AMOS YAHIL

Astronomy Program, State University of New York at Stony Brook

ABSTRACT

A report is presented of the current status of a redshift survey of IRAS galaxies, which maps their density structure over 76% of the sky, to a distance of $100h^{-1}$ Mpc. The peculiar gravitational field is then calculated, assuming that the IRAS galaxies trace the mass. Special attention is paid to correct the Hubble positions of the galaxies self-consistently for their peculiar velocities.

The picture of peculiar gravity which emerges is more complex than had been imagined in earlier parametric models. It is dominated by two large mass concentrations, with modest over-density, one in the direction of Hydra-Centaurus (the "Great Attractor"), and the other around Perseus-Pisces. As a result, the gravitational field bifurcates not far from the position of the Local Group.

A comparison between the predicted peculiar velocities and the observed ones shows good overall agreement, confirming the gravitational origin of the peculiar velocities. It is planned to tackle the remaining differences mainly by extending the redshift survey down to $|b| \geq 5$, and by calculating peculiar velocities nonlinearly.

1. INTRODUCTION

This paper reports on the current status of a program to map the density structure of the universe to $\sim$ 10,000 km s^{-1}, through a redshift survey of tracer galaxies detected by the Infrared Astronomical Satellite (IRAS). The initial investigations of the distribution of these galaxies (Yahil, Walker, and Rowan-Robinson 1986; Meiksin and Davis 1986), used only their positions and fluxes. These studies showed a dipole anisotropy in the surface brightness of the IRAS galaxies that was coincident, within the errors, with that of the cosmic microwave background (CMB). This suggested both that the IRAS galaxies traced the mass, and that the peculiar gravitational field was due to density perturbations within the volume surveyed by the IRAS galaxies.

Unfortunately, owing to the broad luminosity function, fluxes are poor distance indicators, and offer only a limited description of the density structure. In order to improve this situation, a redshift survey was launched to obtain individual distances for the IRAS galaxies. Collaborators in this effort have included M. Davis, J. Huchra, M. Strauss, J. Tonry, and A. Yahil. While the results from this ongoing project are not yet final, the redshifts available to date have already significantly modified our picture of the universe around us, superseding the previous studies based on fluxes. This paper, and the complementary one by Strauss and Davis (1988), are reports on the current status of the project.

The IRAS redshift survey is not different in principle from the initial RSA (Sandage and Tammann 1981) or CfA (Huchra *et al.* 1983) optical surveys. The big difference is in the volume sampled. The current IRAS survey covers 76% of the sky, and densities can be determined reasonably accurately to a distance of $100h^{-1}$ Mpc ($H_0 = 100h$ km s^{-1} Mpc^{-1}). Observations are now underway, which would increase the sky coverage to 87%. This should be compared with the RSA sample, which covered 50% of the sky to $40h^{-1}$ Mpc, and the CfA catalogue, which covered 20% of the sky to $80h^{-1}$ Mpc.

The structure of this paper is as follows. The selection criteria of the IRAS sample are delineated in Sec. 2. The calculation of the density structure, presented in Sec. 4, is preceded by a discussion of the luminosity function in Sec. 3. Special attention is paid to the underlying assumption of a universal luminosity function, which is shown to be consistent with the data. The calculation of the gravitational field is taken up in Sec. 5, assuming that the IRAS galaxies indeed trace the matter. In performing this calculation, care is taken not to assume a smooth Hubble flow, with galaxies at distances given by their redshifts. Instead, a new method is devised, in which distances are obtained by correcting the redshifts self-consistently for the peculiar velocities induced by the peculiar gravity, which itself depends on the corrected distances. The essence of the paper are the results of the calculations of Sec. 4 and 5, which are presented in Sec. 6 in the form of maps of both density and gravity. The detailed predictions of the calculation can then be confronted with observations, by comparing the inferred peculiar velocities with the observed ones. This is done in Sec. 7, using the compiled data on peculiar velocities presented at this conference by Faber and Burstein (1988).

2. SAMPLE

From the observational point of view, the IRAS galaxies are ideal tracers, since they are homogeneously detected over most of the sky, and their fluxes are unaffected by galactic extinction. The present sample covers 76% of the sky. Excluded are a band $|b| < 10°$ around the galactic plane, a strip in which the IRAS coverage was incomplete, and a few patches of known "cirrus" infrared emission.

Candidates for the redshift survey included all the sources in the Second IRAS Point Source Catalogue (PSC), with high quality 60μ flux $S_{60} > 1.936$ Jy, and satisfying the color criterion $S_{12} < S_{60}/3$ (Meiksin and Davis 1986). This condition, corresponding to a spectral inedx $\alpha < -0.68$, cannot be used for $S_{60} < 0.75$ Jy, since 12μ fluxes are not available below 0.25 Jy, but the problem does not arise with our higher flux limit. Yahil *et al.* (1986) used a somewhat less restrictive condition, $S_{25} < 3S_{60}$, corresponding to $\alpha < +1.25$, which can be used down to the 60μ limit of the PSC at 0.5 Jy. In the flux range of our redshift survey, however, this more liberal color criterion yields only a few percent more candidates, and was not used.

Prior to the the redshift survey, a major problem was contamination of the sample by infrared sources in our own interstellar medium, the so-called "cirrus". These sources have spectra that are similar to those of external galaxies, and are, therefore, included in the list of candidates generated by the above algorithm. While they are mainly confined to the plane, they do extend to higher galactic latitude in several directions of molecular cloud complexes, such as Ophiuchus and Orion, or of other galaxies of the Local Group. The existence of this cirrus contamination has prompted several attempts to use more sophisticated and restrictive color criteria to exclude them (Harmon, Lahav, and Meurs 1987; Rowan-Robinson 1988). In our redshift survey the cirrus problem was overcome by individual inspection of all the candidates on sky survey prints. This method is by far the most reliable one, since a nearby galaxy can easily be identified, even through a large amount of extinction. In any case of doubt, a spectrum of the object was obtained.

We have been so encouraged by the success of the identification program, that we decided to extend the survey to the previously excluded patches of cirrus, and down to $|b| \geq 5°$. When this survey is completed, we will have covered 87% of the sky. I have myself performed all the identifications of the candidates in these regions. My confidence in the identification is based on three factors: (1) I personally found it rather easy to make the identifications; (2) sources classified as cirrus were invariably clustered together in areas of larger obscuration; and (3) sources classified as galaxies, or questionable, were more smoothly distributed, with approximately the same sky surface density as the well identified galaxies at high galactic latitude. A quantitative test of my identifications will be available shortly for a part of the sample, where a "blind" Arecibo survey is taking spectra of all the IRAS galaxy candidates (Dow *et al.* 1988).

The above procedure for selecting galaxy candidates, as well as the subsequent determination of the luminosity function (see Sec. 3), presuppose accurate IRAS fluxes. There are several causes of concern in this regard. First, the PSC is known to underestimate the fluxes of extended sources. This would create a bias against large nearby extended galaxies, and may seriously affect our estimate of density in the Virgo supercluster. In order to overcome this difficulty, we asked Tom Soifer and Elizabeth Smith of the Infrared Processing and Analysis Center (IPAC) to addscan all the sources flagged as extended in the PSC, and used these addscanned fluxes. We also used the co-added fluxes of Rice *et al.* (1988) for galaxies with angular diameters greater than 10′. Secondly, we have noticed that the correlation coefficient flag in the PSC, which measures how well a point source template fits the scans, is anticorrelated with the ratio of addscanned to PSC fluxes. There are over a thousand sources which are not flagged as extended, but have poor correlation coefficients, and their addscanning is now in progress. Initial indications are that the problematic sources are virtually all cirrus objects. Finally, errors result from intensity enhanced detector responsivity ("hysteresis"), when crossing the galactic plane. This effect, which is expected to be important only at very low galactic latitudes, is now under investigation.

Our current redshift survey, which does not yet include those new areas of higher optical extintion, consists of 2244 galaxies, whose sky distribution is shown in Fig. 1, in both galactic and supergalactic coordinates. The excluded areas are also shown. The first impression is that the sky distribution of the galaxies is fairly smooth. Well known nearby clusters can be identified, but the contrast with their surroundings is far less marked than in similar plots of optically selected galaxies. Fig. 2 shows the distribution of the IRAS galaxies with distance, and compares it with that predicted for a homogeneous universe, with the luminosity function determined in Sec. 3. The biggest deviation is an over-density in the Virgo supercluster, but otherwise no obvious clustering is seen in this distribution, which is an average over all the 76% of the sky which we cover.

The smooth distributions in Figs. 1 and 2 might be construed to indicate that the survey extends beyond the "local" irregularities, and the volume it encompasses begins to approximate a "fair sample" of the universe. Based on our calculation of the gravitational field (see Sec. 5), we believe that this is indeed the case. The visual impression, however, is somewhat misleading, because of the absence of early-type galaxies in the IRAS sample. We are therefore seriously undersampling the high density regions, which are known to have a higher fraction of early-type galaxies (Dressler 1980, Postman and Geller 1984). This point is taken up again in Sec. 4.

3. LUMINOSITY FUNCTION

The mapping of the density structure is performed by counting galaxies as a function of position. In order to convert these counts into densities, it

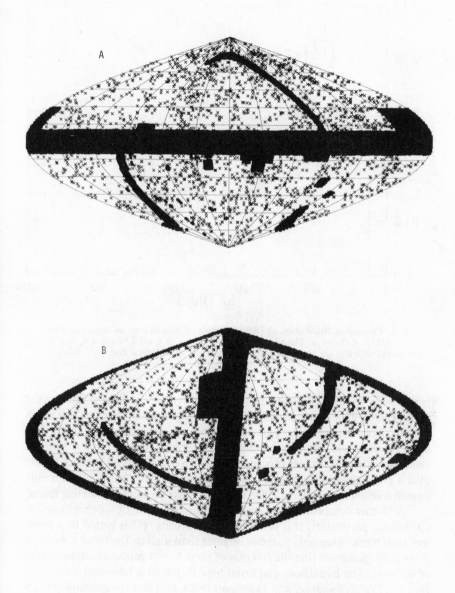

FIG.1. Sky map of the current IRAS redshift survey, showing both the area covered and the distribution of observed galaxies: (a) in galactic coordinates, (b) in supergalactic coordinates. Meridians start at 0° on the left of the plot, and end at 360° on the right.

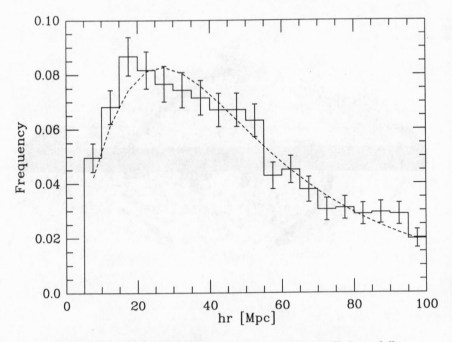

FIG.2. Frequency distribution of IRAS redshifts, averaged over all observed directions (76% of the sky). The observations are shown as a solid histogram, and the expected distribution for a homogeneous universe as a dashed line.

is necessary to take into account the luminosity function. In a flux-limited survey there will be more galaxies per unit volume at closer distances, for which the flux from less luminous galaxies is above the observational cutoff, than at larger distances, for which only the brightest galaxies make it into the catalogue. The extravagant procedure, in which one rejects all galaxies fainter than a given absolute luminosity, is totally impractical. One is left with only a small fraction of the galaxies, and statistical noise becomes the limiting factor.

A further complication arises if the luminosity function depends on environment, particularly if it is a function of density. (This would be a problem even when using only galaxies brighter than a given luminosity, because there is no guarantee that the fraction of these bright galaxies is independent of density). The hypothesis employed here is that of a universal luminosity function (Yahil, Sandage, and Tammann 1980), i.e., that the luminosity function is everywhere identical in *shape,* and differs from one location to another only in *normalization.* Indeed, it is this normalization which determines the local density (see Sec. 4). This hypothesis, therefore, needs to be confirmed, at least *a posteriori.*

The determination of the luminosity function in a density-independent fashion follows the method of Sandage, Tammann and Yahil (1979), except that the functional form of the luminosity function due to Schechter (1976), is replaced by the *cumulative* luminosity function

$$\Phi(L) = C \left(\frac{L}{L_*}\right)^{-\alpha} \left(1 + \frac{L}{\beta L_*}\right)^{-\beta}, \tag{1}$$

with its corresponding *differential* luminosity function

$$\phi(L) = -d\Phi(L)/dL = \left(\frac{\alpha}{L} + \frac{\beta}{\beta L_* + L}\right) \Phi(L). \tag{2}$$

The *conditional* luminosity function, for a galaxy at distance r, is therefore

$$f(L/r) = \begin{cases} \phi(L)/\Phi(L_m), & \text{if } L \geq L_m; \\ 0, & \text{otherwise,} \end{cases} \tag{3}$$

where $L_m(r) = 4\pi r^2 \nu_{60} S_m$ is the minimum luminosity that can be seen at distance r above the flux limit S_m. (The frequency $\nu_{60} = 5 \times 10^{12}$ Hz is used to convert flux into "luminosity".)

This probability function is density-independent, i.e., it is not a function of the normalization constant C in eq. (1), which cancels in the ratio $\phi(L)/\Phi(L_m)$. It also has a unit integral, and is therefore the correct probability function to use in a maximum-likelihood fit (Sandage *et al.* 1979), which is equivalent to minimizing

$$\Lambda = -2 \sum_i \ln f(L_i/r_i), \tag{4}$$

with respect to the three parameters which define the shape of the luminosity function: α, β, and L_*. The minimum value of Λ is arbitrary, but the deviations from the minimum follow the usual χ^2 rules, and provide error estimates for the parameters.

Our best fitted values are listed in Table 1. Fits are presented both for "redshift space", in which distances are not corrected for peculiar velocities, and for the case in which a self-consistent correction to distance was applied (see Sec. 5). Virgo galaxies—defined to be all galaxies within 10° of M87 with velocities smaller than 2500 km s^{-1}—were not included in any of the fits to the luminosity function. Neither were galaxies with distances smaller than $5h^{-1}$ Mpc or larger than $100h^{-1}$ Mpc.

It is useful to generalize the conditional luminosity function of a single galaxy to that of for a *set* of galaxies with distances $r_1, \ldots, r_N$:

$$f(L/r_1, \ldots, r_N) = \sum_{i=1}^{N} f(L/r_i), \tag{5}$$

where the distribution function is here normalized to an integral of N. Any combination of galaxies can be studied using this probability function, as long as the galaxies are not selected by luminosity. It is possible to test for the universality of the luminosity function by considering galaxies at different distances, densities, morphologies, or even colors (provided the colors are available without limits on luminosity). For example, Fig. 3 is a check of the dependence of the derived luminosity function on distance. The top panel shows the entire sample, while the lower ones break up the counts into distance bins. The fits of eq. (5) to the data (dashed lines), with the χ^2 values given in the figure, show that there is no Malmquist-like bias with distance. Note that the same fit is used in all the panels.

TABLE 1

FITTED PARAMETERS OF THE LUMINOSITY FUNCTION

	α	β	L_* $[h^{-2} L_\odot]$	C $[h^3 \text{ Mpc}^{-3}]$
Redshifts Only	0.60 ± 0.08	1.83 ± 0.15	$(3.5 \pm 1.0) \times 10^9$	5.96×10^{-3}
Corrected Distances	0.55 ± 0.08	1.92 ± 0.16	$(3.6 \pm 0.9) \times 10^9$	5.59×10^{-3}

4. Density

The calculation of local density follows the method of weights developed by Yahil *et al.* (1980) and Davis and Huchra (1982). For a flux-limited sample of galaxies with a universal luminosity function, the conversion from number counts to density is obtained by assigning each galaxy a distance-dependent weight

$$w_i = C/\Phi(L_m). \tag{6}$$

Beyond a certain distance, the density derived in this manner becomes subject to large statistical noise, and is no longer useful. A cutoff distance of $100h^{-1}$ Mpc was, therefore, imposed. The normalization constant C in eq. (6) is determined by requiring the sum of the weights to equal the surveyed volume. The mean weighted density per unit volume is then unity by construction.

In order to calculate the gravitational field, it is necessary to make some assumption about the distribution of galaxies in the masked 24% of the sky which was not surveyed. This is most easily accomplished by adding to the

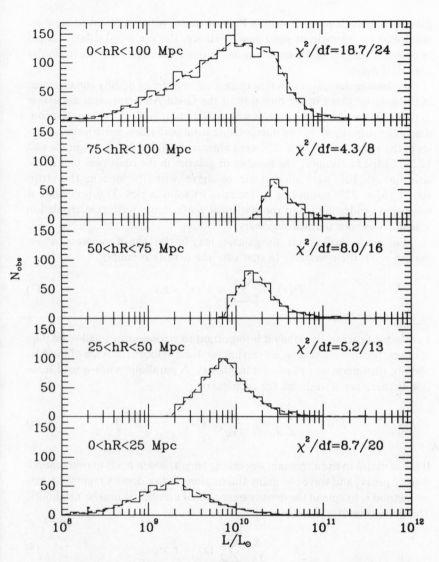

FIG.3. Distributions of infrared luminosity at different distances, compared with the expected function. All the data are fit to the same luminosity function given in Table 1. The quality of the fit is indicated by the χ^2 (measured for bins containing 5 or more galaxies). There is no Malmquist-like bias with distance.

observational data a simulated set of galaxies in the masked region. Initially, a spatially homogeneous distribution was assumed, i.e., random positions, and distances drawn from the distribution shown in Fig. 2. The number of simulated galaxies was made to agree, within Poisson noise, with the observed number, multiplied by the ratio of solid angles. Hence, the simulated galaxies could be assigned weights in the same way as the observed ones, resulting in similar statistical noise.

Following the realization that there may be a strong density enhancement in the galactic plane in the direction of the Great Attractor, and discussion to that effect at this conference, it was decided to modify the galaxy distribution in the strip $|b| < 10°$ by interpolating from both sides. Specifically, galaxies in the strips $10° < |b| < 22°$ were binned in cells of $30°$ in longitude and $12.5h^{-1}$ Mpc in distance. The number of galaxies in the equivalent bins in the strip $|b| < 10°$ were then made to agree with the one in the strips $10° < |b| < 22°$, multiplied by the ratio of solid angles. This procedure is somewhat crude, but given the statistical noise, more refined interpolation schemes were not deemed necessary.

For many applications, the galaxies may be thought of as points, whose masses equal their weights. In that case the density is simply

$$D(\mathbf{r}) = \sum_i w_i \delta^3(\mathbf{r} - \mathbf{r}_i). \tag{7}$$

For the purposes of this study it is important to smooth over small-scale fluctuations, because the large-scale gravitational and velocity fields are of interest, and not the interaction of nearest neighbors. A parabolic point-spread function is therefore introduced for each galaxy

$$D(\mathbf{r}) = \sum_i w_i \frac{15}{8\pi a_i^3} \left(1 - \frac{(\mathbf{r} - \mathbf{r}_i)^2}{a_i^2}\right). \tag{8}$$

It is not useful to use a constant smoothing length, which tends to over-smooth density peaks, and leave too many fluctuations in low density regions. A better method is to spread the density over a fixed number of nearest neighbors. The definition of a_i used here is

$$a^2_i = \frac{5}{3k} \sum_{j=1}^k (\mathbf{r}_i - \mathbf{r}_j)^2, \tag{9}$$

with k = 10. (For Virgo galaxies, which were all placed at the center of the cluster, k was set to the number of Virgo galaxies plus 10). This procedure treats all densities in the same way, and should therefore introduce a minimal bias in density.

An underlying assumption of the entire procedure is the existence of a universal luminosity function. The density calculation would need to be completely re-evaluated if the luminosity function proved, for example, to b density-dependent. It is possible to test for a density bias as for a distance bias Fig. 4 shows a breakup of the observed counts into density bins with the same

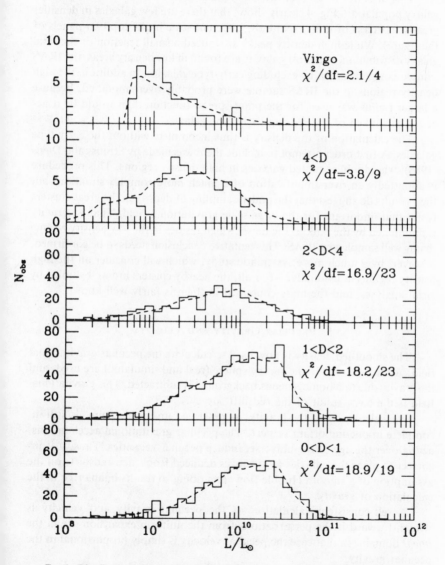

FIG. 4. Distribution of infrared luminosity at different densities (analogous to Fig. 3). The absence of a density-dependent bias supports the assumption of a universal *infrared* luminosity function, but there remains the problem of undersampling of early-type galaxies in rich clusters of galaxies.

fit shown in Fig. 3. Again, the χ^2 show no significant deviations. Thus, the assumption of a universal *infrared* luminosity function appears to be consistent with the data.

This does not mean that the IRAS galaxies are fair tracers of the entire galaxy population. Fig. 4 clearly shows that there are few galaxies in densities greatly in excess of unity. (Note the different scale in the top two panels of the figure). While high density peaks are indeed a small fraction of the total galaxy distribution, and most galaxies are found in low density areas, the IRAS sample exaggerates this by excluding early-type galaxies. In addition, the high density regions in the IRAS sample were probably over-smoothed, because a larger radius was used for the point-spread function than would be if account was taken of the missing early-type galaxies.

The calculation of the density is thus incomplete without the early-type galaxies. A first order attempt to include them was made by Strauss and Davis (1988), who double counted galaxies in high density regions. This procedure is admittedly an oversimplification of a much more complex situation, but their results do suggest that the under-sampling of density in the great clusters is not that important for the large-scale gravitational field. That is, gravity is dominated by the bulk of the mass, which resides in the low density regime that is well sampled by IRAS. This tentative conclusion needs to be confirmed, however, by a much more systematic study, which will consider all the high density regions individually. After all, the nearby clusters are well studied by virial analyses, and the mass contained in them is fairly well known.

5. Peculiar Gravity and Velocity

The smoothed density is now used to calculate the peculiar gravitational field. All the galaxies in the survey sphere (real and simulated) are used, and the gravity due to a homogeneous background is subtracted. This gravitational field is the basic result of the redshift survey.

The calculation of the gravity, as described up to this point, is still incomplete in one important respect. The peculiar gravitational accelerations, acting over the age of the universe, induce peculiar velocities. Thus, it is incorrect to place the galaxies at distances deduced from their redshifts on the assumption of a smooth Hubble flow, and doing so results in an error in the calculation of gravity.

A self-consistent calculation of the peculiar gravity and velocity is straightforward if the perturbations from the uniform expansion are in the *linear* domain. In that case the peculiar velocity is simply proportional to the peculiar gravity

$$\mathbf{u}_i = \frac{2}{3} H_0^{-1} \Omega_0^{-0.4} \mathbf{g}_i(\mathbf{r}_i). \qquad (10)$$

This can be combined with the usual relation between distance and the observe radial velocity in the Local Group frame (Yahil *et al.* 1977)

$$v_i = H_0 r_i + (\mathbf{u}_i - \mathbf{u}_{LG}) \cdot \hat{\mathbf{r}}_i, \tag{11}$$

to yield a closed implicit set of equations for the true distances r_i. These equations have only one free parameter, the cosmological density parameter, Ω_0. They are independent of the Hubble constant, except as a scaling factor to convert km s^{-1} to Mpc.

Given the observed velocities v_i, and a value of Ω_0, eqs. (10) and (11) can therefore be solved for the distances r_i of the galaxies. This was done iteratively, starting by assuming the distances to be Hubble ones, and using eq. (10) to obtain a first estimate of the peculiar velocities. These were then used to obtain new distances from eq. (11). (The angular coordinates of the galaxies on the sky were, of course, unchanged). The process was repeated, with an updated luminosity function and density distribution determining a corrected **g** at each iteration. Convergence to an accuracy ~ 20 km s^{-1} was obtained after a few iterations. Numerical oscillations were damped by setting the new distance in each iteration to the average of the previous one, and the one suggested by the current iteration.

The choice of Ω_0 needs some clarification. The models presented at the conference were for $\Omega_0 = 1$. This value was chosen because the *magnitude* of the derived peculiar velocity of the Local Group was then of the order of its velocity with respect to the CMB, 610 km s^{-1} (Lubin and Villela 1986). With the new interpolation scheme for the galactic plane strip, this was no longer so. The procedure was, therefore, modified by changing Ω_0 in each iteration, with the resultant rescaling of all peculiar velocities, so that the magnitude of the peculiar velocity of the Local Group became exactly 610 km s^{-1}. The final value of the cosmological density parameter was $\Omega_0 = 0.5$, significantly lower than earlier determinations from infrared galaxies (Yahil *et al.* 1986; Villumsen and Strauss 1987; Lahav, Rowan-Robinson, and Lynden-Bell 1988), and closer to the value deduced from the optical dipole (Lahav 1987; Dressler 1987; Lahav *et al.* 1988).

However, this estimate of Ω_0 should be viewed as very preliminary, because the predicted peculiar velocity of the Local Group is toward $l = 231°$ and $b = 48°$, which is 35° away from the *direction* of the CMB, $l = 272°$ and $b = 30°$ (Lubin and Villela 1986). It cannot be argued that Ω_0 has been properly determined, until this difference in directions is understood.

A major fault could be the assumption of linear perturbations. In the spherical case, the correct peculiar velocity is smaller than the linear approximation by a factor $\simeq \langle D \rangle^{-1/4}$, where $\langle D \rangle$ is the mean density in the sphere interior to the galaxy in question (Yahil 1985). Linear theory therefore *over-estimates* the effect of high density regimes on the peculiar velocities. The Virgo supercluster is one region where the linear calculation might thus be in error (although this is offset by the undersampling of early-type galaxies in the central cluster). Lowering the contribution of the Virgo supercluster will bring the peculiar velocity of the Local Group into better agreement with the direction of the CMB.

In fact, N-body calculations show that, in the *nonlinear* regime, peculiar velocities and accelerations are typically misaligned by $\sim 25°$ (Villumsen and Davis 1986), comparable to the discrepancy found here. A careful comparison of the predicted peculiar velocity field with the observed one will therefore probably require a self-consistent nonlinear calculation of the gravitational field. Such a calculation cannot use only the positions of the galaxies at the present epoch, as is done in the linear approximation. Instead, a guess has to be made of their positions at an earlier epoch, say $z = 10$, when linear perturbations might be a better approximation, and integrated nonlinearly to the present epoch, using an N-body code. We are now in the process of implementing such a scheme, but the results presented here still use the linear approximation.

Even in the nonlinear calculation, it will be impossible to determine the original positions of galaxies that are now in virialized clusters, where phase-mixing has completely obliterated any memory of initial conditions. In fact, the suspicion is that any shell crossing will make it very difficult to determine which galaxy came from where. Fortunately, the IRAS galaxies avoid the virialized regions of high density. Fig. 5 shows a breakup of the observed pairwise separation in redshift space, Δv_{ij}, into a radial (line-of-sight) component π, and a tangential component σ. (Only one component of σ is shown, by projecting it along a random azimuthal axis.) In such a scatter diagram, virialized velocities are seen as the familiar "finger of God" concentration of pairs for which $\pi \gg \sigma$. No such effect is seen. The same conclusion can be drawn from Fig. 6, which shows a random distribution of the angle α which Δv_{ij} makes with the line-of sight (Turner and Sargent 1977).

Another question of major interest is how far one needs to integrate over density perturbations for the gravitational acceleration to converge. Put differently, where are the perturbations responsible for the gravity? Fig. 7 addresses this question by plotting the *cumulative* peculiar velocities due to concentric shells around the Local Group, both their absolute values and their three Cartesian coordinates in supergalactic coordinates. (See Sec. 6 for the coordinate convention used here.) It is seen that the bulk of the peculiar velocity of the Local Group is generated within $40h^{-1}$ Mpc, although there is a slow growth beyond that distance (Vittorio and Juszkiewicz 1987). The same picture emerges from Fig. 8, which plots the *differential* contributions of the concentric shells to the peculiar velocity of the Local Group. Poisson error bars are shown here as well, so the reader may obtain a sense of the ratio of signal to shot noise in the calculation of gravity.

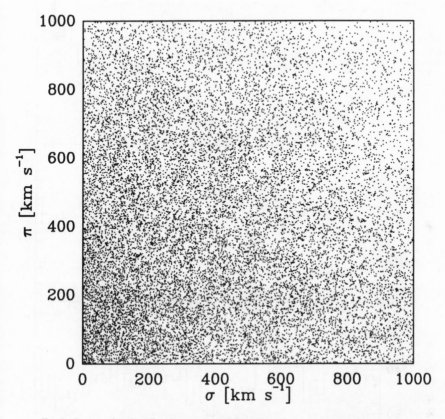

FIG.5. Scatter plot showing the breakup of the pairwise separation in redshift space, Δv_{ij}, into a radial (line-of-sight) component π and a tangential component σ. (Only one component of σ is shown, by projecting it along a random azimuthal axis.) The lack of a "finger of God" concentration of pairs, for which $\pi \gg \sigma$, shows that the IRAS galaxies avoid virialized clusters.

If the gravitational acceleration acting on the Local Group is largely due to perturbations within the local supercluster ($r < 40h^{-1}$ Mpc), then it cannot be dominated by the more distant Great Attractor, whose contribution to gravity should not converge until well beyond that distance. The IRAS redshift survey does see the Great Attractor as a large complex in the direction of Hydra-Centaurus, but it is largely counteracted by the Perseus-Pisces complex in the opposite direction. The detailed maps presented in Sec. 6 show that, in fact, the gravitational field bifurcates close to the position of the Local Group.

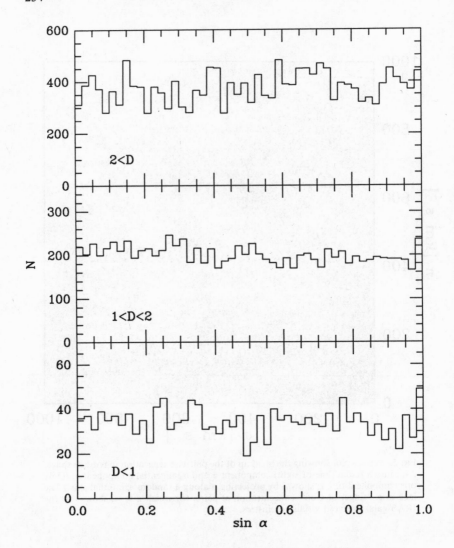

FIG.6. Distribution of the angle α which Δv_{vj} makes with the line-of- sight. The randomness of this distribution, and the lack of a concentration around $\alpha = 90°$, is another demonstration of the paucity of virialized IRAS galaxies.

6. MAPS

A major difficulty in the study of any 3-d vector field is visualization. In an effort to overcome the problem, this section present a series of maps of the density structure and gravitational field from different vantage points.

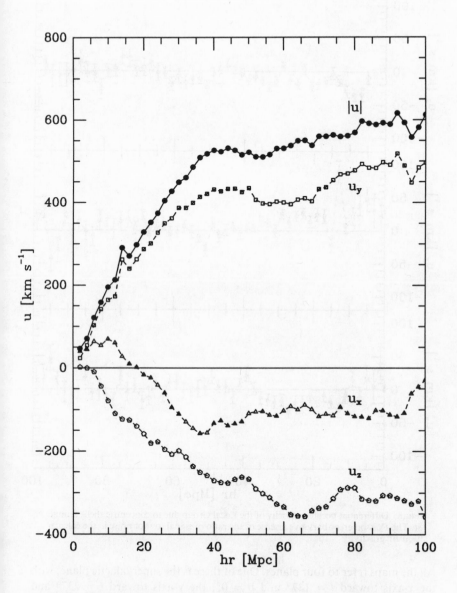

FIG. 7. Cumulative peculiar velocity acting on the Local Group due to concentric shells around it. The cartesian components are in supergalactic coordinates. The bulk of the peculiar velocity of the Local Group is generated within $40h^{-1}$ Mpc. It is therefore *not* dominated by the Great Attractor, whose center is at $42h^{-1}$ Mpc, and whose contribution should not converge until well beyond that distance.

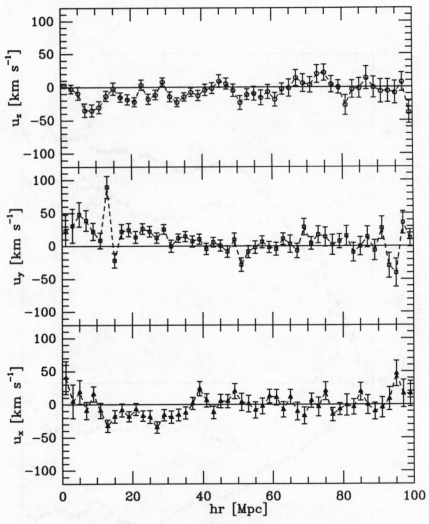

FIG.8. Differential peculiar velocity of the Local Group due to concentric shells around
it. The Poisson error bars give a sense of the ratio of signal to shot noise in the calculation of gravity.

All the maps refer to four planes. One of these is the supergalactic plane, with
the x-axis toward $l = 137°$ and $b = 0°$, the y-axis toward $l = 227°$ and
$b = 84°$, and the z-axis, normal to the plane, toward $l = 47°$ and $b = 6°$.
The Local Group is at the origin of the plot. The planes of the other plots
are obtained by rotating by 45° around the x-axis, which is thus identical in
all plots. The galactic coordinates of the z-axis are marked on each plot.

Fig. 9 shows the *smoothed* density structure in the four planes. The contours are spaced logarithmically at intervals of 2 decibels. Contours for which $D \geq 1$ are shown as solid lines, and the ones for which $D < 1$ as dashed lines. The contour plots are of thin slices, because the point-spread functions of the galaxies allow the calculation of density at any point in space. In reality, however, the density represents some mean density above and below the plane with scale height depending on location, as explained in Sec. 4. Careful inspection of the figures shows all the familiar nearby clusters of galaxies. Most of the galaxies, however, are in large-scale agglomerates with relatively low density, and these dominate gravity. One such large concentration is in the direction of the Great Attractor, along the (common) negative x-axes of all the plots. There is, however, also a significant over-density in the opposite direction, toward Perseus-Pisces. Of additional note is a big void beyond Fornax, in a direction roughly opposite to that of Virgo.

The gravitational field, calculated in Sec. 5, is presented in Fig. 10. All galaxies (both real and simulated) that are within $\pm 22.5°$ of the plane of the plot are shown as vectors, whose lengths are proportional to the components of their peculiar gravities in the plane. While the overall scale of the vectors is arbitrary, their relative lengths and directions correspond to the gravitational acceleration. Fig. 10 clearly shows that the two large-scale centers of attraction are indeed the Great Attractor and Perseus-Pisces. The gravitational field shows a distinct bifurcation between these two large complexes, not far from the position of the Local Group. In addition, there is strong gravity toward Virgo, but this local field looks far from spherical symmetry.

The major conclusion from Figs. 9 and 10 is that the local gravitational field is fairly complex, and is not adequately described by parametric models of the sort used to date. First, the Virgocentric field is not spherical. (Incidentally, the peculiar velocity of the Local Group relative to Virgo in this calculation was 410 km s^{-1}, of which the component in the direction of Virgo was 330 km s^{-1}). Secondly, the quadrupolar approximation of the shear field (Lilje, Yahil, and Jones 1986) is valid over a limited range in distance. Finally, the Great Attractor is not the only large-scale source of pull; Perseus-Pisces and the void beyond Fornax also play their role. In fact, this was already anticipated by Lilje *et al.* (1986), who showed that the eigenvalues of the quadrupole shear tensor were not in the ratio $-1 : -1 : +2$, that would be expected for a single Great Attractor.

The inadequacy of the parametric models is understandable in view of the limitations of the data from which they were derived. Previous determinations were limited in sky coverage, and in the case of the RSA catalogue also in depth. Purely kinematic models (Lynden-Bell *et al.* 1988, Faber and Burstein 1988), on the other hand, are based on the inherently incomplete information contained in the one observable (radial) component of peculiar velocities. They may also suffer from some incompletion in sky or depth

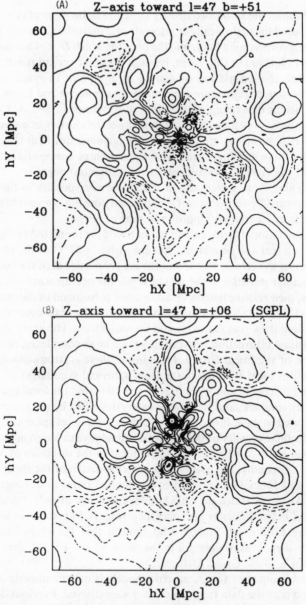

FIG.9. Smoothed density structure in four planes. The contours are spaced logarithmically at intervals of 2 decibels. Contours for which $D \geq 1$ are shown as solid lines, and ones for which $D < 1$ as dashed lines. The orientation of the planes is such that the x-axis always points in the supergalactic x direction, $l = 137°$ and $b = 0°$. The direction of the z—axis, normal to the plane, is shown at the top of each frame.

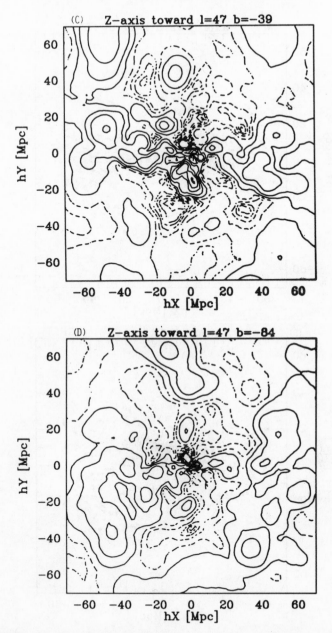

Frame (b) is the supergalactic plane (SGPL). The largest mass agglomerates are seen to be two gigantic perturbations of relatively small over-density. The Great Attractor is along the negative x-axis, and the Perseus-Pisces complex is along the positive x-axis. There is also a void beyond Fornax, roughly opposite the direction of Virgo (the highest density peak).

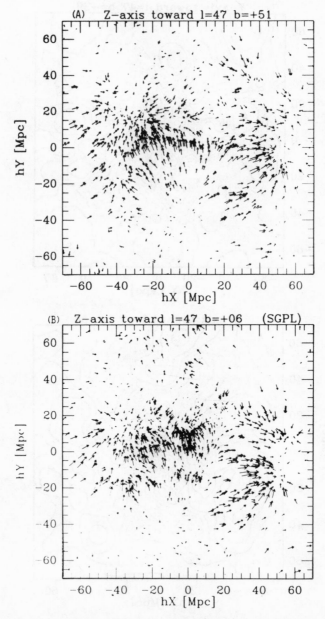

FIG.10. The gravitational field predicted by the IRAS galaxy distribution. All galaxies (real and simulated) that are within ± 22.5° of the planes of Fig. 9 are shown as vectors, whose lengths are proportional to the components of their peculiar gravities in the plane. (The positions of the galaxies are at the tails of the vectors.) The overall scale of the vectors is arbitrary, but their relative lengths and directions correspond

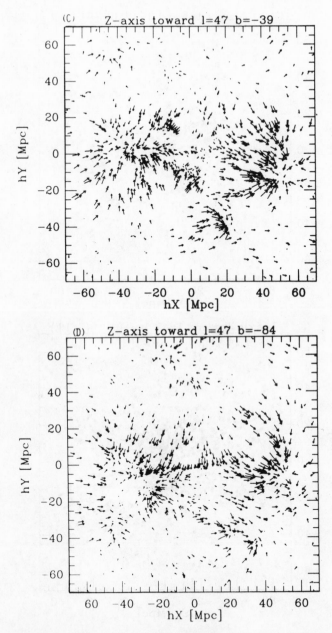

to the gravitational acceleration. The dominance of the two large mass concentrations is clearly seen, resulting in a bifurcation of the gravitational field close to the Local Group. The Virgo supercluster also exerts a significant pull, but its gravity is far from the spherical model frequently adopted in the past.

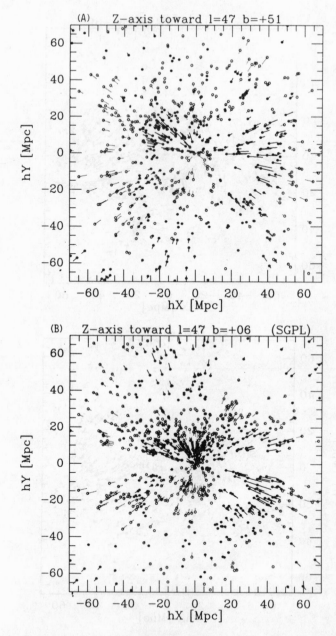

FIG.11. Predicted radial components of the peculiar velocities of the observed IRAS galaxies, which can be compared directly with the observations presented in Figs. 6-8 of FB in this volume. Galaxies with receding peculiar velocities (in the frame of the CMB) are marked by filled circles and solid lines, and the ones with approaching

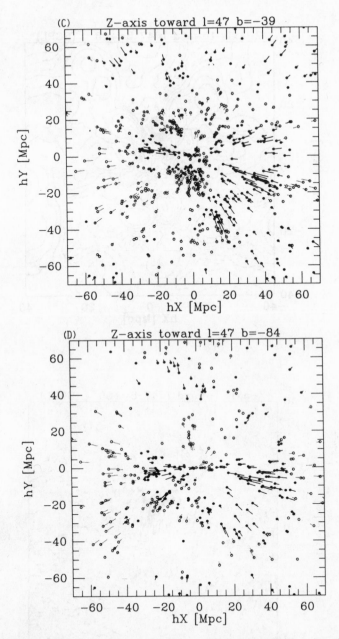

peculiar velocities are denoted by open circles and dashed lines. The location of each galaxy, after correction for peculiar velocity, is at the center of the circle, and the length of the line is the magnitude of the radial component of the peculiar velocity.

(A)
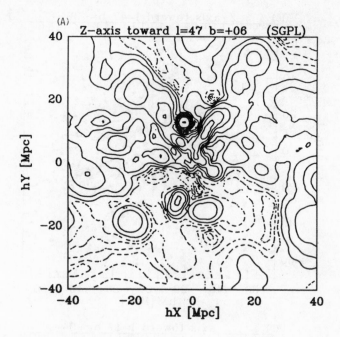

(B)
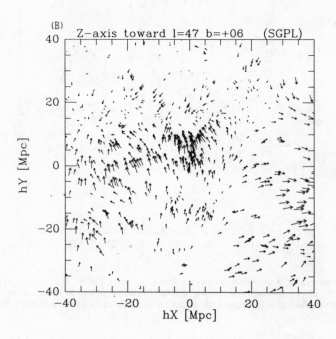

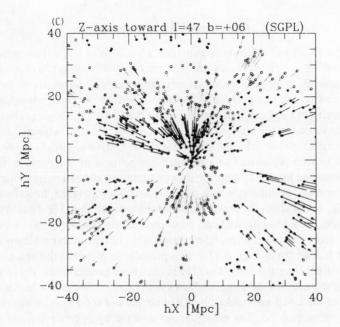

FIG.12. Blow-up of frames (b) of Figs. 9-11 (the supergalactic plane). The scale of Figs. 12a, b, c is close to that of Fig. 8 of FB. Comparison of Fig. 12c with Fig. 8 of FB shows good overall agreement, demonstrating the gravitational origin of the peculiar velocities. There is lack of agreement in the first quadrant, in the region of Ursa Major.

coverage; paucity of data in the direction of Perseus-Pisces, and at larger distances in the direction of the Great Attractor are particular problems.

The time has therefore perhaps come to replace the parametric models by detailed 3-d tables of density, such as provided by the IRAS redshift survey, and future surveys of equivalent sky coverage and depth. These surveys make detailed predictions of the peculiar velocity field, which can then be confronted with observations. An initial attempt in this direction is made in the next section.

7. COMPARISON WITH OBSERVATIONS

The IRAS redshift survey provides a 3-d density map, from which one can *predict* the entire complex peculiar velocity field, subject to only one free parameter, Ω_0. This prediction can be confronted with hundreds of measured radial components of peculiar velocities, as well as the three components of the velocity of the Local Group relative to the CMB. The basic premise, that peculiar velocities are induced by gravity, can therefore be subjected to a very over-constrained test. The comparison with the CMB was already made in Sec. 6. This section is devoted to a detailed comparison of the peculiar velocity field predicted by IRAS, with the superb summary and re-analysis of existing data, presented at this conference by Faber and Burstein (1988, henceforth FB).

Fig. 11 translates the vector peculiar velocities shown in Fig. 10 into observable radial components, and can be compared directly with Figs. 6 and 8 of FB. In order to facilitate this comparison, Fig. 12 reproduces a blown up version of frames (b) of Figs. 9-11 (the supergalactic plane) to the exact scale of Fig. 8 of FB. In particular, Fig. 12c is the IRAS prediction of Fig. 8 of FB. The notation is also the same. Galaxies with receding peculiar velocities in the frame of the CMB are marked by filled circles and solid lines, while the ones with approaching peculiar velocities are denoted by open circles and dashed lines. The location of each galaxy, after correction for peculiar velocity, is at the center of the circle, and the length of the line is the magnitude of the radial component of the peculiar velocity. Further comparisons are provided in Figs. 13-15, which are the analogues of Figs. 9-11 of FB.

A careful examination of the predicted versus observed peculiar velocities shows good overall agreement. This confirms both that the IRAS galaxies trace the mass, and that the peculiar velocities are induced by gravity. There is just too much correlation between peculiar gravity and velocity for alternative mechanisms to dominate.

There is, however, some disagreement in detail. The most striking difference is in the first quadrant of the supergalactic plane, where the gravity maps predict peculiar velocities moving away from the Local Group, while the data call for velocities toward it. (Recall that all velocities are in the frame of the CMB.) This difference, together with the (possibly related) misalign-

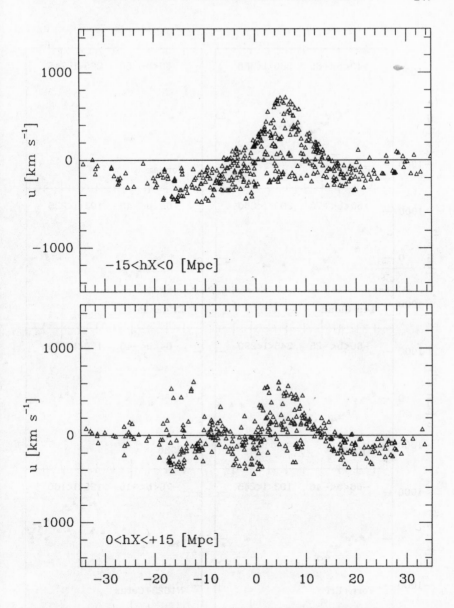

Fig.13. IRAS predictions to be compared with Fig. 9 of FB.

ment of the peculiar gravity acting on the Local Group and its velocity with respect to the CMB, are the major difficulties left to overcome. Four possible corrections to the calculation of gravity come to mind:

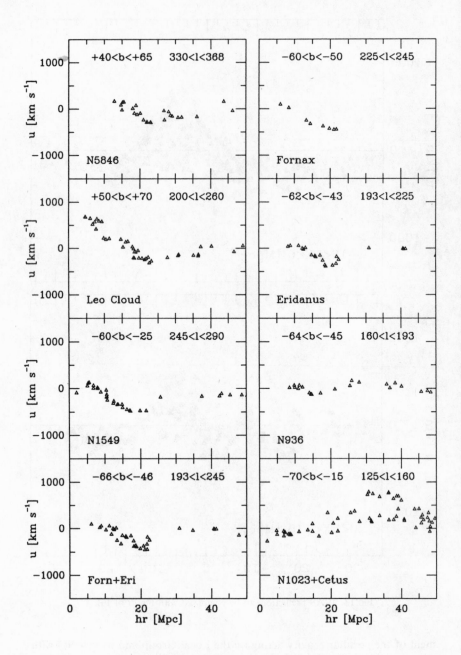

FIG.14. IRAS predictions to be compared with Fig. 10 of FB.

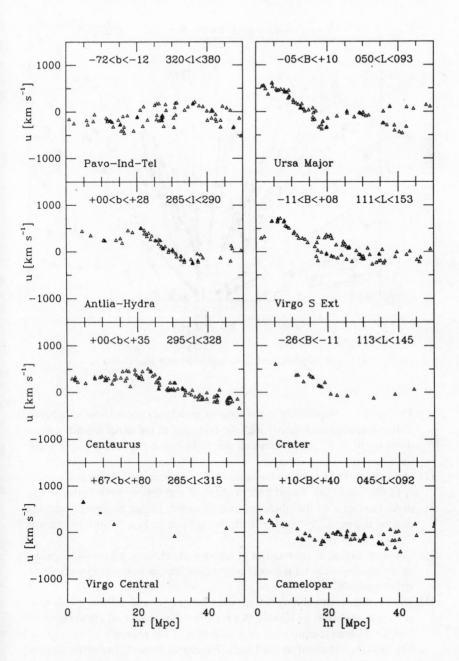

A. YAHIL

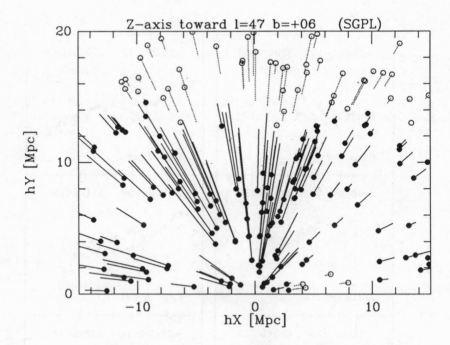

FIG.15. IRAS predictions to be compared with Fig. 11 of FB.

(1) The current interpolation in the zone of avoidance needs to be improved. This will be remedied shortly with the extension of the IRAS redshift survey down to $|b| \geq 5°$. The remaining interpolation, in the strip $|b| < 5°$, will then be much more secure.

(2) Nonlinear effects can lead to misalignment of peculiar gravity and velocity of the magnitude found here (Villumsen and Davis 1986). If this is, indeed, the cause of the observed misalignment, then a nonlinear calculation of the peculiar velocity field, as outlined in Sec. 5, should remove this source of error.

(3) The contribution of the handful of high density clusters, where IRAS galaxies undersample the total galaxy population, can be evaluated by standard virial methods.

(4) The overall agrement between the peculiar gravity and velocity fields suggests that the effect of biasing (Kaiser 1984, Bardeen *et al.* 1986) is small. Significant bias is equivalent to a violation of the principle of superposition, and should manifest itself in the balance of forces between the various mass agglomerates. A quantitative evaluation of the role of biasing is still lacking, however, and some biasing may be called for.

8. Conclusions

The IRAS redshift survey has been the first to provide density tracers with enough sky coverage (76%) and depth ($100h^{-1}$ Mpc) to allow a reliable determination of the local peculiar gravitational field. The infrared luminosity function of the IRAS galaxies appears to be universal, with no bias as a function of density. It is therefore possible to determine density, and hence gravity, using the method of weights of Yahil *et al.* (1980) and Davis and Huchra (1982). This calculation of density and gravity requires a correction of the Hubble distances of the galaxies for peculiar velocities. This has been accomplished by a self-consistent calculation of density, gravity, and peculiar velocity in a linear approximation for gravitational instabilities.

The density structure of the IRAS galaxies, and the resultant peculiar gravity, have been presented as a series of maps. These show that the gravitational field is dominated by two large concentrations ranging over a large volume, but having a modest overdensity. One is the Great Attractor in the direction of Hydra-Centaurus, along the negative supergalactic x-axis; the other is the Perseus-Pisces complex, in roughly the opposite direction in the sky. As a result, the gravitational field bifurcates not far from the position of the Local Group. In addition, a non-spherical Virgo supercluster exerts a significant pull, and there is a void beyond Fornax, roughly in the opposite direction.

The gravitational field is thus far more complex than had been imagined in earlier studies, and it is questionable whether it can be represented by simple parametric models. It is, however, possible to confront the detailed predictions of the IRAS redshift survey with a growing body of data on peculiar velocities of galaxies, summarized at this conference by Faber and Burstein (1988). This comparison shows good overall agreement, and thus confirms the gravitational origin of the peculiar velocities, a major astrophysical result. It also demonstrates that the IRAS galaxies trace the mass, thus providing a limit on biasing, although that is yet to be quantified.

There are, however, also disagreements. The predicted peculiar gravitational acceleration of the Local Group is misaligned by 35° relative to its peculiar velocity with respect to the CMB. Although this is comparable to nonlinear misalignments found in N-body calculations (Villumsen and Davis 1986), it is not yet known whether the linear approximation used here is the cause of the discrepancy in this case. There is also a difference between the predicted and observed peculiar velocities in the first quadrant of the supergalactic plane, in the direction of Ursa Major.

These difficulties may be resolved by a number of improvements planned for the near future, which include: (1) an extension of the survey closer to the galactic plane, down to $|b| \geq 5°$; (2) a nonlinear calculation of the predicted peculiar velocities; (3) a careful evaluation of the effect of high density clusters, where the fraction of IRAS galaxies is smaller than in the general field; and (4) the possible inclusion of some biasing.

Further observations should provide additional constraints. Most needed are peculiar velocities in the direction of Perseus-Pisces, and on the back side of both the Great Attractor and Perseus-Pisces. Ultimately, if and when the observations of density and peculiar velocity are brought into acceptable agreement, there will emerge a reliable determination of the cosmological density parameter, Ω_0.

ACKNOWLEDGEMENTS

It is a great pleasure to thank my collaborators—M. Davis, J. Huchra, M. Strauss, and J. Tonry—for many fruitful exchanges, as well as permission to use the IRAS redshift data prior to our joint publications. Special thanks go to D. Burstein and S. Faber for their useful comments, and for making their preprint available in a timely fashion, allowing a one-to-one comparison of the IRAS predictions and the observed peculiar velocities. T. Soifer and E. Smith were very helpful in providing addscans for hundreds of IRAS sources.

REFERENCES

Bardeen, J. M., Bond, J. R., Kaiser, N., and Szalay, A. S. 1986. *Ap J.* **304**, 15.

Davis, M. and Huchra, J. 1982. *Ap J.* **254**, 437.

Dow, M. W., Lu, N. Y., Houck, J. R., Salpeter, E. E., and Lewis, B. M. 1988. *Ap J Letters.* **324**, L51.

Dressler, A. 1980, *Ap J.* **236**, 351.

———. 1987, preprint.

Faber, S. M. and Burstein, D. 1988. this volume. (FB)

Harmon, R. T., Lahav, O., and Meurs, E. J. A. 1987. *MN.* **228**, 5p.

Huchra, J., Davis, M., Latham, D., and Tonry, J. 1983. *Ap J Suppl.* **52**, 89.

Kaiser, N. 1984. *Ap. J Letters.* **284**, L9.

Lahav, O. 1987. MN. **225**, 213.

Lahav, O., Rowan-Robinson, M., and Lynden-Bell, D. 1988. *MN.* **234**, 677.

Lilje, P. B., Yahil, A., and Jones, B. T. J. 1986. *Ap J.* **307**, 91.

Lubin, P. and Villela, T. 1986. in *Galaxy Distances and Deviations from Universal Expansion.* p. 169. eds. B. F. Madore and R. B. Tully. Dordrecht: Reidel.

Lynden-Bell, D., Faber, S. M., Burstein, D., Davies, R. L., Dressler, A., Terlevich, R. J., and Wegner, G. W. 1988. *Ap J.* **326**, 19.

Meiksin, A. and Davis, M. 1986. *AJ.* **91**, 191.

Postman, M. and Geller, M. J. 1984. *Ap J.* **281**, 95.

Rice, W. *et al.* 1988. *Ap J Suppl.* **68**, 91.

Rowan-Robinson, M. 1988. in *Comets to Cosmology: 3rd IRAS Conference.* ed. A. Lawrence. Berlin: Springer. in press.

Sandage, A. and Tammann, G. A. 1981. *Revised Shapley Ames Catalogue.* Washington: Carnegie Institution.

Sandage, A., Tammann, G. A., and Yahil, A. 1979. *Ap J.* **232**, 352.

Schechter, P. L. 1976. *Ap J.* **203**, 297.

Strauss, M. and Davis, M. 1988. this volume.

Turner, E. E. and Sargent, W. L. W. 1977. *Ap J Letters.* **22**, L3.

Villumsen, J. V. and Davis, M. 1986. *Ap J.* **308**, 499.

Villumsen, J. V. and Strauss, M. A. 1987. *Ap J.* **322**, 37.

Vittorio, N., and Juszkiewicz, R. 1987. in *Nearly Normal Galaxies.* p. 451. ed. S. M. Faber, Berlin: Springer.

Yahil, A. 1985. in *The Virgo Cluster of Galaxies.* p. 359. eds. O. G. Richter and B. Binggeli, Garching: European Southern Observatory.

Yahil, A., Sandage, A., and Tammann, G. A. 1980. *Ap J.* **242**, 448.

Yahil, A., Tammann, G.A., and Sandage, A. 1977. *Ap J.* **217**, 903.

Yahil, A., Walker, D., and Rowan-Robinson, M. 1986. *Ap J Letters.* **301**, L1.

THE PECULIAR VELOCITY FIELD
PREDICTED BY THE DISTRIBUTION OF IRAS GALAXIES

MICHAEL A. STRAUSS and MARC DAVIS

Departments of Astronomy and Physics, University of California, Berkeley

ABSTRACT

We have recently completed a redshift survey of $\sim$ 2300 *IRAS* galaxies selected uniformly over 76% of the sky, and with a characteristic depth of $\sim$ 4000 km s^{-1}. The sky coverage is unprecedented, allowing us to map the mass distribution in the local universe in great detail, under the assumption that *IRAS* galaxies trace the mass. In particular, as the distribution of matter around any nearby galaxy is known, we can predict its peculiar velocity from linear perturbation theory. Our own peculiar acceleration points within 22° of our microwave velocity vector, and the majority of the material inducing our acceleration is located within a redshift of 3000 km s^{-1}. The existence of a Great Attractor in the galactic plane with a power-law radial density distribution that dominates our motion is inconsistent with the convergence of the *IRAS* peculiar acceleration. However, we measure an overdensity in the sphere reaching to us centered on the Great Attractor of 40%, consistent with the prediction of Lynden-Bell *et al.* (1988). The center of mass of this overdensity is displaced $\approx$ 500 km s^{-1} towards us.

The *IRAS* galaxy redshift distribution is completely consistent with that of the observed optical galaxies in the direction of the Great Attractor. *IRAS* predicts large bulk flows with coherence length $\sim 20h^{-1}$ Mpc. The *IRAS* velocity field qualitatively reproduces the recent observations of peculiar velocities of spiral and elliptical galaxies. In particular, the substantial outflows reported for the Hydra-Centaurus-Pavo-Indus region are also qualitatively in agreement with the velocity field expected from the distribution of *IRAS* galaxies. Although we cannot rule out the existence of excess mass in the galactic plane in the direction of the Great Attractor, it is not needed to explain the observed peculiar velocities.

1. INTRODUCTION

The *IRAS* database has in many ways answered the dreams of those who wish to study the large-scale structure of the distribution of galaxies in the local universe. Peebles (1980) shows in linear perturbation theory that our peculiar velocity is directly proportional to the dipole moment of the matter distribution around us, the constant of proportionality depending only on the value of the density parameter Ω. In order to measure the dipole moment directly, one needs a redshift survey of a sample of galaxies covering the entire sky, free from systematic biases between the northern and southern hemispheres. The *IRAS* Point Source Catalog contains some 25,000 galaxies (Soifer, Houck, and Neugebauer 1987) selected in a uniform way over the sky. Furthermore, galactic extinction is negligible in the infrared, so a galaxy catalogue may be compiled from the Point Source Catalog (PSC) with a well-defined flux limit over all parts of the sky that are not limited by confusion.

Meiksin and Davis (1986), and Yahil, Walker, and Rowan-Robinson (1986) were the first to extract galaxy catalogs from the PSC, and showed that the angular dipole moment of the galaxy distribution points within 30° of the peculiar velocity vector of the Local Group, as inferred from the dipole anisotropy of the Cosmic Microwave Background. The implication is that the *IRAS* galaxies trace the mass that gives rise to our peculiar velocity. We have measured redshifts for a sample of objects extracted from the PSC based on the criteria of Meiksin and Davis (1986): high flux quality at 60μm, $F_{60}/F_{12} > 3$, and $F_{60} > 1.936$ Jy. This last condition is imposed to keep the sample to a manageable size. Objects within ten degrees of the galactic plane, as well as a few regions of high-latitude star formation, are excluded to avoid excessive cirrus contamination. A preliminary discussion of our results has appeared in the proceedings of IAU Symposium 130 (Strauss and Davis 1988, hereafter SD88); this paper is an update of that report, and is complementary to the paper of Amos Yahil (1988) in these proceedings. In Sec. 2 we discuss recent changes in our sample. In Sec. 3, we detail the calculation of peculiar velocities and show the effect they have on our peculiar acceleration. A detailed comparison of the *IRAS* density field with the Great Attractor model of Lynden-Bell *et al.* (1988) is presented in Sec. 4. We show full sky maps of the peculiar velocities and compare them to peculiar velocities measured by Aaronson *et al.* (1982), Burstein *et al.* (1987), and Rubin (1988). Sec. 5 contains our conclusions.

2. REFINING THE GALAXY CATALOGUE

There have been several important developments since the writing of SD88. A major worry at that time was the problem of extended sources, objects of sufficiently large angular diameter to be resolved by the *IRAS* beam. The PSC fluxes were determined by fitting the scan across a source to a template of the expected response to a point source, resulting in systematically underestimated fluxes for resolved objects. Most extended sources are flagged as such in the PSC; we sent a list of *all* such objects satisfying our color criterion, regardless of flux, to Tom Soifer and Elizabeth Smith of IPAC, who had them ADDSCANed. This involves creating the median of all scans crossing a given source, and integrating the result between zero-crossings. The results are encouraging; 86 objects were added to our list and the fluxes of 400 others were corrected. As we indicated in SD88, increasing the fluxes of the galaxies flagged as extended makes a big difference in the calculation of our peculiar gravity.

We have also received from Walter Rice co-added *IRAS* fluxes for galaxies with angular diameters greater than 10′ (Rice *et al.* 1988). We have obtained a machine-readable copy of the Point Source Catalog, Version 2.0, and have updated our galaxy sample accordingly. With these changes, our sample consists of 2285 galaxies with fluxes greater than 1.936 Jy at 60μm, of which 29 still require observation.

We discovered a strong anti-correlation between the correlation coefficient (CC), a measure of the goodness-of-fit of the point source template to a scan, and the ratio of ADDSCAN to PSC flux; that is, sources with poor CC have PSC fluxes that are systematically underestimated. There are almost 1400 objects in the PSC satisfying our color criterion with poor CC, but which are not flagged as extended. The ADDSCANing of these sources is in progress; preliminary indications are that virtually all of these sources are associated with foreground cirrus, and thus it is unlikely that these sources present any systematic problem with our catalog.

Finally, the *IRAS* detectors suffered hysteresis after passing over regions of large flux density, in particular after passing over the galactic plane. We are in the process of assessing how large an effect this is.

3. Calculation of and Correction for Peculiar Velocities

Given the distribution of *IRAS* galaxies around us, and armed with the assumption that they trace the mass on the large scale, we can calculate our peculiar acceleration due to the inhomogeneity of that distribution. As explained above, linear theory then relates that directly to our peculiar velocity, allowing us to obtain an estimate for Ω. Because of the large sky coverage of the present sample, we can calculate the gravitational acceleration of other points in space, as we know the distribution of matter around them. In practice, we do the following. From the observations we calculate the number density and selection function of galaxies in the survey, using the methods of Davis and Huchra (1982). The selection function is simply the fraction of the luminosity function sampled at any given redshift in a flux-limited sample. With this we fill the 24% of the sky not covered by the survey (principally the region within 10° of the galactic plane) with random galaxies with the same number density and selection function, yielding a galaxy catalogue with true 100% sky coverage. Nearby dwarf galaxies not observable to at least 500 km s^{-1} are deleted from the analysis, but otherwise the catalogue is flux limited. Each galaxy in the sample is labelled with the value of the selection function at that distance. The peculiar velocity of a point P in the sample is then estimated as:

$$V_P = \frac{H_0 \Omega^{0.6}}{4\pi n_l} \sum_i \frac{1}{\phi(r_i)} \frac{\mathbf{r}_i - \mathbf{r}_P}{|\mathbf{r}_i - \mathbf{r}_P|^3}, \tag{1}$$

where r_i is the vector to galaxy i, $\mathbf{r}_P$ is the vector to galaxy P, n_1 is the galaxy density calculated as in Davis and Huchra (1982), and $\phi(r_i)$ is the selection function at the distance of the galaxy i. The observer is at the origin. Note that the right-hand side of equation (1) is independent of H_0. Thus given a value of Ω, we can estimate the velocity flow field in the local universe. In all of the following, we shall set Ω equal to 1 for simplicity.

Strictly speaking, equation (1) is correct only when the sum extends over all of space, while our survey has a finite depth. We have found (SD88) that our own peculiar acceleration converges within 4000 km s^{-1}, so we carry out the sum in Equation (1) for all galaxies for which 400 km s^{-1} < $|\mathbf{r}_i - \mathbf{r}_P|$ < 5000 km s^{-1}. As our sample becomes very sparse for redshifts greater than 10,000 km s^{-1} we do not compute peculiar velocities for test particles more distant than r_P = 5000 km s^{-1}. The small scale cutoff in the sum is a smoothing intended to eliminate small scale nonlinear behavior, where Equation (1) will not apply. Note that our sample is flux limited, not volume limited, and the shot noise errors in the computed velocity increase with distance, becoming quite substantial by a redshift of 5000 km s^{-1}.

The first thing we may do with this technique is to correct our measured redshifts for the peculiar velocities of the galaxies in our sample. We initially assume pure Hubble flow, and thus put each galaxy at the distance indicated by its velocity. The one exception to this is that galaxies within 6° of the center of the Virgo cluster, with redshifts less than 2500 km s^{-1}, are all positioned at the distance of Virgo; no further Virgocentric correction is applied. The peculiar velocity of each galaxy in our sample within 5000 km s^{-1} is calculated using Equation (1), and the redshift is corrected accordingly. This will change the density field, of course, and we recalculate the peculiar velocity field, continuing until the process converges. We find that using the average of the peculiar velocity of a given galaxy found in the previous two iterations as the initial guess, the algorithm converges to an accuracy of typically 20 km s^{-1} per galaxy within four iterations. During this procedure one must update the estimate of the selection function, which also changes slightly. In

Figure 1, we show how this process affects the calculation of our peculiar acceleration. The quantity plotted is $V\Omega^{-0.6}$ as a function of the distance to which the sum in equation (1) is carried out. This is the analogue to Figure 2 of SD88. The open symbols show our peculiar acceleration due to our sample when each galaxy is placed at its redshift distance, while the solid symbols show the effect of correcting for peculiar motions of each galaxy, and our own motion. The net effect is really quite minor. The amplitude of the acceleration is $V\Omega^{-0.6} \approx 500$ km s^{-1}, and is approximately constant from 3400 km s^{-1} outward after correcting for random shot noise. This amplitude would imply $\Omega = 1.2$ from Equation (1), although we should point out that we had to assume a value for Ω to calculate the peculiar velocities; there is some circular reasoning involved.

The cumulative direction with respect to the microwave velocity averages 22° between 3400 km s^{-1} and 7000 km s^{-1} radius, but shot noise from the

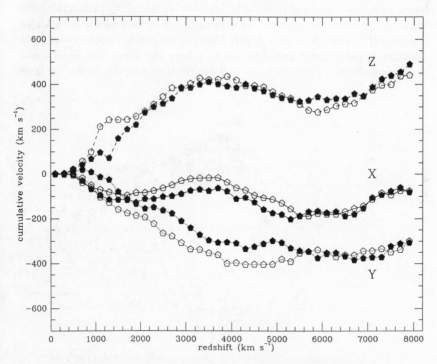

FIG. 1. The three components of the cumulative peculiar gravity. X points toward the galactic center, Y is in the galactic plane and Z is toward the North Galactic Pole. The open symbols result assuming pure Hubble flow. The solid symbols result after correcting each galaxy's distance for its inferred peculiar gravity and for our own motion.

dilute sampling causes the cumulative direction to begin to random walk for radii beyond 7000 km/s. We have not yet determined how to perform the minimum variance extraction of the "asymptotic" direction and amplitude of our acceleration in the presence of this shot noise, particularly since the expected convergence behavior depends on the power spectrum of large-scale perturbations.

We show the radial peculiar velocity field directly in Figures 2a and 2b, in galactic coordinates. In Figure 2a, the radial peculiar velocities (motion relative to the comoving frame) of all galaxies in our sample within 2000 km s^{-1} are indicated; those objects undergoing outflow relative to the comoving frame are indicated with solid symbols, while those pointing towards us are indicated with open symbols. The number of sides on the symbols indicates the magnitude of the peculiar velocity; triangles are for objects with $V_r < 200$ km s^{-1}, squares for $200 < V_r < 400$ km s^{-1}, and so on. The peculiar velocities of the local galaxies exhibit a strong dipole field which points approximately in the direction of our own peculiar motion, indicating that these galaxies are flowing coherently, and that we are taking part in this same motion. Smaller-scale motions are evident in the vicinity of the Ursa Major complex ($l = 150°$; $b = 60°$). Figure 2b shows the peculiar velocities for those galaxies between 2000 and 4000 km s^{-1}; here the flow field breaks into several incoherent zones and is not well described by a dipole. Notice the large

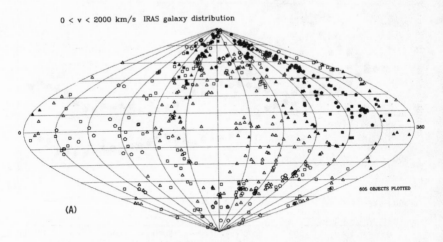

0 < v < 2000 km/s   IRAS galaxy distribution

(A)

360

805 OBJECTS PLOTTED

FIG. 2a. The whole-sky distribution in galactic coordinates of *IRAS* galaxies with redshift v < 2000 km s^{-1}. Open symbols represent galaxies with negative inferred radial peculiar velocity v_r relative to the comoving frame; solid symbols represent galaxies with $v_r > 0$. Triangles plot galaxies with $|v_r| < 200$ km s^{-1}; squares are for galaxies with $200 < |v_r| < 400$ km s^{-1}; pentagons, hexagons, etc. are defined in an obvious sequence.

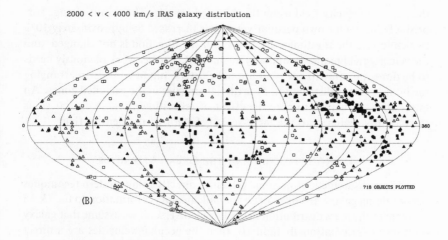

2000 < v < 4000 km/s IRAS galaxy distribution

FIG. 2*b*. The whole-sky distribution in Galactic coordinates of *IRAS* galaxies with redshifts 2000 < v < 4000 km s^{-1}. The symbols are the same as in Figure 2*a*.

outward velocities in the Hydra-Centaurus-Pavo-Indus region ($270° < l < 360°$, $-30° < b < 30°$). This is the region in which large positive peculiar velocities have been reported for the elliptical and spiral galaxies (Lynden-Bell *et al.* 1988; Aaronson *et al.* 1988), and at least qualitatively the *IRAS* maps are consistent with this expectation. A second region of outflow is in the foreground of the massive Perseus supercluster ($90° < l < 200°$; $-45° < b < 15°$). The band of galaxies with $|b| < 10°$ is part of the homogeneous random sample added to bring the survey to full sky coverage.

All of this discussion assumes that *IRAS* galaxies trace the mass distribution. However, we know that this is not true in at least one important respect. *IRAS* galaxies are mostly dusty, late-type systems, and thus our sample is deficient in elliptical and S0 galaxies. The cores of rich clusters are overdense in early-type galaxies, so our sample systematically under-represents the matter distribution in the rich clusters. The well-known nearby cluster cores; Virgo, Hydra, Centaurus, Perseus, Pavo, show lower contrast and are less conspicuous in *IRAS* than in optically selected catalogs. For example, within the 6° radius centered on the Virgo cluster to a limiting redshift of 2400 km s^{-1}, the number contrast relative to background in the CfA optical survey is twice that seen in the *IRAS* survey. We are currently experimenting with schemes to add the missing early galaxies to the *IRAS* sample. One simple scheme is to add galaxies to cluster cores until the density contrast matches that of the CfA survey. Thus we doubled the 33 galaxies within 6° of the core of Virgo, and the 8 galaxies within 3° of the center of Centaurus. We then further iterated the peculiar velocity field using this boosted sample. The results change surprisingly little. Our own peculiar velocity is increased by approximately 10% (which lowers the Ω estimate by 16%), but the direction is not changed, and the velocity field maps are similarly unchanged. Further work obviously needs to be done, not only to boost additional clusters, but to consider rearranging the homogeneous distribution of artificial galaxies in the excluded zones. An interesting test will be to note the effect of repositioning these objects so as to smoothly match onto the large-scale behavior of the density field.

4. IRAS GALAXIES AND THE GREAT ATTRACTOR

The measurement of peculiar velocities directly using modern techniques for obtaining galaxy distances offers complementary information to the IRAS estimate of the mass distribution in the local Universe. If we assume that galaxy motions are gravitationally induced, then the peculiar velocities are a direct measure of the inhomogeneities of the density distribution, regardless of the distribution of galaxies. Thus in principle, the measurement of peculiar velocities may lead to new insights into the relative distribution of dark and luminous matter and the nature of the bias (see Davis and Efstathiou 1988). It then becomes particularly important to compare the predictions of the IRAS galaxies with direct measurements of peculiar velocities.

Certainly a prime motivation for the present conference, and a central topic of much of the discussion, is the model of Lynden-Bell et al. (1988, hereafter collectively called the Seven Samurai; see also Faber and Burstein 1988) of a so-called Great Attractor (hereafter GA). The large outward peculiar velocities seen in the clusters of Centaurus, Pavo, and Telescopium suggest that they lie directly in front of a huge mass concentration. The model is quite specific; this GA is centered at a redshift of 4,350 km s^{-1}, at galactic coordinates $l = 307°$, $b = +9°$, and has a density distribution that drops as the square of the distance from the center. There are several questions that IRAS can hope to answer vis-à-vis the GA model. Do we see any evidence for such a structure in the distribution of galaxies in our sample? Is it possible that there is a very large amount of material hidden by the galactic plane? In particular, we wish to address the Seven Samurai's claim that the GA is responsible for essentially the entire component of our peculiar velocity in that direction. Do we see any evidence for this in the distribution of IRAS galaxies? On the face of things, the answer is no; Figure 1 shows that our peculiar acceleration converges rather nicely by 4000 km s^{-1}, pointing within 22° of our peculiar velocity vector. In Figure 3 we quantify this statement. The points plotted are the amplitude of the induced peculiar acceleration, after correcting for shot noise. The convergence at 4000 km s^{-1} is particularly clear here. The smooth curves are the expected induced acceleration from a GA at $b = 0°$, $v = 4200$ km s^{-1}, with a power-law density distribution $\varrho \propto r^{-\gamma}$, for various

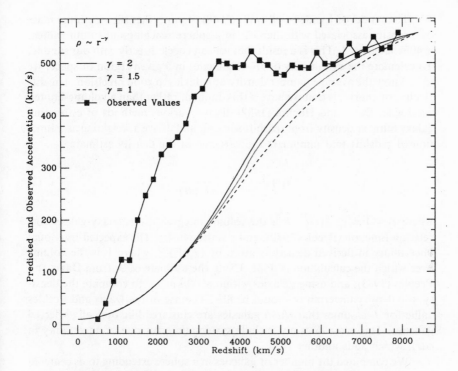

FIG. 3. The plotted squares show the amplitude of our peculiar acceleration as a function of redshift, as calculated from the fully iterated catalogue. A correction for shot noise has been made, so this is not simply the quadratic sum of the components in Figure 1. The smooth curves are the expected peculiar acceleration due to a GA placed at a redshift of 4200 km s^{-1}. The density distribution around the GA was taken to be a power law. The GA was centered squarely on the galactic plane ($b = 0°$), and the material within 10° of the plane was ignored, so as to duplicate the observational situation. All curves are normalized to the same peak value.

values of γ (see Faber and Burstein 1988; they find $\gamma = 1.7$; they also give the GA a core radius, which has not been modeled here.) An excluded zone around the galactic plane twenty degrees thick has been included in the calculation, to mimic our observational situation. All the curves have been normalized to the same peak value. Although putting the GA in the center of the plane minimizes its effect, the smooth curves all rise steeply from 3000 to 6000 km s^{-1}, in sharp contrast to what is observed. It is true that adding excess mass in the vicinity of the GA to our sample will pull the *IRAS* acceleration vector closer to the peculiar velocity vector. However, Figure 3 shows that a model of the GA *which dominates our motion* is not consistent with our results.

The model of Lynden-Bell *et al.* (1988) requires that the fractional mass overdensity associated with the GA, in a sphere reaching out to our radius, is 40% for $\Omega = 1$. This is a result that we can check directly with our sample. To calculate the overdensity of *IRAS* galaxies in a given region of space, we must know the average number density of galaxies to good precision. The difficulty, of course, is that we have a flux-limited, rather than a volume-limited catalogue. Davis and Huchra (1982) discuss various methods of estimating galaxy number density from redshift surveys, and derive a weighting as a function of redshift that minimizes the variance of the density estimate:

$$w(r) = \frac{1}{1 + \bar{n}J_3\phi(r)} , \qquad (2)$$

where $J_3 \equiv 4 \pi \int_0^\infty \xi(r)r^2 \, dr$ is the volume integral of the galaxy-galaxy correlation function (Peebles 1980), and $\bar{n}$ is the density. The expected fractional uncertainty in derived density is given by $(J_3/V)^{1/2}$, where V is the volume over which the calculation is done. Using the estimate of J_3 from Davis and Peebles (1983), and using galaxies within 8000 km s^{-1} to compute the density, the density uncertainty should be 8%. The use of the Davis and Peebles value for J_3 assumes that *IRAS* galaxies are clustered like optically selected galaxies. This error does not include systematic error due to uncertainty in $\phi(r)$, which adds another 10%.

We computed the number of galaxies in a sphere extending to us centered on the position of the GA. The fully iterated distribution was used, in which we filled the galactic plane uniformly, and the counts were weighted according to Equation (2). We found $(\delta\varrho/\varrho)_{GA} = 0.4 \pm 0.2$, a large value indeed. The error quoted is that due to the uncertainty in the background density alone. This overdensity is very insensitive to the value of J_3 used in Equation (2), and does not depend on any absolute luminosity limit placed on the sample. Much work remains to characterize this overdensity and compare it with the Seven Samurai model. The center of mass of the overdensity within this sphere is centered approximately 500 km s^{-1} from the GA in our direction, which can be thought of as the mass-weighted mean position of the rich clusters in the vicinity: Hydra, Centaurus, Pavo, Indus, etc.

Dressler (1988) has done a redshift survey of ESO galaxies in the vicinity of the GA, and shows in his Figure 2 (reproduced in Faber and Burstein 1988) the redshift distribution of 890 galaxies in the region 290° $< l < 350°$, $-35° < b < 45°$, $|b| > 10°$. Two large peaks are seen in the distribution; one at 3000 km s^{-1} is due to the Centaurus cluster, the other is centered at 4500 km s^{-1}, which Dressler cites as direct evidence for the GA.

Figure 4 shows the redshift distribution of the 236 *IRAS* galaxies in the same region of sky as the Dressler survey. *IRAS* is a much more dilute sample than that of the ESO galaxies, but the agreement between the histograms is

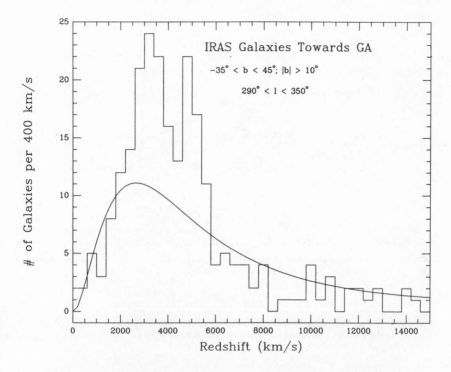

FIG. 4. The radial distribution of *IRAS* galaxies in the region
$290° < l < 350°$; $-35° < b < 45°$; $|b| > 10°$. Compare this plot to Figure
2 of Dressler (1988). The smooth curve plots the expected distribution if the
galaxies were homogeneously distributed with mean density as determined over
the entire *IRAS* sample.

remarkable, implying that the *IRAS* galaxies trace the distribution of the ESO
galaxies very well in this region. The smooth curve is the expected distribution for a homogeneous universe with a mean density as given by the average
over the full *IRAS* sample. Dressler plots the redshift distribution in four
separate zones of galactic latitude within his survey, and again the distribution of *IRAS* galaxies within each region matches his plots within the statistical
uncertainty. From Figure 4 it is apparent that the density contrast in the GA
region is approximately 2 ± 1, similar to the value inferred by Dressler, whose
control sample was less well defined. However, there may be a large amount
of excess hidden in the plane; *IRAS* has the potential to probe the galaxy
distribution even closer to the galactic plane. We are in the process of identi-
fying and obtaining redshifts for *IRAS* galaxies in the range $5° < |b| < 10°$,
which will increase our sky coverage to 85%.

Can the distribution of *IRAS* galaxies in the present survey explain the
large peculiar velocities reported in the literature which motivated the GA

AHM measured peculiar velocities

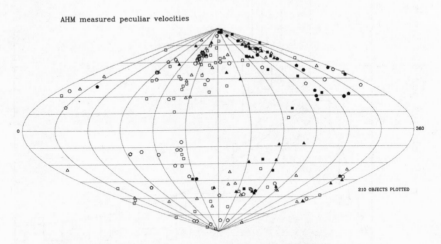

FIG. 5a. The peculiar velocities (relative to the comoving frame) measured using the IRTF distance indicator for the data of AHM.

IRAS predicted velocities, AHM sample

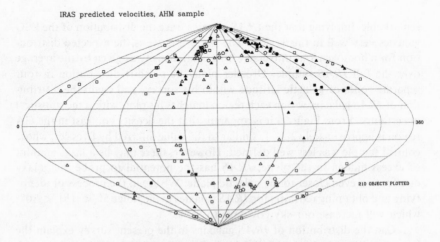

FIG. 5b. The *IRAS* inferred peculiar velocities for the AHM galaxies.

EGALS measured peculiar velocities

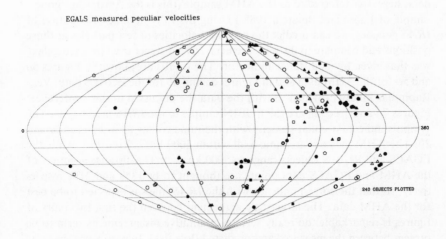

FIG. 6*a.* The peculiar velocities measured by Burstein *et al.* for the EGALS sample.

IRAS predicted peculiar velocities, EGALS sample

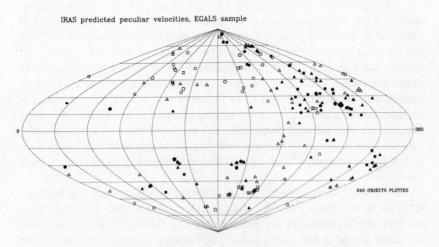

FIG. 6*b.* The *IRAS* inferred peculiar velocities for the EGALS galaxies.

model? David Burstein has kindly sent us the data for the best quality subset of the Seven Samurai program galaxies (EGALS) as well as the subset of the Aaronson *et al.* (1982) spirals having the best optical magnitude and diameter data, hereafter referred to as the AHM sample (this is the Aaronson "good" sample of Faber and Burstein 1988.) Using the fully iterated distribution of *IRAS* galaxies, we can predict the peculiar velocities of test particles at those positions and compare to the measurements. The distance used for each galaxy was that given by its distance indicator. The results are shown in Figures 5*a* and 5*b* for the AHM data and in Figures 6*a* and 6*b* for the EGALS data. Vera Rubin has kindly provided us with the data she discussed in this conference; Figures 7*a* and 7*b* compare the RUBIN sample to the *IRAS* predictions. The symbols indicate peculiar velocity with the same coding as in Figures 2*a* and 2*b*. The AHM galaxies are almost all within 3000 km s^{-1} redshift, while the EGALS and RUBIN objects range to 5000 km s^{-1}. The distance accuracy of the AHM and EGALS data points are thought to be 13% and 21% respectively, so that the correspondence with the *IRAS* maps is expected to be best for the AHM data. The large-scale agreement between the first two pairs of figures is remarkable; no really serious qualitative disagreements seem to be present between the measured and predicted flow field. In particular, the large motions seen in the vicinity of the nominal GA are reproduced very well. The RUBIN sample also gives qualitative agreement with the *IRAS* expectations, but the scatter is somewhat larger. These plots all use the unboosted *IRAS* sample to compute the velocity field; using the boosted sample increases the *IRAS* predicted velocity of many galaxies but changes the sign of relatively few. We are currently studying alternative schemes of boosting the influence of the cluster centers.

The agreement between the *IRAS* predictions and the various observations is made more quantitative in Figures 8—10, which are scatterplots of observed *vs.* predicted peculiar velocities for each of the three data sets. There is a reasonable correlation between the two quantities, with correlation coefficients listed on each figure. Note that the EGALS sample would match the predictions better if the *IRAS* velocities were increased a factor of 2, corresponding to an inferred $\Omega = 3.0$! In Figures 8 and 9, much of the scatter in the correlation is caused by a few outlying points, implying either that these are galaxies with unusual properties, thus rendering their measured peculiar velocities meaningless, or perhaps that they are attracted by matter not well-traced by our sample. Clearly, it will be very interesting to investigate these possibilities.

Error analysis on these graphs is somewhat of a nightmare. The horizontal error in the absence of systematic problems can be derived simply from the fact that the distances are measured to a given fractional accuracy. For the depth of the typical AHM galaxy, this is approximately 300 km s^{-1} rms error. The vertical error bar of each point, assuming the *IRAS* galaxies trace

RUBIN measured peculiar velocities

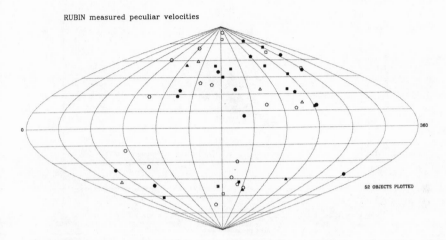

FIG. 7a. The peculiar velocities measured by Rubin and collaborators for the RUBIN sample.

IRAS predicted peculiar velocities, RUBIN sample

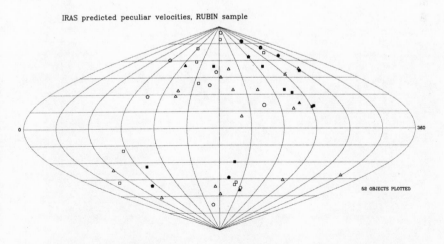

FIG. 7b. The *IRAS* inferred peculiar velocities for the RUBIN galaxies.

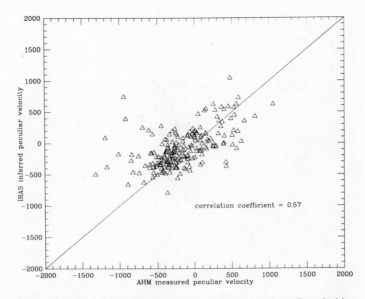

FIG. 8. The scatter plot of AHM measured *vs. IRAS* inferred peculiar velocities
for the AHM galaxies. No attempt at compensating for the underweighted
cluster centers has been made.

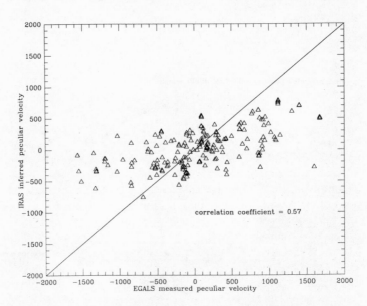

FIG. 9. The same as Figure 8, but for the EGALS sample.

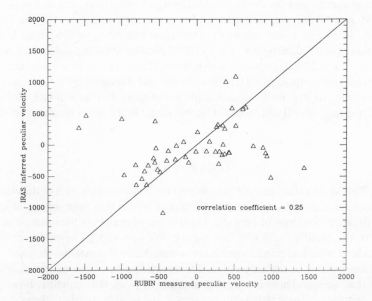

FIG. 10. The same as Figure 8, but for the RUBIN sample.

the mass and that deviations from linear theory are negligible (!?), is comprised of three terms: the shot noise in the computation of the *IRAS* acceleration which is a strong function of the distance of the point, the neglect of the acceleration due to galaxies within 400 km s^{-1} of the galaxies, and the shear in the velocity field given that we have computed the acceleration exactly at the distance given by the secondary estimator which is uncertain at the 10-20% level. Given all that, the 57% covariance in Figure 8 is probably as good as could be hoped for, although it may be that much of the scatter is due to certain regions in which *IRAS* galaxies are poor tracers of the mass distribution. The plots of the EGALS data and the RUBIN data (Figure 9 and 10) show a similar trend but larger scatter because the distance estimators are not quite as precise and the galaxies are more distant. The correlation coefficients for the latter two cases are 0.57 and 0.25 respectively. A preliminary experiment of adding more material in the galactic plane toward the GA direction did *not* improve this scatter, implying that the absence of such a mass concentration in our data is not the source of the scatter. Further work will attempt to quantify this statement.

An alternative analysis is simply to measure the scatter in the diagram of velocity width or dispersion versus absolute luminosity, both with and without corrections for the flowfield. Doing a simple linear regression on the AHM sample for galaxies with $M_H < -19$, assuming pure Hubble flow, results in an rms scatter about the regression line of $\sigma_M = 0.64$ mag. Using distances from redshifts corrected by $IRAS$ peculiar velocities reduces this scatter to $\sigma_M = 0.50$ mag, which is a significant effect. The fit is not substantially improved with a quadratic IRTF law. Faber and Burstein (1988) find a scatter of 0.40 mag or less for the same sample in the context of their model, perhaps implying that there is still room for improvement in the $IRAS$ predicted velocity field.

5. CONCLUSIONS

The $IRAS$ galaxy sample has allowed us to reach a new level in the discussion of peculiar velocities in the universe. For the first time we are able to ask detailed questions of how the distribution of matter in the universe gives rise to the peculiar velocities that we see. We now can go beyond simplistic models of spherical infall to perfectly symmetric isothermal clusters, and can explain motions in the universe as due to the totality of the matter distribution that we see. There is a terrific amount of work that remains, however. The problem of the missing cluster cores in the sample is rather irksome, and our simple attempt at compensating can certainly be improved. We need to quantify our errors in peculiar velocity determinations; the $IRAS$ sample is a very dilute sample of galaxies, and we suspect that we are largely dominated by shot noise in the calculations of peculiar gravity. Quantifying this and other effects will take large Monte-Carlo and N-body simulations. Detailed quantitative comparisons of observed and predicted peculiar velocities in various regions of the sky remain to be done; points of disagreement between the two may teach us interesting things about the relative distribution of $IRAS$ galaxies and dark matter.

At present, we can state several disparate conclusions with regard to the GA model; (a) The convergence of our peculiar acceleration by 4000 km s^{-1} is inconsistent with the GA model; (b) The addition of a density enhancement in the direction of the GA center would bring the $IRAS$ dipole vector closer to the microwave dipole vector; (c) There is a density enhancement in $IRAS$ galaxies of 40 $\pm$ 20% in a sphere centered on the GA position and reaching out to our radius; the center of mass of the overdensity is 500 km s^{-1} from the GA position, displaced in our direction; (d) The redshift distribution of $IRAS$ galaxies in the region agrees very well with that of optically selected galaxies; a large enhancement is seen at $\sim$ 4000 km s^{-1}; (e) We find good agreement between our predicted peculiar velocities and those measured by other workers, *without* the addition of extra material in the plane. In particular, we see coherent flow within 2000 km s^{-1}, and large outflow in Centaurus and Pavo-Indus.

Thus we find ourselves halfway towards an understanding of the GA model. We suspect that the origin of the confusion is the Seven Samurai's assumption that the GA is smooth and spherically symmetric; we look forward to studying this in detail, and we predict that a picture that we all can agree upon will emerge over the next year or two.

Finally, it is time to ask detailed questions of how our results compare with predictions of various cosmological scenarios; the exciting possibility exists that the comparison of predicted and measured flow fields will be able to discriminate between competing theories.

Perhaps the most gratifying aspect of this work to come out of this workshop is the sense that it has inspired those hard-working observers who measure galaxy distances to continue their work with renewed effort. The call for more and better distances is stronger than ever; we look forward to the comparison with new data sets.

ACKNOWLEDGEMENTS

We thank our collaborators John Huchra, Amos Yahil, and John Tonry for permission to use our data before publication. Avery Meiksin is thanked as always for his many helpful suggestions and comments on this project. We thank Dave Burstein for supplying us with the Seven Samurai and AHM data in machine-readable form, and Vera Rubin for her data. MAS gratefully acknowledges the support of a Berkeley Graduate Fellowship. This research was supported by the NSF and NASA.

REFERENCES

Aaronson, M., Bothun, G., Budge, K., Dawe, J., Dickens, R., Hall, P., Mould, J., Murray, J., Persson, E., Schommer, R., and Wright, A. 1988. in *Large-Scale Structures of the Universe*. eds. J. Audouze, M.-C. Pelletan, and A. Szalay. Dordrecht: Kluwer Academic Publishers, p. 185.

Aaronson, M., Huchra, J., Mould, J. R., Tully, R. B., Fisher, J. R., van Woerden, H., Goss, W. M., Chamaraux, P., Mebold, U., Siegman, B., Berriman, G., and Persson, S. E., 1982. *Ap J Suppl.* **50**, 241.

Burstein, D., Davies, R. L., Dressler, A., Faber, S. M., Stone, R. P. S., Lynden-Bell, D., Terlevich, R. J., and Wegner, G. A. 1987. *Ap J Suppl.* **64**, 601.

Davis, M. and Huchra, J. 1982. *Ap J.* **254**, 437.

Davis, M. and Peebles, P. J. E. 1983. *Ap J.* **267**, 465.

Davis, M. and Efstathiou, G. 1988. this volume.

Dressler, A. 1988, *Ap J.* **329**, 519.

Faber, S. and Burstein, D. 1988. this volume.

Lynden-Bell, D., Faber, S. M., Burstein, D., Davies, R. L., Dressler, A., Terlevich, R. J., and Wegner, G. 1988. *Ap J.* **326**, 19. (Seven Samurai).

Meiksin, A., and Davis, M., 1986. *AJ.* **91**, 191.

Peebles, P. J. E. 1980. *The Large-Scale Structure of the Universe*. Princeton: Princeton University Press, Sec. 14.

Rice, W. L., Persson, C. J., Soifer, B. T., Neugebauer, G., and Kopan, E. L. 1988. *AP J Suppl.* **68**, 91.

Rubin, V. 1988. this volume.

Soifer, B. T., Houck, J. R., and Neugebauer, G. 1987. *Ann Rev Astron Astrophys.* **25**, 187.

Strauss, M. A., and Davis, M. 1988. in *Large-Scale Structures of the Universe*. eds. J. Audouze, M.-C. Pelletan, and A. Szalay. Dordrecht: Kluwer Academic Publishers, p. 191 (SD 88).

Yahil, A. 1988. this volume.

Yahil, A., Walker, D., and Rowan-Robinson, M. 1986. *Ap J Letters,* **301**, L1.

IV

SMALL-SCALE MICROWAVE FLUCTUATIONS

SMALL AND INTERMEDIATE SCALE ANISOTROPIES
OF THE MICROWAVE BACKGROUND RADIATION

A. N. LASENBY

Mullard Radio Astronomy Observatory, Cavendish Laboratory

and

R. D. DAVIES

Nuffield Radio Astronomy Laboratories, Jodrell Bank

ABSTRACT

Anisotropies of the Cosmic Microwave Background Radiation (MWB) have, for some years now, been one of the key observational constraints against which all theories of galaxy formation and evolution have had to be tested. In this contribution, we should like to review some of the latest results in this field and perhaps raise some questions arising from them, pertinent to the main theme of this Study Week, namely the organization of structure and velocities on large scales. This is a particularly exciting period in MWB anisotropy studies, with advances in equipment, and the work of many groups around the world, bringing us closer to discovery level across a wide range of angular scales. In particular, recent results from a twin-horn experiment in Tenerife (Davies *et al.* 1987), provide evidence for a possible detection of anisotropy on a scale of 8°, although, as will be indicated below, much work still needs to be done in order to eliminate other possible non-cosmological origins for this signal. It is these new results from several different groups we want to concentrate on, and the discussion will be ordered in terms of increasing angular scale, as follows: Sec. 1. Latest small-scale VLA and Jodrell Bank results ($\theta \lesssim 1'$); Sec. 2. Owens Valley results versus the Uson and Wilkinson experiment ($2' \lesssim \theta \lesssim 7'$); Sec. 3. Tenerife results on intermediate scales ($5° \lesssim \theta \lesssim 8°$).

Then in Section 4 we consider future experimental prospects — what *is* needed to determine $\Delta T/T$ on small and intermediate angular scales? In Section 5 we offer some thoughts on the current relation between theory and experiment.

1. Latest Small-Scale VLA and Jodrell Bank Results

To set the scene for subarcminute scale MWB anisotropies, we recall that the angular scale subtended by a spherical density perturbation of mass M at the epoch of recombination is given by

$$\theta \approx 9' (\Omega^2 h)^{1/3} (M/10^{15} M_\odot)^{1/3}$$

where $h = H_0/50$ km s^{-1} Mpc^{-1}.

Anisotropies on subarcmin scales, therefore, correspond to *galactic* sized masses — $10^{12} M_\odot$ and below. However, on these scales, a variety of processes operate during the epochs up to and including recombination to render such primordial perturbations more or less invisible, at least on current theories. Thus if anisotropies were discovered on subarcmin scales, a more natural interpretation for them would be in terms of secondary effects associated with the later, presumably non-linear, stages of galaxy formation. For example, an integrated Sunyaev-Zeldovich effect from hot gas in the cores of young galaxy clusters (Cole and Kaiser 1988) or (coupled with a Doppler scattering effect) from the early stages of non-linear growth occuring in hydrodynamical theories of galaxy formation (Ostriker and Vishniac 1986) could provide quite sizeable (10^{-5} to 10^{-4} in $\Delta T/T$) fluctuations in the MWB, on scales up to a typical core size, $\sim 1'$, and with a substantial covering fraction with respect to the sky. Note that such fluctuations would not be expected to have the Gaussian signature associated with primordial perturbations.

Two instruments have been used recently in the search for these fine-scale anisotropies; the Very Large Array (VLA) and the Jodrell Bank MkIA-MkII interferometer. At the VLA an exciting position was reached about 18 months ago, with the two independent groups involved both claiming an excess of "noise" near the centre of their maps, correlated with the primary beamshape. With interferometer observations (which are necessary at radio freqencies in order to achieve the fine angular resolution), receiver and atmospheric noise should be spread uniformly over the entire sky plane, and one can look for microwave background fluctuations by looking for a component of the noise that instead peaks towards the centre, where the primary beam envelope means one has greatest sensitivity to real structures on the sky. Unfortunately, imperfections in the interferometer electronics also tend to lead to excess noise near the centre of the map, but by a combination of improvements to the VLA hardware, and much hard work by the two groups aimed at identifying and removing such sources of systematic error, the situation was reached (see e.g. Kellermann *et al.* 1986) where it was clear that there were excess signals near the map centre, due to astronomy rather than electronics. These had a surprisingly large amplitude, $\Delta T/T \sim 10^{-4}$, not easily accounted for by extrapolation of the radio source counts known at higher levels down to the flux levels probed with the new measurements (Martin and Partridge 1987).

However, as revealed by Fomalont and Kellermann at the "ΔT/Tea" meeting at Toronto in May 1987 and Wilkinson at IAU 130 in June 1987, it is clear that in the field used by Fomalont *et al.*, some of the excess is due to radio source emission from a cluster of faint galaxies (detected via a CCD) near the field centre, and that the rest can be accounted for as due to the interaction of the CLEAN algorithm, used in making the final maps, with effects due to weak sources near the confusion limit. Briefly, the missing short spacings occuring in any interferometer system (the minimum baseline on the VLA is 45 m corresponding to a scale of ~ 4.5 arcmin at the 5 GHz frequency of the observations) mean that the synthesized beam is surrounded by an extended negative bowl, so that confusing sources can give negative responses, which look like negative-going microwave background fluctuations, as well as positive ones. This would be accounted for analytically once the source counts (and source size distribution) were known at the appropriate flux levels, except that interaction of the CLEAN algorithm with an initial dirty map containing extended features (the negative bowls and possibly the sources) is nonlinear and unpredictable. Instead, the Fomalont group found that full scale numerical simulations had to be carried out, with simulated data produced and then CLEANed using the VLA package. This revealed that the excess fluctuations *could* be accounted for with reasonable extrapolations of the source counts from higher levels. Thus their results have now become upper limits to $\Delta T/T,$ as follows (Fomalont, "ΔT/Tea" meeting, Toronto, 1987):

Scale	2σ Upper limit
10"	$< 4.4 \times 10^{-4}$
17"	$< 1.6 \times 10^{-4}$
30"	$< 1.1 \times 10^{-4}$
50"	$< 5.0 \times 10^{-5}$

The limit on the 50" scale is in fact very competitive with the best single-dish results so far, and was achieved in far shorter integration time (100 hours versus 174 out of 445 scheduled hours for Uson and Wilkinson [1984a,b] for example), showing that once the new class of problems they introduce has been understood and dealt with, interferometers may well be the best way forward for the future in pushing down $\Delta T/T$ limits (more on this in Section 4).

The final results from the other group using the VLA for these measurements (Partridge *et al.* 1988) should be available shortly. However, they have recently analyzed their data for polarization anisotropies, and have published interesting upper limits on these, at around the 10^{-4} level on a scale of 18″ to 160″ (Partridge *et al.* 1988).

Although care is taken to reduce the effects of systematic errors on the VLA, it would obviously be of interest to have measurements available from an independent system, not so prone to the combination of correlator offsets and dish-dish crosstalk which has made VLA observations difficult. Such a system is the MkIA-MkII broadband interferometer at Jodrell Bank (Padin *et al.* 1987). This operates at 5 GHz with a bandwidth of 380 MHz and telescope spacing of 425 m. Observations totalling 315 hours have been made in 1985 and 1987, resulting in limits to $\Delta T/T$ of about $(8 - 10) \times 10^{-4}$ on scales from 6″ to 50″ (see Figs. 1 and 2, taken from Waymont *et al.* 1988). The novel feature of these observations lies in the improved method of analysis used throughout, which for the first time applies likelihood techniques to interferometer observations. This means that we can give a precise definition of the 'angular scale' on which results are given — it is the coherence angle $\theta_c = (-C''(0)/C(0))^{-1/2}$ for the assumed sky covariance function $C(\theta)$ (see Section 2) and our limits are derived for the rms $[(C(0))^{1/2}]$ of an assumed Gaussian covariance function:

$$C(\theta) = C(0) \exp \left(-\frac{\theta^2}{2\theta_c^2} \right).$$

This gives a slightly pessimistic view of the sensitivity of the experiment, since previous results (e.g. from the VLA) have only been given in terms of limits on the sky rms when already convolved with the interferometer synthesized beam. We have attempted a deconvolution to set limits to the *intrinsic* rms and hence our results will appear to be higher (by a factor $\sim \sqrt{2}$) from this cause. Note that problems of interaction of CLEAN with weak confusing sources do not arise. The fields were chosen, from VLA observations, to be blank at the levels required for the MWB experiment, and no CLEANing was carried out, except for a single source far from the main beam, which was removed via direct manipulations with the uv data.

A further novel feature, is that for the first time we have set limits directly to the power spectrum of microwave background fluctuations (Figs. 1a and

(A) Limits on the power spectrum

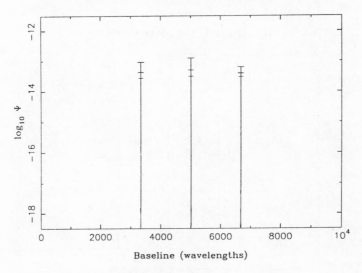

(B) Limits on Gaussian spectrum

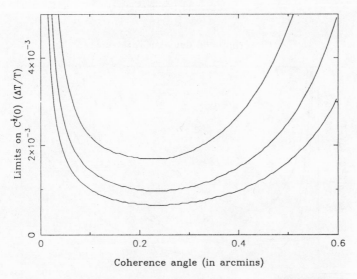

FIG. 1. Limits to small scale MWB anisotropies from Jodrell Bank MkIA-MKII observations made in 1985:

(a) Limits to the power spectrum Ψ as a function of baseline measured in wavelengths. The horizontal bars correspond to 90, 95 and 99% confidence limits respectively (working up from the bottom).

(b) Limits to $\sqrt{C_0}$ as a function of coherence angle θ_c for an assumed Gaussian form of the spectrum. The curves correspond to 90, 95 and 99% confidence limits.

(A) Limits on the power spectrum

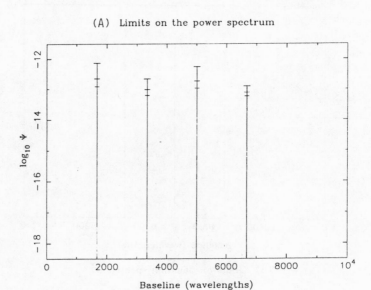

(B) Limits on Gaussian spectrum

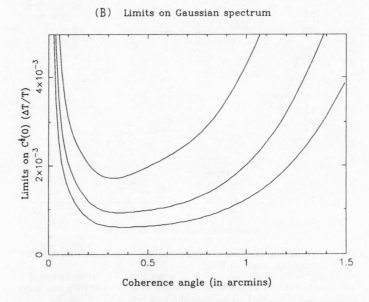

FIG. 2. Same as for Fig. 1a and b except the limits are for the 1987 data. A lower declination field was used so that smaller baselines would be included. Note the increased angular scale coverage in (b).

2a). The ultimate goal of observations of MWB anisotropy is to discover the true power spectrum (fluctuation power versus angular scale) of the microwave sky, and compare it with the theoretical predictions of, for example, cold dark matter (CDM) models. An interferometer provides an opportunity of doing this directly, since the response on a given baseline is directly proportional to the Fourier amplitude at the angular scale corresponding to the baseline. We employ a likelihood technique to provide confidence intervals for the value of the power-spectrum at 3 or 4 coarsely spaced bins of baseline length. For these small angular scales, the relation between 'baseline', measured in wavelengths — call this k — and the spherical harmonic component l more usually used to represent the power spectrum, is given by $l = 2\pi k$. With this change to the labeling of the x-axis, Figs. 1(a) and 2(a) can now be compared directly with predictions for the power spectrum such as those in e.g. Bond and Efstathiou (1987, their Figures 7a and b). If this comparison is carried out, one sees that, as expected on these scales, the theoretical predictions for primordial anisotropy are well below the upper limits, here by orders of magnitude, reflecting the inappropriateness of trying to look for CDM type fluctuations on arcsecond scales. However, we see the importance of the present results as consisting in the way they demonstrate a completely new method of getting from the observations to what we really want to know — the power spectrum. It should be stressed again that this really is model independent — the 'parameters' varied in the likelihood function are the trial values of the power spectrum at the centres of the defined bins, so apart from the unavoidable discretization of the spectrum, it is completely general in its form. We hope to apply such methods to the interferometer experiments with the 5-km telescope, and planned Very Small Array, currently being worked upon at Cambridge (see Section 4).

2. Recent Owens Valley Results and the Uson and Wilkinson Experiment

Readhead and collaborators at Caltech (see for example Readhead *et al.* (1988) and Lawrence *et al.* (1988)), have recently announced first results from a programme of observations using the Owens Valley 40 m telescope at 20 GHz. They have measured the temperature of a central spot relative to the mean of two outside positions at each of 8 centres on a circle 1° from the North Celestial Pole (NCP). The observing configuration involves rapid switching between two beams, which at a slower rate are alternately positioned over the central spot. Each beam has a FWHM of 110″ and the beamthrow is 7.′15. The temperatures they observe are:

RA of Field Centre	ΔT microKelvin
1^h	-57 ± 31
3^h	18 ± 30
5^h	-26 ± 24
7^h	144 ± 25
9^h	20 ± 23
11^h	-21 ± 23
13^h	-18 ± 29
15^h	-32 ± 35

The field at 7^h appears to include a known radio source (and the value is also highly variable), hence this field is discarded before subsequent analysis. A matter of great interest is to be able to compare these results with those from Uson and Wilkinson using the NRAO 140 foot telescope, again at 20 GHz (Uson and Wilkinson 1984a and b). They also used a combination of beam-switching and telescope motion to create a triple beam pattern and had a beam-width of $90''$ and a beamthrow of $4'.5$. The numbers for the twelve Uson and Wilkinson fields are (Wilkinson, private communication):

Field Number	ΔT microKelvin
1	68 ± 128
2	-23 ± 125
3	73 ± 105
4	-68 ± 131
5	-106 ± 138
6	68 ± 169
7	290 ± 181
8	-130 ± 151
9	167 ± 182
10	189 ± 176
11	-54 ± 101
12	102 ± 135

A useful method of comparing these results, and in particular of seeing what angular scales they set strictest limits to (e.g. will these be in the ratio of the beamthrow or the beamsize?), is by means of the likelihood function.

Again assuming the Gaussian form given in Section 1 for the true sky autocovariance function (acf), and that each field is independent from the others (this comes closest to being untrue in the Owens Valley case, where the minimum interfield distance is $\approx 30'$), we have for the likelihood function

$$L(\mathbf{X}|C_0, \theta_c) \propto \prod_{i=1}^{n} \frac{1}{(\sigma_i^2 + \sigma_s^2)^{1/2}} \exp \left\{ \frac{-X_i^2}{2(\sigma_i^2 + \sigma_s^2)} \right\}.$$

Here $X_i \pm \sigma_i$ $(i = 1, \ldots, n; n = 7$ or $12)$ corresponds to the numbers given in the respective tables and σ_s — that part of the rms scatter in the observations due to a real signal from the sky — is derived as follows. Consider the effect of smearing in a single Gaussian beam, with dispersion σ $(\equiv \frac{\text{FWHM}}{2\sqrt{2 \ln 2}})$. We have $C(\theta) \mapsto C_M(\theta)$ where in the Gaussian sky case

$$C_M(\theta) = \frac{C_0 \theta_c^2}{2\sigma^2 + \theta_c^2} \exp \left\{ \frac{-\theta^2}{2(2\sigma^2 + \theta_c^2)} \right\}.$$

Now consider the combination of beamswitching and telescope wagging which gives the triple beamshape. If θ_b is the beamthrow and T_C, T_E, T_W the "central, east, and west" temperatures respectively, we have

$$\Delta T = \frac{1}{2} \{(T_C - T_E) - (T_W - T_C)\} = T_C - \frac{1}{2} (T_E + T_W).$$

and thus

$$\sigma^2 = \text{Var}(\Delta T) = \frac{3}{2} C_M(0) - 2C_M(\theta_b) + \frac{1}{2} C_M(2\theta_b).$$

One can plot contours of likelihood in the (C_0, θ_c) plane and look either for a peak (corresponding to a detection of anisotropy) or, failing this, locate the angular scale of maximum sensitivity, where the contours fall (from a maximum at $C_0 = 0$) most rapidly with C_0. Fig. 3 shows such a contour plot for the data of Uson and Wilkinson. There is no peak away from the origin and the scale of maximum sensitivity is $\theta_c = 1\rlap{.}'8$, A plot of a slice through the likelihood contours at $\theta_c = 1\rlap{.}'8$ is shown in Fig. 4, where the point where the likelihood ratio falls to 1/20th of its peak value (at 0) is shown. This is 156 μK. Without wanting to call this value of $\sqrt{C_0}$ a 'confidence limit', which begs several questions, we do want to take such a value as representative of the sensitivity of an experiment on a given angular scale. As an aside, it may be worth pointing out that having constructed the likelihood function one then has several choices as to how to use it to construct limits at a given level of

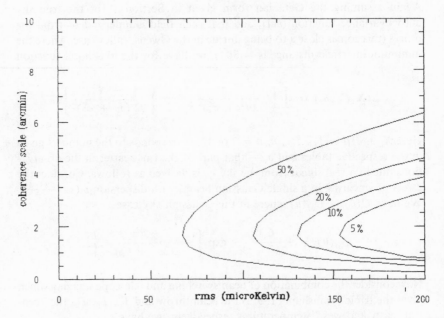

FIG. 3. Contours of likelihood in the ($\sqrt{C_0}$, θ_c) plane for the Uson and Wilkinson data. The percentages are expressed relative to the peak value, which occurs along the axes.

confidence. Currently all such methods — Bayesian, frequentist and fiducial — seem to contain serious problems and inadequacies, but it is arguable that the method with least difficulties, providing a suitable prior can be found, is the Bayesian method. In this spirit, the limits given for the Jodrell Bank results in Section 1 were constructed via a Bayesian integration of the area under the likelihood function multiplied by a prior we believe most suitable for the purpose. In the present context, however, simple comparison of the values of the likelihood functions themselves is sufficient — see Kaiser and Lasenby (1988).

The corresponding contour plot for the Readhead *et al.* (1988) data is shown in Fig. 5. We see immediately that the angular scale of greatest sensitivity is $\theta_c \approx 2.5$, larger than Uson and Wilkinson, (note that the corresponding FWHM is 5.9), with a corresponding 1/20th likelihood point of 56 μK, about three times lower than in the Uson and Wilkinson case. Looking at the 1/20th point of the likelihood as a function of angular scale and thus (essentially by turning the contours on their sides) obtaining a plot of $\sqrt{C_0}$ versus θ_c, a comparison of Fig. 3 with Fig. 5 yields a comparison of the Owens Valley results versus Uson and Wilkinson over a full range of angular scales,

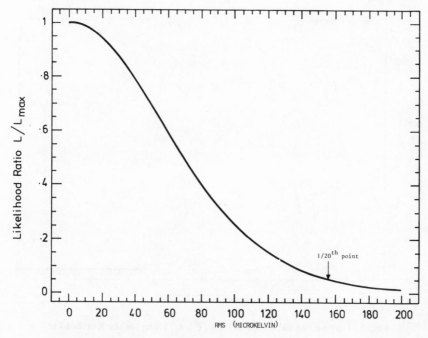

FIG. 4. Slice through the Uson and Wilkinson likelihood contours at the coherence scale of maximum sensitivity, $\theta_c \approx 1.'8$.

and shows that the Readhead *et al.* results *do* represent a very significant advance on the already powerful Uson and Wilkinson upper limits.

Readhead *et al.* (1988) themselves employ a Bayesian method of setting upper limits, with a uniform prior, and find $\Delta T/T \lesssim 1.5 \times 10^{-5}$ at 95% confidence, on a coherence scale of $\approx 2.'5$ [Lawrence has recently (private communication) revised this to 1.9×10^{-5}.] We shall return to the significance of these results in Section 5.

3. Tenerife Results on Intermediate Angular Scales

We should start this section by emphasizing straightaway the collaborative nature of this work, which is carried out jointly with Bob Watson, Ted Daintree and John Hopkins of Jodrell Bank, and John Beckman and Raphael Rebolo of the Instituto de Astrofisica de Canarias. We have been looking at intermediate to large angular scales in the MWB, with resolutions so far of 8° and 5°. It is worth saying that one expects the MWB fluctuations on these scales to be essentially primordial and that the fluctuations in matter density they are associated with will not yet have gone non-linear. This means

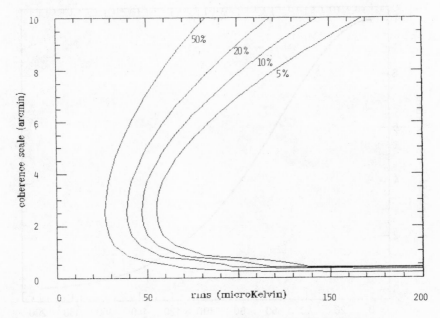

FIG. 5. Contours of likelihood in the ($\sqrt{C_0}$, θ_c) plane for the Readhead *et al.* data.

that on these scales one gets a direct picture of the long wavelength tail of the primordial perturbation spectrum with no complicated physics intervening (including no erasure by possible later reionization, which is not expected to affect scales $\geq 3° - 5°$ on *any* model). Referring to the equation for angular scale in Section 1, we see that the corresponding mass scales are of order 10^{20} $M_\odot$, and therefore there is no immediate evidence (from the structures we see around us) that there should be MWB fluctuations on these scales. However, current theories for galaxy formation, e.g. CDM and baryon isocurvature models (Peebles 1987), *do* predict a long wavelength tail to the perturbation spectrum with $\Delta T/T$'s perhaps 5×10^{-6} in an 8° beam for a CDM model, and somewhat higher (because of the different spectral index needed) in baryon isocurvature models. All such values are strongly dependent on Ω and the power law slope *n* of the primordial perturbation spectrum, and the *detection* of anisotropy on intermediate scales, and knowledge of how it varies as a function of angular scale, would enable one in principle to 'read off' the values of Ω and *n*, and check the details of galaxy formation theory in a quite unprecedented way. The only comparable information we have comes from the peculiar velocity field on large scales, since this is also expected to correspond to perturbations still in the linear regime.

The experimental set-up used for the Tenerife observations consists of a 10. 4 GHz twin horn drift scan system with a wagging mirror. One can adjust the inclination of the mirror to control the declination being surveyed, and then as the earth rotates the triple beam difference $\Delta T = T_C - \frac{1}{2}(T_E + T_W)$, formed as usual via the combination of beam-switching and azimuthal wagging, is recorded automatically at 80 second intervals, building up a 24 hour RA coverage. Data taken when the Sun is above horizon are discarded, which means that observations of a particular declination strip have to be spaced over a year in order to give complete coverage. The analysis of a section of data at Dec 40° has already been published (Davies *et al.* 1987) and we want here to concentrate on the analysis of the *total* data set, which consists of coverages as shown in the following table:

Declination	Number of Scans
45°	22
40°	74
35°	22
25°	21
15°	16
5°	26
0°	57
−5°	12
−15°	25

(Note that there are two independent receivers, and therefore the number of days' observations is the number of scans divided by 2.)

Stacking these implies an rms error on the two regions with greatest coverage (Decs 40° and 0°) of ~ 220 μK per 1° bin of RA. Preliminary versions of the stacked scans for all declinations are shown in Fig. 6. The two galactic plane crossings (one strong, one weak) are clearly visible, illustrating vividly why it is necessary to restrict the analysis for MWB fluctuations to well away from the plane.

The analysis reported previously was of the lowest noise section of the Dec 40° data, from 12^h to 16.6^h in RA, with $|b| > 35°$ throughout. The same likelihood technique as described in Section 2 was used, except that here each "field" overlaps with its neighbours to a very considerable extent, since the beams are ~ 8° wide, but the data are represented by 1° bins. This means one has to compute the likelihood function via a multinormal distribution having a broad diagonal band down the covariance matrix. This turns out still to be feasable however and likelihood contours having a strong peak away

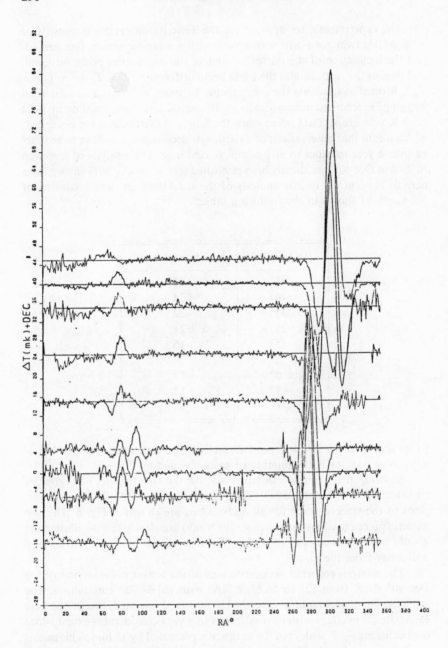

FIG. 6. Preliminary versions of the stacked scans for all the 8° data. The second difference ΔT is plotted against Right Ascension, with the vertical spacing indicating the relative Declinations.

from the origin in the (C_0, θ_c) plane were found (Davies *et al.* 1987) indicating the *detection* of anisotropy, of whatever origin, for this region. The level given was $\sim 3 \times 10^{-5}$ in $\Delta T/T$ in an 8° beam.

It must be recorded here now that there is a slight worry about whether the significance (i.e. height of the likelihood peak) was overestimated in the original analysis due to non-independence of the successive 80 sec data points which go into a 1° RA bin for a given day, perhaps due to large-scale atmospheric irregularities causing offsets with coherence time > 80 secs as they pass across the instrument. Taking such points as statistically independent results in an *under*estimation of the true scatter in the data, and hence an *over*estimation of the significance of any peaks away from zero in the likelihood function. This is currently being investigated, but it appears so far that with proper account taken of possible correlations between these points (by estimating scatter exclusively from variations day to day), the location of the peak in the likelihood contours for the Dec 40° data is not affected, although its significance is reduced.

Another problem with the data is slowly varying baselines, different from day to day but varying on timescales of several hours. These are apparent on visual inspection of some of the data sets and typically have amplitudes of 1 — 2 mK. The instrument is not sensitive to structures on the sky having these scales, so this is not a problem for the 'detection', but the worry is that they make a properly weighted stacking of the data difficult (the results given in Davies *et al.* (1987) were for this reason *un*weighted averages). In order to combine *all* the data from the different declinations properly and deal with the baseline problem, a program has been written which carries out a simultaneous χ^2 minimization with respect to (a) a baseline for each day (represented via Fourier components, with a minimum period of 8 hours) and (b) a two-dimensional model of the microwave sky, from $-30°$ to $+60°$ in declination. Currently it uses a maximum entropy regularizing function as a constraint on the sky model. This is to overcome non-uniqueness problems in the deconvolution of the triple beam pattern and the fact that in some places there is a 10° gap between neighbouring declinations. The results so far look very interesting (Fig. 7a), and suggest that at least as far as the maximum entropy reconstruction goes, there is definite evidence for real structure in the data (as already found statistically via the likelihood technique). Fig. 7a is the result of using the method with the 8° data from declinations 45°, 40° and 35° simultaneously. To help in understanding this grey-scale image (shown with logarithmic levels), also shown, in Fig. 7b, is the equivalent plot for the 408 MHz all-sky survey (Haslam *et al.* 1982), kindly provided in digital form by Dr. J. Osbourne of Durham University. It is clear that the structures near the galactic plane repeat well, but that the variations in the 10.4 GHz data

(a)

8 degree - Dec 40

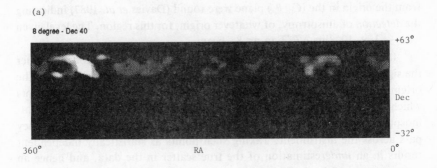

+63°

Dec

−32°

360°   RA   0°

(b)

90°

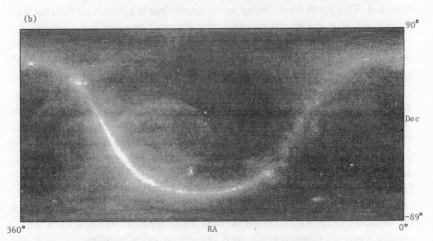

Dec

−89°

360°   RA   0°

408 MHz All Sky Survey - 1 degree resolution

(c)

5 degree - Dec 40

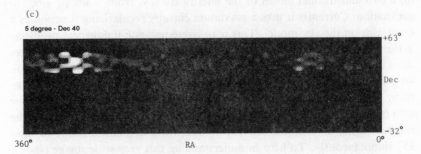

+63°

Dec

−32°

360°   RA   0°

FIG. 7. (a) 2-dimensional reconstruction of the 10.4 GHz microwave sky at Dec 40°, using the 8° data.
(b) 408 MHz all sky continuum survey in the same rectangular projection.
(c) 2-dimensional reconstruction of the 10.4 GHz microwave sky at Dec 40°, using the new 5° data.

at high galactic latitudes do not correspond to what one would predict from the lower frequency data.

Two possibilities immediately present themselves aside from the interpretation that one is looking at a picture of primordial perturbations in the MWB: (a) the maximum entropy reconstruction is producing spurious features, corresponding to an overinterpretation of the 'noise', or (b) the galactic spectral index is *not* uniform, and varies sufficiently to produce the structure seen. Clearly (b) is partially true, in that the heights of galactic plane crossings indicate a .408 — 10.4 GHz index of 2.3-2.4, while in the Dec 0° case, if the high latitude areas had an index this low, we would see much larger fluctuations in the Tenerife data than are observed. These considerations put a lower limit of 2.8-2.9 to the high latitude spectral index. Such a range — from ~ 2.3 at the plane to ~ 2.8 at high latitudes — is consistent with the spectral indices found by Lawson *et al.* (1987), working from lower frequency surveys. It is difficult to repeat the exercise at Dec 40° since the high latitude region here is remarkably flat in the 408 MHz map and the same lower limits to the high latitude index do not arise. Put differently, we believe the possible galactic component is fairly small here (Davies *et al.* 1987). However, to pursue this further, it is obviously necessary to have some data at intermediate frequencies, and to this end observations are being made at Jodrell Bank using a 5 GHz twin horn interferometer system operating in the same drift scan mode as the switched beam system at Tenerife, and with a resolution of ~ 2°. Preliminary results from this are already becoming available, and while they are encouraging for a non-galactic origin of the fluctuations seen at Dec 40°, it is still too early to give a definitive answer. Ultimately, the best way to prove the cosmological origin of the 10.4 GHz fluctuations will be to work at higher frequencies, where the galaxy will unquestionably have sunk to low enough levels, in both thermal and non-thermal emission, to be negligible (see Section 4).

As regards (a), the only satisfactory way is to repeat the experiment to obtain an independent data set, and this has now been done, at a resolution of 5° rather than 8° (achieved by extending the horns, to make them longer and broader. Note the latter has the disadvantage of increasing the sensitivity to discrete radio sources, and these start to be someting that must be taken into account in assessing the 5° results. This will be carried out in future work to be published, but the results presented below have not been corrected for this). The 5° data set has comparable coverage and temperature sensitivity to the 8° data. The results of the χ^2 fitting program for four declinations near 40° (Decs 45°, 40°, 37°.5 and 35°) are shown in Fig 7c. Clearly the resolution is different, but there is an exciting similarity to Fig 7a in the position of the main features in the reconstructuion, suggesting that these features are not mere artifacts of the maximum entropy process. Again, these results are preliminary only, but they point forward to perhaps a new direction in MWB anisotropy studies, where we can start discussing individual features on maps, rather than statistical analyses only.

Before leaving this area, one should mention the interesting agreement between the levels of microwave anisotropy indicated by these studies (upper limits to the primordial anisotropy if the galactic contribution is significant) and the 1981 results of Melchiorri *et al.* who worked with a mm/sub-bolometer system, with 6° beams. Both imply $\Delta T/T$ levels of $\sim 3 \times 10^{-5}$, but in the Melchiorri case also, significant contamination from galactic emission (at their wavelengths due to dust) is probably present, although they attempted to subtract this. Direct comparison of the two data sets is difficult due to the completely different beam switching geometries and scanning techniques, but if a 2-dimensional equivalent single beam reconstruction could be formed from the Italian data, perhaps by methods similar to those used to produce Fig. 7, then the comparison could be very significant.

4. FUTURE EXPERIMENTS

Obviously many groups around the world are currently working on new experiments and here only a limited list will be attempted, comprising experiments related to work we are involved in at our respective institutions. However, two broad lessons for the future can perhaps already be drawn. The first of these is that on small (arcminute to 1°) scales, single dishes are beginning to approach the limits of their usefulness and it is likely that interferometers will be the key to future breakthroughs to sensitivity levels of a few $\times 10^{-6}$ (rather than 10^{-5}) in $\Delta T/T$. One pressing reason for this is that studies as sensitive as those at Owens Valley are now beginning to approach the 'confusion limit', where signals from discrete, weak radio sources, at a level of ~ 1 per beam, begin to limit further progress. An interferometer can avoid this, by providing a tuneable range of angular scales, in which discrete radio sources can be filtered out (on the basis of their small angular size in relation to the MWB fluctuations sought) and sensitivity to fluctuations retained, provided the interferometer has enough short baselines. At Cambridge, the 5-km telescope is being enhanced in sensitivity by a factor 20 (via a combination of cooled front ends and broadband electronics) and the dishes moved closer together to turn it into a premier instrument for observing MWB fluctuations on sub-arcmin and arcmin scales. Exciting results can be expected from this, particularly perhaps in the area of Sunyaev-Zeldovich astronomy (Saunders 1988), in the near future. For work on slightly larger scales, a few arcmin up to maybe 2°, Cambridge is proposing a completely new instrument, the Very Small Array (VSA), which would consist of 10-20 elements in a dense pack configuration, with maximum baseline 12 m, operating at 15 GHz. The angular scales involved here come within the province of the second "lesson", namely that we will see an increasing trend towards experiments on intermediate scales where (a) the currently popular theories indicate the $\Delta T/T$ fluctuation power should be largest (e.g. the coherence scale for CDM fluctuations is ~ 15 arcmin) and (b) as already sketched in Section 3, the physics involved becomes simpler as one goes up in angular scale, with fewer complicated non-linear processes separating us from the genuine primordial perturbations.

On these larger scales, several Jodrell Bank experiments are planned as follow-ups to the 8° and 5° Tenerife experiments. These are: (1) A 10.4 GHz ~ 2° resolution experiment, based on small paraboloids rather than horns, and already in place in Tenerife. [This is in fact an adaptation of the equipment originally used by Mandolesi *et al.* (1986)]; (2) A 15 GHz ~ 4° resolution switched beam twin horn experiment expected to come into operation in Tenerife late 1988; (3) A 30 Ghz ~ 3° resolution switched beam twin horn experiment sited initially in Tenerife, but with a possible future move to the drier observing conditions of Antartica.

It is via instruments (2) and (3) in particular, that the possible fluctuations found at 10.4 GHz can be confirmed as cosmological in origin or not.

5. RELATION TO THEORY

Several aspects of the consequences for theory of anisotropies at a level ~ 3×10^{-5} in $\Delta T/T$ in an 8° beam have already been discussed in Davies *et al.* (1987). Here we just want to highlight what seem to be some pertinent additional current issues in MWB anisotropy studies.

1. The recent Berkeley/Nagoya result (Matsumoto *et al.* 1988), indicating a radiation excess in the Wien region of the MWB spectrum may be about to change completely our picture of conditions at intermediate redshift ($z = 10 — 100$). This will have important consequences for FIR anisotropies and, more relevant to the radio studies discussed here, it may mean we have to consider seriously scenarios in which the universe is reionized completely at early epochs. This would then point attention strongly towards those theories (e.g. the recent baryon isocurvature models) in which reionization is a natural ingredient.

2. Given that reionization occured, how does it affect e.g. the Readhead *et al.* and Uson and Wilkinson results? The recently reported 'Vishniac effect' is exciting in showing us that on small enough scales, fluctuations are actually boosted by the reionization (see contribution by Efstathiou, 1988).

3. In the context of bayron-dominated models with reionization, the joint constraints on small scales (e.g. Readhead *et al.* and Uson and Wilkinson) and intermediate scales (Davies *et al.* and Melchiorri *et al.*) seem close to ruling out $\Omega < 1$ universes completely (again see Efstathiou contribution). In this case, if we want to retain a baryon-dominated cosmology presumably we would have to accept "non-standard" nucleosynthesis, in which there is a segrega-

tion of protons and neutrons following the quark-hadron phase transition (see Applegate and Hogan 1985) in order to avoid previous nucleosynthesis constraints on light element production, which implied $\Omega_{baryon} \lesssim 0.1$.

4. If the Davies *et al.* and Melchiorri *et al.* results are taken as detections of primordial anisotropy, then what are the consequences for predictions of large-scale streaming motions? Obviously this is a key question for this Study Week. Our previous impression has been that both the reported large-scale motions and the MWB anisotropy (if confirmed) would be too large for the CDM picture, and in fact a severe problem for it. This still seems to be true for the MWB anisotropies. However, if what Nick Kaiser says is correct (Kaiser and Lahav 1988), namely that the Great Attractor picture and its associated velocity field are compatible with CDM, then that presumably means that the MWB anisotropy 'detections' at the 3×10^{-5} level on scales of $6° - 8°$ are *too large for the velocity fields we observe*. What would be very helpful is a relatively model independent route from a measurement of MWB anisotropy (on intermediate scales) to a prediction for rms velocities on a given scale, and vice versa. Since this contribution was given, a paper by Juszkiewicz *et al.* (1987) has appeared, providing a means of setting at least lower bounds to each quantity in terms of the other. This apparently confirms that the Davies *et al.* result is bigger than the minimum required by a factor ~ 10, if the coherence scale and amplitude for the velocity streaming as found from the elliptical galaxy studies (Dressler *et al.* 1987) are correct. A coherence scale approximately 3 times larger however, $\sim 180 \, h^{-1}$ Mpc (h here in units of 100 km s^{-1} Mpc^{-1}), would dramatically alter the situation, leading to minimum fluctuations at about the level seen.

5. Finally, will structures such as the Great Attractor or, on the other hand, large voids, about which we have heard much this week, themselves be directly evident in the MWB at detectable levels (rather than via their effect, in surrogate forms, in terms of their partners at earlier epoch)? This is an old subject (see e.g. Rees and Sciama 1968) but also a promising line of enquiry for the future.

ACKNOWLEDGEMENTS

We would again like to thank the many colleagues and collaborators whose work has made the Tenerife observations possible, among them Bob Watson, John Hopkins, Ted Daintree, John Beckman, Jorge Sanchez-Almeida, and Raphael Rebolo.

REFERENCES

Applegate, J.H. and Hogan, C.J. 1985. *Phys. Rev.* **D31**, 3037.

Bond, J.R. and Efstathiou, G. 1987. *MN.* **226**, 655.

Cole, S. and Kaiser, N. 1988. Submitted to *MN.*

Davies, R.D., Lasenby, A.N., Watson, R.A., Daintree, E.J., Hopkins, J., Beckman, J., Sanchez-Almeida, J., and Rebolo, R. 1987. *Nature.* **326**, 462.

Dressler, A., Faber, S.M., Burstein, D., Davies, R.L., Lynden-Bell, D., Terlevich, R.J., and Wegner, G. 1987. *Ap J.* **313**, L37.

Efstathiou, G. 1988, this volume.

Haslam, C.G.T., Salter, C.J., Stoffel, H., and Wilson, W.E. 1982. AA *Suppl.* **47**, 1.

Juszkiewicz, R., Górski, K., and Silk, J. 1987. *Ap J.* **323**, L1.

Kaiser N. and Lahav, 0. 1988, this volume.

Kaiser, N. and Lasenby, A.N. 1988. Preprint.

Kellermann, K.I., Fomalont, E.B., Weistrop, D., and Wall, J. 1986. in J.-P. Swings, editor, *Highlights of Astronomy — Volume 7,* p. 367.

Lawrence, C.R., Readhead, A.C.S., and Meyers, S.T. 1988. *The Post Recombination Universe.* eds. N. Kaiser and A.N. Lasenby (NATO ASI), in press.

Lawson, K.D., Mayer, C.J., Osbourne, J.L., and Parkinson, M.L. 1987. MN **225**, 307.

Mandolesi, N., Calzolari, P., Cortiglioni, S., Delpino, F., Sironi, G., Inzani, P., De Amici, G., Solheim, J.-E., Berger, L., Partridge, R.B., Martenis, P.L., Sangree, C.H., and Harvey, R.C. 1986. *Nature.* **319**, 751.

Martin, H. and Partridge, R.B. 1987. Preprint.

Matsumoto, T., Hayakawa, S., Matsuo, H., Murakami, H., Sato, S., Lange, A.E., and Richards, P.L. 1988. *Ap J.* In press.

Melchiorri, F., Melchiorri, B.O., Ceccarelli, C., and Pietranera, L. 1981. *Ap J.* **250**, L1.

Ostriker, J.P. and Vishniac, E.T. 1986. *Ap J.* **306**, L51.

Padin, S., Davis, R.J., and Lasenby, A.N. 1987. *MN.* **224**, 685.

Peebles, P.J.E. 1987. *Ap J.* **315**, L73.

Readhead, A.C.S., Lawrence, C.R., Meyers, S.T., and Sargent, W.L.W. 1988. *The Post Recombination Universe* eds. N. Kaiser and A.N. Lasenby (NATO ASI). In press.

Rees, M.J. and Sciama, D.W. 1968. *Nature.* **217**, 511.

Saunders, R.D.E. 1988. *The Post Recombination Universe* eds. N. Kaiser and A.N. Lasenby (NATO ASI). In press.

Uson, J. and Wilkinson, D.T. 1984a. *Ap J.* **283**, 471.

_____. 1984b. *Nature.* **312**, 427.

Waymont, D.K., Lasenby, A.N., Davies, R.D., Davis, R.J., and Padin, S. 1988. In preparation.

EFFECTS OF REIONIZATION
ON MICROWAVE BACKGROUND ANISOTROPIES

G. EFSTATHIOU

Institute of Astronomy, Cambridge

SUMMARY

Star formation, or other sources of energy, could have reionized the intergalactic medium at high redshifts. Primary anisotropies in the microwave background would be erased on small angular scales, but new fluctuations would be generated as a direct result of the peculiar motions of the scatterers. We summarize the key physical effects responsible for these secondary fluctuations and show that the new anisotropy limits from Owens Valley set stringent constraints on purely baryonic models with Gaussian isocurvature initial conditions.

1. INTRODUCTION

Limits on the temperature fluctuations in the microwave background radiation have proved to be powerful constraints on theories of galaxy formation. On large angular scales ($\theta \gtrsim 5°$), anisotropies are generated by spatial fluctuations in the curvature (the Sachs-Wolfe effect, Sachs and Wolfe 1967), and by fluctuations in the entropy (the "isocurvature" effect, Efstathiou and Bond 1986). These are insensitive to the recombination history, so we can set tight limits on the shape of the primordial fluctuation spectrum (see *e.g.* the article by Vittorio 1988) without worrying too much about energy injection into the intergalactic medium (IGM), except insofar as this might modify the matter distribution on small scales thereby affecting the normalization of the initial fluctuations.

On angular scales $\lesssim 1°$, the main source of anisotropy arises from the scattering of photons off moving electrons. Inferences from small-scale anisotropy limits are, therefore, extremely sensitive to the assumed recombination history. Most authors have assumed that recombination occurs at the usual time of $z \sim 1000$ (Peebles 1968) and that the IGM remains neutral thereafter. Yet the lack of a Lyα absorption trough in quasar spectra implies that the IGM must have been reionized at redshifts $z \gtrsim 4$. If reionization of the IGM happened early enough, the small-scale anisotropy pattern generated at $z \gtrsim 1000$ (here called the "primary" anisotropies) would have been severely modified.

The possibility that secondary reheating has affected fluctuations in the microwave background radiation deserves to be taken seriously for at least the following reasons:

1. Suppose that the IGM remained fully ionized; the optical depth from Thomson scattering between redshift z and the present epoch would be

$$\zeta = \frac{c\sigma_T H_0}{4\pi G m_p} \, (1 - Y_{He}/2) \, (\Omega_B/\Omega^2) \, (2 - 3\Omega + (1 + \Omega z)^{1/2} \, (\Omega z + 3\Omega - 2)), \quad (1)$$

where Y_{He} is the abundance of Helium (by mass), Ω is the cosmological density parameter at the present epoch, and Ω_B is the contribution of baryons to Ω. In dark matter dominated universes, the redshift z_c at which $\zeta = 1$ can be quite high, e.g., if $\Omega \approx 1$, $\Omega_B \approx 0.1$, $h \approx 0.75$, then $z_c \approx 50$; but for baryon dominated models z_c can be substantially lower. Clearly there is no requirement that energy be injected into the IGM at high redshits ($z \gg z_c$) for reionization to have substantially modified the primary anisotropy pattern.

2. The IGM need not have been heated to a high temperature ($T_e \gg 10^4$K). Correspondingly, there are no general contraints on reionization (at interesting redshifts) based on either the energy outlay or on spectral distortions of the cosmic microwave background.

3. Non-linear structures must have formed before z_c to reheat the IGM (e.g. by photoionization due to massive stars). This would be quite natural in some models of galaxy formation (e.g. isocurvature baryon models).

4. In some specific models (e.g. pancake, or scale-invariant cold dark matter models), the first non-linear structures are expected to form at recent epochs, so reionization at $z \gtrsim z_c$ might seem implausible. However, this depends critically on the assumption of a pure power law initial spectrum extending to small scales. Reionization of the IGM could have occurred if "primordial seeds" (e.g. cosmic strings) caused a small fraction of the baryons to collapse into stars at $z \gtrsim z_c$.

One might imagine that the main effect of reionization at $z > z_c$ would be to erase anisotropies on all scales smaller than the horizon length at z_c (subtending an angle $\theta_c \sim (\Omega/z_c)^{1/2}$). However, just as peculiar motions at $z \sim 1000$ generate small-angle anisotropies, peculiar motions at $z \sim z_c$ can generate significant secondary anisotropies. These secondary temperature fluctuations can be analysed in considerable detail without recourse to complicated numerical calculation. In Section 2 we review computations of velocity-induced temperature fluctuations using first-order perturbation theory. These are generally far below current upper limits for universes with $\Omega_B \gtrsim 0.1$. However, Vishniac (1987) has recently shown that second-order contributions can dominate over the first-order effects on arcminute scales; this effect is reviewed in Section 3.

Throughout, we will apply the results to baryon isocurvature models. Peebles (1988) has mentioned some of the attractive features of this type of model (see also Peebles 1987a). Firstly, baryons are assumed to dominate the mass density of the universe; thus $\Omega \approx \Omega_B$ and there is no need to appeal to exotic forms of dark matter such as axions or gravitinos. Secondly, the value of Ω_B deduced from dynamical studies (e.g. Davis and Peebles 1983, Bean et al. 1983) is in reasonable agreement with the low value implied by the standard theory of primordial nucleosynthesis (Yang et al. 1984). This may indicate that all of the "missing mass" is baryonic and that the universe is open. Thirdly, for certain choices of initial conditions, isocurvature baryon models produce peculiar velocity fields with a large coherence length (Peebles 1987b), consistent with observations of coherent flows reported by, for example, Rubin et al. (1976) and Dressler et al. (1987).

Unfortunately, there is no known physical mechanism that could give rise to entropy fluctuations with an acceptable spectrum. The most likely outcome of an early inflationary phase would be a flat universe with scale invariant perturbations. In a baryon isocurvature model, these initial conditions would lead to large angle temperature fluctuations well in excess of the observational limits (see equation 31 below). Of course, the inflationary picture could be wrong since it is based on a bold extrapolation of particle physics to extraordinarily high energies. If we abandon current ideas on inflation, we are allowed considerable freedom in the choice of initial conditions. For example, to define a baryon isocurvature model, we need to specify the following free parameters and functions: the background cosmology is unconstrained, so we can vary Ω_B, the Hubble constant H_0 and the cosmological constant Λ; the initial spectrum of entropy fluctuations need not be a power-law, but even if it were, the spectral index n and the amplitude are free parameters; we know nothing about the statistics of the initial fluctuations, we have no reason to assume that they are Gaussian; the shape of the post-recombination fluctuation spectrum depends on the history of the IGM. Peebles (1987a) has applied the term "minimal" to the baryon isocurvature model, presumably because hypothetical dark matter particles are not needed, but it should be recognised that a great deal of new physics is required to fix the initial conditions.

There is such a large family of baryon isocurvature models that it would be difficult to test each one against observations using time consuming techniques such as N-body simulations. Nevertheless, these models are worth investigating because it would be extremely important if we could demonstrate that gravitational instability in a purely baryonic universe could not be made to work. Calculations of the microwave background anisotropies provide an inexpensive method of narrowing the options. The calculations of Efstathiou and Bond (1987) show that low density baryon isocurvature models with a standard recombination history are convincingly excluded by anisotropy limits on arcminute scales. This is a weak constraint since, as argued above, reionization is quite likely in these models. We will show below that the new anisotropy limits obtained by Readhead and collaborators at Owens Valley set strong constraints even if the IGM were reionized at $z \geq z_c$.

2. First-Order Velocity-Induced Anisotropies in Reionized Models

The perturbation to the photon distribution function may be expressed as $\delta f_T(\mathbf{q}, \mathbf{x}, t) = T(\partial \bar{f}/\partial T)\Delta_T/4$, where $\bar{f}$ is the Planck function corresponding to temperature T and $\mathbf{q}$ is the comoving photon momentum. The Fourier transform of the equation describing radiative transfer is

$$\dot{\Delta}_T + ik\mu\Delta_T = \sigma_T \bar{n}_e a (\Delta_{T0} - \Delta_T + 4\mu v_B), \quad \mu = \hat{\mathbf{k}}.\hat{\mathbf{q}}, \qquad (2)$$

where a is the cosmological scale factor, dots denote differentiation with respect to conformal time $\tau = \int dt/a$ and Δ_{T0} is the isotropic component of $\Delta_T(k, \mu, \tau)$. In writing equation (2) we have assumed that Thomson scattering is isotropic and we have ignored gravitational source terms on the *rhs* which are unimportant for the short wavelength perturbations of interest here. Furthermore, in this Section we ignore spatial variations in the electron density n_e, *i.e.* we have used the spatial average $\bar{n}_e$ in (2); we shall see in the next Section that fluctuations in n_e can lead to important second order contributions to the small angle anisotropies (Ostriker and Vishniac 1986, Vishniac 1987). The baryon peculiar velocity v_B is related to the fractional perturbation in the baryon density by the equation of continuity

$$v_B = -\frac{\dot{\delta}_B}{ik}. \qquad (3)$$

Since z_c is generally $\ll 150(\Omega h^2)^{1/5}$ (*i.e.* we can ignore Compton drag) but $> (1/\Omega - 1)$ (*i.e.* $a \propto \tau^2$), we may assume that δ_B follows the usual growth rate for linear perturbations in an $\Omega = 1$ universe, $\delta_B \propto \tau^2$.

The solution to (2) is given by

$$\Delta_T(\tau, \mu, k) = \int_0^\tau [\Delta_{T0} + 4\mu v_B]_{\tau'} \, g(\tau, \tau') e^{ik\mu(\tau' - \tau)} \, d\tau', \qquad (4)$$

where

$$g(\tau, \tau') = [\sigma_T \bar{n}_e a]_{\tau'} \, exp \, [-\int_{\tau'}^\tau \sigma_T \bar{n}_e a \, d\tau],$$

and the terms in square brackets are evaluated at τ'. Solutions to (2) in the short-wavelength limit $k\tau \gg 1$ have been discussed by a number of authors (*e.g.* Sunyaev 1978, Davis and Boynton 1980, Silk 1982) who ignored the isotropic term Δ_{T0} on the *rhs*, and by Kaiser (1984) who showed that the isotropic term in (4) contributes a $\Delta T/T$ of similar magnitude to the baryon velocity term.

The zeroth moment of (2) gives

$$\Delta_{T0} = \int_0^\tau \Delta_{T0} g(\tau, \tau') j_0(k(\tau - \tau')) \, d\tau' - \int_0^\tau 4v_B i g(\tau, \tau') j_1(k(\tau - \tau')) \, d\tau'. \tag{5}$$

The integrands in (5) give significant contributions if $k(\tau - \tau') \lesssim 1$, while for $k(\tau - \tau') \gg 1$ the spherical bessel functions oscillate between positive and negative values. For short wavelengths ($k\tau \gg 1$), the slowly varying terms g, Δ_{T0} and v_B may be taken out of the integrals and replaced with their values at τ. It is this "localized" nature which allows a simple analytic solution for short wavelength perturbations (Kaiser 1984). The first integral in (5) is negligible, since it is order $\Delta_{T0}\zeta_\lambda$, where ζ_λ is the optical depth across one wavelength ($\lambda = 2\pi/k$). The isotropic component is thus

$$\Delta_{T0} \approx - \left[\frac{4v_B i}{k} \right]_\tau g(\tau, \tau), \tag{6}$$

and the solution (4) may be written

$$\Delta_T(k, \mu, \tau) \approx \left[\frac{8\delta_B}{(k\tau_c)^2} \right]_{\tau_c} e^{-ik\mu\tau} \int_0^\tau \tau' [g(\tau', \tau') + ik\mu] g(\tau, \tau') e^{ik\mu\tau'} \, d\tau', \tag{7}$$

where we have used equation (3) and assumed $\delta_B \propto \tau^2$.

Our goal is to compute the observed temperature pattern on the sky. This can be described statistically by the temperature autocorrelation function,

$$C_T(\theta) = \langle \Delta T/T(\mathbf{q}, \mathbf{x}, \tau_o) \, \Delta T/T(\mathbf{q}', \mathbf{x}, \tau_o) \rangle, \quad \cos\theta = \hat{\mathbf{q}}.\hat{\mathbf{q}}'.$$

Generally, the background radiation is observed with a finite beam-width which we will approximate as a Gaussian of width σ. A convenient expression for the smoothed autocorrelation function $C_T(\theta, \sigma)$, valid for small angles $\theta \ll 1$, has been given by Doroshkevich, Sunyaev, and Zeldovich (1978),

$$C_T(\theta, \sigma) = \frac{V_x}{32\pi^2} \int_0^\infty \frac{1}{2} \int_{-1}^{+1} |\Delta(\tau, k, \mu)|^2 \, J_o(kR_c\theta(1 - \mu^2)^{1/2})$$
$$\exp(-(kR_c\sigma)^2(1 - \mu^2)) \, d\mu k^2 \, dk, \tag{8}$$

where

$$R_c = L_c \sinh(\tau_0/L_c), \; L_c = H_0^{-1}(1 - \Omega)^{-1/2}a_0^{-1},$$

$\tau \gg \tau_c$, and the perturbations are assumed to be periodic in a large box of volume V_x. Equation (8) can be simplified by noting the following points: (i) For free-streaming photons (neglecting gravity), the quantity $|\Delta_T(k, \mu, \tau)|^2$ is conserved. Provided scattering is unimportant at $\tau \ll \tau_0$, which is generally true (see equation (1)), we may choose any arbitrary epoch τ in (8). (Another consequence of $\tau_c \ll \tau_0$ is that the angle-length relation ($\theta \sim 1/(kR_c)$) is independent of τ_c). (ii) The quantity $|\Delta_T(k, \mu, \tau)|^2$ is a highly peaked function with width $\mu \sim 1/(k\tau_c)$. This can be seen immediately from (7), since the integral is approximately the Fourier transform of a broad function of width $\sim \tau_c$. This simply expresses the fact that for short wavelength plane waves, $k\tau_c \gg 1$, many wavelengths will lie across the last-scattering shell (and so their contributions to $\Delta T/T$ will cancel) unless the wavenumber **k** is oriented almost at right angles to the direction of the observer. We can therefore approximate (8) as

$$C_T(\theta, \sigma) \approx \frac{V_x}{32\pi^2} \int_0^\infty W_T^2(k)J_o(kR_c\theta)\exp(-(kR_c\sigma)^2)k^2 \, dk, \quad (9a)$$

where

$$W_T^2(k) = \frac{1}{2} \int_{-1}^{+1} |\Delta_T(k, \mu, \tau_0)|^2 \, d\mu. \quad (9b)$$

Applying Parseval's theorem to (7) we find

$$W_T^2(k) = \left[\frac{64\delta_B^2(\tau_c)}{(k\tau_c)^5}\right]\pi I_T, \quad (10a)$$

$$I_T \approx \int_0^\infty (G_I - dG_v/dx)^2 \, dx, \; x \equiv \tau'/\tau_c, \quad (10b)$$

and

$$G_I = (\tau/\tau_c)g(\tau, \tau)g(\tau_0, \tau)\tau_c^2,$$

$$G_V = (\tau/\tau_c)g(\tau_0, \tau)\tau_c.$$

The integral (10b) can be evaluated analytically for many interesting ionization histories. For example, if all of the baryonic material remains ionized $\tilde{n}_e \propto a^{-3}$, then

$$g(\tau_0, x) \approx 3/\tau_c x^{-4}e^{(-1/x^3)}, \quad (11a)$$

and so

$$I_T \approx 81 \int_0^\infty x^{-8} e^{(-2/x^3)} = \frac{27}{2^{7/3}} \Gamma(7/3) = 6.38. \qquad (11b)$$

It is also easy to verify that in a fully ionized universe, retaining only the velocity term G_v reduces I_T by a factor of about 2 while retaining only the isotropic term increases I_T by a factor of about 2. As noted by Kaiser (1984), the velocity and isotropic terms tend to cancel in the full integral (10b).

We will now apply these results to purely baryonic universes with isocurvature initial conditions (Peebles 1987a). The initial perturbation in the entropy per baryon is assumed to be a power law, $|S_\gamma|^2 \propto k^n$. The power spectrum for linear baryon density fluctuations at the present epoch is related to $S_\gamma(k)$ by a transfer function $T(k)$,

$$P(k, \tau_o) = \langle |\delta_B(k, \tau_o)|^2 \rangle = T^2(k) |S_\gamma(k)|^2.$$

On scales smaller than about the horizon length at the epoch when radiation and matter have equal densities ($k_{equ}^{-1} \approx 5 \, (\Omega_B h^2)^{-1} \, \text{Mpc}$), $P(k)$ retains the initial power law form $\propto k^n$, but on scales much larger than k_{equ}^{-1}, $T(k) \propto k^2$. Transfer functions for various baryon isocurvature models are plotted by Peebles (1987b) and Efstathiou and Bond (1987). We therefore assume

$$|\delta_B(k, \tau_c)|^2 = Ak^n, \quad k \gg k_{equ} \gg 1/\tau_c, \qquad (12)$$

and so from (10a) $W_T^2 \propto k^{(n-5)}$. From equation (8), we see that the correlation function at zero lag diverges if $n < 2$, i.e. the contribution to $C_T(0)$ is dominated by long wavelength perturbations and so is sensitive to the detailed shape of the transfer function $T(k)$ and to the behaviour of long wavelength perturbations with $k\tau_c \sim 1$. However, the sky variance measured by "triple-beam" arrangements, such as that used in the Uson and Wilkinson (1984) experiment, is related to $C_T(\theta, \sigma)$ by

$$\left(\frac{\Delta T}{T}\right)_\theta^2 = 2[C_T(0) - C_T(\theta)] - \frac{1}{2} [C_T(0) - C_T(2\theta)], \qquad (13)$$

(Bond and Efstathiou 1984) which is, of course, insensitive to long wavelengths and converges if $n > -2$. Evaluating (13), we find

$$\left(\frac{\Delta T}{T}\right)_\theta^2 = \frac{2I_T}{\pi \tau_c^5} \, A V_x (R_c \sigma)^{(2-n)} \varphi^2(n, \theta/(2\sigma)), \qquad (14a)$$

where the function φ^2 is

$$\varphi^2(n, x) = \sum_{m=2}^{\infty} \frac{(-1)^m}{(m!)^2} \Gamma(n/2 - 1 + m)x^{2m}[2^{(2m-2)} - 1], \quad (14b)$$

and is plotted in Figure 1 for three values of the spectral index n. For $\chi \gg 1$, the assymptotic behaviour of φ^2 is

$$\varphi^2(n, x) \to \frac{\Gamma(n/2 + 1)}{\Gamma(2 - n/2)\, n\, (2 - n)} \frac{4}{}[1 - 2^{-n}]x^{(2-n)}, \quad x \gg 1, n \neq 0,$$
$$\qquad\qquad (14c)$$
$$\to 2x^2\ln 2 \qquad x \gg 1, \quad n = 0.$$

To normalize the amplitude of $\Delta T/T$, we match the second moment of the matter autocorrelation function

$$J_3(x_o) = \int_0^{x_o} \xi x^2 \, dx, \qquad (15a)$$

at $x_o = 10h^{-1}$ Mpc, to the value $J_3(10h^{-1} \text{ Mpc}) \approx 270 \, h^{-3} \text{ Mpc}^3$, deduced using the galaxy correlation function of the CfA redshift survey (Davis and Peebles 1983). Thus

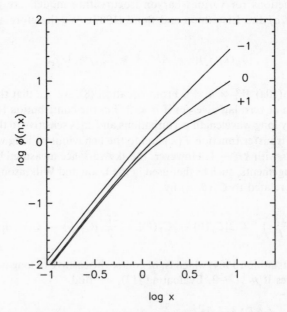

Fig. 1. The function $\varphi(n, x)$ defined in equation (14b) plotted for three values of the spectral index n, $n = -1, 0, +1$.

$$AV_x = \frac{2\pi^2}{p_n^2} \frac{J_3(x_o)x_o^n n}{\Gamma(n+2)\sin(n\pi/2)} \left[\frac{D(\tau_c)}{D(\tau_o)} \right]^2, \tag{15b}$$

where $D(\tau)$ is the growth factor for linear perturbations (Peebles 1980, equation 11.16) and the factor p_n is unity if the power spectrum is a pure power law. Values of p_n are tabulated by Efstathiou and Bond (1987) for various baryon isocurvature models, but even if $\Omega = 1$ and $n \sim -2$, $p_n \lesssim 2$. Notice that equations (14a) and (15b) imply the approximate scaling

$$(\Delta T/T)_\theta \propto h^{-1/6} \, \Omega^{(3n-5)/6}.$$

Thomson scattering can induce a net linear polarization in the microwave background fluctuations (Kaiser 1983). A linear polarization amplitude Δ_P may be defined in an analogous way to the total amplitude Δ_T and obeys the equation

$$\dot{\Delta}_P + ik\mu\Delta_P = \sigma_T \bar{n}_e a [1/2(1 - P_2(\mu)) (\Delta_{P0} - \Delta_{T2} - \Delta_{P2}) - \Delta_P]. \tag{16}$$

If $k\tau_c \gg 1$, the quadrupole component Δ_{T2} is the dominant source term in (16), the polarization amplitudes being smaller by a factor of $\sim \zeta_\lambda$. The second moment of (7) gives

$$\Delta_{T2} \approx - \left[\frac{8\delta_B}{(k\tau)^3} \right]_{\tau_c} \frac{\pi}{4} \, \tau_c \, \frac{d[\tau g(\tau, \tau)]}{d\tau}. \tag{17}$$

Thus,

$$W_P^2 = \frac{9\pi^3}{4} \frac{\delta_B^2(\tau_c)}{(k\tau_c)^7} I_P \tag{18a}$$

where

$$I_P \approx \int_0^\infty G_P^2 dx, \; G_P \equiv \tau_c^2 \frac{d[xg(\tau, \tau)]}{dx} g(\tau_o, \tau). \tag{18b}$$

For constant ionization fraction,

$$I_P \approx (27)^2 \int_0^\infty x^{-16} e^{(-2/x^3)} dx = (27)^2/4.$$

Equations (18) show that the polarization fluctuations induced by optically thin perturbations are negligible, in agreement with the arguments given by Kaiser (1984).

In Table 1, we list values for $(\Delta T/T)_\theta$ for several baryon isocurvature models for the experimental set-up used by Uson and Wilkinson ($\theta = 4.5'$, $\sigma \approx 0.64'$). Typically, the predicted anisotropies are of order 10^{-6} and are well below the level of present experiments on arcminute scales. The first-order anisotropies would only approach the current upper limits if Ω were much less than ~ 0.1 and $n \lesssim -1$. These results agree qualitatively with the conclusions of other workers (e.g. Sunyaev 1978, Kaiser 1984).

TABLE 1

FIRST ORDER VELOCITY INDUCED ANISOTROPIES*

	$\Delta T/T$ ($\times 10^{-6}$)		
n	$\Omega = 0.1$	$\Omega = 0.2$	$\Omega = 1.0$
$+1$	0.7	0.7	0.5
0	1.3	0.9	0.3
-1	3.8	1.7	0.4

* Computed for the Uson and Wilkinson (1984) experiment for baryonic models with h = 0.5.

An analysis of the regime $k\tau_c \sim 1$ requires a full numerical solution of the Boltzman equation. In Figure (2a), we show the behaviour of $W_T^2(k)$ for two baryon isocurvature models with $\Omega = 0.2$ and $h = 0.5$. The solid line shows $W_T^2(k)$ for a model with standard recombination history while the dotted line shows what happens if the matter stays fully ionized (Efstathiou and Bond 1987). The k^{-5} tail predicted by equation (10) for the latter case is shown clearly in Figure (2b). The prominent peaks in Figure (2a) arise from Doppler contributions at wavenumbers $k \sim 2\pi/\tau_{rec} \sim 10^{-2}\,\mathrm{Mpc}^{-1}$ for standard recombination and $k \sim 2\pi/\tau_c \sim 10^{-3}\,\mathrm{Mpc}^{-1}$ in the fully ionized case. A prominent peak at $k\tau_c \sim 1$ is a characteristic feature of the numerical solutions.

Figure 3 shows a similar comparison between two adiabatic, scale-invariant cold dark matter models with $\Omega = 1$, $\Omega_B = 0.1$, $h = 0.5$. The solid lines show the case of standard recombination (see also Bond and Efstathiou 1987) and the dotted lines correspond to a universe in which the baryons never recombine. As in Figure 2, the fully ionized model displays a large peak in both W_T^2 and W_P^2 at wavenumbers $k \sim 2\pi/\tau_c \sim 4 \times 10^{-3}\,\mathrm{Mpc}^{-1}$. In fact, the values of the correlation functions at zero lag for these models (normalized according to the "mass traces light" assumption described above) are $C_T^{1/2}(0) = 3.5 \times 10^{-5}$, $C_P^{1/2}(0) = 2.0 \times 10^{-6}$ (standard recombination) and $C_T^{1/2}(0) = 2.3 \times 10^{-5}$, $C_P^{1/2}(0) = 1.7 \times 10^{-6}$ (fully ionized). Reionization has indeed erased the temperature fluctuations from short wavelengths, but the velocity

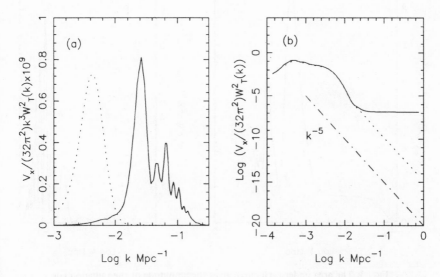

Fig. 2. Figure 2a shows $k^3 W_T^2(k)$ (including only first-order velocity induced anisotropies) for two baryon isocurvature models with $n = 0$, $\Omega = 0.2$, $h = 0.5$. Results for standard recombination are shown by the solid line while the dotted line shows results for a fully ionized universe. For clarity, the solid line has been divided by 25. Figure 2b shows $W_T^2(k)$ on a log-log plot for the fully ionized model shown in Figure 2a. The dotted line shows the k^{-5} tail predicted by equation (10a) describing first-order velocity induced anisotropies. The second-order velocity induced anisotropies described in Section 3 give rise to the constant offset (for $n = 0$) in $W_T^2(k)$ at large $\mathbf{k}$.

induced secondary fluctuations at $k \sim 1/\tau_c$ lead to *rms* temperature and polarization fluctuations which are almost as large as in the case with standard recombination. From equation (18) we could have anticipated relatively large polarization fluctuations at $k \sim 1/\tau_c$. Polarization amplitudes at the level of a few percent, coherent over angles $\sim (\Omega/z_c)^{1/2}$, are characteristic of reionized models. Since polarization is a direct result of Thomson scattering, the coherence angle for linear polarization provides a measure of the width of the last scattering shell. A polarization pattern coherent on scales $\gtrsim 10' \Omega^{1/2}$ would provide very strong evidence that reionization had occurred (Hogan, Kaiser, and Rees 1982, Efstathiou and Bond 1987).

The results of this Section may seem a little depressing; the anisotropies predicted for experiments with beam throws of $\sim$ arcminutes are extremely small and would in any case be difficult to observe given the expected confusion at cm wavelengths from faint radio sources (Danese, de Zotti, and

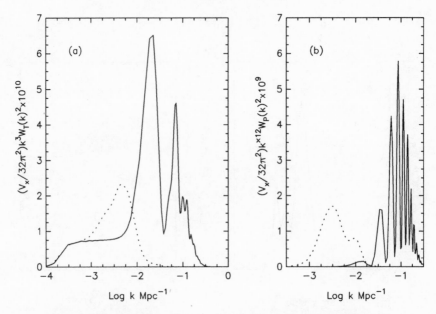

FIG. 3. The area under each curve gives the amplitude of the radiation correlation at zero lag $(C(0))$ for a scale-invariant adiabatic cold-dark matter universe with $\Omega = 1$, $\Omega_b = 0.1$, $h = 0.5$. The solid lines show results for the case of standard recombination and dotted lines show what happens if the IGM stays fully ionized. Figure 3a shows the integrand for $C_T(0)$ while Figure 3b shows the integrand of the polarization autocorrelation function $C_P(0)$.

Manolesi 1983). Substantial secondary anisotropies are generated for relatively large wavenumbers, of size comparable to the horizon at z_c, and these could be observed by experiments with beam widths of $\sigma \lesssim 1°$ and beam throws $\gg \sigma$. However, such experiments have not been performed. Fortunately, there is yet another physical effect, related to the velocity induced anisotropies analysed here, which can produce larger fluctuations on arcminute scales and may well allow us to exclude a wide class of purely baryonic models. This is described in the next Section.

3. SECOND-ORDER VELOCITY-INDUCED ANISOTROPIES

The first-order effects described above give a small contribution to $\Delta T/T$ on small angular scales; as we have shown in the previous section, the temperature perturbation in linear theory from a single plane wave is small because the contributions from peaks and troughs cancel apart from the slow evolution of the peculiar velocity and optical depth in the light travel time across a wavelength. Ostriker and Vishniac (1986) and Vishniac (1987) have pointed out that second-order effects can, under certain circumstances, dominate over the first-order anisotropies. The physical origin of Vishniac's effect can be understood as follows. If the electron density n_e is spatially variable, then we may see a net $\Delta T/T$ on a patch of the sky which depends on the motions of the scatterers along the line-of-sight and on the fluctuation in electron density within the beam. To compute the effect on $\Delta(k, \mu, \tau)$ we are therefore required to perform a "mode-coupling" sum over products of the baryon velocity and the fractional perturbation in n_e.

The equation of radiative transfer, before Fourier transforming, is

$$\dot{\Delta}_T + \gamma_\alpha \partial \Delta_T/\partial x^\alpha = \sigma_T n_e a [\Delta_{T0} + 4\gamma_\alpha v^\alpha - \Delta_T], \tag{19}$$

where the γ_α are the direction cosines defined by the photon momentum $\hat{q}$. If we define the fractional perturbation in the electron density δ_e by

$$n_e = \bar{n}_e(1 + \delta_e), \tag{20}$$

and Fourier transform (19) with $\mathbf{k}$ along the z-axis, we recover equation (2) together with additional second-order terms arising from the fluctuating part of n_e. The most important of these second-order terms is retained in the following equation

$$\begin{aligned}
\dot{\Delta}_T(k) + i\mu k \Delta_T(k) &= \sigma_T \bar{n}_e a [\Delta_{T0}(k) + 4\mu v_B(k) - \Delta_T(k)] \\
&\quad + \sigma_T \bar{n}_e a [4\hat{\gamma} \cdot \sum_{k'} v_B(\mathbf{k}') \delta_e(\mathbf{k} - \mathbf{k}')].
\end{aligned} \tag{21}$$

If we now split $\Delta_T(k)$ into first-and second-order pieces,

$$\Delta_T(k) = \Delta_T^1(k) + \Delta_T^2(k),$$

the solution for $\Delta_T^2(k)$ is,

$$\Delta_T^2(k, \mu, \tau) \approx 4 \int_0^\tau [\hat{\gamma} \cdot \sum_{k'} v_B(\mathbf{k}') \delta_e(\mathbf{k} - \mathbf{k}')] g(\tau, \tau') e^{ik\mu(\tau' - \tau)} \, d\tau'. \tag{22}$$

For simplicity, we will assume that $\delta_e(\mathbf{k}) = \delta_B(\mathbf{k})$; this is reasonable on scales where the matter fluctuations are linear if most of the baryonic material remains in a diffuse ionized form. Symmetrizing the sum in (22) we obtain

$$\begin{aligned}
\Delta_T^2(k, \mu, \tau) \approx -\frac{4}{i} e^{-i\mu k\tau} \bigg[&\frac{1}{\tau_c} \sum_{k'} \delta_B(\tau_c, \mathbf{k}') \delta_B(\tau_c, \mathbf{k} - \mathbf{k}') \hat{\gamma} \cdot \mathbf{D}(\mathbf{k}, \mathbf{k}') \\
&\int_0^\tau (\frac{\tau}{\tau_c})^3 g(\tau, \tau') e^{ik\mu(\tau')} \, d\tau' \bigg],
\end{aligned} \tag{23}$$

where

$$\mathbf{D}(\mathbf{k}, \mathbf{k}') = \left(\frac{\mathbf{k}'}{|\mathbf{k}'|^2} + \frac{\mathbf{k} - \mathbf{k}'}{|\mathbf{k} - \mathbf{k}'|^2} \right).$$

As for equation (7), the integral in (23) may be approximated as a Fourier transform $\hat{G}^2(k\mu\tau_c)$. Assuming random phases,

$$|\Delta_T^2(k)|^2 = 32 \sum_{\mathbf{k}'} |\delta_B(\tau_c, \mathbf{k}')|^2 |\delta_B(\tau_c, \mathbf{k} - \mathbf{k}')|^2 (\hat{\gamma}.\mathbf{D}(\mathbf{k}, \mathbf{k}'))^2 \frac{|\hat{G}^2(k\mu\tau_c)|^2}{\tau_c^2}.$$

Thus, the contribution to W_T^2 (equation 9b) may be written

$$W_T^2$$
$$= \frac{32\pi}{k\tau_c^3} I_1 \frac{V_x}{(2\pi)^3} \int |\delta_B(\tau_c, \mathbf{k}')|^2 |\delta_B(\tau_c, \mathbf{k} - \mathbf{k}')|^2 (\hat{\gamma}.\mathbf{D}(\mathbf{k}, \mathbf{k}'))^2 \, d^3\mathbf{k}',$$
$$\text{(24a)}$$

where

$$I_1 = \frac{1}{2\pi} \int_{-1}^{+1} |\hat{G}^2(k\mu\tau_c)|^2 \, d(k\mu\tau_c). \tag{24b}$$

The integral in (24a) may be expressed in terms of the power-spectrum of the baryon perturbations at τ_c, $P(k, \tau_c)$, and one of the angular integrals may be performed analytically giving,

$$W_T^2 = \frac{4V_x P^2(k, \tau_c)}{\pi \tau_c^3} I_1 I_2(k) \tag{25a}$$

where

$$I_2(k)$$
$$= \int_0^\infty \int_{-1}^{+1} dy\,d\mu \, \frac{(1 - \mu^2)(1 - 2\mu y)^2}{(1 + y^2 - 2\mu y)^2} \frac{P[k(1 + y^2 - 2\mu y)^{1/2}, \tau_c]}{P(k, \tau_c)} \frac{P[ky, \tau_c]}{P(k, \tau_c)}.$$
$$\text{(25b)}$$

The integral I_1 may be evaluated using Parseval's theorem,

$$I_1 \approx \int_0^{\tau_o} (\tau/\tau_c)^6 (g(\tau_o, \tau)\tau_c)^2 \, d(\tau/\tau_c). \tag{25c}$$

If all of the matter remains ionized, (25c) gives

$$I_1 \approx 9\Gamma(4/3)/2^{1/3} \approx 6.38. \tag{26}$$

The second-order contributions to the polarization may easily be computed following the procedures outlined in Section (2) and above. The result is

$$W_P^2 = \frac{9\pi V_x P^2(k, \tau_c)}{64\tau_c^3(k\tau_c)^3} I_3 I_2(k),$$ (27a)

where

$$I_3 \approx \int_0^{\tau_o} (\tau/\tau_c)^6 (g(\tau, \tau)g(\tau_o, \tau)\tau_c^2)^2 \, d(\tau/\tau_c)$$ (27b)

and is approximately equal to

$$I_3 \approx 27/4$$ (28)

for constant ionization fraction. As with the first-order contribution to the polarization described in Section 2, the second-order anisotropies do not have a significant polarization signature because $k\tau_c \gg 1$ on the scales of interest.

For the baryon isocurvature models, the integral I_2 does not diverge at long wavelengths for $n < -1$ because the transfer function falls steeply for $k \lesssim k_{equ}$. For short wavelengths I_2 is approximately constant, so W_T^2 in (25a) will vary as a power-law $\propto k^{2n}$. The radiation correlation function may be evaluated from (9a)

$$C_T(\theta, \sigma) = (AV_x)^2 \frac{I_1 I_2}{16\pi^3 \tau_c^3} \frac{\Gamma(n + 3/2)}{(R_c\sigma)^{(2n+3)}} \, {}_1F_1(n + 3/2; 1; -\theta^2/4\sigma^2).$$ (29)

If $n > -1.5$ the contribution to $C_T(0, \sigma)$ is dominated by short wavelength perturbations and converges for a finite beam width σ. For $\theta \gg \sigma$, the degenerate hypergeometric function in (19) varies as $\theta^{-(2n+3)}$ for $n \neq -0.5$ and as $\exp(-\theta^2/4\sigma^2)$ for $n = -0.5$ (there is a sign change in the assymptotic behaviour at $n = -0.5$). Thus if $n > -1.5$ a double, or triple beam experiment will essentially measure uncorrelated noise and the amplitude of the temperature fluctuations will be determined by the size of the beam width. Note that equation (29) implies the approximate scaling

$$C_T^{1/2}(0, \sigma) \propto h^{5/6} \Omega^{(2+3n)/3}.$$

However, the actual scaling with Ω in the results described below is somewhat weaker than implied by this relation (this is because $C_T^{1/2}$ depends on the square of the growth factor D).

Readhead and collaborators have recently made sensitive measurements at Owens Valley which are potentially capable of detecting the fluctuations predicted by (29) (Hardebeck, Lawrence, Moffel, Myers, Readhead and Sargent, in preparation; see also Sargent 1987). A full discussion of this experiment should properly be postponed until their final analysis is published, but we can get a good idea of what their results imply as follows. Their experimental arrangement resembles the triple beam arrangement of Uson and Wilkinson; the temperature difference measured for each of eight fields is the difference between the temperature at a field point and the mean temperature in two reference beams covering arcs of 30° on the sky at 7.15′ on either side of each field point. Their beam width is $\sigma \sim 0.78'$ (1.8′ FWHM), and their experiment suggests an upper limit of 1.5×10^{-5} at the 95% confidence level (Readhead, private communication). If $n > -1.5$, we can neglect correlations on scales of order of the beam throw; furthermore, since the signal in the reference beams is averaged over a relatively large area it should be a good approximation to assume that their sky variance estimate is given by $\langle (\Delta T/T)^2 \rangle \sim C_T(0, \sigma)$. In Table 2 we list values of $\sqrt{C_T(0, \sigma)}$ computed from (29) for their beam width. The results are typically an order of magnitude larger than those listed in Table 1 and in many cases exceed the Owens Valley limit.

TABLE 2

SECOND ORDER VELOCITY INDUCED ANISOTROPIES*

| | $C^{1/2}(0, \sigma)$ ($\times 10^{-5}$) | | | |
n	$\Omega = 0.1$	$\Omega = 0.2$	$\Omega = 0.4$	$\Omega = 1.0$
0	0.1	0.3	0.9	3.0
-0.5	0.2	0.4	1.1	1.7
-1	0.9	1.1	1.7	1.4

* Computed for a beam width $\sigma = 0.78'$ for baryonic models with h = 0.5.

4. DISCUSSION

Figure 4 shows a schematic plot of the constraints on n and Ω implied by the Owens Valley results. The contours in the figure correspond approximately to the positions where $\Delta T/T = 1.5 \times 10^{-5}$ from the first-and second-order anisotropies discussed above. As Tables 1 and 2 show, the predictions are fairly slowly varying functions of the parameters n and Ω, so the region excluded by the contours should not be taken too literally. For example if we had plotted contours at twice the Owens Valley limit, we would hardly exclude any of the area shown in Figure 4. A more detailed statistical comparison with the observations (e.g. using maximum likelihood) is required to establish accurate limits on the models. Figure 4 has been plotted for $h = 0.5$; the scal-

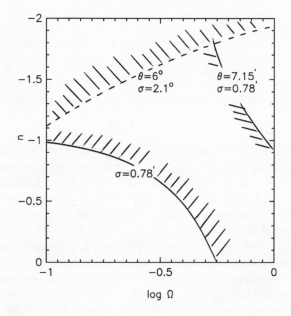

FIG. 4. Diagram showing approximate constraints on baryon-dominated isocurvature models (with $h = 0.5$). The shaded regions show areas excluded by limits on microwave backround anisotropies. The dotted line shows the region excluded by limits on angular scales of $6°$ ($\Delta T/T \lesssim 5 \times 10^{-5}$, Melchiorri *et al.* 1981), while the solid lines show regions excluded by the recent Owens Valley experiment ($\Delta T/T \lesssim 1.5 \times 10^{-5}$).

ing derived in Section 3 shows that that the anisotropies would have nearly twice the amplitude if $h = 1$.

The dashed line in Figure 4 shows the constraints imposed by the experiment at $6°$ (interpreted as an upper limit of $\Delta T/T < 5 \times 10^{-5}$) performed by Melchiorri and collaborators (Melchiorri *et al.* 1981). For wavenumbers $k < k_{equ}$, the temperature perturbations can be written in terms of the initial entropy fluctuation spectrum,

$$|\Delta_T(k, \mu, \tau_o)|^2 = (8/5)^2 |S_\gamma(k)|^2, \tag{30}$$

where 4/3 of the factor 8/5 comes from the initial entropy perturbation and the remaining 4/15 comes from the Sachs-Wolfe effect arising from the gravitational potential fluctuations induced by the motion of the matter relative to the radiation. From equation (8), we derive the following expression for the temperature autocorrelation function at large angular scales,

$$C(\theta, \sigma) = \frac{\sqrt{\pi}}{25} \frac{1}{(R_C\sigma)^{(n+3)}} \frac{1}{(fp_n)^2} \frac{J_3(x_o)x_o^n n}{\Gamma(n + 2)\sin(n\pi/2)}$$

$$\Gamma((n + 3)/2) \frac{\Gamma(-(n + 1)/2)}{\Gamma(-n/2)}$$

$$_1F_1[(n + 3)/2; 1; -\theta^2/4\sigma^2], \quad n < -1, \quad \sigma > \tau_c/R_c \sim (\Omega/z_c)^{1/2},$$

$$(31)$$

where the factor f is the total linear growth factor for short wavelength perturbations ($T(k) = f$, $k \gg k_{equ}$ and is tabulated by Efstathiou and Bond (1987). (If $n > -1$, the amplitude of $C_T(\theta, \sigma)$ for $\theta \sim \sigma$ depends on the way that $|\Delta_T(k, \mu \tau_o)|^2$ departs, at short wavelengths, from equation (30)). Equation (31) shows that spectral indices $n \leq -2$ lead to excessive anisotropies at large angular scales.

Lasenby (1988) has described results of an experiment performed at Tenerife using a frequency 10.4GHz and a beam throw of 8.3° (beam width $\sigma = 3.5°$). The results indicate a *positive* detection of anisotropies at the level $\Delta T/T \sim 4 \times 10^{-5}$ (Davies *et al.* 1987). It is extremely important to determine whether this is a detection of intrinsic fluctuations in the microwave background, rather than radio continuum emission from the Galaxy. This is currently being checked by Lasenby and collaborators with further experiments at angular scales of 5° and at frequencies of 5GHz. In fact the Melchiorri *et al.* (1981) experiment also detected fluctuations at the level of $\Delta T/T = 4.1 \pm 0.7 \times 10^{-5}$. Since they used a bolometer with a wide spectral coverage in the infrared (500 — 3000μm; see Fabbri *et al.* 1982 for experimental details), we have conservatively interpreted this measurement as an upper limit of 5 $\times 10^{-5}$ because it is quite likely that the detection is caused by dust in the Galaxy. However, it is interesting that the detection is in qualitative agreement with the Tenerife results.

The Tenerife experiments lead to very similar constraints as those for the Melchiorri *et al.* (1981) experiment shown in Figure 4. If these two experiments have detected primordial anisotropies in the background radiation, then the results apparently conflict with the Owens Valley upper limits unless $\Omega_B \approx 1$ and $n \approx -1.5$. If we really are forced to high Ω_B, then the baryon isocurvature model loses many of its attractive features; a high value of Ω_B is incompatible with standard primordial nucleosynthesis and conflicts with local determinations of Ω unless galaxies are more clustered than the mass distribution. There are ways of avoiding this conclusion, for example, we have assumed throughout that the IGM stays fully ionized from a redshift $z_* > z_c$ to the present. But if a large fraction of the the ionized IGM condensed into dark matter, or into high density neutral clumps, at a redshift $z' > z_c$, then the second order anisotropies would be reduced by about $(z_c/z')^{5/4}$.

The physical effects described in this article illustrate some of the interesting consequences of velocity fields at high redshift. In a reionized universe Thomson scattering off moving electrons produces several potentially observable effects. The clearest indicator of reionization would be a polarization pattern that is coherent over scales $\gtrsim 30'$. The second order anisotropies described in Section 3 have a characteristic "spiky" pattern and so should be readily distinguishable from the much smoother pattern expected in models with a standard recombination history. As we have shown, the secondary fluctuations can be quite large and could be detected by modern experiments.

ACKNOWLEDGEMENTS

I thank my $\Delta T/T$ collaborator, Dick Bond, for many interesting discussions and contributions. I am grateful to Anthony Readhead for supplying me with details of the Owens Valley experiment.

REFERENCES

Bean, A.J., Efstathiou, G., Ellis, R.S., Peterson, B.A., and Shanks, T. 1983. *MN.* **205**, 605.

Bond, J.R. and Efstathiou, G. 1984. *Ap J.* **285**, L45.

_____. 1987. *MN.* **226**, 655.

Danese, L., de Zotti, G., and Mandolesi, N. 1983. *AA.* **121**, 114.

Davies, R.D., Lasenby, A.N., Watson, R.A., Daintree, E.J., Hopkins, J., Beckman, J., Sanchez-Almeida, J., and Rebelo, R. 1987. *Nature.* **326**, 462.

Davis, M. and Boynton, P.E. 1980. *Ap J.* **237**, 365.

Davis, M. and Peebles, P.J.E. 1983. *Ap J.* **267**, 465.

Doroshkevich, A.G., Zeldovich, Ya. B., and Sunyaev, R.A. 1978. *Soviet Astron.* **22**, 523.

Dressler, A., Faber, S.M., Burstein, D., Davies, R.L., Lynden-Bell, D., Terlevich, R.S., and Wegner, G. 1987. *Ap J.* **313**, L37.

Efstathiou, G. and Bond, J.R. 1986. *MN.* **218**, 103.

_____. 1987. *MN.* **227**, 33p.

Fabbri, R., Guidi, I., Melchiorri, F. and Natale, V. 1982. in *Proceedings of the Second Marcel Grossman Meeting on General Relativity.* p. 889. ed. R. Ruffini. North-Holland.

Hogan, C.J., Kaiser, N., and Rees, M.J. 1982. *Phil Trans R Soc Lond A.* **307**, 97.

Kaiser, N. 1983. *MN.* **202**, 1169.

_____. 1984. *Ap J.* **282**, 374.

Lasenby, A. 1988. this volume.

Melchiorri, F., Melchiorri, B.O., Ceccarelli, C. and Pietranera, L. 1981. *Ap J.* **250**, L1.

Ostriker, J.P. and Vishniac, E.T. 1986. *Ap J.* **306**, L51.

Peebles, P.J.E. 1968. *Ap J.* **153**, 1.

_____. 1980. *The Large-Scale Structure of the Universe.* Princeton: Princeton University Press.

_____. 1987a. *Ap J.* **315**, L73.

_____. 1987b. *Nature.* **327**, 210.

_____. 1988. this volume.

Rubin, V.C., Thonnard, N., Ford, W.K., and Roberts, M.S. 1976. *AJ.* **81**, 719.

Sachs, R.K. and Wolfe, A.M. 1967. *Ap J.* **147**, 73.

Sargent, W.L.W. 1987. *Observatory.* **107**, 235.

Silk, J. 1982. *Acta Cosmologica.* **11**, 75.

Sunyaev, R.A. 1978. in *IAU Symp. 79, The Large-Scale Structure of the Universe.* p. 393. eds. M.S. Longair and J. Einasto. Dordrecht: Reidel.

Uson, J.M. and Wilkinson, D.T. 1984. *Ap J.* **277,** L1.

Vishniac, E.T. 1987. *Ap J.* **322,** 597.

Vittorio, N. 1988. this volume.

Yang, J., Turner, M.S., Steigman, G., Schramm, D.N., and Olive, K.A. 1984. *Ap J.* **281,** 493.

V

THEORY: *AB INITIO*

COHERENCE OF LARGE-SCALE VELOCITIES

ALEXANDER S. SZALAY

Department of Physics and Astronomy, The Johns Hopkins University
and Department of Atomic Physics, Eötvös University, Budapest

ABSTRACT

The radial components of the galaxy peculiar velocities are related to primordial density fluctuations via a tensor window function. For a given shell of galaxies the radial velocity as a function of direction can be expanded in multipoles, the $l = 1$ term corresponding to the bulk motion, the $l = 2$ to a quadrupole anisotropy of the velocity field. The $l = 0$ 'breathing mode', even if present, is unobservable; it would be absorbed into the local value of the Hubble constant.

We derive the distribution of the bulk velocity of such a shell and compare it to the results obtained with the usual scalar window function. The estimated velocity and the statistical scatters are significantly affected by sampling anisotropies in the galaxy selection function. The observed bulk velocity is most likely to be close to the galactic plane, aligned with the dipole anisotropy of the galaxy distribution. This effect is nongravitational, purely a result of nonuniform sampling.

We estimate the rms values of the various multipole moments of the velocity field and generate some Monte-Carlo realizations assuming the cold dark matter (CDM) spectrum. Generally the CDM velocities are too low to be compatible with the current observations.

1. INTRODUCTION

In the last few years several groups attempted to determine the bulk motion of a relatively large region (about 10 - 50 Mpc in radius) of the universe centered on us (Rubin *et al.* 1976a, b, Hart and Davies 1982, de Vaucouleurs and Peters 1984, Aaronson *et al.* 1986, Burstein *et al.* 1986, Collins *et al.* 1986).

These regions are large enough to be well described by linear theory, since nonlinear motions within the region cancel out, only the collective part remains (Clutton-Brock and Peebles 1981, Kaiser 1983). Using an independent distance determination to subtract the Hubble-flow from the observed redshift, the radial component of the peculiar velocity can be determined for each object. Distances are calculated from secondary distance indicators, different for elliptical galaxies (Burstein *et al.* 1986) and spirals (Collins *et al.* 1986). These methods usually carry systematic errors of the order of 10-20 percent.

The net center-of-mass (CM) velocity of the observed region can be determined from the raw data by using variants of the least squares fit or maximum likelihood techniques. This can be compared to the primordial fluctuation spectrum directly or indirectly (Kaiser 1988). The results have large systematic errors partly due to uncertainties in the distance estimators, partly due to the poor statistics.

Motion on such a large scale is linear; therefore, it can be related to the primordial linear fluctuation spectrum in a straightforward way as follows. The fluctuations are described by a dimensionless density fluctuation field and the peculiar velocity field relative to the Hubble flow. Both are functions of the comoving spatial coordinate x and time t. They can be expressed in terms of their Fourier transforms:

$$\delta(x, t) = \int d^3x \, e^{ikx} \, \delta_k \; ; \; v(x, t) = \int d^3k \, e^{ikx} \, v_k, \tag{1}$$

where δ_k has random phases and $P(k) = |\delta_k|^2$ is the power spectrum. From the continuity equation we obtain

$$v_k = -i(H_0 af) \, \frac{k}{k^2} \, \delta_k = \hat{k} v_k, \tag{2}$$

with $H_0 f = (\dot{D}/D)$ and $\hat{k} = k/k$. $H_0 = (\dot{a}/a)$ is the Hubble-constant, a is the expansion factor, and D is the growing solution of the linear equation on δ_k (Peebles 1980). Given a selection function describing the observed region one can predict the distribution of the CM velocity, based upon the fluctuation spectrum. The quantity used for this prediction is the rms (root mean square) value, which is compared to the observed velocity.

The expected value of the CM velocity for a given region is the integral of $v(x)$ over the selected area, described by a selection function $\Psi(s)$, normalized to $\int d^3s \, \Psi(s) = 1$,

$$\tilde{U}(x_0) = \int d^3s \, \Psi(s)v(s) = \int d^3k \, v_k \, e^{ikx_0} \, \tilde{W}(k), \tag{3}$$

using $x = x_0 + s$, where x_0 is the position of the observer, and s is the relative distance. This derivation assumes that we can measure the individual peculiar velocities $\mathbf{v}(\mathbf{s})$ in full, which is not the case. All theoretical calculations of this kind have also assumed an isotropic $\Psi(s)$, whose Fourier transform is the scalar window function $\tilde{W}(k)$:

$$\tilde{W}(k) = \int d^3\mathbf{s}\ \Psi(\mathbf{s})e^{i\mathbf{k}\mathbf{s}} = 4\pi \int_0^\infty ds\ s^2 \Psi(s)\ j_0(ks). \tag{4}$$

Here $j_0(ks)$ is the spherical Bessel function of 0th order. The dispersion $\langle \tilde{U}^2 \rangle$ can be expressed with these quantities.

$$\langle \tilde{U}^2 \rangle = \int d^3\mathbf{k}\ |v_{\mathbf{k}}|^2\ \tilde{W}(k)^2. \tag{5}$$

This isotropic rms value has been the one compared to the observed velocity so far. We show below that significant deviations occur from this simple prediction from experiment to experiment due to the way the observations and the data analysis were carried out, and we suggest that theoretical predictions should match the particular observations individually.

2. THE TENSOR WINDOW FUNCTION

The observed peculiar velocities are only the radial components of the full velocity of the ith galaxy in the sample : $v^i = \hat{s}^i \mathbf{v}^i$, with errors σ_i. First let us consider the vector sum of the projected velocities

$$\mathbf{V}(\mathbf{x}_0) = \frac{\Sigma_i v^i \hat{s}^i / \sigma_i^2}{\Sigma_i\ 1/\sigma_i^2} = \langle v_r(\mathbf{s})\hat{s}\rangle = \int d^3\mathbf{s}\ \Phi(\mathbf{s})\hat{s}v_r(\mathbf{s}). \tag{6}$$

$\hat{s} = \mathbf{s}/s$ is a vector of unit length parallel to $\mathbf{s}$, and $\Phi(\mathbf{s})$ is the *generalized selection function*, containing the errors as well. The errors are non-negligible, they mostly come from the distance estimation, so they are proportional to distance, s^i. They can be well approximated as $\sigma_i^2 = \sigma_f^2 + \Delta^2 s_i^2$ (Lynden-Bell *et al.* 1988). This will modify the radial selection function, more heavily weighting smaller scales. When we determine $\mathbf{V}$ in a Cartesian coordinate system, we need a tensor window-function, similar to the one discussed by Grinstein *et al.* (1987) arising from the projector $\hat{s}_\alpha\hat{s}_\beta$:

$$V_\alpha = \int d^3\mathbf{s}\ \Phi(\mathbf{s})\hat{s}_\alpha\hat{s}_\beta v_\beta(\mathbf{x}) = \int d^3\mathbf{k}\ e^{i\mathbf{k}\mathbf{x}_0}\ W_{\alpha\beta}(\mathbf{k})v_{\mathbf{k}\beta} \tag{7}$$

$$W_{\alpha\beta}(\mathbf{k}) = \int d^3\mathbf{s}\ \Phi(\mathbf{s})e^{i\mathbf{k}\mathbf{s}}\ \hat{s}_\alpha\hat{s}_\beta. \tag{8}$$

Let us consider first a completely isotropic selection function $\Phi(s)$. We can evaluate $W_{\alpha\beta}$ in this case:

$$W_{\alpha\beta}(\mathbf{k}) = 4\pi \int_0^\infty ds\, s^2\, \Phi(s) \left\{ \frac{1}{3}\, \delta_{\alpha\beta}[j_0(ks) + j_2(ks)] - \hat{k}_\alpha \hat{k}_\beta\, j_2(ks) \right\}, \quad (9)$$

significantly different from the window function in Eq. (4).

The center of mass velocity $\mathbf{U}$ can be determined from the observed radial velocities using a least squares fit:

$$\chi^2 = \sum_i \frac{(v^i - \mathbf{U}\hat{s}^i)^2}{\sigma_i^2} \quad \text{leading to} \quad \mathbf{U} = \mathbf{M}^{-1}\mathbf{V}. \quad (10)$$

$\mathbf{M}$ is a weighted projection matrix:

$$M_{\alpha\beta} = \sum_i \frac{\hat{s}_\alpha^i \hat{s}_\beta^i}{\sigma_i^2} \Big/ \sum_i 1/\sigma_i^2 = \int d^3s\, \Phi(s)\hat{s}_\alpha \hat{s}_\beta = W_{\alpha\beta}(0). \quad (11)$$

For an isotropic selection function $M_{\alpha\beta}$ is diagonal, and the elements are $\frac{1}{3}$, so only the diagonal part of $W_{\alpha\beta}$ contributes to the dispersion of U:

$$\langle U^2 \rangle = \int d^3\mathbf{k}\, |v_\mathbf{k}|^2\, W(k)^2 \quad (12)$$

$$W(k) = 4\pi \int_0^\infty ds\, s^2\, \Phi(s)\, (j_0(ks) - 2j_2(ks)). \quad (13)$$

This window function will differ from the scalar one by the presence of $-2j_2(ks)$. In the case of an isotropic sample this is the correct window function for calculating the dispersion of the observed bulk velocity. Its deviation from the istotropic scalar $\bar{W}(k)$ can be seen in Fig. 1a. We estimate the ratio of the dispersions due to the radial projection to be $\langle U^2 \rangle / \langle \bar{U}^2 \rangle \approx 0.75$, with a mild dependence on the radius of the Gaussian selection function shown on Fig. 1b., calculated using the CDM spectrum.

3. EFFECT OF ANISOTROPIES

The assumption of isotropy is generally not correct due to clumpy distribution of galaxies, galactic extinction, and fluctuations on the scale of the survey. The anisotropies will influence the measured bulk velocity, since the radial velocities only measure a projected part of the bulk flow. Here we take the sampling anisotropies explicitly into account, and show that they have a strong effect, summarizing a more extended discussion (Regös and Szalay 1988). We expand the galaxy selection function in multipoles:

$$\Phi(s) = \frac{1}{\sqrt{4\pi}} \sum_{lm} \phi_{lm}(s)\, Y_{lm}(\Omega_s). \quad (14)$$

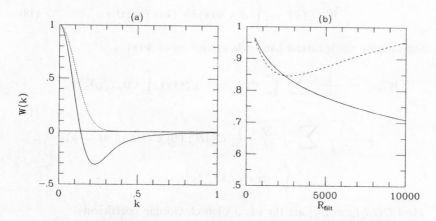

FIG. 1. (a) Deviation of the tensor window function (solid line) from the scalar (broken line) for an isotropic Gaussian radial selection (width R). (b) The solid line shows the correction to be applied to the scalar velocity prediction to obtain the tensor one. The broken line shows the same, but the window function also includes the distance error weights.

The normalization is $\int d^3s\ \Phi(s) = \int ds\ s^2 \phi_{00}(s) = 1$. The hermiticity of the expansion requires that $\phi_{lm} = (-1)^m \phi^*_{l,-m}$. From now on we will use spherical coordinates. In this system the unit vectors are

$$\hat{s}_\mu = \sqrt{\frac{4\pi}{3}}\ Y_{1\mu}(\Omega_s)\ ;\ \hat{k}_\mu = \sqrt{\frac{4\pi}{3}}\ Y_{1\mu}(\Omega_k). \tag{15}$$

The component of the 'vector-sum' velocity $\mathbf{V}$ corresponding to $\mu = 0, \pm 1$ can be expressed with the help of the tensor window function. Let us expand $V_\mu(\mathbf{x}_0)$ in multipoles and determine the coefficients.

$$V_\mu(\mathbf{x}_0) = i(H_0 af) \int d^3k\ \frac{\delta_\mathbf{k}}{k}\ e^{i\mathbf{k}\mathbf{x}_0} \int d^3s\ \Phi(s)\hat{s}_\mu(\hat{s}^*_\nu \hat{k}_\nu)e^{i\mathbf{k}\mathbf{s}}. \tag{16}$$

Using the Rayleigh expansion one can write

$$V_\mu(\mathbf{x}_0) = 4\pi(H_0 af) \int d^3k\ \frac{\delta_\mathbf{k}}{k}\ e^{i\mathbf{k}\mathbf{x}_0} \sum_{L,M} i^L Y_{LM}(\Omega_\mathbf{k})$$
$$\int d^3s\ \Phi(s)\hat{s}_\mu j'_L(ks) Y^*_{LM}(\Omega_s). \tag{17}$$

Here $j_L(ks)$ is the spherical Bessel function of order L, and $j'_L(ks)$ is the derivative with respect to ks. One can define the tensor window function in spherical coordinates as

$$W_{LM,\mu}(k) = \int d^3s \; \Phi(s)\hat{s}_\mu j_L'(ks) Y_{LM}^*(\Omega_s). \tag{18}$$

Substituting the spherical harmonic expansion of $\Phi(s)$

$$W_{LM,\mu} = \frac{1}{\sqrt{3}} \sum_{l,m} \int_0^\infty ds \; s^2 \, \phi_{lm}(s) j_L'(ks) \int d\Omega_s \, Y_{LM}^* Y_{1\mu} Y_{lm} =$$

$$= \frac{1}{\sqrt{4\pi}} \sum_l \sqrt{\frac{2l+1}{2L+1}} \; C(1lL; 00) \, C(1lL; \mu \, M - \mu) \tag{19}$$

$$\int_0^\infty ds \; s^2 \phi l, M_{-\mu}(s) j_L'(ks).$$

Here $C(l_1 l_2 l_3; m_1 m_2)$ are the usual Clebsch-Gordan coefficients.

For a given selection function one can easily express the expectation value

$$\langle V_\mu V_\nu \rangle = 4\pi (H_0 af)^2 \int d^3k \, \frac{|\delta_k|^2}{k^2} \sum_{L,M} W_{LM,\mu}^* W_{LM,\nu}. \tag{20}$$

The bulk motion components U_μ are correlated as

$$\langle U_\mu^* U_\nu \rangle = (M^{-1})_{\mu\varrho}^* (M^{-1})_{\nu\eta} \langle V_\varrho^* V_\eta \rangle. \tag{21}$$

Now let us calculate $\mathbf{M}$ in this system. We find that only the $l = 0$ and $l = 2$ terms are non-vanishing, thus $\mathbf{M}$ depends only on the quadruple anisotropy of the generalized selection function. Here $a_{lm} = \int ds \, s^2 \phi_{lm}(s)$.

$$M_{\mu\nu} = \frac{1}{\sqrt{4\pi}} \sum_{lm} \int_0^\infty ds \; s^2 \phi_{lm}(s) \int d\Omega_s \, Y_{lm} \frac{4\pi}{3} Y_{1\nu}^* Y_{1\mu}$$

$$= \frac{1}{3} \left[\delta_{\mu\nu} - \sqrt{2} a_{2,\nu-\mu} C(211; \nu - \mu, \mu) \right]. \tag{22}$$

3.1 - Statistical (sampling) Effect

The anisotropies have an important effect on the statistical scatter in the value of $\langle U_\mu^* U_\nu \rangle$ due to the discrete sampling and the error in the radial velocities. This dispersion can most easily be visualized in the extreme, when there are galaxies only at the north and south galactic poles. In this case we can determine the component of the bulk motion along the axis, but there is no information on the components in the galactic plane. Any quadrupole anisotropy due to galactic extinction will have such an effect, so the component in the plane has always a larger scatter than the axial one. Here we calculate its magnitude. First we define the mean statistical scatter of the velocity errors as

$$\langle \sigma^2 \rangle = \left(\frac{1}{N} \sum_{i=1}^{N} \frac{1}{\sigma_i^2} \right)^{-1}. \tag{23}$$

The variance matrix of the bulk velocity components becomes

$$\text{Var} \, (U_\mu^* U_\nu) = \frac{\langle \sigma^2 \rangle}{N_e} (M^{-1})_{\mu Q}^* (M^{-1})_{\nu \eta} M_{\eta Q} = \frac{\langle \sigma^2 \rangle}{N_e} (M^{-1})_{\mu \nu}^*, \tag{24}$$

where N_e is the effective degrees of freedom. For an isotropic selection function we obtain the trivial expression

$$\text{Var}(U_\mu^* U_\nu) = 3\delta_{\mu \nu} \frac{\langle \sigma^2 \rangle}{N_e}. \tag{25}$$

If we try to simulate the galactic extinction with a rotationally symmetric quadrupole a_{20}, $\mathbf{M}$ is still diagonal, but

$$\text{Var} \, (|U_0|^2) = 3 \frac{\langle \sigma^2 \rangle}{N_e} \frac{1}{1 + 2a_{20} / \sqrt{5}}$$

$$\tag{26}$$

$$\text{Var}(|U \pm 1|^2) = 3 \frac{\langle \sigma^2 \rangle}{N_e} \frac{1}{1 - a_{20} / \sqrt{5}}.$$

3.2 - Systematic effects

Even if there were no statistical errors ($\langle \sigma^2 \rangle = 0$), i.e. we measured each radial velocity totally accurately or we have a very large number of galaxies, the anisotropy in the selection function would still affect the expectation value $\langle U_\mu^* U_\nu \rangle$.

Below we calculate the contribution of various multipole moments ϕ_{lm} to the tensor window function. It turns out that, for reasonable selection functions, the single most important contribution arises (not surprisingly) from the dipole anisotropy of galaxy distribution. We should stress, however, that this is not a gravitational but a sampling effect.

If we consider a single dipole, we can choose our coordinate system by $\phi_{1\mu} = \delta_{\mu 0} \phi_{10}$, the dipole is along the z-axis. One can simplify notation below by using

$$w_{lm;L}(k) = \int_0^\infty ds \, s^2 \phi_{lm}(s) j_L(ks) \tag{27}$$

for the Lth Bessel-moment of the lm term in the expansion. The tensor window function becomes

$$
W_{LM,\mu} = \frac{\delta_{M\mu}}{\sqrt{12\pi}} \left[\delta_{L1} w_{00;1} + 3\delta_{L0}\delta_{M0} w_{10;0} + \delta_{L2} \sqrt{\frac{4}{5}}\, w_{10;2} \right]. \quad (28)
$$

Calculating the expectation value

$$
\langle V_\mu^* V_\nu \rangle = \frac{1}{3} \left[\delta_{\mu\nu} (H_0 a f)^2 \int d^3k\, \frac{|\delta_k|^2}{k^2} \right.
$$

$$
\left. \left(|w_{00;1}|^2 + 3\delta_{\mu 0}|w_{10;0}|^2 + \frac{4}{45}|w_{10;2}|^2 \right) \right]. \quad (29)
$$

Normalizing the various moments to the integral of $|w_{00;1}|^2$, the term corresponding to the isotropic tensor window function is

$$
A_{lm;L} = \frac{\int d^3k\,(|\delta_k|^2/k^2)|w_{lm;L}|^2}{\int d^3k\,(|\delta_k|^2/k^2)|w_{00;1}|^2}. \quad (30)
$$

Assuming that there is no quadrupole, we obtain the coefficients K_ν, the correction to be applied to the results of calculations based upon the assumption of isotropy:

$$
K_\nu = \frac{\langle |U_\nu|^2 \rangle}{\langle |U_\nu|^2 \rangle_{isotr}} = 1 + \frac{4}{15} A_{10;2} + \delta_{\nu 0}\, 9A_{10;0}. \quad (31)
$$

For a numerical evaluation of the coefficients, we assume that the radial dependence of the selection function ϕ_{10} is the same as the overal ϕ_{00}, the dipole amplitude is D, the case of nonuniform sampling,

$$
\phi_{00}(r) \propto e^{-(r/R)^2/2} \,/\, (\sigma_f^2 + \Delta^2 r^2)\,;\, \phi_{10}(r) = D\phi_{00}(r). \quad (32)
$$

We use the adiabatic cold dark matter spectrum to get the coefficients. Only the $L = 0$ term has a significant contribution, the $L = 2$ becomes negligible: $A_{10;0} = 2.44D^2$, $A_{10;2} = 0.25D^2$. Thus the dipole moment will change the correction considerably:

$$
K_\nu = 1 + 0.067\, D^2 + \delta_{\nu 0}\, 21.96\, D^2. \quad (33)
$$

Even a small $D = 0.2$ dipole anisotropy will increase the dispersion of the observed velocity component along the direction of the dipole by a factor of 2. This effect is also coming from the nonuniform sampling. A real dipole anisotropy of the galaxy distribution would cause additional effects. Besides, another complication will arise from the use of magnitude limited samples, as discussed by Fall and Jones (1976).

3.3 - Applications to the Elliptical Galaxy Sample

We used the Dressler *et al.* (1987) elliptical galaxy sample to estimate the magnitude of these effects. We take a Gaussian of width $R = 3000$ km/s as the radial selection for both multipoles and use an error model of

$$\sigma_i^2 = \sigma_f^2 + \Delta^2 r_i^2 / \sqrt{N_i}, \tag{34}$$

where $\sigma_f = 300$ km/s is the field-dispersion, $\Delta = 0.23$ is the error of the distance determination, and N_i is the multiplicity assigned to groups as in Lynden-Bell *et al.* (1988). The effective degrees of freedom in this case will not be integer, $N_e = 252.7$. Calculating the mean scatter for the sample we obtain

$$\langle \sigma^2 \rangle = 1.8 \, \Delta^2 R^2 = 1.8 \times 690^2 (\text{km/s})^2. \tag{35}$$

We calculated the various multipole moments of the selection function numerically. The quadrupole moment became $a_{20} = 0.44$, the various dipole components are $a_{10} = 0.156$, $a_{11} = (-0.027) + i(-0.213)$. This dipole has an amplitude $D = 0.342$, points to $l = 262.7^\circ$, $b = 27.2^\circ$, very close to the direction of the microwave anisotropy. The distribution of galaxies and the dipole direction is shown on Fig. 2. In order to simplify we transform the results to rectangular coordinates (U_x, U_y, U_z) in the galactic system.

The sampling variance in the velocities becomes in units $(\text{km/s})^2$

$$\text{Var}(U_\alpha U_\beta) = \begin{pmatrix} 12720 & 0 & 0 \\ 0 & 12720 & 0 \\ 0 & 0 & 7280 \end{pmatrix}. \tag{36}$$

The variance matrix of the bulk velocity, using the CDM spectrum, no biasing, normalizing to the isotropic tensor velocity dispersion $\langle U^2 \rangle$ becomes

$$\text{Var}(U_\alpha U_\beta) = \frac{1}{3} \langle U^2 \rangle \begin{pmatrix} 1.04 & -0.26 & 0.13 \\ -0.26 & 3.25 & -1.03 \\ 0.13 & -1.03 & 1.44 \end{pmatrix}. \tag{37}$$

The trace of the matrix is $1.91 \langle U^2 \rangle$, corresponding to a total rms velocity of $1.38U$, equal to 430 km/s for $\langle U^2 \rangle = (310 \text{ km/s})^2$.

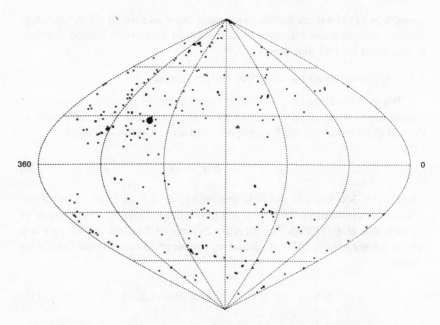

FIG. 2. The distribution of the elliptical galaxy sample of Burstein *et al.* (1986). The large black mark shows the direction of the dipole moment, $D = 0.342$, $l = 267.7^o$, $b = 27.2^o$.

4. MULTIPOLE EXPANSION OF THE RADIAL VELOCITY

If we select a relatively thin shell of galaxies, then the radial velocity on the surface of the sphere is a scalar function of the angles α, δ. It can be expanded in terms of spherical harmonics:

$$V_r(\Omega_s) = \sum_{l,m} V_{lm} Y_{lm}(\Omega_s), \tag{38}$$

with $V_{lm} = (-1)^m V^*_{l,-m}$. For that particular shell we can calculate the ensemble average of the multipole moment correlation matrix

$$\langle V^*_{lm} V_{l'm'} \rangle \, \delta_{ll'} \, \delta_{mm'} \, 4\pi \, (H_0 a f)^2 \int d^3 \mathbf{k} \, \frac{|\delta_k|^2}{k^2} \, |W_l(k)|^2 = \delta_{ll'} \, \delta_{mm'} \, 4\pi \, u_l^2, \tag{39}$$

where

$$W_l(k) = \int d^3 \mathbf{s} \, \Phi(s) j_l'(ks). \tag{40}$$

In the first part of this paper we already considered the $l = 1$ term corresponding to the dipole 'bulk' motion. There is a monopole $l = 0$, corresponding to an overall uniform expansion or inward flow, caused by the local over- or under-density, which is absorbed into the local value of the Hubble constant. The presence of this term can possibly be observed by taking larger and larger shells.

If we have two separate shells, described by their respective window functions $W_l^{(a)}(k)$ and $W_{l'}^{(b)}(k)$, the correlation of the velocity coefficients is given by

$$\langle V_{lm}^{(a)*} V_{l'm'}^{(b)} \rangle = \delta_{ll'} \, \delta_{mm'} \, 4\pi (H_0 a f)^2 \int d^3 \mathbf{k} \; \frac{|\delta \mathbf{k}|^2}{k^2} \; W_l^{(a)*}(k) \; W_{l'}^{(b)}(k). \quad (41)$$

The relative magnitude of the u_l cofficients will determine the coherence, or patchiness of the large-scale velocity field. This coherence can also be described by using the multipole coefficients: a smooth velocity field has a cutoff for the higher harmonics. Here we suggest that this patchiness carries significant information about the shape of the fluctuation spectrum, independent of the normalization. In Fig. 3 we plot the coefficients u_l^2 as a function of l for the CDM spectrum, using $R = 3000$ km/s, weighted with errors, normalized to a mass-variance of 1 for a top-hat radius of 800 km/s. It is obvious that the $l = 0$ monopole term is the largest, and the others are falling off quite rapidly. For this normalization,

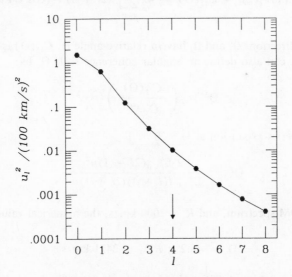

FIG. 3. The dependence of the velocity multipole moments u_l^2 on the harmonic number l, using the CDM spectrum, normalized to unit top-hat mass variance at 800 km/s, no biasing. The sample is approximated by a Gaussian selection of $R = 3000$ km/s radius, and error weighting. $\sigma_f = 300$ km/s, and $\Delta = 0.23$. The arrow indicates the angular coherence scale of the CDM picture.

$$u_0^2 = 2.147$$
$$u_1^2 = 1.032 \tag{42}$$
$$u_2^2 = 0.188$$

in units of $(100 \text{ km/s})^2$. One should note that these are the dispersions for each respective $|V_{lm}|^2$. The rms 'monopole' velocity is about 150 km/s, about 5% distortion on the Hubble flow. The dispersion of the isotropic bulk motion is $\langle U^2 \rangle = 9u_1^2 = (310 \text{ km/s})^2$. For the quadrupole, there are 5 components of the irreducible tensor, and dispersion of the anisotropy in the quadrupole components can be estimated to be $(337 \text{ km/s})^2$, comparable to the bulk motion prediction.

Currently, in Lynden-Bell *et al.* (1988) there is a discussion of a quadrupole anisotropy of $0.1 - 0.2$ in the Hubble flow for the various samples (300 - 600 km/s). Besides, the velocity field shows remarkable coherence, on linear scales of $500 - 2000$ km/s.

5. Angular Velocity Correlations

A good way to quantify the coherence of the velocity fields is to calculate the angular velocity correlation function. From the multipole expansion

$$\langle V_r(1)V_r(2) \rangle = C_{12}(\Theta) = 4\pi \sum_{l=0}^{\infty} (2l + 1) P_l(\cos \Theta) u_l^2 \tag{43}$$

where the directions Ω_1 and Ω_2 have a relative angle Θ. $C_{12}(\Theta)$ is shown on Fig. 4. One can also define an angular coherence scale Θ_0 by

$$\Theta_0^2 = \left(\frac{C_{12}(\Theta)}{C_{12}''(\Theta)} \right)_{\Theta = 0}. \tag{44}$$

Using the series expansion at $\Theta = 0$

$$\Theta_0^2 = \frac{2 \sum_{l=0}^{\infty} (2l + 1)u_l^2}{\sum_{l=0}^{\infty} l(l + 1)(2l + 1)u_l^2}. \tag{45}$$

For the CDM spectrum, and $R = 3000$ km/s, the numerical value becomes

$$\Theta_0 = 46.2^o \; ; \; R\Theta_0 = 2421 \text{ km/s} \tag{46}$$

corresponding to $l = 4$. This is indeed very close to the observed patchiness of the velocity field. Note, that this quantity does not depend on the absolute normalization of the density fluctuation spectrum, nor on the amount of biasing.

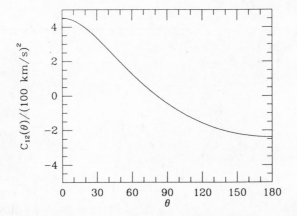

FIG. 4. The angular velocity correlation function, in units of $(100 \text{ km/s})^2$.

The angular correlation function for the elliptical galaxy data has been calculated numerically (Burstein and Szalay 1988). This function, smoothed in $10°$ bins is plotted in Fig. 5.

6. MONTE-CARLO REALIZATIONS

Using the multipole expansion, we generate some images of the angular distribution of the radial velocities. For a given selection function we know the dispersion $|V_{lm}|^2$ of the expansion coefficients, and the velocity field is described by Gaussian statistics. Each V_{lm} is a random complex number,

$$V_{lm} = \sqrt{2\pi} \; u_l(x_{lm} + iy_{lm}) \qquad (47)$$

where x_{lm} and y_{lm} are normal Gaussian random numbers, with unit dispersions. The hermiticity requires that

$$V_{l,-m} = (-1)^m V_{lm}^* = (-1)^m \sqrt{2\pi} \; u_l(x_{lm} - iy_{lm})$$
$$V_{lo} = \sqrt{4\pi} \; u_l' x_{lm} \qquad (48)$$

Then we just add up the series, to a harmonic number $l = 10$, and generate equal area plots of the resulting velocity field. These are shown on Fig. 6. Also for comparison we smoothed the original elliptical galaxy data with various Gaussian filter lengths and these results are shown in Figs. 7 and 8.

7. CONCLUSION

We have made several improvements over the usual ways of calculating large-scale motions in the universe, taking into account that observers measure

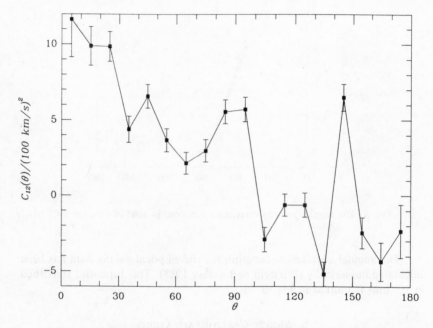

FIG. 5. The angular velocity correlation function in 10° bins, estimated from the elliptical galaxy sample (Burstein and Szalay 1988).

radial velocities, resulting in a tensor window function. Furthermore, we have calculated the twofold effect of selection anisotropies in the galaxy distribution previously not considered and we have shown that they have important consequences on the bulk motions.

The fluctuation spectrum causes a large-scale coherence in the velocity field, quantified by the multipole components of the projected velocities. The use of radial projection mixes these modes corresponding to different l, and a nonuniform sampling results in nonseparable errors. With an anisotropic sample the spherical harmonics become nonorthogonal, and the usual methods of least squares or maximum likelihood yield systematically biased estimations of the bulk motion. The major effect comes from the dipole and the quadrupole anisotropy: the direction of the velocity will be correlated with the dipole moment of the galaxy distribution and statistical uncertainties align perpendicular to the quadrupole. This will cause the measured bulk velocity to be preferentially close to the galactic plane and aligned with the dipole anisotropy, very close to what the actual observations show, although there is a real density enhancement in that direction.

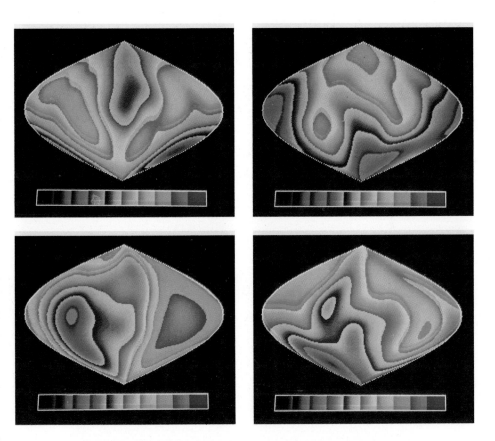

FIG. 6. Realizations of the velocity field in the unbiased CDM scenario, where the survey size is limited to 3200 km/s, and the selection function is error weighted, normalization is as above (see Fig. 3). A total of 16 different random cases have been created, the false colors span the range of − 1000 km/s through 1000 km/s, and the small bins with intensity modulation are 200 km/s wide. Red colors correspond to redshift, blue to blueshift. Out of the 16 models created about 2 have a large dipole of about 600 km/s in magnitude, like the case of the Great Attractor, and several others have much smaller velocities. With a biasing factor of a few, such a large dipole would be extremely unlikely.

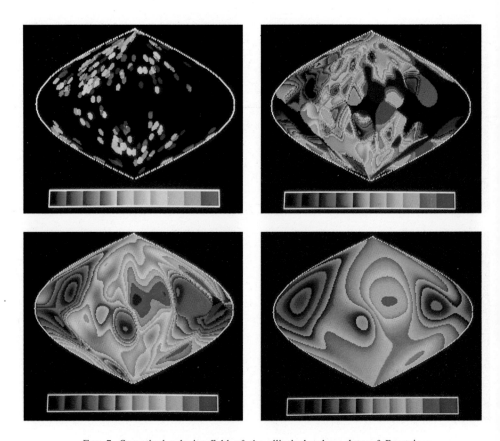

FIG. 7. Smoothed velocity field of the elliptical galaxy data of Burstein *et al.* (1986) with distances out to 8000 km sec^{-1}. Each galaxy has been smoothed with a Gaussian filter of a radius 1,5, 10, and 20 degrees respectively. The Gaussian was cut off at four filter radii. The regions with no galaxies within that distance were left blank. Velocities are coded as in Fig. 6.

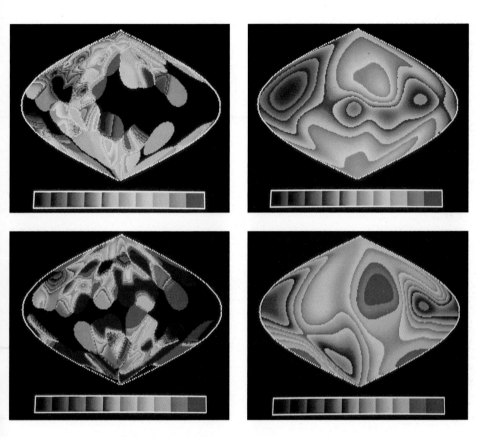

FIG. 8. Smoothed velocity field of the galaxies in the elliptical set: in the top two panels are the near galaxies ($V < 3200$ km sec^{-1}); in the bottom two panels are the far galaxies ($3200 < V < 8000$ km sec^{-1}). Smoothing lengths are 5 and 20 degrees; other details are as in Fig. 7. Note that the dipole in both the near and far fields is accompanied by a monopole term of comparable magnitude.

The radial velocity field observed over the whole sky can also be expanded in terms of spherical harmonics, and each coefficient has a different physical meaning. The monopole term describes a uniform distortion in the expansion, the dipole is the bulk motion, the quadrupole appears as an anisotropy of the Hubble flow. The general coherent appeareance of the velocity field can be well described by an angular coherence scale. Using the CDM spectrum to estimate this angle for the elliptical galaxy sample we obtain an excellent agreement, $\Theta_0 = 46.2^o$, independent of normalizations. The shape of the correlation function (the higher multipole moments) carries interesting information on the shape of the power spectrum around the 10 Mpc scales. On the whole, the standard normalization of the CDM picture gives somewhat low velocities, but the relative ratios of the bulk motion to quadrupole and the correlation length are quite acceptable.

Acknowledgements

The author would like to acknowledge useful conversations with Enikö Regös, Dave Burstein, Vera Rubin, and Nick Kaiser. This work was supported by a grant of the Hungarian Academy of Sciences (OTKA) in Hungary, and by the National Science Foundation of the United States.

REFERENCES

Aaronson, M., Bothun, G., Mould, J., Huchra, J., Schommer, R.A. and Cornell, M.E. 1986. *Ap J.* **302**, 536.

Burstein, D., Davies, R.L., Dressler, A., Faber, S.M., Lynden-Bell, D., Terlevich, R.J., and Wegner, G. 1986. in *Galaxy Distances and Deviations from Universal Expansion* (NATO ASI Series) ed. B. Madore and B. Tully. p. 123. Dordrecht: Reidel.

Burstein, D. and Szalay, A.S. 1988. in preparation.

Clutton-Brock, M. and Peebles, P.J.E. 1981. *AJ.* **86**, 1115.

Collins, C.A., Joseph, R.D., and Roberts, N.A. 1986. *Nature.* **320**, 506.

de Vaucouleurs, G. and Peters, W.L. 1984. *Ap J.* **287**, 1.

Dressler, A., Lynden-Bell, D., Burstein, D., Davies, R.L., Faber, S.M., Terlevich, R.J., and Wegner, G. 1987. *Ap J.* **313**, 42.

Fall, S.M. and Jones, B.J.T. 1976. *Nature.* **262**, 457.

Grinstein, B., Politzer, H.D., Rey, S.J., and Wise, M.B. 1987. *Ap J.* **314**, 431.

Hart, L. and Davies, R.D., 1982. *Nature.* **298**, 191.

Kaiser, N. 1983. *Ap J Letters.* **273**, L17.

———. 1988. *MN.* **231**, 149.

Lynden-Bell, D., Faber, S.M., Burstein, D., Davies, R.L., Dressler, A., Terlevich, R.J., and Wegner, G. 1988. *Ap J.* **326**, 19.

Peebles, P.J.E. 1980. *The Large Scale Structure of the Universe.* Princeton: Princeton University Press.

Regös, E. and Szalay, A.S. 1988. *Ap J.* submitted.

Rubin, V.C., Thonnard, N., Ford, W.K. Jr., Roberts, M.S., and Graham, J.A. 1976a. *AJ.* **81**, 687.

Rubin, V.C., Thonnard, N., Ford, W.K. Jr., and Roberts, M.S. 1976b. *AJ.* **81**, 719.

THEORETICAL IMPLICATIONS
OF COSMOLOGICAL DIPOLES

N. KAISER and O. LAHAV

Institute of Astronomy, Cambridge

ABSTRACT

We have compared the dipole anisotropy of the IRAS sky and the peculiar velocity bulk-flow solutions with the cold dark matter (CDM) predictions using linear theory. The model predictions fit very nicely with the observations; the χ^2 statistics are perfectly acceptable, and the likelihood analysis using these probes of large-scale structure gave essentially identical normalisation parameters as those we favour from considerations of clustering on much smaller scales.

We have developed and applied an improved version of Gott's test using the correlations between a number of angular dipoles and bulk flow solutions, rather than the traditional implementation using only the flux weighted dipole and the Local Group. We obtain a value for the bias parameter for IRAS galaxies of $b \simeq 1.5$, corresponding to an Ω_0 of around 0.4.

Following the modeling of the peculiar velocity field, we have also made simple models with one or more spherical perturbations. These models, in which roughly 300 km/s of our peculiar motion derives from inhomogeneity at distances between about 40 and 80 Mpc/h, are compatible with the observed dipoles, and also reproduce the apparent *lack* of acceleration seen at these distances in redshift space. The acceleration from distant perturbations in these models is similar to that predicted in CDM.

1. INTRODUCTION

In the cold dark matter model (hereafter CDM), it is assumed that at some very early time there were Gaussian isentropic density fluctuations $\Delta(\mathbf{r}) \equiv (\varrho(\mathbf{r}) - \bar{\varrho})/\bar{\varrho}$ with the 'Harrison-Zeldovich' constant curvature spectrum.

With the additional assumptions that the universe has critical density and is dominated by a cold collisionless particle, the power spectrum $P(k) \equiv \langle \Delta(k)\Delta^*(k) \rangle$ emerging in the matter dominated era can be calculated (e.g. Bond and Efstathiou 1984), and this gives a complete statistical description of the density and associated velocity field.

We wish to confront this model with observations of large-scale structure. One approach is to use estimates of the power spectrum or autocorrelation function, which effectively measure the rms fluctuations. On small scales these can be determined to high precision, but as we approach the depth of the survey we no longer sample enough volume to determine the properties of the ensemble. There are, however, certainly statistically significant detections of density fluctuations at very large scales from the dipole moments of galaxy counts for instance, and there are analogous detections of large-scale peculiar velocities. Our goal here is to extract the theroretical predictions for these observations. We would like to make clear at the outset that, while we will consider only *dipole* moments of the various fields, we do not assume that the dipole moments alone provide a good description of the inhomogeniety around us. The dipoles are simply a convenient set of statistics for which it seems to us to be reasonable to calculate the probability distribution function and compare with the theory as though the statistics had been chosen *a priori*. We could in principle perform a similar analysis to calculate the probability distribution function for the coefficients of a Great Attractor model for instance, but since the form of this model was chosen *a posteori* we would find it more difficult to interpret the results.

Before we can make these predictions it is necessary to specify two parameters which determine the amplitude of the fluctuations. The first parameter controls the amplitude of the *mass* fluctuations, for which an arbitrary but convenient measure is the rms fluctuation in spheres of radius 8 Mpc/h, and which we denote by σ_ϱ. Another parameter is the bias parameter b which describes the amount by which the fluctuations in galaxy number density are amplified relative to the underlying matter concentrations. These parameters are constrained, albeit rather roughly, from considerations of mass and galaxy clustering on smaller scales. Assuming that clusters of galaxies correspond to rare high peaks of the initial mass fluctuations when filtered on a suitable scale, we find (see e.g. Bardeen *et al.* 1986) that $\sigma_\varrho \simeq 0.6 - 0.7$. The greatest uncertainty here is in the determination of the masses of clusters from velocity dispersions. Since the rms fluctuations in galaxies in 8 Mpc/h spheres, $b\sigma_\varrho$, is roughly unity, this would indicate $b \simeq 1.4 - 1.8$.

With these normalisation parameters set, the model has great predictive power and is vulnerable to a multitude of observational tests, particularly those which probe structure on scales much larger than the scale where the normalisation is performed. An added bonus is that fluctuations on such scales should be very well described by linear theory, and the primordial Gaussian statistics should be preserved.

If we imagine observers throughout the universe each of whom makes a dipole measurement with identical selection procedures, then, in this model, the results form a Gaussian random vector field. For a single dipole, the statistical distribution is very simple. We can, however, generate a number of dipoles by splitting the data into flux and distance intervals, and hopefully improve the power of the test. These dipoles are necessarily quite strongly correlated with each other, so they must be interpreted with care. We shall adopt two approaches. First we shall attempt an *absolute* assessment of the CDM model with the normalisation as given above. To this end we calculate the rms values for the individual dipoles. This simple test enables one to see if there are any obvious discrepancies. Then, in order to properly incorporate the correlations and changes in direction we calculate a 'chi-squared' statistic. Secondly, as a rather crude first attempt to make a *relative* assessment of a range of theoretical models, we perform likelihood analysis with the normalisation parameters as free variables. The hope here is that if we live in a world in which there is more power at large scales than in CDM-world, then this analysis should give larger values for the normalisation parameters than those determined above. We appreciate that there are widely differing views on how to do statistics, particularly for problems like this. We hope that the combination of statistics used here will have some appeal to both 'frequentists' and 'Bayesians'.

As well as asking whether the amplitudes of the observed dipoles are compatible with the very specific CDM model, one can also use these data to test the more fundamental underlying assumption that the fluctuations in galaxy number density trace the mass fluctuations (albeit in a linearly biased manner). Traditionally, this test has been applied using the flux-weighted dipole of the galaxy counts, which gives an estimate of the net acceleration acting on the Local Group, and this can be compared with the Local Group motion relative to the microwave frame. There are two issues here. One can ask whether the velocity is parallel to the apparent acceleration — as predicted (in linear theory) in all models where structure grows by gravitational instability, though not in the case of the explosive scenario for instance. If these vectors are at least approximately parallel then it is also interesting to look at their ratio. Before 'biasing' became popular, this ratio was interpreted as a determination of Ω. In the present context we *assume* $\Omega \equiv 1$ and interpret the ratio as a measure of the bias parameter b. An appealing feature of this test is that it appears to give a determination of b which is independent of any assumption about the spatial distribution of sources of acceleration. Realistically, considerations of the finite number density of galaxies and flux limits in the catalogues tend to decouple the estimates of acceleration and velocity somewhat, and to estimate the magnitude of the error requires a model to tell us which wavelengths contribute significantly to the acceleration. We find that, in CDM-world at least, the scatter is perhaps disappointingly large using the standard choice of dipoles, but that a considerable improvment can be made making use of other dipoles which have now become available. Crudely speaking, just as an idealised flux-weighted dipole correlates very well with the motion of the Local Group, there are other dipoles which correlate well with the motion of the Local Supercluster or of even larger regions. These can be measured quite accurately now, and give an estimate of b (or $b\Omega^{-0.6}$ according to ones taste) with considerably smaller statistical uncertainty. An additional advantage, though one which is harder to quantify, is that in using the motion of large aggregates rather than the Local Group one can be much more confident in applying linear theory.

A largely independent view of large-scale structure is provided by redshift surveys. Initial estimates (Strauss and Davis 1987; Yahil 1988) of the acceleration vector from the 2 Jy IRAS redshift survey seemed to indicate that the source of the acceleration is very localised. This conflicts strongly with the view of a rather deep source for the bulk of the Local Group motion coming from studies of departures from Hubble flow. A loophole which may remove this apparent discrepancy is that no allowance was made for the effect of peculiar velocities, and these are expected to have a profound effect on the clustering pattern in redshift space (Kaiser 1987). We use a variety of models to explore this possibility and show that this effect does indeed seem to largely resolve the apparent inconsistency. We also calculate the statistical uncertainty in the acceleration vector in the CDM model, and compare with the precision of the acceleration estimate obtained from the angular dipoles.

The implications of all the observations we analyse below have been considered before in one way or another. An important feature of our analysis is that we attempt to model as realistically as possible the actual observations. We show that for each observation there corresponds a unique 'window function' defined unambiguously by the weighting scheme adopted by the observer. The observation can be written as a linear convolution of the density or velocity field and the window, plus an independent random 'noise' component deriving from the discreteness of galaxies, statistical errors in peculiar velocities or whatever. We have also attempted to model this statistical error realistically, and include this in our comparison with the CDM predictions. We find that for the velocity observations in particular, previous theoretical studies have tended to overestimate the effective depth of the surveys, and this led to the misleading impression that our universe is more inhomogeneous on large scales than CDM-world. Similarly, the anisotropy of the IRAS sky was initially thought to be indicative of density fluctuations on scales of order 100 Mpc/h (Rowan-Robinson *et al.* 1986), again with exciting implications for cosmogonical theories. Our analysis will show that the simple dipole statistics at least, are in as good agreement with the CDM predictions as could be expected. Whether the same will prove true for more detailed representations of the data we cannot say. Also, the perhaps more interesting question of which alternative models are *incompatible* with observations is only touched on slightly here.

Perhaps the most unrealistic feature of our analysis is that we assume that in all of the observations we have full and and uniform sky coverage; this approximation is not strictly necessary, and zones of obscuration etc. can be allowed for within the framework set out below. The assumption of spherical symmetry does, however, greatly simplify the formulae, and is not a bad approximation for most of the data we consider.

The layout is as follows. In Sec. 2 we consider the dipole moments of the IRAS galaxy counts. In this section we set out the details of the theoretical model and illustrate the technique used for constructing the probability distribution function for a real observation. In Sec. 3 we analyse in a completely analogous manner the dipole moments of the peculiar velocity field. In Sec. 4 we combine these data, as described above, to determine the bias factor. These sections are an extract from the more complete exposition of Kaiser and Lahav (1988) where fuller detail is given and the dipoles obtained from optical catalogues are also considered. In Sec. 5 we turn to the interpretation of the apparent acceleration vector as inferred from the 2 Jy IRAS redshift survey. Finally we summarise our results, discuss their relationship to other probes of large-scale structure and sketch our plans to extend these analyses in the future.

2. THE ANGULAR DIPOLES

In this section we will consider dipole moments of the angular counts of galaxies. In Sec. 2.1 we state the model for the matter density field, and for the distribution of galaxies. In Sec. 2.2 we calculate the probability distribution function (pdf) for a single dipole moment, and in Sec. 2.3 we calculate the pdf for a collection of dipoles. We obtain expressions for the likelihood as a function of the normalisation parameters of the model. In Sec. 2.4 we apply these results to the IRAS dipoles, and also consider some alternative models.

2.1 - The Matter Density Field and the Galaxies

We assume that the Fourier components of the density field are drawn from a Gaussian distribution with $\langle \Delta(\mathbf{k})\Delta^*(\mathbf{k}) \rangle = P(k)$. The field

$$\Delta(\mathbf{r}) \equiv \frac{\varrho(\mathbf{r}) - \langle \varrho \rangle}{\langle \varrho \rangle} = \frac{V_u}{(2\pi)^3} \int d^3k \, \Delta(\mathbf{k}) \, e^{-i\mathbf{k} \cdot \mathbf{r}} \qquad (2.1)$$

is therefore a statistically homogeneous and isotropic random field which is periodic in the large but arbitrary volume V_u.

Writing $P(k) = P_0 F(k)$, the specific form we adopt is a fit to the CDM spectrum with initially adiabatic fluctuations and with $\Omega_0 = 1$ and $h = 0.5$ as given by Bond and Efstathiou (1984):

$$F(k) = k \{1 + [11.55 \ k + (5.70 \ k)^{3/2} + (3.24 \ k)^2]^{1.25} \}^{-1.6}. \quad (2.2)$$

The unit of length here and throughout is 1 Mpc/h, or the distance corresponding to a recession velocity of 100 km/s.

We express the normalization in terms of the variance of density in a 'top-hat' sphere of radius a = 8 Mpc/h:

$$\sigma_\varrho^2 \equiv \left\langle \left(\frac{\Delta M}{M} \right)^2 \right\rangle = \frac{1}{(2\pi)^3} \int d^3k \ P(k) \ |U(ka)|^2 \quad (2.3)$$

where $U(ka) = 3 \ j_1(ka)/(ka)$, and $j_1(y) \equiv \sin y/y^2 - \cos y/y$.

We assume that there is a universal luminosity function $\Phi(L)$ and that the galaxies 'fairly sample' the linearly biased density field $1 + \Delta_{galaxies} = 1 + b\Delta$. Thus, if we describe this point process by occupation numbers $n(\mathbf{r}_i, L_j)$ for microscopic cells in $\mathbf{r}$ and L space which are 0 or 1 if the volume element $d^3r \ dL$ with labels i, j is empty or contains 1 particle respectively, then the probability of occupation is

$$P(\text{galaxy in cell } \mathbf{r}_i, L_j) = \langle n(\mathbf{r}_i, L_j) \rangle \propto (1 + b\Delta(\mathbf{r}_i)) \cdot \Phi(L_j). \quad (2.4)$$

The inclusion of the parameter b is motivated by current thinking about biased models for galaxy formation in which production of galaxies is modulated by longer wavelength modes. Various schemes have been proposed to effect biasing, but the property shared by all of these is that for small amplitude waves one obtains a linear bias. More generally one would let b be a function of luminosity or some other attribute of the galaxies, but we will not explore that possibility.

The 'universal luminosity function' assumption according to which luminosities are assigned purely at random is highly idealised, and is almost certainly unrealistic. This assumption is important in determining the statistical noise in the observations. If, as seems at least plausible, environmental effects systematically perturb the luminosities of galaxies, then one would expect the simple 'Poissonian' estimate used below to underestimate the true noise. We have made tests to see if this is important by analysing the differences between dipoles obtained from IRAS and optical samples with similar depth. The constraint is unfortunately rather weak, and while we found no strong indication of excess fluctuations, we would warn against too literally interpreting dipoles for which the shot noise term is appreciable.

2.2 - Modeling a Single Dipole

Given a flux limited catalogue which lists positions and fluxes we can construct a dimensionless dipole moment which we define to be

$$\mathbf{D} = \frac{3 \sum w(S_q) \, \hat{\mathbf{r}}_q}{\int_{S_{min}}^{S_{max}} dS \, N(N) \, w(S)} \tag{2.5}$$

where the sum is over all galaxies with flux S in the range $S_{max} > S \geq S_{min}$. $N(S)$ is defined such that $\int_S^\infty N(S) \, dS = N_0 (S/S_{lim})^{-3/2}$, where N_0 is the total *expected* number of galaxies brighter than S_{lim} (the flux limit of the catalogue) in the entire sky. The choice of weighting function $w(S)$ is at the disposal of the observer.

In terms of the occupation numbers defined above,

$$\mathbf{D} = \frac{3 \sum_{i,j} w(L_j/4\pi r_i^2) \, \hat{\mathbf{r}}_i \, n(\mathbf{r}_i, L_j)}{\int dS \, N(S) \, w(S)} \tag{2.6}$$

where the sum is now over microcells. We now write $\mathbf{D}$ as the sum of two terms; the first we call the 'theoretical' dipole $\mathbf{D}^{th}$, which is the value that would be obtained in a world where galaxies are so numerous that they approach the continuum limit, while the second $\mathbf{D}^{sn}$ represents the sampling 'shot noise' error arising from the discreteness of the galaxies. For the former we have

$$\mathbf{D}^{th} = \langle \mathbf{D} \rangle = \frac{3 \sum_{i,j} w(L_j/4\pi r_i^2) \, \hat{\mathbf{r}} \, \langle n(\mathbf{r}_i, L_j) \rangle}{\int dS \, N(S) \, w(S)} . \tag{2.7}$$

Using 2.4, converting the sums to integrals, and shifting the spatial origin so the observer lives at $\mathbf{r}_0$, gives

$$\mathbf{D}^{th}(\mathbf{r}_0) = 3 \, b \int d^3r \, W(r) \, \Delta(\mathbf{r}_0 + \mathbf{r}) \, \hat{\mathbf{r}}, \tag{2.8}$$

where the radial window is

$$W(r) \propto \int_{4\pi r^2 S_{min}}^{4\pi r^2 S_{max}} dL \, w(L/4\pi r^2) \, \Phi(L), \tag{2.9}$$

where $\Phi(L)dL$ is the luminosity function, and we normalise $W(r)$ so that $\int d^3r \, W(r) = 1$.

From the convolution theorem we find

$$\mathbf{D}^{\text{th}}(\mathbf{k}) = W^*(k) \, \hat{\mathbf{k}} \, \Delta(\mathbf{k}) \qquad (2.10)$$

where

$$W^*(k) \equiv -12\pi i \int dr \, r^2 \, W(\mathbf{r}) \, j_1(kr). \qquad (2.11)$$

From Parseval's theorem we find:

$$\langle \mathbf{D}^{\text{th}} \cdot \mathbf{D}^{\text{th}} \rangle \equiv \frac{1}{V_u} \int d^3r_0 \, \mathbf{D}^{\text{th}} \cdot \mathbf{D}^{\text{th}} = \frac{b^2}{(2\pi)^3} \int d^3k \, P(k) \, W(k) W^*(k). \qquad (2.12)$$

To obtain the total variance we must add the shot noise variance, which turns out to be

$$\langle \mathbf{D}^{\text{sn}} \cdot \mathbf{D}^{\text{sn}} \rangle = \frac{9}{N_0} \frac{\int_{S_{\min}}^{S_{max}} dS \, S^{-5/2} \int_{S_{\min}}^{S_{max}} dS \, w^2(S) \, S^{-5/2}}{\left(\int_{S_{\min}}^{S_{max}} dS \, w(S) \, S^{-5/2} \right)^2}. \qquad (2.13)$$

The final result then is that the observation defined by equation 2.5 is a single sample of a zero-mean, statistically isotropic, Gaussian vector field $\mathbf{D}$ with variance

$$\langle \mathbf{D} \cdot \mathbf{D} \rangle = \langle \mathbf{D}^{\text{th}} \cdot \mathbf{D}^{\text{th}} \rangle + \langle \mathbf{D}^{\text{sn}} \cdot \mathbf{D}^{\text{sn}} \rangle. \qquad (2.14)$$

2.3 - Modeling N-dipoles

One can learn more from the data by splitting it into flux bins; and in the limiting case of very fine bins one would be using full information contained in the run of the dipole with flux.

The multivariate probability distribution of a set of M measurements of Gaussian variables a_i is

$$P(a_1, a_2, \ldots a_M) \, d^M a = \frac{d^M a}{\sqrt{(2\pi)^M |A|}} \exp -\frac{1}{2} \sum_i \sum_j a_i \, A_{ij}^{-1} \, a_j. \qquad (2.15)$$

where $A_{ij} \equiv \langle a_i a_j \rangle$. Here we have N dipoles, defined by N different weighting functions, each with 3 components labelled α, β etc. By virtue of the assumed spherical symmetry of the sample we have $A_{(i\alpha),(j\beta)} = \delta_{\alpha\beta} C_{ij}$ with

$$C_{ij} \equiv \langle D_{i\alpha}^{\text{th}}(\mathbf{r}) D_{j\alpha}^{\text{th}}(\mathbf{r}) \rangle + \langle D_{i\alpha}^{\text{sn}}(\mathbf{r}) D_{j\alpha}^{\text{sn}}(\mathbf{r}) \rangle, \qquad (2.16)$$

where no summation over α is implied, and where

$$\langle D_{i\alpha}^{\text{sn}}(\mathbf{r}) D_{j\beta}^{\text{th}}(\mathbf{r}) \rangle = \delta_{\alpha\beta} \, \frac{1}{3} \, \frac{b^2}{(2\pi)^3} \int d^3k \, P(k) \, W_i(k) \, W_j^*(k) \quad (2.17)$$

and

$$\langle D_{i\alpha}^{\text{th}}(\mathbf{r}) D_{j\beta}^{\text{sn}}(\mathbf{r}) \rangle = \frac{3\delta_{\alpha\beta}}{N_0} \frac{\int dS \, N(S) \int dS \, N(S) \, w_i(S) \, w_j(S)}{\int dS \, N(S) \, w_i(S) \int dS \, N(S) \, w_j(S)} . \quad (2.18)$$

The final probability distribution for the $D_{i\alpha}$ can then be written as

$$P(D_{i\alpha}) d^{3N} D = \frac{d^{3N}D}{(2\pi)^{3N/2}|C|^{3/2}} \exp - \frac{1}{2} \sum_i \sum_j C_{ij}^{-1} \sum_\alpha D_{i\alpha} D_{j\alpha} .$$
$$(2.19)$$

This formula contains, via 2.17, the normalisation parameter $b\sigma_\varrho$, and gives, on inserting actual measurements, the likelihood function $L(b\sigma_\varrho) = P(\text{data}|b\sigma_\varrho)$. We will also look at the χ^2 statistic that appears in the argument of the exponential.

2.4 - IRAS Dipoles

As our data source we use a colour-selected IRAS sample (Meurs and Harmon 1988; Harmon, Lahav, and Meurs 1987) at 60μ, which is based on version II of the IRAS *Point Source Catalog*. The advantages of this sample are its good sky coverage (more than 75%) and its "objective" selection of galaxies, free from association with optical galaxies. For this IRAS sample we use a double power law luminosity function at 60μ (Lawrence *et al.* 1986):

$$\Phi(L) dL = CL^{-\alpha} \left[1 + \frac{L}{L_*\beta} \right]^{-\beta} dL \qquad (2.20)$$

with $\alpha = 1.8$, $\beta = 1.7$ and $L_* = 1.58 \times 10^{10} \, h^{-2} \, L_\odot$.

Fig. 2.1a shows the IRAS radial window functions (eq. (2.9)-(2.10)) for flux limits $0.7 - 10.0$ Jy. The solid line corresponds to a number weighted window ($w(S) = 1$) and the dashed line corresponds to a flux weighted window ($w(S) = S$). The ordinate chosen is $r^3 W(r)$, which is proportional to the net weight per logarithmic distance interval, and this peaks at about 10000 km/sec.

Fig. 2.1b shows the corresponding k-space window functions (eq. (2.11)). The ordinate shows $k^2 W^2(k)$ for the number and flux weighted windows and

IRAS WINDOW FUNCTIONS

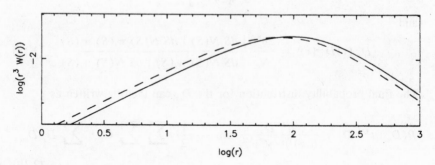

FIG. 2.1a. IRAS radial window functions (weighted by the cube of the distance), as calculated from the IRAS luminosity function (eq. 2.9) for a flux range 0.7-10.0 Jy. The curves correspond to number weighted (solid line) and flux weighted (dashed line) windows. The plot indicates that in a uniform universe a galaxy with a flux in this range has the highest probability to be found at a distance of about 100 Mpc/h. Note the similarity of the two curves.

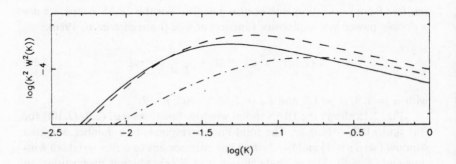

FIG. 2.1b. The Fourier transform of the curves in Fig. 2.1a according to eq. 2.11 (weighted by the square of the wave-number). The dashed-dotted line shows k F(k), where F(k) is the CDM power-spectrum (eq. 2.2). The curves show the contribution to the integral 2.12, for the theoretical dipole.

the dashed-dotted line shows $k\,F(k)$, where $F(k)$ is the CDM power spectrum given in (2.2). By plotting these expressions we can see the contribution of various wavelengths to the integral (2.12). It is clear from these figures that for this sample, the two weighting schemes give very similar windows, and therefore probe very similar depths. Consequently, measuring both dipoles will give little more information than either one alone.

Windows can however be constructed which do give better resolution in depth, and which therefore have greater power to discriminate between theories. Fig. 2.2 shows the windows for 5 IRAS differential flux slices, where the slices boundaries are 0.7, 1.2, 2.0, 3.5, 5.9, 10.0 Jy (see also Table 2.2). The solid curve corresponds to the fainter slice (0.7-1.2 Jy). The other curves are shifted to smaller distances in increasing order of their fluxes.

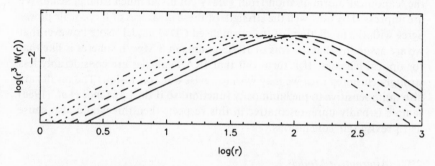

5 IRAS WINDOW FUNCTIONS

FIG. 2.2. IRAS radial window function for 5 differential flux slices (see Table 2.2). The solid line corresponds to the faintest slice, which samples the most distant volume. The other curves are shifted to smaller distances in increasing order of their fluxes.

Details of the construction of the dipoles are given in Kaiser and Lahav (1988). The results are given in columns 1-6 of Table 2.1. In column 7 we show the predicted dipoles assuming $b\sigma_\varrho = 1$. The predictions and the observations are quite similar. Both increase with increasing flux, but the predictions do so more rapidly than the data. This is interesting, since it is an indication that the data would fit better to a model with somewhat more power at larger scales than the CDM model - though in order to fit the normalisation condition for 8 Mpc/h spheres the spectrum would have to turn up again at small scales. To see quantitatively whether these dipoles are consistent with CDM we calculated the statistic $\chi^2 \equiv \Sigma_i \Sigma_j C_{ij}^{-1} \Sigma_\alpha D_{i\alpha} D_{j\alpha}$. This should be roughly Gaussian distributed with mean $3N$ (= 15 in this case) and variance $6N$. The result was 15.0 exactly, clearly a bit of a fluke, but equally clearly showing that the dipoles are quite compatible with CDM.

TABLE 2.1
FLUX WEIGHTED IRAS DIPOLES

S_{min}	S_{max}	D_x	D_y	D_z	D_{obs}	D_{pred}
0.7	1.2	-0.008	-0.131	0.049	0.140	0.110
1.2	2.0	-0.003	-0.103	0.089	0.136	0.154
2.0	3.5	0.053	-0.179	0.147	0.238	0.212
3.5	5.9	-0.102	-0.182	0.081	0.224	0.295
5.9	10.0	-0.177	-0.090	0.168	0.260	0.401

We have also calculated the likelihood versus $b\sigma_\varrho$ as shown by the strong solid line in Fig. 2.3. The ML solution is $b\sigma_\varrho = 1.1$, which agrees well with the 'canonical' normalisation from galaxy counts at much smaller scale. We are impressed by how well the strength of these measures of large-scale power agree with the predictions of the normalised CDM model. Note however that we are assuming that the rms of IRAS counts in 8 Mpc/h spheres is like that for optical galaxies. If it turns out that IRAS galaxies are considerably less strongly clustered then we may wish to revise our conclusion. The results are also quite sensitive to the luminosity function, so if the Lawrence *et al.* (1986) sample is badly unrepresentative in this respect, then this would also cause us to revise our conclusions.

2.5 - Alternative Models

An alternative to the random Gaussian fields considered so far is to model the perturbation as a single localised overdense region. Figure 2.2 is very useful here as it shows graphically how changing the distance of the region producing the dipole affects the run of **D** with flux. If the perturbing region is situated at $r < 20$ Mpc/h then the high flux dipole should be 4.5 times that for the lowest flux, whereas we observe a factor 2 difference. Setting the perturbation ar $r = 60$ Mpc/h gives equal dipole for these flux bins, and for $r > 200$ Mpc/h the low flux dipole is 8 times larger than the highest flux bin - quite different from that observed. The best fitting single distance perturber lies at $r \simeq 40$ Mpc/h. Our result differs from the conclusion of Villumsen and Strauss (1987) who used the same 'delta-function' model and obtained a best fit for a perturber at 17.5 Mpc/h. They used the Meiksin and Davis (1986) sample which is based on different selection criteria and has poorer sky coverage than that used here, so perhaps the difference in results stems from this. Our result is also somewhat in disagreement with the result of Lahav, Rowan-Robinson, and Lynden-Bell (1988), who concluded that both the IRAS and optical dipoles are generated on scales smaller than 4000 km/sec. Again, their samples

5 IRAS dipoles – CDM spectrum

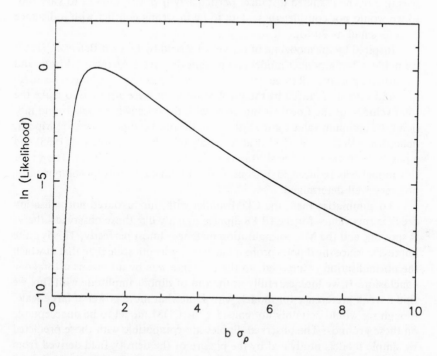

FIG. 2.3. The likelihood function versus $b\sigma_\varrho$ (where b is the bias parameter and σ_ϱ is the normalization), using 5 IRAS flux dipoles (see Table 2.2). The curve peaks at $b\sigma_\varrho = 1.1$.

and model (a "shell model") are different, and that may be a cause for discrepancy.

Our preferred distance $r \simeq 40$ Mpc/h is reminiscent of the Great Attractor picture. We have made a more detailed calculation of these 5 dipoles for a spherically symmetric r^{-2} density enhancement profile centred at 40 Mpc/h and which induces a retardation of the Hubble flow at our position of 600 km/s in the direction of the Local Group - microwave background motion. If we assume a b of unity the dipoles come out too small, but with $b = 1.5$ the predicted dipole amplitudes are in very good agreement with those observed. The directions of the dipoles are in poorer agreement - clearly, a single attractor model predicts that all dipoles should be parallel - but it is encouraging that we see at least rough agreement in the scale of the dominant perturbation in our vicinity as determined from these very different techniques. The value of $\chi^2 \equiv \Sigma(D_{\text{obs}} - D_{\text{pred}})^2/D_{\text{sn}}^2$ was 13, with 15 degrees of freedom. We experimented with other distances and found that χ^2 was disappointingly insensitive to the assumed distance, particularly if b was allowed to vary too, so we would not consider these data to be in strong conflict with a distance even as small as 20 Mpc/h.

Inspired by the modeling of the velocity field by Lynden-Bell *et al.* (1988) we made a 3-component model comprising the Great Attractor, Virgo, and a third component. All three components had the same r^{-2} profile. The direction and velocity induced by the third component were adjusted to make the total velocity of the Local Group agree with the microwave background motion. The optimum value for the distance to the third component was 18 Mpc/h, in accord with the idea of a Local Anomaly of Faber and Burstein (1988) and a possible effect of the Local Void (Lynden-Bell and Lahav 1988). The χ^2 for this model was reduced to 10, but again, the distance to this component is not very well determined.

To summarise then, the CDM model with our favoured normalisation predicts rms values for the IRAS dipoles roughly like those observed - the χ^2 is spot on, and the ML normalisation matches almost perfectly. This is quite impressive since the dipoles probe a much larger length scale than that at which the normalisation is imposed, so the outcome was by no means a foregone conclusion. If we look carefully at the run of dipole amplitude with flux we find a weak preference for models with somewhat more power at large scale, though we would certainly not consider the CDM model to be unacceptable on these grounds. The observed dipoles are compatible with those predicted by simple models motivated by the picture of the density field derived from quite independent studies of the peculiar velocity field, though clearly the real universe is more complicated than any of these simple models.

3. PECULIAR VELOCITIES

We now consider dipole moments of the line of sight peculiar velocity field - often called 'bulk flow solutions'. The analysis here is formally very similar to that of the previous section. One important difference, however, is that while the angular counts constrain the product $b\sigma_\varrho$, the velocity data constrain the *mass* fluctuations σ_ϱ alone.

The velocity field $\mathbf{v}(\mathbf{r})$ and $\Delta(\mathbf{r})$ are connected by the equation of continuity $\nabla \cdot \mathbf{v} = -\partial\Delta/\partial t$, or for $\Omega_0 = 1$, in terms of Fourier components

$$\mathbf{v}(\mathbf{k}) = \frac{-iH_0\mathbf{k}}{k^2} \Delta(\mathbf{k}). \tag{3.1}$$

This formula gives the velocity of a particle with respect to the uniformly expanding frame defined by the matter at large distances. Operationally this can be defined to be the frame in which the microwave background dipole anisotropy vanishes.

3.1 - Window Function for a Velocity Dipole

Let us assume that we are given a catalogue of direction vectors $\hat{\mathbf{r}}_q$, distance estimates and associated distance error estimates r_q, σ_q and redshifts corrected to the microwave frame z_q. An estimate of the line of sight peculiar velocity of the qth galaxy is $u_q \equiv cz_q - r_q$. We can construct a distance weighted dipole of the u_q which is analogous to equation 2.5:

$$\mathbf{U} \equiv \frac{3 \sum w(r_q) \, \hat{\mathbf{r}}_q \, u_q}{\int d^3r \, n(r) \, w(r)} \, . \tag{3.2}$$

As with the angular dipoles, the weighting scheme $w(r)$ is as yet unspecified. Note that for the case of uniform Hubble expansion, $\mathbf{U}$ vanishes, and for a pure 'bulk flow', $\mathbf{U}$ simply measures the flow velocity.

Neglecting for the moment the uncertainty in the distance estimates, and assuming that the galaxies observed are approximately spherically distributed around the sky with a radially varying number density $n(r)$, we can write this 'theoretical' $\mathbf{U}$ (see 2.7) as a convolution of the peculiar velocity field $\mathbf{v}$:

$$\mathbf{U}^{\text{th}} (\mathbf{r}_0) = 3 \int d^3r \, W(r) \, (\hat{\mathbf{r}}.\mathbf{v}(\mathbf{r}_0 + \mathbf{r})) \, \hat{\mathbf{r}}, \tag{3.3}$$

where the window $W(r)$ is simply a normalised version of the radial weight distribution $n(r)w(r)$:

$$W(r) = \frac{n(r)w(r)}{\int d^3r \, n(r) \, w(r)} \, . \tag{3.4}$$

Applying the convolution theorem we obtain

$$\langle \mathbf{U}^{\text{th}} \cdot \mathbf{U}^{\text{th}} \rangle = \frac{H_0^2}{(2\pi)^3} \int d^3k \, W^2(k) \, P(k), \tag{3.5}$$

where

$$W(k) = \frac{12\pi}{k} \int dr \, r^2 \, W(r) \, j_2(kr), \tag{3.6}$$

and we have defined $j_2(x) \equiv 1/2 \int_{-1}^{1} d\mu \, \mu^2 \, e^{ix\mu}$ (note however that this differs from the usual definition of the spherical Bessel function).

With 3-dimensional velocity data, one would compute

$$\mathbf{U}(\mathbf{r}_0) = \int d^3r \, W(r) \, \mathbf{v}(\mathbf{r}_0 + \mathbf{r}),$$

and the corresponding k-space window would be the standard Fourier integral rather than the j_2 integral above. The main difference is that the use of line of sight velocities increases the acceptance to high-k modes, and thereby increases the variance (Kaiser 1988).

As before, we must include the 'shot noise' velocity that would arise even in a universe with zero peculiar velocity. If the rms error in the line of sight peculiar velocity u for a galaxy at estimated distance r is $\sigma(r)$, then the shot-noise variance in the dipole is

$$\langle \mathbf{U}^{sn} \cdot \mathbf{U}^{sn} \rangle = 9 \int d^3r \; W^2(r) \; \sigma^2(r). \tag{3.7}$$

If the error is dominated by scatter in the 'Tully-Fisher' relation, or whatever analogue thereof is used, then the fractional error is constant with $\sigma(r) = fr$, where $f \simeq 23\%$ for the best methods, and if an inverse variance weighting scheme is used, this becomes

$$\langle \mathbf{U}^{sn} \cdot \mathbf{U}^{sn} \rangle = \frac{9f^2}{\Sigma r^{-2}}. \tag{3.8}$$

Just as for the dipole moments in Sec. 2, if we work in 1st order perturbation theory then the shot noise variance 3.7 and the theoretical variance 3.5 simply add to give the final total variance.

We can construct the probability density function for a set of N bulk flow solutions (defined by N different weighting functions $w(r)$) just as in Sec. 2. First one must construct the window $W(k)$ for each shell. One then proceeds to calculate the theoretical covariance matrix elements:

$$\langle U_{i\alpha} U_{j\beta} \rangle^{th} = \frac{1}{3} \frac{H_0^2 \delta_{\alpha\beta}}{(2\pi)^3} \int d^3k \; P(k) \; W_i(k) \; W_j^*(k). \tag{3.9}$$

Add to these the shot noise matrix elements

$$\langle U_{i\alpha} U_{j\beta} \rangle^{sn} = 3f^2 \delta_{ij} \delta_{\alpha\beta} / \sum r^{-2}, \tag{3.10}$$

which we have again taken to be diagonal since we will not consider overlapping samples. Finally, after inverting the resulting matrix $C_{ij} \equiv \langle U_{i\alpha} U_{j\alpha} \rangle^{th} + \langle U_{i\alpha} U_{j\alpha} \rangle^{sn}$, we have

$$P(U_{i\alpha}) d^{3N}U = \frac{d^{3N}U}{(2\pi)^{3N/2} |C|^{3/2}} \exp -\frac{1}{2} \sum_i \sum_j C_{ij}^{-1} \sum_\alpha U_{i\alpha} U_{j\alpha} \tag{3.11}$$

(see equation 2.19).

3.2 - The Data vs Predictions

Dressler *et al.* (1987) have given a streaming solution of the form 3.3, for the galaxies with estimated distances $r < 60$ Mpc/h, with a weighting function which, for large r, varies inversely as the variance; $w(r) \propto 1/r^2$. This weighting scheme minimises the shot noise fluctuations, which seems a desirable property, but does not necessarily give the maximum signal to noise or maximal cosmologically significant information. The radial distribution of number of galaxies and weight are shown in Figures 3.1 and 3.2. These figures are very informative; note how the weight is concentrated at small distances; this results in a much broader window function in k-space than if one imagined the weight to be uniformly distributed in a sphere of radius 60 Mpc/h. Indeed, it is apparent that with so little weight at large distances, the choice of outer boundary is essentially irrelevant. We stress that in order to model a real observation it is necessary to know how the weight in the solution is distributed in radius. If a theoretical window (Gaussian sphere or whatever) is adopted with a scale length tied to the outer boundary of the sample then the results are likely to be misleading.

In a more recent paper, Lynden-Bell *et al.* (1988) have given bulk flow solutions for a variety of shells around us. In Table 3.1 we give the rms predictions for streaming solutions with $1/r^2$ weighting and with upper and lower limits to estimated distances shown. The predictions given correspond to an 'unbiased' normalisation $\sigma_\varrho = 1$. We have also included in this table the prediction for the Local Group motion, which we model as the motion of a point.

TABLE 3.1

FIVE *BULK FLOW* SOLUTIONS AND THEIR EXPECTATION VALUES

Sample	U_x	U_y	U_z	$|U|$	U_{th}	U_{sn}	U_{tot}	
1	− 20	− 532	277	600	990	— —	990	local group
2	362	− 486	112	616	431	90	440	E's 0-3200
3	194	− 132	− 3	235	268	251	376	E's 3200-8000
4	61	− 349	− 13	355	627	31	628	AHMST 0-3000
5	139	− 398	15	420	590	29	591	E's + AHMST < 3200

If we consider a biased model with $\sigma_\varrho = 0.7$ for instance, then the Local Group motion lies right on the rms prediction; the distant sample is still a little lower than the prediction; but the solution for the galaxies in the range 0-32 Mpc/h is now about twice the rms prediction, which is starting to look embarrassing. Before condemning this model on the basis of this single observation one should pause and consider the following points. First, one should

N(r) for E-galaxy sample

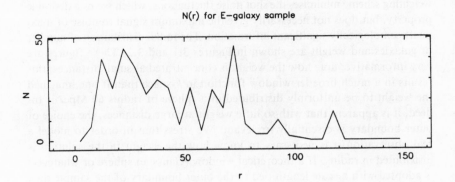

FIG. 3.1. The radial distribution of the sample of elliptical galaxies of Lynden-Bell *et al.* (1988). Each galaxy is equally weighted.

Weight distribution for E-galaxies

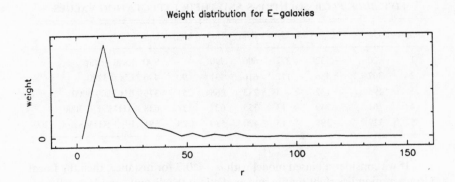

FIG. 3.2. The radial distribution of ellipticals, where each galaxy is weighted by the inverse of its variance. Since the variance increases as the square of the distance, the distribution is peaked at a much smaller distance than the unweighted distribution in Fig. 3.1.

note that the elliptical galaxies give the largest motion for this dist ance range (616 km/s), while the solution of Aaronson *et al.* (1982, hereafter AHM.T) for their nearby spiral sample gives the much smaller value of 355 km/s (li..e 4 of Table 3.1), as do the results of de Vaucouleurs and Peters. Seco .d, o..e should ask how the division between the shells was obtained; in our str tistical model we implicly assume that such parameters are set *a priori,* but cle-rly that is a fiction and there is the possibility for bias here. Third, but closely related to the second point, one should ask oneself how many quasi-independer . tests of the model could have been constructed, and whether to have fou .d one 2-sigma vector is really unacceptable. We have also included as the 5th line in Table 3.1 a weighted combination of the ellipticals and the spirals which has $|U| = 420$ km/s.

It is interesting to note that the 'far field' elliptical sample has a motion which is not detected above the noise level. The initial response to these observations was that they ruled out CDM because they demonstrated coherent flows on very large scales (e.g. Vittorio *et al.* 1986; Bond 1987). With the data separated into near and far field, we can see clearly that even the unbiased CDM model predicts an rms velocity which is now, if anything, *larger* than that observed in the far field. It is only fair to add that there have been other claims of much larger motions for distant spirals, and quite possibly these may be problematic for theories like CDM, in which the rms velocity falls inversely as the sample depth at large distances. If these observations hold up then this would be very exciting.

As with the angular dipoles, an interesting quantitative way to see whether these data, taken together, are consistent with the models is to calculate χ^2 (i.e. twice the argument of the exponential in 3.11). We took 3 dipoles as the data. The first is the Local Group motion; the second is the weighted combination of the near field ellipticals and the AHMST spirals; the third is the far field ellipticals. We consider these to be the best data available for the task. With our preferred normalisation $\sigma_\varrho = 0.6$, χ^2 came out to be 8.7, whereas the expectation is 9. Again, the fact that this is so close to the expectation can only be a fluke, but this does not alter the conclusion that these data are compatible with CDM.

In Figure 3.3 we show the likelihood function $L(\sigma_\varrho)$ using again the 3 dipoles. The most likely value $\sigma_\varrho \simeq 0.6$ agrees very well with our canonical biased CDM model, and even a smaller amplitude of $\sigma_\varrho \simeq 0.5$ would not be ruled out. The form of these functions is very similar to those obtained from the angular dipoles. It is important to note that the likelihood falls quite slowly for large values of σ_ϱ - reflecting the fact that it is easier to reconcile a low observation with a high rms prediction than *vice versa* - so we would not exclude models with somewhat more power at large scales than CDM. As with the dipole results however, we are struck by how well the predictions of the CDM model agree with the observations.

3 STREAMING SOLUTIONS

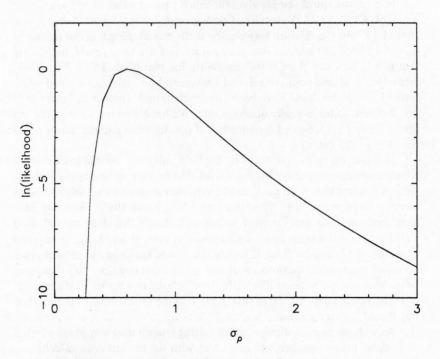

FIG. 3.3. The likelihood function versus the normalization parameter σ_ϱ, using 3 velocity vectors: the motion of the Local Group, a motion of a nearby (< 32 Mpc/h) shell of ellipticals and spirals, and a distant (32-80 Mpc/h) shell of ellipticals. The most likely value is $\sigma_\varrho = 0.6$. Note the asymmetry of the likelihood function.

4. ANGULAR DIPOLES, BULK FLOWS, AND THE BIAS PARAMETER

In the previous two sections we looked at the angular dipoles and bulk flow solutions and asked whether they were compatible with the CDM predictions. Here we wish to use these data together to test for the correlations expected between these vectors under the hypothesis that structure is growing by gravitational instability, and to determine the bias parameter b which appears as a constant of proportionality.

The basic relation between the peculiar velocity and the peculiar acceleration is (Peebles 1980):

$$\mathbf{v} = H_0 \, f(\Omega_0) \, \mathbf{g} \tag{4.1}$$

where $f(\Omega_0) \simeq \Omega_0^{0.6}$ and

$$\mathbf{g} = \frac{1}{4\pi} \int d^3r\, \Delta(r)\, \frac{\mathbf{r}}{r^3}. \qquad (4\ 2)$$

The test, which we understand was originally due to Gott, has most co mo⌐-ly been applied (e.g. Yahil *et al.* 1980, Davis and Huchra 1982, Yah⁻¹ *et al.* 1986, Lahav 1987) using the flux weighted angular dipole and the velocity of the Local Group. The idea here is that since both flux and acceleration vary as $1/r^2$, the flux weighted dipole should look very much like equaton 4.2, or ly with $\Delta_{galaxies}$ in place of the mass fluctuation Δ, and give an estimate of ɔ times) the acceleration g.

The limitations of this test arise from the flux limits that must realistically be imposed and from the shot noise resulting from the finite number of galaxies. With no flux limits, the window function for a flux weighted dipole would vary as $1/k$ for all wavelength. We see from Figure 2.1b that this ideal is not realised in practice for the IRAS sample, and that $kW(k)$ is strongly dependent on wavelength. The lower flux limit is unavoidable. The upper limit could in principle be increased, but the price paid is that the shot noise becomes unacceptably large. The result then is that a realistic dipole will not sample the acceleration from all shells equally, and consequently any constraint on b will depend on where one assumes the acceleration originates. Here we will obtain a best estimate of b, and of the statistical uncertainty, under the hypothesis that the fluctuations are Gaussian with the CDM spectrum.

We first consider the traditional application of the test using $\mathbf{D}_F$ and $\mathbf{U}_{LG}$. We find the uncertainty to be quite large. The reasons for this are twofold. In the CDM-world there are big fluctuations in the distance from which the acceleration arises. Because the shape of the window function differs from the idealised flux weighted $1/k$ form this converts to a big uncertainty in b. Second, because one is attempting to measure the acceleration acting on a point, the graininess of the nearby galaxies inevitably gives a big shot noise term. Additionally, there is the underlying worry as to how well the Local Group motion is described by linear theory, though that is much harder to quantify.

In Sec. 4.2 we look at a modified test that gives reduced uncertainty. The basic idea here is that by giving less weight to the high flux galaxies the shot noise will be reduced. Now, this will also attenuate the acceptance of the window to high-k modes still further, and this will degrade the correlation of this dipole with the Local Group motion, so little, if any, gain in performance will be achieved. However, the correlation with other bulk-flow solutions (which also have window functions with high-k modes strongly attenuated) will increase, so if we use these in place of the Local Group motion we should get a more precise test. Crudely, the aim is to find a low noise dipole which measures the acceleration acting on the Local Supercluster, or some other large region for which the peculiar velocity has been measured. The specific problem we solve is: given a particular bulk-flow solution, what is the flux-weighting scheme that results in the optimum dipole for estimating b. This question of what is the 'best' dipole to measure the acceleration was discussed by Lynden-Bell and Lahav (1988). Clearly any such optimisation is model dependent; the solution given here is contingent on the assumption of Gaussian fluctuations. By using many bulk-flow solutions, a whole family of tests can be constructed which are independent to the extent that the windows sample different regions of k-space. Finally, in Sec. 4.3 we apply all tests simultaneously by means of multivariate likelihood analysis to obtain our best estimate of b.

4.1 - Flux Weighted Dipole and the Local Group Velocity

The joint probability distribution $P(\mathbf{D}_F, \mathbf{U}_{LG})$ is a 6-variate zero-mean Gaussian with covariance matrix elements

$$C_{11} = \frac{1}{3} \langle \mathbf{U}_{LG} \cdot \mathbf{U}_{LG} \rangle = \frac{H_0^2}{3(2\pi)^3} \int d^3k \; P(k) \; W_U(k) \; W_U^*(k)$$

$$C_{12} = \frac{1}{3} \langle \mathbf{U}_{LG} \cdot \mathbf{D}_F \rangle = \frac{bH_0}{3(2\pi)^3} \int d^3k \; P(k) \; W_U(k) \; W_D^*(k) \quad (4.3)$$

$$C_{22} = \frac{1}{3} \langle \mathbf{D}_F \cdot \mathbf{D}_F \rangle = \frac{b^2}{3(2\pi)^3} \int d^3k \; P(k) \; W_D(k) \; W_D^*(k) \; + \; \text{shot noise}$$

where $W_D(k)$ is given by equation (2.12), with the shot noise variance given by equation (2.15), and, treating the local Group as a point, we have $W_U = 1/k$. We ignore shot noise for the Local Group velocity.

The likelihood function $L(b) \equiv P(\mathbf{D}_F, \mathbf{U}_{LG} | b, ...)$ is shown in Figure 4.1 for the IRAS flux dipole given in Table 2.1 and for a Local Group velocity of 600 km/sec towards galactic coordinates ($l = 268°$, $b = 27°$). We have set $\sigma_\varrho = 0.7$. The curve peaks at $b_{max} = 1.5$. For $\sigma_\varrho = 1.0$ we find $b_{max} = 1.3$. These estimates are very similar to those obtained in Sec. 2.5 by matching the 5-dipole IRAS amplitudes with the predictions in the Great Attractor model.

An illuminating way of quantifying the uncertainty in b in the method is to use the conditional probability $P(\mathbf{U}_{LG} | \mathbf{D}_F, ...)$ which is Gaussian with mean

$$\hat{\mathbf{U}}_{LG} = a\mathbf{D}_F, \quad (4.4)$$

where

$$a = \langle \mathbf{U}_{LG} \cdot \mathbf{D}_F \rangle / \langle \mathbf{D}_F^2 \rangle \quad (4.5)$$

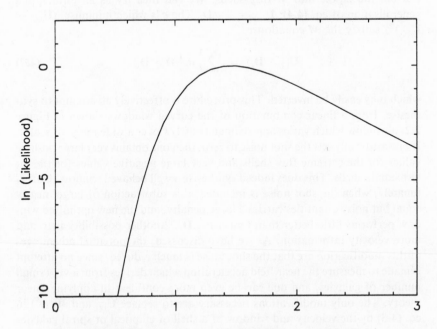

FIG. 4.1. The likelihood function versus the bias parameter b, using the velocity vector of the Local Group and an IRAS flux dipole. The normalization is set to be $\sigma_\varrho = 0.7$. The curve peaks at $b = 1.5$, but it is rather broad.

and with variance

$$\langle (\mathbf{U}_{\mathrm{LG}} - \bar{\mathbf{U}}_{\mathrm{LG}})^2 \rangle = \langle 3\mathbf{U}_{\mathrm{LG}}^2 \rangle \left[1 - \frac{\langle \mathbf{U}_{\mathrm{LG}} \cdot \mathbf{D}_{\mathrm{F}} \rangle^2}{\langle \mathbf{U}_{\mathrm{LG}}^2 \rangle \langle \mathbf{D}_{\mathrm{F}}^2 \rangle} \right]. \quad (4.6)$$

We can say that a measurement of a flux-dipole $\mathbf{D}_F$ *predicts* a velocity $a\mathbf{D}_F$ with uncertainty given by (4.6). A good estimator would be one that has $E_U^2 \equiv \langle (\mathbf{U}_{\mathrm{LG}} - \bar{\mathbf{U}}_{\mathrm{LG}})^2 \rangle / \langle \mathbf{U}_{\mathrm{LG}}^2 \rangle \ll 1$. Unfortunately, with this choice of dipoles the coupling is quite weak because of the poor shape of W_D and the shot noise. For the IRAS dipole we find $E_U = 35\%$.

We will now consider some modifications which result in a more tightly coupled combination of dipoles.

4.2 - Alternative Schemes

One possibility is to vary the flux weighting scheme. As in Section 2 we can bin the dipole into M flux shells. We can then try as an estimator a generalisation of eq. (4.4): $\bar{U}_{LG} = \Sigma_i a_i D_i$. The a_i's which minimise $\langle(U_{LG} - \bar{U}_{LG})^2\rangle$ satisfy the M equations:

$$\langle U_{LG} \cdot D_j \rangle = \sum_i a_i \langle D_i \cdot D_j \rangle \qquad (4.7)$$

which may easily be inverted. This procedure is effectively an attempt to synthesise, from a linear combination of the curved windows shown in Figure 2.2, a window which varies approximately as $1/k$ over a wide range of scales. If one artificially sets the shot noise to zero, then one obtains very large positive values for the extreme flux shells and very large negative values for the intermediate shells. This does indeed synthesise a well behaved window. Unfortunately, when the shot noise is included, this substraction of large, nearly equal but noisy quantities carries a large penalty, and the new optimised window performs little better than that using D_F. Another possibility is to add more velocity information. As we have discussed, the potential advantages of this modification are that the shot noise is much reduced, since no attempt is made to measure the near field acceleration which derives from a very small number of galaxies, and one can be even more confident in applying linear theory. The only modifications necessary are to replace U_{LG} and $W_U(k)$ in eq. (4.3) by the velocity and window of a shell of elliptical or spiral galaxies given in Table 3.1. Rather than present the results for a number of quasi-independent bulk-flows, we prefer to proceed directly to the more general likelihood function which uses all the data together rather than special linear combinations.

4.3 - Multi-dipole Likelihood Function

We use 5 IRAS flux dipoles (Table 2.2) and 3-velocity shells (from Table 3.1) to construct the likelihood function.

$$L(b) = P(D_1, ..., D_5, U_1, ..., U_3 | b...). \qquad (4.8)$$

As U_1 we use the Local Group velocity (though because this has a strong correlation matrix element to the noisy high flux dipole it enters the likelood function with little weight), as U_2 we use the combined velocity of the nearby shell (ellipticals and spirals) and as U_3 we use the velocity of the distant ellipticals shell (3200 — 8000 km/sec). Figure 4.2 shows the likelihood function (4.8) for $\sigma_\varrho = 0.7$. The curve peaks at $b_{max} = 1.6$, and the location of the peaks is insensitive to σ_ϱ. See how narrow the curve is, in particular when compared

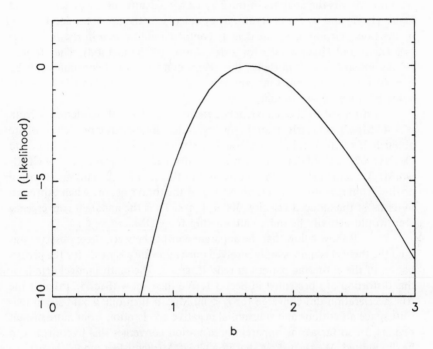

FIG. 4.2. The likelihood function versus the bias parameter b, using the 3 velocity vectors (of Fig. 3.3) and 5 IRAS flux dipoles with $\sigma_\varrho = 0.7$. The curve peaks at $b = 1.6$. Note that the curve is much narrower than the curve of Fig. 4.1.

to Figs. 4.2 and 5.3. This demonstrates how likelihood analysis, which uses all the data simultaneously, has considerably reduced the statistical uncertainty.

5. 'GRAVITY' VECTOR FROM REDSHIFT SURVEYS

In the previous sections we have used angular dipoles of the galaxy counts to estimate the acceleration acting on the Local Group and on larger regions, and we have used the run of dipole with flux to get a handle on the distance from which this acceleration derives. A more or less consistent picture has emerged; the run of **D** with flux agrees approximately with that predicted in the Great Attractor model, and comparing the acceleration with the Local Group velocity (or the more sophisticated test of Sec. 4.3) gives consistent estimates of $b \simeq 1.5$. In addition, the amplitudes of both angular dipoles and peculiar velocities are quite consistent with CDM.

An important piece of evidence which, at first sight, seems to conflict with this view is the analysis of the 2 Jy IRAS redshift survey by Strauss and Davis (1987) and by Yahil (1988). They calculated the apparent acceleration of the Local Group using the density contrast field as seen in *redshift* space (see Davis and Huchra 1982 for a description of the method). They found: (i) that integrating out in shells, this vector converged to a constant value by $r = 40$ mpc/h; and (ii) that comparing this acceleration with U_{LG} gave a value of b very close to unity.

With regard to the 'convergence radius', the lack of acceleration from $r > 40$ Mpc/h is clearly incompatible with the Great Attractor picture where about half of the total acceleration arises from these distances. Strauss and Davis (1987) claimed that what they saw conformed to the predictions of CDM-world. A simple calculation reveals that in this model, the rms acceleration arising from $r > 40$ Mpc/h is about 1/3 of the total rms, or, when expressed in terms of the induced velocity, 300 σ_ϱ km/s. Thus the apparent convergence seen would actually be quite embarassing for CDM.

We will show below that the apparent conflict between these observations (and the theory) is quite simply resolved once allowance is made for the perturbation of the clustering pattern in redshift space. The mathematical details of this distortion are presented in Sec. 5.1. We then show (Sec. 5.2) that if the true acceleration did converge for $r < 40$ Mpc/h then one would see in redshift space a fictitious and substantial negative acceleration from more distant regions. In so far as the apparent acceleration converges this hypothesis can be discounted. We then show that in a Great Attractor-like model the net apparent acceleration for 80 Mpc/h $> r > 40$ Mpc/h is small - much like that observed - even though about 1/2 of the true acceleration arises from this region.

Turning to Gaussian models like CDM, we shall show that if we Fourier analyse the density field we find that each Fourier mode gives a contribution to the apparent acceleration vector which is proportional to the true acceleration, but with an amplitude which is wavelength dependent. This can be analysed in terms of a window function for this observation, which is quite analogous to those used in preceding sections. In Sec. 5.2.3 we construct this window function, and use this to calculate the uncertainty in the estimate of b under the assumption of CDM. We find that the uncertainty is larger than that obtained in Sec. 4 using only angular coordinates and fluxes. This is in part because of the distortion of the clustering pattern and in part because of the large shot noise. This latter source of uncertainty is made particularly large by the choice of a 'semi volume limited' weighting scheme. We discuss ways in which the test can be improved.

5.1 - Distortion of Clustering in Redshift Space

The mapping from real space coordinates **r** to redshift space **s** is just

$$s = r \left(1 + \frac{\hat{r} \cdot (v(r) - v(0))}{r} \right) \qquad (5.1)$$

where $v(0)$ is the observer's velocity relative to the cosmic frame and it is assumed that $H_0 = 1$. From this one obtains (Kaiser 1987) for the linearised apparent density perturbation in redshift space:

$$\Delta_s(r) = \Delta_r(r) - \left(2 + \frac{d\ln\phi}{d\ln r} \right) \frac{\hat{r} \cdot (v(r) - v(0))}{r} - \hat{r} \cdot \frac{dv(r)}{dr}. \qquad (5.2)$$

The origin of these terms can be understood as follows. If v were identically zero then redshifts would exactly measure distances and we would have just the first term $\Delta_s = \Delta_r$. The other terms arise because peculiar velocities distort the pattern of clustering in redshift space and introduce spurious apparent density enhancements. Consider the second term in equation 5.2. From 5.1 we see that $2\hat{r} \cdot (v(r) - v(0))/r$ is just twice the fractional radial displacement of a galaxy in redshift space, so this factor gives the amount by which a volume element is transversely compressed or rarefied assuming $v(r)$ to be constant over the volume element. If, for instance, redshift underestimates the distance to the galaxies in some region, this causes an apparent density enhancement, relative to the density we would measure for the same galaxies with real distance estimates. The logarithmic derivative of the selection function comes in because, in computing Δ_s we compare the number of galaxies with that predicted from the selection function, but evaluated at a perturbed position. (It is interesting to note that if nature were so generous as to provide a population of galaxies with $\Phi(L) \propto L^{-2}$, so $\phi(r) \propto r^{-2}$, this correction terms vanishes. Were this the case one would also be spared the need to collect redshifts, since the sum in equation 5.3 below is then independent of the redshifts. Unfortunately, for real galaxies this does not seem to be the case.)

Finally, the third term is necessary to allow for gradients of the peculiar velocity field, which act to compress or rarefy the volume element in the radial direction.

It is not difficult to modify the analysis to treat a semi-volume limited survey. Such a survey is obtained by taking a redshift-dependent flux limit; $S(s) = S_{lim}$ if $s > s_i$ and $S(s) = (s/s_i)^{-2} S_{lim}$ if $s < s_i$. The result is simply to replace in equation 5.1 the factor $2 + d\ln\phi/d\ln r$ by its value at the boundary of the magnitude limited region.

An interesting consequence of this is that by choosing the boundary appropriately, the second term in equation 5.2 can be made to vanish throughout the volume-limited region. If one had a much deeper redshift survey this might appear to be a very nice way to get around the problems of redshift space distorsions, since provided the volume limited region encompasses the source of our acceleration the 3rd term in 5.2 can easily be corrected for (it simply causes the 'acceleration' measured in redshift space to exceed the true acceleration by a constant factor). Unfortunately, for a 2 Jy survey this would limit the volume limited regime to $r < 20$ Mpc/h which is uninterestingly small.

5.2 - Apparent Acceleration Vector in Redshift Space

Define the 'acceleration' vector:

$$\mathbf{v}_s = \frac{H_0}{4\pi n_1} \sum_i \frac{1}{\phi(s_i)} \frac{\hat{\mathbf{s}}_i}{s_i^2} \tag{5.3}$$

just as in Davis and Huchra (1982), and in Strauss and Davis (1987), and where it is implied that we take $\phi = \phi(s_1)$ in the volume-limited region. Taking the limits of the sum to be spheres in redshift space, and setting $H_0 = 1$ for convenience gives

$$\mathbf{v}_s = \frac{1}{4\pi} \int_{s_{min}}^{s_{max}} d^3s \, \frac{\hat{\mathbf{s}}}{s^2} \, \Delta_s(\mathbf{s}) + \text{shot noise}$$

$$= \frac{1}{4\pi} \int_{s_{min}}^{s_{max}} d^3r \, \frac{\hat{\mathbf{r}}}{r^2} \left(\Delta_r(\mathbf{r}) - \left(2 + \frac{d\ln\phi}{d\ln r} \right) \frac{\hat{\mathbf{r}} \cdot (\mathbf{v}(\mathbf{r}) - \mathbf{v}(0))}{r} \right.$$

$$\left. - \hat{\mathbf{r}} \cdot \frac{d\mathbf{v}(\mathbf{r})}{dr} \right) + \text{shot noise} \tag{5.4}$$

We will presently do the usual decomposition into plane waves, construct the window function for $\mathbf{v}_s$ and proceed as before. There are, however, important conclusions that can be drawn from equation 5.4 directly.

5.2.1 - Apparent acceleration seen by a rocket borne observer

Consider an observer who has a locally generated peculiar motion $\mathbf{v}(0)$ (perhaps generated by a rocket, or by the gravitational influence of nearby matter), but who resides in a universe which is uniform and at rest with respect to the microwave background on larger scales. Outside of the local perturbation, Δ_r and $\mathbf{v}(\mathbf{r})$ vanish, so this observer calculates

$$\mathbf{v}_s = \frac{1}{3} \mathbf{v}(0) \int \frac{dr}{r} \left(2 + \frac{d\ln\phi}{d\ln r} \right)$$

$$= \frac{1}{3} \mathbf{v}(0) \left[\left(2 + \frac{d\ln\phi}{d\ln r} \right)_{s_1} \ln \frac{s_1}{s_{min}} + \ln \frac{\phi(s_{max})s_{max}^2}{\phi(s_1)s_1^2} \right]. \tag{5.5}$$

Thus, the motion of the observer induces a spurious $\mathbf{v}_s$ vector which is of order the true velocity and contains logarithmic divergences. Note that this result is independent of the true value of Ω_0 - it simply results from calculating the 1st order change in equation 5.3 when one changes the inertial frame to which the redshifts are referred. It is hoped that this simple example makes clear that the simple estimator (equation 5.3) cannot in general give a reliable estimate of the acceleration vector. It should also make clear the futility of trying to correct equation 5.3 for Virgo infall for instance, when there are much larger sources of error coming from our uncorrected motion with respect to the galaxies at greater distances.

It is interesting to compare the predicted run of $\mathbf{v}_s$ for this simple 'rocket' model with that observed. Strauss and Davis (1987) found that, integrating out in redshift shells, the acceleration converges by about 4000 km/s, there being negligible contribution from the magnitude limited region from 4000 to 8000 km/s. While no details of the luminosity function adopted by Strauss and Davis (1987) are given, they do tell us that the selection function falls by a factor 10 going from 4000 to 8000 km/s, so using this in equation 5.5 we would predict that in this shell they should have found a *negative* contribution to the dipole of about 1/3 of the observer's velocity, or about 150 km/s since with their correction for Virgo infall the redshifts for distant galaxies used are those that would be seen by an observer moving at 450 km/s with respect to the microwave frame. We do not wish to go into the question of exactly how reliably the data exclude or constrain any such fall in $\mathbf{v}_s$, but it is interesting to note that if one were convinced that $\mathbf{v}_s$ had really converged then the one hypothesis that can be excluded immediately is that we live in a world where the true acceleration converges! One of the many hypotheses that *would* be allowed is that we live in a world with velocities generated by perturbations on a scale much larger than the survey region.

5.2.2 - Apparent acceleration in the Great Attractor-like model

With a specific model for the velocity field one can predict the radial run of $\mathbf{v}_s$ using equation 5.4. The heavy line in Figure 5.1 shows the apparent acceleration build up with increasing redshift for a single attractor model centred at $r = 40$ Mpc/h with an r^{-2} density contrast run producing an infall of 600 km/s at our position (see Lynden-Bell *et al.* 1987). The calculation assumes $b = 1$. One finds that the overall $\mathbf{v}_s$ agrees quite well with the true $\mathbf{v}$ - though this is not generally the case. The light continuous line, the dotted-dashed line and the dotted line correspond to the 3 terms in eq. 5.4 in the order they appear. The growth of $\mathbf{v}_s$ with r is quite different; the appearance is that the result has converged by about 40 Mpc/h, and if anything, there appears to be a net repulsion from the back of the perturbation. The behaviour of $\mathbf{v}_s$ in this model is qualitatively similar to that observed, particularly in the

APPARENT ACCELERATION FROM GT ATTRACTOR

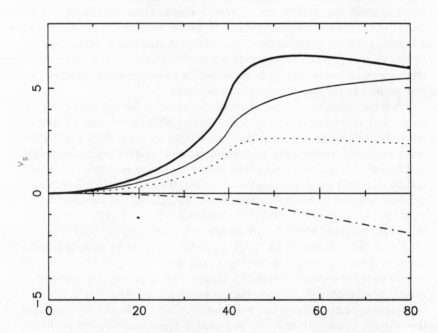

FIG. 5.1. The heavy solid line shows how the apparent acceleration of the
Local Group (in 100 km/sec units) builds up with distance (in Mpc/h) for
a single attractor model placed at 40 Mpc/h (see text). The light continuous
line, the dotted-dashed line and the dotted line correspond to the three terms
of eq. 5.4, in the order they appear. In particular, the light continuous line
shows the true acceleration.

far field, so there is clearly no conflict between the redshift space picture and
the Great Attractor-like model once allowance is made for peculiar velocities.

5.2.3 - Apparent acceleration vector in Gaussian models

While one cannot simply equate v_s and the true acceleration, it is still a
potentially useful statistic. Just as for the observations discussed in the previous
sections, that defined by equation 5.3 corresponds to the convolution of the
density field with a particular window function, and so we can proceed much
as before to calculate the corresponding window function, and calculate the
correlation between v_s and other vectors such as the Local Group motion. For
any particular model, we can then try to compensate for any systematic bias

introduced by the distortion, obtain an improved estimate of b, and quantify the uncertainty.

Doing the usual plane-wave decomposition: $\Delta(r) = \Delta_k e^{i\mathbf{k}\cdot\mathbf{r}}$, $\mathbf{v}(\mathbf{r}) = \hat{\mathbf{k}} v_k e^{i\mathbf{k}\cdot\mathbf{r}}$, with $\Delta_k = (bk/i)v_k$, and performing the integral in 5.3 results in

$$v_s(k) = [b\ W_r(k) + W_s(k)]\ v(k) + \text{shot noise} \tag{5.6}$$

where

$$W_r(k) = j_0(ks_{\min}) - j_0(ks_{\max}) \tag{5.7}$$

and

$$W_s(k) = j_2(ks_{\min}) - j_2(ks_{\max}) + \int_{s_{\min}}^{s_{\max}} \frac{dr}{r}\left(2 + \frac{d\ln\phi}{d\ln r}\right)\left(\frac{1}{3} - j_2(kr)\right) \tag{5.8}$$

where, as before

$$j_2 \equiv \frac{\sin(kr)}{kr} + 2\left[\frac{\cos(kr)}{(kr)^2} - \frac{\sin(kr)}{(kr)^3}\right].$$

We have evaluated these windows for the selection function derived from the Lawrence *et al.* (1986) luminosity function. Figure 5.2 shows the window functions W_r (light continuous), W_s (dashed line), and their sum (heavy line). The actual window differs considerably from the desired window, which in these plots would be the horizontal line $W(k) = 1$. Were the actual window function simply offset vertically this would present no problem, since this would bias the result by a multiplicative factor which can easily be corrected for. The problem revealed by Figure 5.2 is that the actual window for the $\mathbf{v}_s$ estimate is wavelength dependent, so for a realistic power spectrum, this reduces the correlation between $\mathbf{v}_s$ *and* $\mathbf{v}$.

For a quantitative comparison of theory and observation we must of course include the noise in the measurement. We find for the shot noise term

$$\langle \mathbf{v}_s \cdot \mathbf{v}_s \rangle^{\text{sn}} = \frac{1}{N} \int_{s_{\min}}^{s_{\max}} dr\ r^2\ \phi(r) \times$$

$$\left[\int_{s_{\min}}^{s_1} \frac{dr}{r^2\phi(r_1)} + \int_{s_1}^{s_{\max}} \frac{dr}{r^2\phi(r)}\right], \tag{5.9}$$

where s_1 is the boundary of the volume limited region and N is the total number of galaxies in the sample (*not* the number in the semi-volume limited

WINDOW FUNCTIONS FOR STRAUSS DAVIS EXPT

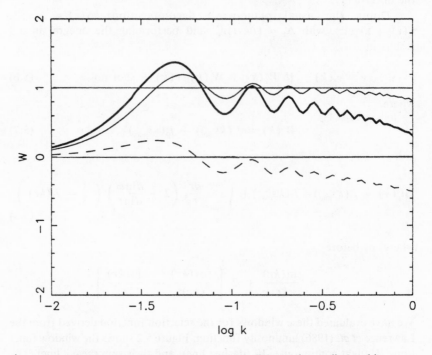

FIG. 5.2. The window functions in real space (light continuous line) and in redshift space (dashed line) and their sum (heavy line). The desired window, which represents the true gravity, is the horizontal line.

sample). This diverges with the lower redshift cuf-off. Taking $s_{min} = 500$ km/s, which seems reasonable, we get $\langle \mathbf{v} \cdot \mathbf{v} \rangle \simeq (340 \text{ km/s})^2$, so the observed signal is only about a factor 2 larger than the rms noise.

We are now in the position to use the $\mathbf{v}_s$ of Strauss and Davis (1987) and the Local Group motion $\mathbf{U}_{LG}$ to esimate b, and, perhaps more importantly, give a realistic estimate of the uncertainty. A good indication of the performance of this test can be obtained by looking at the probability distribution for $\mathbf{U}_{LG}$ conditional on an observation $\mathbf{v}_s$ as used in Sec. 4. With $b = 1$ one finds $\bar{\mathbf{U}}_{LG} \simeq 0.8\mathbf{v}_s$, but the scatter about this is quite large:

$$\langle (\mathbf{U}_{LG} - \bar{\mathbf{U}}_{LG})^2 \rangle \simeq 0.37 \langle \mathbf{U}_{LG}^2 \rangle, \tag{5.10}$$

which, with our favoured normalisation parameter $\sigma_\varrho = 0.7$, gives an rms fluctuation about $\bar{\mathbf{U}}_{LG}$ of about 450 km/s. Since the observed value is about

600 km/s, the total statistical noise in this test is considerable, and arises roughly equally from shot noise and distortion effects. Figure 5.3 shows the likelihood function for b using these data. Not surprisingly it is very broad, though it is interesting that it peaks at a very similar value to that obtained from the angular dipoles in the previous section.

While the performance of this estimator of b (or of Ω_0) is rather poor when compared with the estimate of Sec. 4, there are however several ways to improve the test. The shot noise is large both because of the use of a semi-volume limiting flux cut-off and because the measurement samples the acceleration right up to the observer. Provided one has a good luminosity function estimate, there is no reason for making a semi-volume limited catalogue - unless of course one particlarly wishes to exclude intrinsically faint galaxies, and by not doing this one decreases the shot noise considerably (though tests we have made using alternative forms for the luminosity function suggest that the results

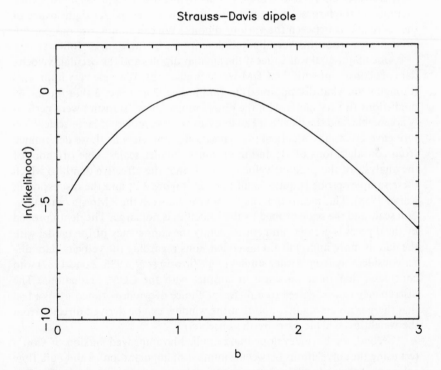

Strauss—Davis dipole

FIG. 5.3. The likelihood function versus the bias parameter b, using the IRAS redshift dipole. The curve peaks at $b = 1.3$. Compare the breadth of this curve to the curves in Figs. 4.1 and 4.2.

are then more sensitive to the choice of luminosity function adopted). With regard to sampling the local acceleration, just as discussed in Sec. 4, there are various advantages in calculating an acceleration which largely ignores the local regions, and which correlates well with the motion of the Local Supercluster, or some other extended region, rather than with the Local Group motion.

Finally, just as in Sec. 4, the formalism we have constructed allows one in principle to use all of the information in the dipoles as a function of distance and luminosity. Ultimately, we hope that by extension of the techniques illustrated here much useful cosmological information can be recovered from the distorted, but 3-dimensional, view of the world provided by redshift surveys.

6. SUMMARY

We have concentrated fairly arbitrarily on *dipole* moments of the line of sight peculiar velocities, and of the counts of galaxies both in flux-angle space and in redshift space. Our choice was motivated in part by common use of these statistics in previous works; in part by the desire to probe density fluctuations on the largest scale, and also because of the special significance of the correlations between the various dipoles. We can summarise our conclusions drawn from this very limited description of the data as follows.

One major goal was to see if the angular dipoles and the peculiar velocity data are consistent with the CDM-world predictions. We were very impressed (though somewhat disappointed) to find in Sec. 2 and Sec. 3 that the model predictions fit very nicely with the observations; the χ^2 statistics were perfectly acceptable, and the likelihood analysis using these probes of large-scale structure gave essentially identical normalisation parameters as those determined from considerations of clustering on much smaller scales. One offshoot of the analysis of the peculiar velocity data is that the effective depth to which these samples probe is quite small (see e.g. Figure 3.2, and the discussion in Kaiser 1988). This means that the 'lever arm' between the 8 Mpc/h normalisation scale and the scale probed by the velocities is not large. The depth probed by the dipoles is greater, and consequently the consistency of the model with the data is more impressive. Our conclusions regarding the velocity data differ considerably from earlier studies (e.g. Vittorio *et al.* 1986, Bond 1987) who concluded that these data are in conflict with the CDM predictions. The discrepancy can be traced to a different choice of window function. We feel that there is no ambiguity, and that the window can be derived directly from the weighting scheme used by the observers.

Second, we have developed and applied an improved version of Gott's test using the correlations between a number of angular dipoles and bulk flow solutions rather than the traditional implementation using only the flux weighted dipole and Local Group. We obtain a value for the bias parameter for IRAS galaxies of $b \simeq 1.5$, corresponding to an Ω_0 of around 0.4. Our analysis differs from previous applications of this type of test in that we make a maximum likelihood estimate of b, contingent on the assumption of the CDM spectrum. The ML estimate of b should be fairly insensitive to the specific spectrum assumed, but the model does play an important role in that it enables us to determine the statistical uncertainty in b. We have applied a very similar analysis to the determination of b (or Ω) from the 2 Jy redshift survey. We find that the statistical uncertainty here is much larger, but improvements are possible.

Third, while our main results assume for the model of the density field the random Gaussian CDM Δ field, we have interspersed these calculations with calculations based on other simpler models using one or more spherical perturbations. These models give a more graphic picture of the density and velocity fields and provide an important link to other studies which have used this type of model extensively. We find a broad degree of consistency between the various observations regarding the distribution in distance of the major components. One way we illustrated this (Sec. 2.5) was to use a single component attractor model centred at 40 Mpc/h, and with amplitude adjusted to give the Local Group microwave background motion. This model reproduces quite nicely the run of angular dipoles with flux, and gave an acceptable chi-squared. Following Lynden-Bell's modeling of the velocity data, we then went on to show that a 3-component model comprising the Great Attractor, Virgo and a third component to make up the microwave velocity provided an even better fit to the dipoles. The distance to the third component was kept as a free parameter, and adjusting gave a best fitting distance of 18 Mpc/h, so this seems to accord with the idea of a 'Local Anomaly' in the velocity field (see Faber and Burstein 1988 and Lynden-Bell and Lahav 1988). Unfortunately, we found that the angular dipole data did not very tightly constrain the depth from which the dominant acceleration derives. We also showed (Sec. 5.2.2) that the simple 1-component model (in which roughly 300 km/s of our peculiar motion derives from inhomogeneity at distances between about 40 and 80 Mpc/h) also reproduces the apparent *lack* of acceleration seen at these distances in redshift space, once the distortion of the clustering in redshift space by peculiar velocities is taken into account. Indeed, the redshift results would seem to preclude the possibility that nearly all of our acceleration derives from smaller distance (Sec. 5.2.1).

These simple models provide at best a very broad brush description of reality; all we claim here is that these very different data sets are all at least *consistent* with models in which a substantial part of our peculiar motion (roughly 300 km/s or so) derives from $r > 40$ Mpc/h. While each line of evidence appears questionable when subjected to close scrutiny, it is encouraging to see a farily coherent picture emerging. On the theoretical side, we presented the simple result that in CDM-world the rms velocity driven by perturbations at $r > 40$ Mpc/h comes to $300\sigma_\varrho$ km/s. This provides a simple way to see that the scale and amplitude of large-scale structure suggested by the data are in reasonable accord with this theoretical model.

7. DISCUSSION

In Sec. 7.1 we discuss the relation of the results obtained above to other probes of large-scale structure; in Sec. 7.2 we consider the question of normalisation; and in Sec. 7.3 we discuss the assumption that $\Omega = 1$, and some implications for galaxy formation. Finally, in Sec. 7.4 we consider the outlook for future extensions of the type of observations described above.

7.1 - Other Probes of Large-Scale Structure

There are various other probes of large-scale structure that provide important independent tests of models like CDM. On very large scales there are constraints on the level of the diffuse X-ray backround (Shafer 1983) and from the isotropy of the microwave background (Kaiser and Silk 1986 and references therein). Mostly, these have given null results, and have been compatible with the rather low values predicted in CDM-world: on large scales the rms fluctuation amplitude varies as λ^{-2}, falling to around 10^{-4} at the Hubble length scale. A notable exception to these null results is the microwave anisotropy detection claimed by Lasenby (1988) and his collaborators. This anisotropy is about an order of magnitude larger than that predicted in CDM-world with our choice of normalisation. If this is confirmed at a higher level of significance, then this would be a very strong indication of much more power at large scales than provided by CDM.

Another very important quantitative probe of large-scale structure is provided by the clustering analysis of the Abell cluster catalogue. There are two lines of evidence here; on one hand we have the angular correlation function estimates from the complete catalogue (now augmented by the southern clusters), and on the other we have the 3-dimensional clustering studies of the much shallower subsample for which redshifts are available. The advantages and disadvantages of these approaches are to a large extent complementary; the formal error bars on the redshift survey estimates of ξ are large and the signal falls below noise at a lag of at most $\simeq 30$ Mpc/h. Note that for a clustered population the realistic error bars are larger than those for a Poissonian distribution, roughly by a factor $1 + 4\pi n J_3$, (see e.g. Kaiser 1986), so the realistic error bars are several times larger than the Poissonian formula at this scale. More graphically, it is apparent from the pair plots of Bahcall and Soneira (1983) that the number of independent clumps is quite small, so it is unlikely that we are seeing a large enough sample to ensure representative results. The formal errors for the deeper angular surveys (Hauser and Peebles 1973) are smaller, and we can be fairly confident that we are seeing a representative sample, but there are other sources of error or bias that might creep in here. Taken at face value the estimates obtained appear to be consistent and give a correlation length of around 25 Mpc/h. This correlation strength is much greater that that predicted in CDM. In this theory the clustering amplitude *is* predicted to be greatly enhanced over the amplitude of the mass correlation function (e.g. Kaiser 1984), but the predicted correlation length is still only around 14 Mpc/h, almost a factor of two below that observed.

Before concluding that CDM is ruled out it is worth critically examining these observations. One possible problem is the spurious clustering that should arise because of Abell's selection criteria. Since he defined clusters to be regions on the sky where the projected density of galaxies was particularly high, one would expect to find an enhancement of pairs of clusters at small angular separation where the physical neighbours of one cluster are counted as members of the other and vice versa. The result is a spurious contribution to the correlation function as a function of the separation of the clusters perpendicular to the line of sight, so this effect would produce an elongation of the clustering pattern along the redshift axis as is seen in the redshift survey. This effect has been explored in some detail by Dekel (1988). To model this effect we have calculated the spurious clustering for a set of spherically symmetric clusters which are laid down in a Poissonian manner, and which have profiles modeled on the cluster-galaxy cross-correlation function. After projecting this onto the sky and selecting an apparent richness limited cluster sample, we get a correlation function which is proportional to the angular cluster-galaxy cross-correlation function, and which falls to unity for projected separation of about 7 Mpc/h (though this is quite sensitive to the assumed cross-correlation function, the cluster richness distribution etc., and is therefore quite uncertain). This strength of correlation would not strongly perturb the estimate of the correlation length from the shallow redshift survey, but would be expected to strongly contaminate the deeper angular surveys. This is very worrying, since it is on these samples that we rely for statistically significant results.

Another source of spurious clustering is expected to arise from the manner in which Abell determined the background density for each plate. Lucey (1983) points out that the background correction becomes large for the deeper samples, so one would expect to find an excess of clusters on plates where the background correction is abnormally low and *vice versa*. Since Abell gives few details of how this background was derived it is difficult to model this effect. It is worth noting, however, that both this and the projection effect mentioned above would be predicted to give a scaling with depth R which is intermediate between that expected for true spatial clustering; $w(\theta) \propto R^{-2}$, and that for a distance independent angular pattern $w \propto R^0$, and so would be quite difficult to detect from scaling tests on the deep samples.

Estimating effects like these accurately is very difficult, and it is easy to think of numerous other sources of spurious clustering, or indeed to speculate on how Abell might have judiciously corrected for such effects. An alternative approach is to use the redshift survey data directly to see if the clustering pattern is really isotropic, and if not, to correct for non-spatial clustering. Bahcall and Soneira (1983) presented plots of the excess pair counts as a function of redshift separation for fixed bins of angular separation and showed that these fall at large Δz. This was held to be evidence that we are seeing true spatial clustering. However, since the pair counts for a uniform distribution also fall off at large Δz, this conclusion is questionable. If one looks at the correlation function which is given by the ratio of the excess pair counts to that for a random distribution (Sutherland 1988), then this shows no sign of falling, and this would suggest a very strong component of non-spatial clustering. Puzzlingly, the effect is seen even for large angular separation of 20 degrees or so, and this would not be predicted by either of the effects mentioned above. Sutherland has tried to correct for this contamination in a manner similar to that used in quasar clustering studies where the 'raw' value of $1 + \xi$ is divided by the mean $1 + \xi$ for a bin with the same perpendicular separation but much larger Δz. The corrected ξ thus obtained looks quite spherically symmetric, has a much lower amplitude, and appears to be quite compatible with the CDM predictions. It is still questionable whether Sutherland's corrected estimate is representative; it is, after all, still based on the redshift survey sample, but his results certainly cast a shadow of doubt over the robustness of the conventional estimates. Sutherland and Efstathiou (1988) have subsequently tested for the reality of the clustering by comparing estimates of $w(\theta)$ derived from inter and intra-plate pairs respectively. They find a significant discrepancy, casting further doubt over the reality of the apparent cluster clustering, so we feel it is premature to discard CDM to the scrap heap as yet. Cluster-clustering does seem a promising area of vulnerability for CDM, and the prospect of a large X-ray selected cluster sample from ROSAT is very exciting.

7.2 - Normalisation

Our normalisation of the fluctuations in galaxy counts is fairly direct and conventional. Another commonly used way to quote the normalisation is in terms of the J_3 integral. Our choice in very similar, but uses a different integral over the 2-point function (see e.g. Peebles 1980, Sec. 36). One might question however, our assumption that the 2-point function for IRAS galaxies is the same as for optical galaxies. If it were shown convincingly that the correlation length is much smaller then we would consider the dipoles to be evidence for more power at large scale. Another minor problem is the assumption that linear theory is applicable for the 8 Mpc/h normalisation. If this were shown to be badly in error we would again have to revise our conclusions.

Our choice of normalisation $\sigma_\varrho \simeq 0.6$—0.7 for the mass fluctuations is more controversial. It has come to our notice (Efstathiou, private communication) that the normalisation used for N-body experiments is roughly $\sigma_\varrho = 0.35$. This is about a factor 2 smaller than our preferred value, so with their choice of normalisation the peculiar velocities would be quite a problem for CDM. The results of the N-body experiments look in many ways quite realistic, and a very interesting question is whether with our normalisation the results would be unacceptable. It seems unlikely that considerations of galaxy clustering would usefully discriminate since to a large degree one can play off a larger amplitude mass fluctuation against a smaller level of biasing. The most effective way to discriminate between these different normalisation options is through comparison of peculiar velocity observations.

If we assume a normalisation which is a factor 2 larger than used in the numerical experiments then we have to reinterpret the results using the redshift scaling $z \rightarrow 2z + 1$, so a snapshot of the particle positions at the 'present' would be interpreted as a snapshot of $z = 1$ for instance, and the scaling for physical velocities $V \rightarrow \sqrt{2}V$, so all rotation velocities would increase by about 40%. It is arguable that in many respects these changes would be beneficial; the epoch of formation of halos that one would wish to identify with galaxies and perhaps quasars would be pushed back to much higher redshift, and there is some indication that the velocity dispersion of clusters in the simulations are on the low side (White *et al.* 1987), so this would also improve. The Press-Schechter (1974) formula for the distribution of haloes provides a direct link between properties of non-linear clusters at the present epoch and smaller objects of a similar density contrast at early times. One infers from the CDM spectrum that *if* 1% of the present universe is in rich clusters with mean separation 50 Mpc/h and rotation velocities 1000 km/s, then since the rms fluctuations rise by about a factor 5 going to a mass scale 1000 times smaller, we should see at $1 + z \simeq 5$ about 1% of the total mass in objects with mean separation 5 Mpc/h and rotation velocities $\sqrt{1 + z}(M/M_{cluster})^{1/3} \times 1000$ km/s, or $\simeq 220$ km/s. These look very much like the properties of real galaxies. Perhaps improved studies of rich clusters will give lower rotation velocities, but if this value is appropriate then this would imply quite an early epoch for the formation of objects with abundance and internal velocities like real galaxies. Other observational tests that would be very useful to resolve the normalisation question are studies of high velocity dispersion clusters at higher redshift from X-ray studies and possibly from gravitational lensing, since the rapid evolution expected for these object is quite sensitive to the normalisation adopted.

7.3 - Biasing Galaxy Formation

We have assumed throughout that the galaxies trace a biased field which has fluctuations which are amplified relative to the underlying mass fluctuations. The main motivation for this is that if Ω really is unity then some sort of biasing of galaxy formation towards dense regions seems to be necessary in order to reconcile the high L/M of clusters compared to the L/M for the universe as a whole. We now discuss the plausibility or otherwise of biased galaxy formation. The discussion here is largely quantitative and any discussion of the physics of galaxy formation is necessarily speculative. We will make the fairly conventional assumption that galaxies form in the dark matter potential wells, and this at least is amenable to quantitative calculations. Here we will use the Press-Schechter (1974) formalism for the evolution of dark matter condensations in CDM-world (Cole and Kaiser 1988).

The basic idea that we shall appeal to is that the long wavelength perturbations present in the CDM spectrum modulate the formation of halos of smaller mass scale. (We shall use the word halo here to denote dark matter condensations in general). If one accepts that to a useful approximation one can associate the halos which have just formed at any epoch with overdense regions in the initial state (e.g. Press and Schechter 1974, Bardeen *et al.* 1986) then the effect of long wavelength modes is simply to modulate the collapse times of the halos - perturbations will turn around and virialise earlier in a region subject to a positive long wavelength perturbation.

This type of analysis has been applied to rich clusters (Kaiser 1984), which it seems appropriate to identify with very massive halos forming at the present epoch. Since, in CDM and other hierarchical pictures the mass scale of clustering increases as time proceeds, the modulation of collapse times by long waves converts to a spatial modulation of the number density of clusters selected according to some minimum mass threshold. This calculation gives a very strong amplification of cluster-clustering, though as we have discussed, this may still be too small. This analysis, where we effectively take a 'snapshot' and analyse the spatial distribution of clusters at that instant of time, can readily be extended to lower mass halos using the Press-Schechter (1974) or analogous formulae, and one finds that the bias decreases in strength, with a negative bias for objects at the low mass end of the spectrum. The borderline unbiased objects are those where $M^2 n(M)$ peaks, so these objects contain the greatest fraction of the mass of the universe per log interval of cluster mass.

Whether this 'snapshot' calculation is relevant to galaxies depends on how one imagines galaxies form. One not unreasonable picture for the formation of disk galaxies is that halos with roughly flat rotation curves form containing gas which is heated to the virial temperature on collapse. As time goes on the gas can cool from progressively larger radii and this gas contracts within the dark halo to form the disk. Aspects of this type of formation picture have been explored by many authors (e.g. White and Rees 1987, Fall and Efstathiou 1980, Gunn 1982) and have become something of a paradigm. An appealing feature of this model in the context of CDM is that with the gas density implied by baryosynthesis the gas in a $V_{rot} = 220$ km/s halo can just cool by the present at a radius of about 100 kpc, so would plausibly make a reasonable disk after a factor 10 decrease in radius. In this picture, galaxy formation would be an ongoing process, the 'snapshot' result should be applicable, and so one would expect at best a rather weak positive bias for the most massive spirals and a negative bias for small spirals. Worse still, since the effect of a positive long-wavelength 'swell' is to increase the characteristic virial temperature of the halos, this acts to impede the cooling, and results in a net antibias for the total amount of gas which has cooled, and therefore most naturally in a net antibias for the total number of stars. A closely related problem is that if one approximates the cooling time - temperature dependence as a power law, $t_{cool} \propto \varrho^{-1} T^{\beta}$, and if one assumes that the gas is initially distributed like the dark matter, then one finds that the mass of gas which has cooled by the present varies with the halo temperature as $M_{cool} \propto T^{(3-\beta)/2}$. Taking $\beta \simeq 1$ as appropriate for galactic temperatures, one would naively predict a luminosity - rotation velocity relation $L \propto V^2$ unlike the observed relation $L \propto V^4$. The relation between this result and the antibiasing becomes apparent when one notes that the halo mass is proportional to the cube of the velocity, so if one has $L \propto V^\gamma$, with $\gamma < 3$, then L/M is a decreasing function of mass. Thus, in a region subject to a positive long wave perturbation so the characteristic halo mass is increased, one would expect the net L/M to be decreased. Finally, one would also expect an additional antibias to arise in this model if the gas is stripped from the halo in a dense region like a rich cluster. All in all, it is very hard to see how a strong positive bias would result in this kind of picture.

Another type of scenario in which the 'snapshot' calculation might be relevant is if galaxy formation is abruptly terminated, perhaps by some kind of negative feedback. If this occurs while the number density of 'galaxies' is still rapidly rising, or equivalently, when only a very small fraction of the mass is in these objects, then the 'galaxies' would be quite analogous to rich clusters at the present epoch and a strong positive bias seems much more promising. The weakest feature of this scenario is that no obvious candidate exists for the hypothetical feedback process which is invoked solely to effect the bias and reconcile the theoretical prejudice for a closed universe with observations.

Without a feedback to abruptly switch off galaxy formation it seems inappropriate to apply the 'snapshot' calculation. To be sure, at any particular instant one can find a population of objects which are strongly biased towards regions where the long-wave swell is positive, but this is only because in regions which are underdense on large scales the analogous objects are destined to collapse just a little later. If there is to be a positive bias then it must be that the delayed collapse somehow reduces the luminosity of the galaxy, but how might such an effect arise?

Before embarking on what might otherwise seem to be a course of idle speculation it is illuminating to consider for a moment what one would expect to see in an unbiased universe - i.e. a universe in which the stellar luminosity generated in a dark matter condensation is independent of the time of formation and is simply proportional to the mass. When CDM first became popular it was argued that such a 'what-you-see-is-what-you-get' model would nicely reproduce the observed luminosity - velocity relations for galaxies (e.g. Blumenthal *et al.* 1984). The argument is that on galaxy scales the spectral index $n \equiv d\log P(k)/d\log k$ is close to -2, so for the '1-sigma' perturbations, for instance, we have $\Delta\varrho/\varrho \propto M^{-1/6}$. The final density of the system at collapse varies as the cube of the initial perturbation amplitude, or as $M^{-1/2}$. Invoking the virial theorem then gives $M \propto V^4$. While the resemblance of this to the observed $L - V$ relation is encouraging, when one looks at the expected scatter about the mean $M - V$ relation the result is less pleasant. The internal velocity varies as $\sqrt{\nu}$, where ν is the initial overdensity in units of the rms. Consider, for instance, the 1 and 2-sigma perturbations on a given mass scale. These have by assumption the same L, but have V^4 differing by a factor 4. A little more quantitatively, one can translate the Gaussian distribution for ν into the scatter for V at constant L. One finds that 2.5 log V^4 has variance 2.2, which is roughly 5 times larger than the scatter of 0.45 magnitudes for the Tully-Fisher relation. Since this presumably contains substantial observational uncertainty we see that the intrinsic $M - V$ correlation for Gaussian fluctuations is much broader than the intrinsic $L - V$ relation for real galaxies. Perhaps, as suggested by Faber (1982) the different morphological types are stratified along lines of constant ν, possibly because of systematic dependence of the angular momentum on ν. If angular momentum, or more generally some factor depending on the shape of the perturbation were the discriminator for morphological type, then one would still have a potential problem. Long wavelength modes have a negligible effect on the shape of much smaller perturbations, so the galaxies would still be unbiased by such modes. If we take the objects at constant ν and subject them to a positive swell they will collapse earlier and have a higher V^4 at the same L, and consequently, using these objects to determine the distance scale, one would infer a smaller Hubble constant using galaxies in dense environments as compared to field galaxies, and as far as we know no such effect is seen.

This 'unbiased' model, therefore, seems to predict that we should see a broad scatter about the $L - V$ relation, and systematic environmental dependence of the $L - V$ intercept, essentially because the random and systematic variations in collapse time expected in this model will change V but not L. There are two possible ways to resolve this dilemma. First, we could stick with $L \propto M$ and argue that the V we observe for stars is not the V for the dark halo which is what we have used above. Second, one can keep $V_{stars} \simeq V_{halo}$, and assume that in perturbations with the same mass, the efficiency of star formation is greater the earlier is the collapse time. Following this option we are led back (though now on rather firmer empirical grounds) to ask: why should an earlier collapse time give a brighter galaxy?

A positive long wave swell will increase the density, the virial temperature, the pressure, and the collapse redshift. Higher density tends to increase the cooling rate, but the increase in temperature nearly balances this, so the ratio of cooling time to dynamical time is hardly altered by a long wave swell. More promising is the idea that one can form more stars per unit mass of gas in a deeper potential well. Such a dependence has been suggested by Larson (1974), and the idea has more recently been revived by Dekel and Silk (1986) in the context of CDM. Such a dependence might plausibly arise if supernovae are efficient at expelling gas from shallow potential wells. A related possibility is that the increased pressure might modify the initial stellar mass function and perhaps reduce the fraction of massive stars, resulting in more moderate mass stars being produced before the supernovae blow off the gas. Yet another possibility is that speeding up the collapse allows more stars to form before the supernovae go off.

Whatever physics is operating, if we take seriously the idea that it is the velocity dispersion of the potential well which determines the luminosity, and not the amount of gas available, then a positive bias seems inevitable. Quite how strong this bias is depends on when galaxies are assumed to form. Using the Press-Schechter formalism (with the normalisation proposed in Sec. 1) one can calculate the number density of objects with $V_{rot} > 220$ km/s, for instance. This reaches a maximum around $(1 + z) = 3$, and is biased up and down by long wavelength modes with a b value of 1.4-1.5. If one uses the simple spherical collapse model to estimate the initial density contrast for recently virialised systems such as rich clusters, then one obtains about a factor 2 enhancement for the L/M of such systems relative to the global L/M. Thus, the bias we find is substantial, though these simple estimates are somewhat lower than seem to be required to reproduce the virial analysis.

7.4 - *Future Prospects*

As the data we have looked at have failed to falsify CDM, it is interesting to consider the outlook for more powerful tests. One possibility would be to repeat the analysis with much deeper samples. The fluctuation amplitude predicted in CDM does fall quite fast as one probes deeper. Unfortunately, the fundamental statistical errors fall more slowly, so quite rapidly one will reach the point where the CDM predictions fall below the noise level, and then only models which produce much larger fluctuations than CDM can be effectively tested. For the angular dipoles, the predicted dipole falls asymptotically with depth as R^{-2}, whereas the shot noise falls as $R^{-3/2}$. For the velocity, the rms prediction falls as R^{-1}, and the noise falls as $R^{-1/2}$. Since the observations we have considered are at best a few sigma detections and agree with the CDM predictions, it is clear that the possibility for extension is unlikely to be fruitful.

A more hopeful outlook is to use a more complete description of the data, rather than using just the dipole moments. The formalism described here can readly be extended to a more or less complete description of the data (subject only to the constraint that one use only representations of the data that are well described by linear theory). The more general analysis for the peculiar velocity field can be found in Kaiser (1988), and we are currently developing the analysis for counts in cells of galaxies. While the mathematics can be straightforwardly generalised, the problem is deciding which representation of the data provides the best compromise between loss of information and the requirement that linear theory be applicable.

Generalising to e.g. counts in cells rather than dipoles enables us to extract more information from the currently available data, but at the price of probing smaller scales. This also solves the problem of how to deal with the zone of avoidance or catalogue boundaries. For deeper samples this type of analysis will enable us to sample many independent volumes, similar to the volumes sampled by the dipoles we have used here, and that will greatly improve the power of the test. For deeper peculiar velocity samples the analogous benefit will be much less because of the increase in uncertainty.

There is one very important benefit to be obtained from deeper peculiar velocity studies. These will enable the importance of systematic errors due to environmental effects to be quantified. The peculiar velocities of galaxies other than our own are determined rather indirectly; one assumes that the Tully-Fisher relation or whatever is universal. Systematic dependence of internal properties of galaxies on environment must be important at some level, and will tend to inflate the estimates of the peculiar velocities inferred. Some tests for these effects can be made within the currently available data, but the most convincing test would be to determine the 'velocities' of a sample of much more distant clusters since one would then expect any real peculiar velocities to be negligible (though this betrays some theoretical prejudice on the part of the authors), and one would then measure the real total statistical uncertainty. Such a program would be arduous, but would not require a full sky survey.

Major increases in the data available will take some time. On a shorter timescale we might anticipate revisions to the luminosity function or perhaps the appropriate normalisation, which would change our conclusions. For Gaussian models the theoretical formalism provides a promising methodology for analysing the large-scale structure data base. The theoretical questions that seem to us most interesting here are: Does the structure appear to be growing by gravitational instability? If so, how strongly biased are the various types of galaxies? How do the data constrain the space of theoretical models for the initial power spectrum?

REFERENCES

Aaronson, M., Huchra, J., Mould, J., Schechter, P.L., and Tully, R.B. 1982. *Ap J.* **258**, 64. (AHMST)

Bahcall, N.A. and Soneira, R.M. 1983. *Ap J.* **270**, 20.

Bardeen, J.M., Bond, J.R., Kaiser, N., and Szalay, A., 1986. *Ap J.* **304**, 15.

Blumenthal, G.R., Faber, S.M., Primack, J.R., and Rees, M.J. 1984. *Nature.* **311**, 517.

Bond, J.R., 1987. in *Nearly Normal Galaxies, The Eighth Santa-Cruz Summer Workshop.* ed. S.M. Faber, p. 388. New York: Springer-Verlag.

Bond, J.R. and Efstathiou, G., 1984. *Ap J.* **285**, L45.

Cole, S. and Kaiser, N. 1988. in preparation.

Davis, M. and Huchra, J. 1982. *Ap J.* **254**, 437.

Dekel, A. 1988. this volume.

Dekel, A. and Silk, J. 1986. *Ap J.* **303**, 39.

Dressler, A., Faber, S.M., Burstein, D., Davies, R.L., Lynden-Bell, D., Terlevich, R.J., and Wegner, G. 1987. *Ap. J.* **313**, L37.

Faber, S.M. 1982. in *Astrophysical Cosmology.* ed. H.A. Bruck, G.V. Coyne, and M.S. Longair. Vatican City State: Pontifical Academy, p. 191.

Faber, S.M. and Burstein, D. 1988. this volume.

Fall, M.S. and Efstathiou, G. 1980. *MN.* **193**, 189.

Gunn, J. 1982. in *Astrophysical Cosmology.* ed. H.A. Bruck, G.V. Coyne, and M.S. Longair. Vatican City State: Pontifical Academy, p. 233.

Harmon, R.T., Lahav, O., and Meurs, E.J.A. 1987. *MN.* **228**, 5p.

Hauser, M.G. and Peebles, P.J.E. 1973. *Ap J.* **185**, 757.

Kaiser, N. 1984. *Ap J.* **284**, L9.

_____. 1986. *MN.* **219**, 795.

_____. 1987. *MN.* **227**, 1.

_____. 1988. *MN.* **231**, 149.

Kaiser, N. and Lahav, O. 1988. preprint.

Kaiser, N. and Silk, J. 1986. *Nature.* **324**, 529.

Lahav, O. 1987. *MN.* **225**, 213.

Lahav, O., Rowan-Robinson, M., and Lynden-Bell. 1988. *MN.* **234**, 677.

Larson, R.B. 1974. *MN.* **169**, 229.

Lasenby, A. 1988. this volume.

Lawrence, A., Walker, D., Rowan-Robinson, M., Leech, K.J., and Penston, M.V. 1986. *MN.* **219**, 687.

Lucey, J.R. 1983. *MN.* **204**, 33.

Lynden-Bell, D. and Lahav, O. 1988. this volume.

Lynden-Bell, D., Faber, S.M., Burstein, D., Davies, R.L., Dressler, A., Terlevich, R.J., and Wegner, G. 1988. *Ap J*. **326**, 19.

Meiksin, A. and Davis, M. 1986. *AJ*. **91**, 191.

Meurs, E.J.A. and Harmon, R.T. 1988. preprint.

Peebles, P.J.E. 1980. *The Large Scale Structure of The Universe*. Princeton: Princeton University Press.

Press, W.H. and Schechter, P. 1974. *Ap J*. **187**, 425.

Rowan-Robinson, M., Walker, D., Chester, T., Soifer, T., and Fairclough, J. 1986. *MN*. **219**, 273.

Shafer, R.A. 1983. Ph.D. thesis, University of Maryland.

Strauss, M.A. and Davis, M. 1988. in *Large-Scale Structures of the Universe*, eds. J. Audouze, M.-C. Pelletan, and A. Szalay. Dordrecht: Kluwer Academic Publishers, p. 191.

Sutherland, W. 1988. *MN*. **234**, 159.

Sutherland, W. and Efstathiou, G. 1988. in preparation.

Villumsen, J.V. and Strauss, M.A. 1987. *Ap J*. **322**, 37.

Vittorio, N., Juszkiewicz, R., and Davis, M. 1986. *Nature*. **323**, 132.

Webster, A., 1976. *MN*. **175**, 71.

White, S.D.M. and Rees, M.J. 1987. *MN*. **183**, 341.

White, S.D.M., Frenk, C.S., Davis, M., and Efstathiou, G. 1987. *Ap J*. **313**, 505.

Yahil, A. 1988. this volume.

Yahil, A., Sandage, A., and Tammann, G.A., 1980. *Ap J*. **242**, 448.

Yahil, A., Walker, D., and Rowan-Robinson, M. 1986. *Ap J*. **301**, L1.

CAN SCALE INVARIANCE BE BROKEN
IN INFLATION?

J. R. BOND

CIAR Cosmology Program, Canadian Institute for Theoretical Physics, Toronto

D. SALOPEK

Physics Department, University of Toronto

and

J. M. BARDEEN

Physics Department, University of Washington

ABSTRACT

Scale invariant fluctuation spectra are the most natural outcomes of inflation. In this paper we explore how difficult it is to break scale invariance naturally. It can be done when there is more than one scalar field. We sketch the effects that varying the expansion rate, structure of the potential surface and the curvature coupling constant have on the quantum fluctuation spectra. We conclude that: (1) power laws different than scale invariant are very difficult to realize in inflation; (2) designing mountains of extra power added on top of an underlying scale invariant spectrum is also difficult; and (3) double inflation leading to a mountain leveling off at a scale invariant high amplitude plateau at long wavelengths is generic, but to tune the cliff rising up to the plateau to lie in an interesting wavelength range, a special choice of initial conditions and/or scalar field potentials is required.

1. INTRODUCTION

In this paper, we discuss the difficulties of designing a density fluctuation spectrum to order using physical processes occurring during an initial inflation epoch. We denote the two-point functions in momentum (comoving wavenumber) space for homogeneous and isotropic random fields by, for example for the density, $\mathbf{P}_{\varrho}(k) \equiv (k^3/(2\pi^2)) \langle|(\delta\varrho/\varrho)\,(k)|^2\rangle$, the power per logarithmic interval of comoving wavenumber k. If the fluctuations are Gaussian then this is all that is required to completely specify the random field. In the inflation picture, quantum fluctuations in scalar fields ϕ generated during the inflation epoch become density fluctuations. If the spatial eigenmodes have zero occupation number and mode-mode couplings can be ignored (linearity is expected since fluctuations are apparently quite small at recombination), then these quantum fluctuations will be like zero point oscillations of sound waves in a crystal — in a Gaussian ground state. Curiously, the long wavelength sound modes of a zero temperature crystal have a Zeldovich scale invariant spectrum, $\mathbf{P}_{\varrho} \propto k^{3+n}$ with $n = 1$, the generic outcome of inflation. However, the way a Zeldovich spectrum arises during inflation depends critically upon the way scalar perturbations in the expanding universe suffer 'Hubble drag' which damps the oscillations for wavelengths smaller than a Hubble distance and freezes out their amplitude for longer wavelengths.

Inflation occurs if the comoving Hubble distance $(Ha)^{-1}$ decreases with time, i.e., $\ddot{a} > 0$; the equation of state of the homogeneous background universe must then obey $p/\varrho < -1/3$, where p is the total pressure and ϱ is the total energy density. This is realizable for a homogeneous scalar field ϕ self-interacting through a potential $V(\phi)$ for which $p = \dot{\phi}^2/2 - V(\phi)$ and $\varrho = \dot{\phi}^2/2 + V(\phi)$, provided the energy density is potential dominated for some period and over some patch of space. As inflation proceeds, $(Ha)^{-1}$ sweeps in, encompassing ever smaller comoving length scales, arresting causal communication across waves with $k^{-1} > (Ha)^{-1}$.

In the inflation picture, a region of size smaller than the Planck length is inflated by a factor $\sim e^{60+N_I}$ to a region (much) larger than the current Hubble length $H_0^{-1} = 3000\ h^{-1}$ Mpc provided the number of e-foldings during inflation is $N_I \gtrsim 70$. Without inflation, the region would have expanded by only a factor $\sim e^{60}$ to less than a micron across by the current time — an aspect of the 'horizon' problem which inflation so successfully solves.

If only one scalar field drives inflation approximate scale invariance, $\mathbf{P}_{\phi}(k) \approx$ constant, for the spectrum of scalar field fluctuations is the natural outcome. This translates into scale invariant fluctuations for the gravitational potential fluctuations $\mathbf{P}_{\Phi}(k) \approx$ constant for adiabatic scalar perturbations, leading to a Zeldovich spectrum for the density. The Poisson-Newton equation $a^{-2}\nabla^2\Phi = 4\pi G\delta\varrho$ for the gravitational potential in an infinite expanding background remains an exact relation in general relativistic perturbation theory if $\delta\varrho$ is the density perturbation in the comoving frame, and $-\Phi$ is Bardeen's (1980) gauge invariant Φ_H variable.

For isocurvature perturbations, the ϕ fluctuations become fluctuations in one of the species of particles present, for example, in axions, $\mathbf{P}_{\varrho}\,(k) \propto \mathbf{P}_f(k)$, where ϱ_A is the axion mass density (isocurvature CDM perturbations) or, possibly, in baryons, $\mathbf{P}_{n_B}(k) \propto \mathbf{P}_f(k)$, where n_B is the baryon number density (*isocurvature baryon perturbations*). In these cases, the total density and gravitational potential fluctuations vanish, $\mathbf{P}_\varrho = \mathbf{P}_\Phi = 0$, at the time when either the axion mass or the baryon number is generated — provided it happens after inflation has ended. The power law $n = -3$ corresponds to scale invariance in these isocurvature cases for the initial density perturbations in the nonrelativistic matter. We refer to a scalar field giving rise to an isocurvature spectrum as an *isocon* in keeping with the terminology *inflaton* for the scalar driving inflation which gives rise to adiabatic fluctuations.

Although very gentle deviations from scale invariance are expected in inflation it is rather difficult to arrange drastic modifications. Here we are interested in exploring the requirements for obtaining power laws different from $n = 1$ for adiabatic perturbations or $n = -3$ for isocurvature perturbations over extended regimes of k-space, or mountains of extra power built on top of an underlying scale invariant spectrum. More generic are two scale invariant spectra of different amplitudes, the larger at longer wavelengths, corresponding to regimes when two different scalar fields drive inflation, with a matching region (a ramp) that depends upon the details of how one scalar takes over from the other. This is double inflation. Tuning the location of the ramp or the mountain or the power law regime to lie in a specific k range requires rather precise conditions to be imposed on the inflationary model. For example, to put modified power at the scale of clusters implies that some special physics was operating at a redshift $z \sim e^{122}$ involving either finely tuning initial conditions for the background scalars or the potential parameters. We discuss these possibilities in more detail in Sec. 2, reviewing work reported in Bardeen, Bond, and Salopek (1987, 1988 [BBS1, BBS2]).

2. RESPONSE OF SCALAR FIELDS TO POTENTIAL CHANGES

There are only a few parameters at our disposal if we wish to modify the spectrum of fluctuations that comes out of inflation. The linearized equations obeyed by the perturbations in the fluctuations $\delta\phi_j(k, t)$ of fields $\phi_j(\vec{x}, t)$, $j = 1, ..., N$, about a homogeneous average $\bar{\phi}_j(t)$ are:

$$\ddot{\delta\phi_j} + 3H\dot{\delta\phi_j} + (k^2 / a^2) \delta\phi_j + \sum_i (m_{ij}^2 + 12\xi H^2\delta_{ij}) \delta\phi_i \quad (2.1a)$$

$$+ \Gamma_j\dot{\delta\phi_j} + (\dot{\bar{\phi}}_j\delta\Gamma_j) + \dot{\bar{\phi}}_j\dot{h}/2 + \bar{\phi}_j\xi\delta R \quad (2.1b)$$

$$= 0.$$

Equation (2.1) is coupled to equations describing the evolution of the gravitational metric and the perturbed total (field plus radiation) energy and momentum density. There are also equations describing the evolution of the background fields, the expansion factor (Friedmann equation) and the background radiation energy density. Details are given in BBS1, BBS2.

For this discussion we shall ignore the terms in the second line, which involve such factors as the rate at which field energy is dissipated into radiation (Γ_j) and the growth of the field perturbations due to gravitational instability (perturbed metric $h = h_b^b$ in the synchronous gauge and perturbed curvature δR). These operate primarily when the perturbations are well outside the Hubble distance and the shape of the fluctuation spectrum is set.

When the wave is still causally connected, $k \gg H a$, with momenta high compared with masses $k \gg |m|a$, a WKB solution of equation (2.1) exhibits the rapid oscillation appropriate for massless scalars with a general decline due to the Hubble drag: $\delta\phi_j \approx (2k)^{-1/2} a^{-1} \exp[-i \int dt k/a]$, where the positive frequency solution has been selected. $\phi_j(k, t)$ has real $\mathrm{Re}(\delta\phi_j)$ and imaginary $\mathrm{Im}(\delta\phi_j)$ parts which must be calculated separately by solving two sets of evolution equations (2.1). The power spectrum for ϕ_j is given by

$$\mathbf{P}\phi_j \, (k, \, t) = \frac{k^3}{2\pi^2} \, ([\mathrm{Re}(\delta\phi_j(k, \, t))]^2 + [\mathrm{Im}(\delta\phi_j(k, \, t))]^2) \, (1 + 2\bar{n}_{jk}). \quad (2.2)$$

Here $\bar{n}_{jk}$ is the average occupation number of mode k for the scalar field ϕ_j. At horizon crossing ($k = H a$) the interior WKB solution referred to above would give $\mathbf{P}\phi_j = [H/(2\pi)]^2 (1 + 2\bar{n}_{jk})$, relating the amplitude to the Hawking temperature $H/(2\pi)$.

To modify the outcome of inflation — assuming linearization is valid which in some cases is debatable — we are only free to modify (1) the occupation number $\bar{n}_{jk}$, (2) the Hubble parameter $H(t) = \dot{a}/a$, which enters in the Hubble drag term $-3H\dot{\delta}\phi_j$, (3) the effective mass matrix $m_{ij}^2 \equiv \partial^2 V/\partial\phi_i\partial\phi_j$, or (4) the curvature coupling constant. ξ enters as an effective term in the mass matrix; it may be positive or negative. It is 0 for minimally coupled fields, 1/6 for conformally coupled fields, and is subject to renormalization so the value could evolve.

It is usually argued that the modes which are currently within our horizon would have had such large values of k/a relative to any characteristic energy scale describing mode occupation that $\bar{n}_{jk}$ can be taken to be zero, so that only zero point fluctuations contribute to the spectrum. Since we are dealing with sub-Planck scale physics it is by no means clear that this assumption is valid. If $\bar{n}_{jk}$ is nonzero then the fluctuation spectrum would be unlikely to be scale invariant, reflecting instead whatever physics would determine the primordial occupation of modes, and the fluctuations would be non-Gaussian.

An approximate solution to equation (2.1) can be used to illustrate the effect of varying $H(t)$ and $m^2(t)$ histories (ignoring the off-diagonal m_{ij}^2, $i \neq j$ terms): outside the horizon $k^{-1} < (Ha)^{-1}$, but before growth due to the perturbed metric occurs, the power spectrum is

$$\mathbf{P}\phi(k) \approx [H(t_k)/(2\pi)]^2 \exp[-\int_{t_k}^{t} dt\, H(t)\,(3 + n(t))], \qquad (2.3a)$$

$$n(t) \equiv \text{Re}\,[-3(1 - 4m^2(t)/(9H^2(t)))^{1/2}], \qquad (2.3b)$$

$$\varrho_\phi \propto \exp[-\int_{t_k}^{t} dt\, H(t)\,(3 + n(t))], \qquad (2.3c)$$

$$a_k \equiv a(t_k) \equiv k/H. \qquad (2.3d)$$

The time t_k and expansion factor a_k when a wavenumber k crosses the horizon are given by eq. (2.3d). Equation (2.3a) is valid provided n > 0 at t_k, and remains valid even if n subsequently vanishes ($2m/3H > 1$). If n is zero at t_k, the prefactor $H^2 t_k$ should be replaced by $H^3\,(t_k)/m(t_k)$. Note that if n is constant, then the exponential term in (2.3a) is simply $[k/(H(t_k)a)]^{3+n}$, a power law in k if H is constant. A number of interesting cases follow from this result.

2.1 - Scale Invariance

If $m^2 \approx 0$, then the spectrum is scale invariant if H is independent of the expansion factor a. The value of a when k first 'crosses' the horizon is a_k. Since $H^2 = (8\pi G/3)\,[\Sigma_j \dot{\phi}_j^2 + V]$, judicious choice of V can lead to structure in $\mathbf{P}_\phi$. However, the most likely case over the ~ 6 orders of magnitude observable k range and the corresponding 14 e-foldings in a is that V will fall gently, leading to only slight deviations from scale invariance, with just a little more power on large scales than on small.

We can distinguish two types of behaviour, depending upon whether the field is driving inflation (in which case it is the inflaton) or its Hubble drag is driven by another field (in which case it is an isocon). The *inflaton* leads to gravitational metric fluctuations Φ proportional to fluctuations in the gauge invariant variable ζ introduced by Bardeen, Steinhardt, and Turner (1983, see also BBS1) which has the advantage of being independent of the spacelike hypersurface upon which the perturbation is measured: $\mathbf{P}_\Phi \propto \mathbf{P}_\zeta \sim (3H(a_k)/\dot{\phi})^2\,\mathbf{P}_\phi(k, a_k)/H^2$. For an *isocon*, for which we take the axion as the generic case, the fluctuations in the axion mass density after the axion mass is generated is $\mathbf{P}_{\varrho A} \propto \mathbf{P}_\phi$. In both cases, approximate scale invariance is the outcome.

An *isocon* which preferentially dissipates into baryons rather than antibaryons can lead to *isocurvature baryon perturbations,* although the δn_B spectrum will be scale invariant if the ϕ spectrum is. Such spectra can be ruled out by their large angle CMB anisotropies which are predicted to be enormous (Efstathiou and Bond 1986, see also Efstathiou 1988).

2.2 - Double Inflation

If we allow for more than one degree of freedom for our scalar fields, then marvelous mountain ranges, valleys and moguls can be envisaged for potential space. One's intuition regarding motion in this space can be utilized, except that the rolling fields are subject to Hubble drag $(-3H\dot{\phi}_j)$ which can result in a terminal velocity (slow-rollover), a phenomenon fundamental to the realization of inflation. Fluctuations $\delta\phi$ are small spreads in the field about the rolling background value $\bar{\phi}$. Fluctuations within the horizon at the specific epoch are still oscillating, while those outside have their shapes frozen in although the amplitudes may change with time. For some purposes it is useful to include those waves outside the horizon with the background field, to allow for a gentle variation from place to place (Starobinsky 1983).

If there are a number of flat directions in potential space, inflation could be a complicated process, modifying the H profile with time and also the form of the mass matrix m_{ij}^2. This will certainly map onto structure in the fluctuation spectrum. However, to transform specific features in potential space to features in a particular range in k-space, we must arrange for the fields to pass through this V structure at a specific range of a values.

Consider first the case of 2 scalars having a potential forming a broad valley with the valley minimum line very gently sloping down towards the origin with somewhat steeper walls rising away from the minimum. This configuration leads to double inflation (Starobinsky 1985, Kofman and Linde 1987, Silk and Turner 1987, BBS1, BBS2). If the 2-dimensional field begins high enough on the wall away from the origin, it will be potential dominated by the wall part, and experience inflation with a large value of H. The field will roll down towards the valley minimum, oscillating in one direction ($m^2 > 0$ so the power in the field in this direction damps away as $a^{-(3+n)}$ as above), while continuing to roll down towards the origin in the other direction with a lower value of H. The field in the second direction is all that is left after the end of inflation. Since it experiences first the high H for long waves as they leave the horizon, then the low H value for short waves, the plateau structure referred to in Sec. 1 is generic. The ramp between the two levels will depend upon the specific form of the potential. To arrange for the location of the ramp to be tuned to an astrophysically interesting scale, the initial location of the field matters. If the field remains near the valley minimum for too long a period then the spectrum within the current Hubble length will only reflect the low value of H and be effectively scale invariant.

Another procedure to adopt is to litter the valley minimum with moguls that will impose structure in the fluctuations of the field direction along the valley minimum. Bardeen (1988) has analyzed this case for a specific choice of mogul potential which leads to non-Gaussian fluctuations. The resulting spectrum becomes independent of initial conditions of the background field, but completely dependent upon mogul emplacement.

2.3 - Mountains

Choosing m^2 positive or negative can give spectra rising or falling with increasing k, but at the expense of exponential decreases or increases in both the background field energy density and in the fluctuation power, assuming inflation is continuing.

If m^2 is fixed, the drop of H with time eventually leads to n reaching zero. The field oscillates coherently with an energy density averaged over an oscillation period decreasing as $\varrho_\phi \sim a^{-3}$ like nonrelativistic matter, with fractional fluctuations $\mathbf{P}_\phi/\varrho_\phi$ being constant. This is the mechanism by which the axion, once it attains its mass when the temperature of the universe is about 200 MeV, behaves like cold nonrelativistic matter. However this occurs *after* inflation, so the a^{-3} law is not devastating.

To have a field whose fluctuations are still of interest we can only have m^2 positive or negative over a limited regime *during* inflation. To shape a mountain of power on top of a scale invariant spectrum, we would want m^2 to begin at 0 to ensure scale invariance at the longest wavelengths, to become positive for the rise, then negative for the drop, finally returning to zero to maintain scale invariance on short wavelengths. However, the scale invariant long wave structure will first decline in amplitude due to the $m^2 > 0$, then rise in amplitude due to $m^2 < 0$. It will require a restrictive class of choices to ensure the long wave part is not too high. Having m^2 become negative first will tend to give the plateau plus ramp structure of double inflation.

2.4 - Power Laws

One way to get extended power laws over some k range is to arrange for n to be constant over the associated range in a_k. This would be possible if m^2 scales with H^2. The natural way for this to occur is to make use of non-zero curvature coupling constants ξ, for then the local spectral index for isocons is $n = -3(1 - 16\xi/3)^{1/2}$. The severe price to pay is the $\sim a^{-(3+n)}$ fall of the energy density in the field throughout inflation. Only for $n = -2$ might we envisage getting the perturbation strength back by relative growth compared with the radiation, and this requires the very special choice $\xi = 5/48$. Even if such a value were to arise for some obscure reason, it would be subject to renormalization. It is easier to contemplate ξ negative with $n < -3$, falling to short wavelengths. We are currently exploring the $\xi < 0$ region and find some interesting features.

Another way to get power laws has been to invoke power law inflation (Lucchin and Matarrese 1985), with the expansion going as $a \sim t^q$, with $q > 1$ to ensure $\ddot{a} > 0$. This necessitates an equation of state yielding fixed $p/\varrho = -1 + 2/(3q)$. For the scalar field which drives power law inflation, $\mathbf{P}_\phi \sim (H/2\pi)^2 \sim k^{(3+n)}$, where $n = -(3q-1)/(q-1) = 3(1-p/\varrho)/(1+3p/\varrho)$.

This is also the k-dependence of $\mathbf{P}_{\zeta}$, hence the density fluctuations subsequently generated would have the power law index $n_\varrho = n + 4$. This always gives $n \leq -3$, $n_\varrho \leq 1$, if inflation is realized, hence there is more power at large scales than the Zeldovich spectrum. A disadvantage of such spectra is that the redshifts of cluster and galaxy formation will not be well separated for power laws with $n_\varrho \leq 0$. Further if n is too negative large angle CMB anisotropies become too large. With an interaction potential of the specific form $V(\phi) = V_0 \exp(-4(\pi/q)^{1/2} \phi/m_P)$ for the inflaton, power law inflation could be realized, with n related to q as given above. Power law inflation could also drive an isocon to develop power law isocurvature perturbations. For example, a Goldstone boson such as the axion would have $\mathbf{P}_\phi \sim k^{3+n}$, with n also as given above. The axion density perturbations developed once the axion mass is generated would have the same power n; such isocurvature CDM spectra can be strongly ruled out by large angle CMB anisotropies (Efstathiou and Bond 1986, Efstathiou 1988).

From this discussion there seems to be little hope that the $-1 \leq n \leq 0$ power law isocurvature baryon spectra advocated by Peebles (1988) will arise within the inflationary paradigm. In any case, these models would have $\Omega = 1$, and smaller values, Ω 0.2 to 0.4, are preferred.

3. Specific Realizations of Broken Scale Invariance

Consider two scalar fields interacting through a chaotic inflation potential containing quadratic and quartic terms:

$$V(\phi_1, \phi_2) = \frac{(m_2^2 + \xi_2 R)\phi_2^2}{2} + \frac{\lambda_2 \phi_2^4}{4} + \frac{(m_1^2 + \xi_1 R)\phi_1^2}{2} +$$
$$+ \frac{\lambda_1 \phi_1^4}{4} - \frac{\nu \phi_1^2 \phi_2^2}{2}. \tag{3.1}$$

If $\nu = 0$, and $\xi_1 = \xi_2 = 0$, then inflation is likely to be double inflation: If, $\lambda \gg \lambda_2$, first ϕ_1 dominates H, then ϕ_2. The effective m^2 of the first field ϕ_1 is always positive. Hence during the second phase its final amplitude would inflate away to an exponentially small value. $\mathbf{P}_{\phi_2}$ develops the ramp plus plateau structure.

The position of the ramp in k-space is controlled by the initial value $\phi_2(t_i)$. In the spirit of chaotic inflation as originally proposed by Linde (1983), $\phi_2(t_i)$ should fluctuate in space on comoving scales similar to the comoving scales on which ϕ_1 fluctuates, with amplitudes which should range up to $\frac{1}{4} \lambda_2 \phi_2^4 \sim m_P^4$. Given that the initial values of ϕ_1 allow the first stage of inflation to take place in some region, the inhomogeneities in ϕ_2 will inflate away (if ϕ_2 is not rough on arbitrarily small scales compared with the initial Planck scale). The value of ϕ_2 at a given location is frozen until the end of the first stage of inflation. Only if the frozen value of ϕ_2 is very precisely a certain value near $4.3 m_P$ will the ramp on the final perturbation spectrum be at an astrophysically interesting scale. If initial values of ϕ_2 do indeed range over $|\phi_2(t_i)| \leq \lambda_2^{-1/4} m_P$ (Linde 1983) and if, as is necessary for the amplitude of the density perturbations, $\lambda_2 \lesssim 10^{-14}$, then the desired range of $\phi_2(t_i)$ would be a tiny fraction of its possible range. Although the probability distribution of $\phi_2(t_i)$ over this range is unknown, it will undoubtedly be necessary to 'fine tune' the initial conditions — as well as the potential parameters — to place the ramp in the desired location, making this version of double inflation rather unattractive in our view.

In any case, making the ramp interestingly large to generate structure will result in a high plateau giving a high amplitude scale invariant spectrum for microwave background fluctuations which would violate the stringent bounds set by the Soviet RELICT experiment discussed e.g., by Bond (1988).

Adding the coupling $\nu \neq 0$ in (3.1) does not aid matters appreciably. Even though one can arrange ν so that the effective m^2 vanishes in the potential valley minimum, Hubble drag forces the background field to lie outside the trough with $m^2 > 0$ always. This is true whether the field starts in the minimum (so there is no double inflation) or outside of it (so there is). Thus ϕ_1 declines exponentially quickly, becoming dynamically unimportant. The $\mathbf{P}_{\phi_2}$ that is left is still of the ramp plus plateau form.

To get enough e-foldings of inflation with a potential for ϕ_2 of the form eq. (3.1), it is necessary that the field start at a value several times the Planck mass. However, the Lagrangian term involving the Ricci scalar R becomes $\mathbf{L}_G = (m_P^2/(16\pi) - \xi\phi^2/2)R$, which can change sign if ϕ is too large. To avoid this catastrophe it is necessary that $\xi_2 \lesssim 0.002$; i.e., that the field be effectively minimally coupled (or have $\xi_2 < 0$). A similar restriction would hold for ξ_1 if we were contemplating large initial values of ϕ_1 as well. Even if ξ_1 did not have to be zero, unless it is quite negative the potential gives a positive effective mass $m_{11}^2 = m_1^2 + 12\xi_1 H^2 + 3\lambda_1\phi_1^2 - \nu\phi_2^2$: the ϕ_1 fluctuations again inflate away.

The addition of cubic interaction terms to the potential (3.1) can give $m^2 < 0$ over a short range. Subsequently however m^2 becomes positive and the ϕ_1 fluctuations inflate away. Bardeen (1988) shows that before they inflate away they could induce non-Gaussian fluctuations in the field driving inflation, ϕ_2. In another approach to the generation of non-Gaussian perturbations, Grinstein and Wise (1987) considered a universe with massive axions which were a very small component of the dark matter, $\Omega_A \ll 1$, but which had very large amplitude fluctuations.

If a (non-chaotic) 'new inflation' potential is chosen, e.g., of form $\lambda_2(\phi_2^2 - \sigma_2^2)^2/4$, with the initial value of the field near the origin, we would still have to tune $\phi_2(t_i)$ to get the ramp in the right place. In this case it is also not clear why ϕ_2 would start close enough to the origin for inflation to be feasible (Masenko, Unruh, and Wald 1984, but see Albrecht and Brandenberger 1985 who partially address this objection).

Starobinsky (1983) showed that the conformal anomaly of massless scalar fields (nonzero trace of the stress-energy tensor due to quantum effects) is o order R^2, where R is the scalar curvature, and this might drive inflation.

However the conformal anomaly terms that were most likely to appear in typical theories were shown to be unlikely to lead to inflationary behaviour. Starobinsky (1985) now considers the R^2 terms independently of the conformal anomaly, parameterizing the gravitational Lagrangian $L_G = (m_P^2 16\pi)(R + R^2/(6M^2))$, by a small mass scale $M \ll m_P$. Another massive scalar field, the "scalaron," was introduced to get this mass scale $M \sim 10^{-5} m_P$. The Friedman equation is now significantly modified over the form with no R^2 term in the Lagrangian. Kofman, Linde and Starobinsky (1985) show that double inflation may follow provided that $M > (\lambda/6\pi)^{1/2} m_P$. Both the scalar field potential and the R^2 term drive the initial inflationary epoch, followed by an era when the scalar field dominates. Again a ramp plus plateau spectrum remains the generic outcome.

Kofman, Linde and Einasto (1987) suggested that the transition to the second phase might proceed by quantum tunnelling through a potential barrier, as in Guth's original 'old' inflation model, leading to bubbles being generated, superimposed upon a universe smoothed during the first inflationary epoch. Such fluctuations would certainly be non-Gaussian, but it would be extremely difficult to arrange for their amplitude to be just right to be useful to explain the large-scale texture we observe now.

We can conclude that no compelling case now exists either for a natural way to break scale invariance either with a mountain-like spectrum built above a scale invariant base, or for power laws differing appreciably from the scale invariant form.

ACKNOWLEGEMENTS

We would like to thank L. Kofman, A. Starobinsky, R, Fakir, M. Mijic, and B. Unruh for informative discussions. The hospitality of J. Monaghan at Monash University while some of this paper was written is gratefully acknowledged by J.R.B. J.M.B. was supported by DOE grant DE-AC0G-81ER40048 at U.W., J.R.B. by a Canadian Institute for Advanced Research Fellowship and a Sloan Foundation Fellowship. J.R.B. and D.S.S. were also supported by the NSERC of Canada.

REFERENCES

Albrecht, A. and Brandenberger, R. 1985. *Phys Rev. D.* **31**, 1225.

Bardeen, J.M. 1980. *Phys Rev D.* **22**, 1882.

_____. 1988. in preparation.

Bardeen, J.M., Steinhardt, P.J., and Turner, M.S. 1983. *Phys Rev D.* **28**, 679.

Bardeen, J.M., Bond, J.R. and Salopek, D. 1987. *Proc. Second Canadian Conference on General Relativity and Relativistic Astrophysics.* ed. A. Coley and C. Dyer. Singapore: World Scientific. [BBS1]

_____. 1988. in preparation. [BBS2]

Bond, J.R. 1988. in *Large-Scale Structures of the Universe.* eds. J. Audouze, M.-C. Pelletan, and A. Szalay. Dordrecht: Kluwer Academic Publishers, p. 93.

Efstathiou, G. 1988, this volume.

Efstathiou, G. and Bond, J.R. 1986. *MN.* **218**, 103.

Grinstein, B. and Wise, M. 1987. preprint.

Kofman, L.A. and Linde, A.D. 1987. *Nuc Phys.* **B282**, 555.

Kofman, L.A., Linde, A.D., and Starobinsky, A.A. 1985. *Phys Lett.* **157B**, 361.

Kofman, L.A., Linde, A.D., and Einasto, J. 1987. *Nature.* **326**, 48.

Linde, A.D. 1983. *Phys Lett.* **129B**, 177.

Lucchin, F. and Matarrese, S. 1985. *Phys Rev D.* **32**, 1316.

Masenko, G., Unruh, W., and Wald, R. 1985. *Phys Rev D.* **31**, 273.

Peebles, P.J.E. 1988, this volume.

Starobinsky, A.A. 1983. In *Quantum Gravity.* p. 103. ed. M.A. Markov and P. West. New York: Plenum.

_____. 1985. *JETP Lett.* **42**, 152.

LARGE-SCALE ANISOTROPY
OF THE COSMIC BACKGROUND RADIATION

NICOLA VITTORIO

Dipartimento di Fisica, Università dell'Aquila

ABSTRACT

In this paper we discuss a number of related issues concerning the large-scale anisotropy of the cosmic background radiation (CBR). We present cold dark matter model predictions for the peculiar acceleration field and for the CBR large-scale anisotropy. We discuss the number and the angular sizes of the hot spots expected in the CBR temperature distribution, under the assumption of Gaussian cosmic background fluctuations. We show that comparing the available CBR observations on large angular scales may constrain the density fluctuation spectral index. We also present an analysis of the Davies *et al.* (1987) data on CBR anisotropy and a Monte Carlo simulation of the experiment.

1. INTRODUCTION

The cosmic background radiation (CBR) temperature anisotropies expected in initially slightly inhomogeneous cosmological models have proved to be one of the most stringent constraints on different galaxy formation scenarios. In fact, CBR anisotropy studies bypass many of the uncertainties associated with the non linear models of small-scale galaxy clustering. Perhaps the most fundamental aspect of the CBR is that it probes epochs ($\sim$ 700,000 years after the Big Bang) and scales ($\gtrsim$ 200 Mpc) of the universe that are otherwise inaccessible.

In particular, the study of the large scale CBR temperature distribution is very important for at least two reasons. On one hand, the large sky coverage, provided by balloon and satellite experiments, ensures we observe a significant sample of the sky. One the other hand, the observational upper limits to the CBR temperature anisotropy can be interpreted independently of the presence or absence of late reheating of the intergalactic medium. In fact, the gravitational potential fluctuations responsible for the CBR temperature anisotropies (Sachs and Wolfe 1967) are independent of the location of the last scattering surface.

A detection of large scale CBR temperature fluctuations would be, of course, of fundamental interest, not only because it would confirm the current ideas on the origin and the evolution of the large-scale structure of the universe, but also because it would provide a direct measure of the amplitude of the initial density fluctuations in the framework of the linear theory. This, in turn, would constrain both the epoch of galaxy formation and the properties of the dark matter in the universe (possible kinds of weakly interacting particle, mass, lifetime, etc.).

The difficulty in detecting CBR temperature fluctuations has raised the question of devising the best observational strategy. For this goal, it is necessary to calculate for different models, not only rms values for the CBR temperature fluctuations on a given angular scale, but also the pattern expected for the CBR temperature distribution. It is thus possible to evaluate in a given cosmological scenario, for a given observational configuration, the probability of detecting CBR temperature fluctuations at a certain level, and then, to define the best experimental configuration and observational strategy.

The standard inflationary scenario (see, e.g., Turner 1987 for a recent review) predicts a flat universe and adiabatic scale invariant density fluctuations. Attention has been drawn in the last years to a very specific model for the formation and evolution of the large-scale structure of the universe, where the mass density is at the present dominated by cold dark matter (CDM; for a review see Blumenthal et al. 1984). This model combines the inflationary predictions with the observational constraints derived by small-scale galaxy clustering, if galaxies are biased to form in the highest peaks of the density field. But in this case there could be a problem with the predicted large-scale peculiar velocity field (Vittorio, Juzkiewicz, and Davis 1986; Juzkiewicz and Bertschinger 1988; but see also Kaiser 1988).

Here we discuss some related issues concerning the large-scale anisotropy of the CBR. The CDM model predictions for the large-scale peculiar acceleration field are presented in Section 2. The expected large-scale pattern of the CBR is discussed in Section 3. Observational upper limits and CDM model predictions for CBR temperature fluctuations are reviewed in Section 4 and Section 5 respectively. An analysis of the Davies et al. data (1987) and a Monte Carlo simulation of the experiment is presented in Section 6. Finally, a brief summary is given in Section 7.

2. CBR DIPOLE ANISOTROPY AND PECULIAR ACCELERATION FIELD

The dipole anisotropy is the only unambiguously detected anisotropy of the CBR. Its amplitude ($\sim 10^{-3}$) is indicative of our motion relative to the

comoving frame and implies a Local Group peculiar velocity of ∼ 600 km
s^{-1}, in a direction which is 45° away from the Virgo Cluster (for a recent
review see Lubin and Villela 1987). Since the Local Group is falling into Virgo
with a velocity of ∼ 250 km s^{-1} (see, e.g., Yahil, 1985), the Virgo cluster as
a whole moves relative to the CBR, with a velocity of ∼ 500 km s^{-1}, in the
general direction of the Hydra-Centaurus Supercluster. Galaxy formation
scenarios can be tested by checking if the expected gravitational field of typical
mass concentrations is sufficient to generate the observed Virgo Cluster peculiar
velocity.

To compare the observations with the model predictions it is important
to determine the minimum depth of a sample of galaxies necessary to define
the comoving frame. If these galaxies trace the Hubble flow, the peculiar
velocities of the Local Group relative to the sample and to the comoving frame
should coincide. Otherwise, a coherent motion of the sample as a whole relative
to the CBR is implied. From a given sample of galaxies, one can, for example, evaluate the acceleration exerted on the Local Group by all the galaxies
in the sample. In order to verify that all the material outside the sample does
not exert a significant acceleration on the Local Group, it is usual to calculate
the cumulative acceleration exerted by galaxies in nested subsamples of increasing radii. If the cumulative acceleration converges to a finite value, then
the inhomogeneities responsible for the dipole anisotropy should be contained
in the sample and the sample itself should be at rest relative to the CBR.

The IRAS Point Source Catalogue provides for the first time a galaxy
sample uniformly selected and with a nearly complete sky coverage. The depth
of the IRAS sample is estimated to be ≲ 200 Mpc. The analysis of this sample has proven to be a powerful probe of the Hubble flow field (Yahil *et al.*
1986, Meiksin and Davis 1985, Villumsen and Strauss 1987, Strauss and Davis
1987). It has been shown that the distribution of galaxies in the catalogue exhibits a dipole anisotropy in reasonable agreement with the direction of the
CBR dipole anisotropy to within ∼ 20° to 30°.

If the IRAS galaxies trace the mass distribution, the gravitational acceleration exerted on the Local Group is proportional to (Yahil *et al.*, 1986)

$$\vec{d} = \frac{3}{4\pi} \int_0^\infty \delta(\vec{x}) \, \frac{\vec{x}}{x^3} \, \Phi(x, R) d^3x, \tag{1}$$

where $\delta(\vec{x})$ is the density fluctuation field and the window function

$$\Phi(x, R) = [1 + \chi^2/(2.4R^2)]^{-2.4}$$

describes the geometry of the sample. The parameter R gives the effective depth
of the sample.

The rms values of $|\vec{d}|$ and of the bulk motion of the sample relative to
be CBR are [Juszkiewicz, Vittorio, and Wyse (1988)]:

$$D^2(r, R) = \frac{9}{2\pi^2} \int_0^\infty dk \, P_s(k) W_g^2(k, R) \tag{2}$$

and

$$v^2(r, R) = \frac{1}{2\pi^2} \int_0^\infty dk P_s(k) \, W_v^2(k, R), \tag{3}$$

where $P_s(k)$ is the power spectrum smoothed on scale r: $P_s(k) = P(k) exp(-k^2 r^2)$. For the explicit expression for W_g and W_v we refer to Juszkiewicz, Vittorio, and Wyse (1988). In the simple case of a volume limited sample, $W_g^2(k, R) = 1 - j_0(kR)$ and $W_v^2(k, R) = 3j_1(kR)/(kR)$, where j_0 and j_1 are the spherical Bessel functions. This shows that W_g acts as a high pass filter and supresses the contribution of long wave ($kR \to 0$) density perturbations. On the contrary, W_v acts as a low pass filter, suppresses the small ($kR \to \infty$) wavelength perturbations. So D and v provide us with information about the density inhomogeneities.

In a CDM dominated universe the density fluctuation spectrum is conveniently described by the following fitting formula: $P(k) = A \, k^n \, [1 + 6.8 \, k \, / \, \Omega_0 + 72 \, (k \, / \, \Omega_0)^{1.5} + 16 \, (k \, / \, \Omega_0)^2]^{-2}$ (e.g., Davis et al. 1985). Here n is the primordial spectral index, Ω_0 is the density parameter, the wavenumber k is measured in Mpc^{-1} and the Hubble constant is assumed to be $H_0 = 50$ km s^{-1}/Mpc. The constant A defines the initial amplitude of the density fluctuations. The spectrum tilts down from the primordial slope ($\propto k^n$) at low k, to k^{n-4} for high k, as a consequence of the reduced growth of fluctuations entering the horizon during the radiation dominated era (see, e.g., Blumenthal et al. 1984). The variance of the density fluctuations is defined as (Peebles 1980)

$$\sigma_0^2(R) = \frac{1}{2\pi^2} \int_0^\infty dk \, k^2 \, P(k) W_m^2(kR). \tag{4}$$

The window function $W_m(kR) = 3j_1(kR)/(kR)$ is introduced because of the averaging over a sharp edged sphere. The quantity R is the typical size of the proto-object, whose rms density fluctuation is $\sigma_0(R)$. The a priori unknown amplitude of the primordial density fluctuation spectrum, i.e. the constant A, is usually fixed by requiring $\sigma_0(8h^{-1}Mpc) = b^{-1}$. The biasing factor b is the ratio of the galaxy counts-to-mass fluctuations on the scale of 8 h^{-1} Mpc. The results below have been calculated with b = 1. If light does not trace mass they must be rescaled by b^{-1}.

Figure 1 shows D(r,R) and v(r,R) as a function of R, for a flat CDM dominated universe with scale invariant density fluctuations. Due to the lack of large-scale power in the model, D(r,R) converges at large scales to a finite value, which implies that density inhomogeneities on scales $\lesssim$ 100 Mpc are

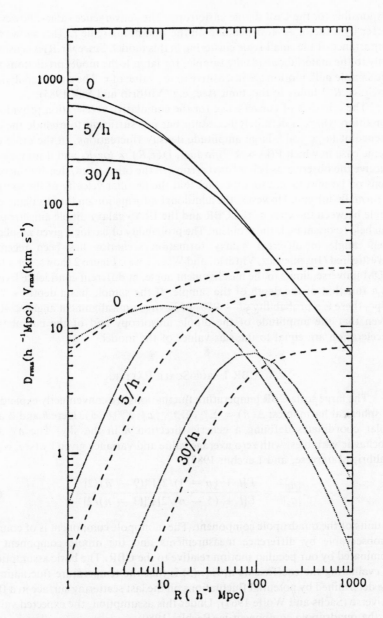

FIG. 1. Rms values of the density moment (dashed lines) and peculiar velocity (continuous line) of a sample defined by the window function $\Phi(x, R)$ (see text). Labels refer to different small-scale cutoffs. The dotted lines show $(d|\vec{d}|/dlog_{10}R)$ vs. R to determine the distance to the dominant fraction of the accelerating material. (From Juszkiewicz, Vittorio, and Wyse 1988).

responsible for the CBR dipole ansiotropy. The convergence values, however, differ by a factor ~ 4, depending on the smoothing scale r. This shows the importance of the small-scale clustering in this model. Since v(r,R) is sensitive only to the material outside the sample, for large R the model predictions for the sample bulk motion are insensitive to the value of r. The simple analytical law $v \propto R^{-1}$ holds in this limit (see, e.g. Vittorio and Silk 1985).

The criterion of convergence for the cumulative acceleration provides a condition which is definitely necessary but not sufficient to exclude the existence of large-scale, large amplitude density fluctuations. In the quite extreme case in which $P(k) \propto k^{-1}$, in fact, $D(r,R) \propto log R$, and it may easily deceive the observer as being convergent. On the other hand, density fluctuations on large-scale are so important that the peculiar velocity of the sample is formally infinite. However, the additional information on the misalignment angle between the apex of the CBR and the IRAS galaxy dipole anisotropies can help to disentangle the problem. The probability of having a given misalignment angle in different galaxy formation scenarios has been recently investigated (Juszkiewicz, Vittorio, and Wyse 1988). Figure 2 shows, for a flat CDM universe, limits to the misalignment angle, at different confidence levels, as a function of the depth of the sample. If the sample has a depth ~ 200 Mpc, there is a probability of ~ 95% of having a misalignment angle < 40°, given that the amplitude of the dipole anisotropy and of the cumulative acceleration are equal to the rms values of the model.

3. CBR LARGE-SCALE PATTERN

The large scale CBR temperature fluctuations are conveninetly expanded in spherical harmonics: $\Delta(\hat{n}) = \Sigma_{l=2}^{\infty} \Sigma_{m=-l}^{m=+l} a_l^m Y_l^m(\theta, \phi)$. Here θ and ϕ are polar coordinates defining a generic direction $\hat{n}$ in the sky. The a_l^m are stochastic variables, with zero average value and variance given by (see, e.g., Fabbri, Matarrese, and Lucchin 1987):

$$\frac{|a_l|^2}{|a_2|^2} = \frac{\Gamma[l + (n - 1)/2]\Gamma[(9 - n)/2]}{\Gamma[l + (5 - n)/2]\Gamma[(3 - n)/2]} \tag{5}$$

in units of the quadrupole component. The monopole component is of course unobservable by difference measurements and the dipole component is dominated by our peculiar motion relative to the CBR. The basic assumption in evaluating the coefficients in Eq. (5) is that the temperature fluctuations are determined by potential fluctuations on the last scattering surface in a flat universe (Sachs and Wolfe 1967). Under this assumption, the expected value of the quadrupole component is (Peebles 1980)

$$|a_2|^2 = \frac{A}{16} \left(\frac{H_0}{c}\right)^{3+n} \frac{\Gamma(3 - n)\Gamma(\frac{3+n}{2})}{[\Gamma(\frac{4-n}{2})]^2\Gamma(\frac{9-n}{2})} . \tag{6}$$

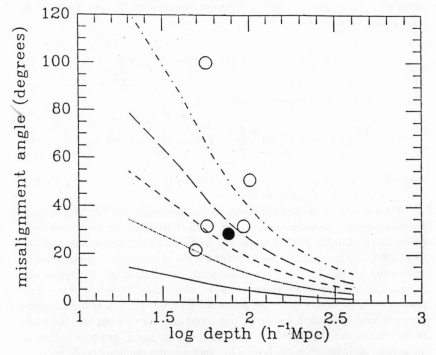

FIG. 2. Confidence levels on the maximum misalignment angle expected in a flat cold dark matter dominated universe: 5% (continuous line), 25%, 50%, 75%, 95% (dotted dashed lines). (From Juszkiewicz, Vittorio, and Wyse 1988).

Here A is the overall amplitude of the density fluctuations and n is the density fluctuation spectral index. The statistics of the CBR temperature field are completely described by the two point angular correlation function $C(\alpha) \equiv \langle \Delta(\hat{n}_1) \cdot \Delta(\hat{n}_2) \rangle$, where $\alpha = \cos^{-1}(\hat{n}_1 \cdot \hat{n}_2)$. It is usual to take into account the finite angular resolution of the antenna by modeling the antenna beam with a Gaussian profile of FWHM = 2.35 σ. Then, the smoothed correlation function is (Scaramella and Vittorio 1988)

$$C(\alpha, \sigma) = \frac{1}{4\pi} \sum_{l=2}^{\infty} |a_l|^2 \, (2l+1) \, P_l \, (\cos \alpha) \, exp \left\{ - \left[\left(l + \tfrac{1}{2} \right) \sigma \right]^2 \right\}$$

(7)

The effect of the beam, as expected, acts as a low pass filter, which severely attenuates harmonics of order $l \gg 1/\sigma$. Knowing the shape of the correlation function allows one to derive from the observations the amplitude of the rms temperature fluctuations, i.e. $C^{1/2}(0, \sigma)$, as a function of n, and, then, to relate directly results obtained in different experiments. In fact, for a single subtraction experiment,

$$\left(\frac{\Delta T}{T}\Big|_{obs}\right)^2 \geq 2\, C(0,\, \sigma)\, [1 - R(\alpha,\, \sigma)], \tag{8}$$

while for a double substraction experiment,

$$\left(\frac{\Delta T}{T}\Big|_{obs}\right)^2 \geq C(0,\, \sigma) \left[\frac{3}{2} - 2\, R(\alpha,\, \sigma) + \frac{1}{2}\, R(2\alpha,\, \sigma)\right] \tag{9}$$

where $R(\alpha,\, \sigma) \equiv [C(\alpha,\, \sigma)/C(0,\, \sigma)]$, and the equality sign holds for a positive detection of CBR temperature fluctuations.

The knowledge of the CBR temperature correlation function also suffices for investigating the large-scale pattern of the CBR temperature distribution (Bond and Efstathiou 1987, Vittorio and Juskiewicz 1987, Scaramella and Vittorio 1988). Two quantities are relevant for the observations: the number of upcrossing regions in the sky and their angular dimension. By upcrossing regions one indicates the regions in the sky where the CBR temperature fluctuation is higher than ν times the rms value [i.e., $C^{1/2}(0,\, \sigma)$]. If these regions are sufficiently abundant and large, one could look for rare but very hot spots in the microwave sky (Sazhin 1985). On the other hand, beam-switching at an angular scale less than the typical hot spot angular diameter could produce a strong reduction in any detectable anisotropy. So, the knowledge of the typical hot spot angular diameter can at least be a guide in designing the observational configuration.

For $\nu \gg 1$, the number of hot spots is well approximated, by (see, e.g., Scaramella and Vittorio 1988):

$$N_{>\nu} = N_*(\sigma,\, n)\, \nu\, exp(- \frac{\nu^2}{2}). \tag{10}$$

The function $N_*(\sigma,\, n)$ is plotted in Figure 3a as function of σ, for different values of the spectral index n. The number of upcrossing regions scales as σ^{-2} for very steep density fluctuation power spectra. This dependence flattens out for negative spectral indices. In fact, lowering n reduces the small scale (relative to the large scale) power. Eventually the finite antenna beam size has pratically no effect. If the primordial fluctuations have a Zeldovich spectrum, the number of upcrossing regions expected in all the sky has an analytical expression (Vittorio and Juszkiewicz 1987):

$$N_{>\nu} = \frac{650}{\sigma^2} \frac{\nu\, e^{-\nu^2/2}}{[-\ln 2\sigma + 3.78]}. \tag{11}$$

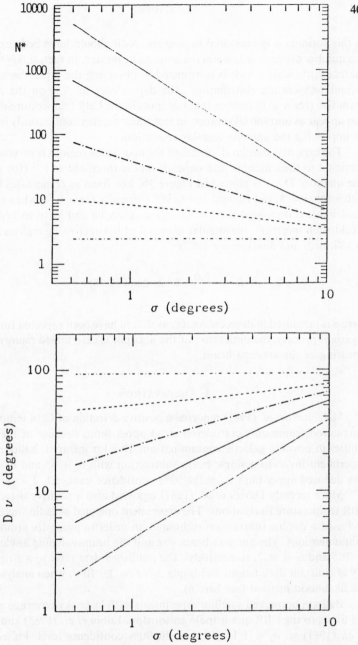

FIG. 3. The quantity N_* (3a), and the hot spot average diameter times the threshold ν (3b), vs. the antenna beam size σ, for different values of the spectral index: $n = 3$ (dotted), $n = 1$ (solid), $n = -1$ (dashed), and $n = -2.9$ (dotted—dashed). (From Scaramella and Vittorio 1988).

In this formula σ is measured in degrees. As it should have been expected, this number depends only upon the antenna beam size. In fact, this is the only characteristic scale which is introduced in observing the otherwise scale invariant temperature distribution. The dependence of $N_{>\nu}$ on the antenna beam size (for $n \gtrsim 0$) implies that hot spots in the CBR temperature distribution appear as unresolved sources: in fact, their number continuously increases on improving the antenna angular resolution.

The expected angular diameter of the upcrossing regions is inversely proportional to their number and depends on the threshold as ν^{-1} (for $\nu \gg 1$). The quantity $D \cdot \nu$ is plotted in Figure 3b. For fixed ν, D increases linearly with σ for $n \simeq 3$, but flattens out to 90° (formally for $\nu = 1$) when only the quadrupole is dominant (i.e., $n < -2$) (Scaramella and Vittorio 1988). For a Zeldovich spectrum, the angular diameter of the upcrossing regions is given by (Vittorio and Juszkiewicz 1987):

$$ D = \frac{5.6}{\nu} \sigma[-ln\ 2\sigma + 3.78]. \tag{12} $$

Here σ is measured in degrees. Again, as should have been expected for a scale invariant process, the dimension of the hot spot is determined mainly by the smearing of the antenna beam.

4. OBSERVATIONS

Melchiorri et al. (1981) reported a positive detection of CBR temperature fluctuation, commonly considered as an upper limit, because of the uncertainties in possible galactic contamination. This far infrared, balloon borne experiment involved a single beam subtraction with $\alpha = 6°$ and $\sigma = 2°.2$. The deduced upper limit is, at the 90% confidence level, $\Delta T/T < 4 \cdot 10^{-5}$.

More recently Davies et al. (1987) reported also a positive detection of CBR temperature fluctuations. The experiment operated at radio wavelengths and used a double subtraction technique, in order to minimize atmospheric contaminations. The antenna beam size and the beamswitching angle were $\sigma = 3°.5$ and $\alpha = 8°.2$, respectively. The published data refer to a strip of the sky at constant declination, and imply $\Delta T/T = 3 \cdot 10^{-5}$, when analysed with the likelihood method (see Sec. 6).

Balloon borne and satellite experiments with large sky coverage give upper limits to the CBR quadrupole anisotropy. Lubin et al. (1985) and Fixsen et al. (1981) set $a_2 < 1.1 \cdot 10^{-4}$ at the 90% confidence level. Fixsen et al. placed also an upper limit to the CBR correlation function: $C(10° < \alpha < 180°, \sigma) < 1.37 \cdot 10^{-9}$. More recently, the RELIC satellite borne experiment (Klypin et al. 1986) set $C(20°, \sigma) < 5.5 \cdot 10^{-10}$ and $a_2 < 4.75 \cdot 10^{-5}$ at the 95% confidence level. Assuming a scale invariant density fluctuation power spectrum, this limit is even more severe: $a_2 < 2.54 \cdot 10^{-5}$. The RELIC experiment sets an upper limit to the amplitude of the octupole component: $a_3 < 9.4 \cdot 10^{-5}$.

In order to have a first comparison among these different experiments we proceed as follows (Scaramella and Vittorio 1988). For a given value of the primordial spectral index, Eq. (8) or (9) allows us to derive from an observational upper limit to (a positive detection of) $\Delta T/T$, an upper limit to (a measure of) $C(0, \sigma)$. Since different experiments operate with different antennas, given a value for $C(0, \sigma)$, Eq. (7) can be used to evaluate $C(0, \sigma')$ for any value of the beam size σ'. So, for each n, one can have four values of $C(0, 3^o.5)$: three upper limits (determined by the Melchiorri et al. 1981, Fixsen et al. 1982, and Klypin et al. 1986 experiments) and one detection (from the Davies et al. 1987 experiment). Comparing these different values of CBR temperature variances (see Figure 4) shows that the Davies et al. (1987) detection is consistent with the Fixsen et al. (1981) and the RELIC upper limits only if $n > 1$. This comparison is fairly independent of the actual value of the density parameter Ω_0, as for $0.2 \le \Omega_0 < 1$, the angles involved are smaller than that subtended by the curvature radius, $\sim 30^o\Omega_0^{1/2}$ (Peebles 1981, 1982).

With a similar strategy, one can predict the quadrupole anisotropy implied by the values of $C(0, \sigma)$, for a given spectral index, deduced from the Melchiorri et al. (1981), Fixsen et al. (1981), and Davies et al. (1987) experiments (see Figure 5). The latter experiment predicts a quadrupole which is higher than that derived by the Melchiorri et al. (1981) experiment for any value of the spectral index n, and it is consistent with the RELIC upper limit only if $n > 1.5$.

5. CDM MODEL PREDICTIONS

In this Section we discuss the large scale CBR temperature anisotropies expected in a CDM dominated universe, where the primordial power spectrum [see Sec. 2] is not necessarily scale invariant. Observational upper limits to the CBR anisotropy can provide bounds on n for a given Ω_0. On intermediate angular scale ($1^o < \alpha < 10^o$), the expected CBR temperature anisotropy, as it would be measured in a single subtraction experiment, is (Vittorio, Matarrese, and Lucchin 1988)

$$\frac{\delta T}{T}\Big|^2_{rms}(\alpha, \sigma) = A \frac{F^2(\Omega_0^{-1} - 1)}{\Omega_0^2} \frac{2}{\pi^{3/2}} \frac{\Gamma(\frac{3-n}{2})}{\Gamma(2 - \frac{n}{2})} \left(\frac{H_0\Omega_0}{2c}\right)^{n+3} \sigma^{1-n}$$

$$\times \sum_{m=1}^{m=\infty} \frac{(-1)^{(m-1)}}{(m!)^2} \Gamma\left(\frac{2m + n - 1}{2}\right) \left(\frac{\alpha}{2\sigma}\right)^{2m}. \tag{13}$$

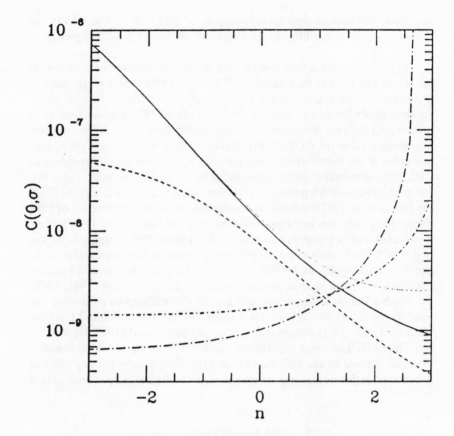

FIG. 4. CBR temperature variance derived from the Melchiorri *et al.* 1981
(dashed), Davies *et al.* 1987 (solid), Fixsen *et al.* 1987 (short dash—dotted),
and the RELIC (long dash—dotted) experiments vs. the spectral index. (From
Scaramella and Vittorio 1988).

The function $F(y) = 2y/[5 + 15y^{-1} + 15\sqrt{1 + y}\,y^{-3/2}ln\,(\sqrt{1 + y} - \sqrt{y})]$
takes into account the reduced growth of fluctuations in an open universe
(Peebles 1981) and the constant A is the overall amplitude of the density fluctuations. If n = 1, the scale invariance of the spectrum reflects the fact that
the expected CBR anisotropy depends only on the ratio α/σ. Figure 6 shows
the rms value of the CBR temperature anisotropy, expected in this case in a
flat universe (Vittorio and Silk 1985). The two fields of view have an angular
radius ~ σ and are separated by an angle α. The expected anisotropy is bigger
for large values of the ratio α/σ. At small values of α/σ the two antennas are
essentially looking to the same region of the sky and the signal is strongly suppressed because of the difference procedure. In Figure 6 the continuous lines

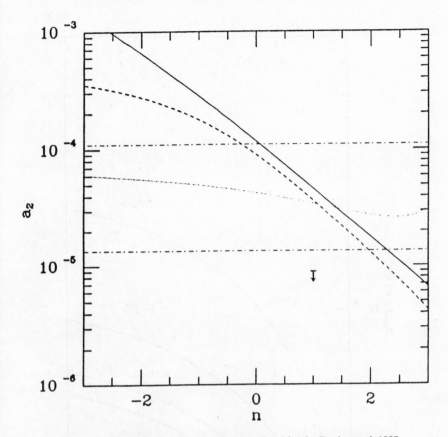

Fig. 5. The rms quadrupole component predicted by the Davies *et al.* 1987 experiment (solid) and the upper limits inferred from the Melchiorri *et al.* 1981 (dashed) and Fixsen *et al.* 1981 (dotted) experiments vs. the spectral index. The upper limits from Lubin *et al.*1985, and Fixsen *et al.*1981 (upper dot—dashed), and the RELIC (lower dot—dashed) experiments are also shown. (From Scaramella and Vittorio 1988).

(labeled c) are the predictions for a CDM dominated universe with scale invariant density fluctuations normalized as described in Sect. 2. If $n \neq 1$, at fixed α/σ, the anisotropy varies as $\sigma^{(1-n)/2}$. A good fit to the numerical result, for $H_0 = 50$ km s^{-1} Mpc^{-1}, is given by (Vittorio, Matarrese, and Lucchin 1988)

$$\frac{\delta T}{T}\Big|_{rms}(6^o, 2^o.2) \simeq 6 \cdot 10^{-5+(0.23\Omega_0-1.15)n}\, \Omega_0^{-1.15}. \tag{14}$$

Consistency with the Melchiorri *et al.* (1981) upper limit requires $n > 0.2$ and $n > 0.6$ for $\Omega_0 = 1$ and $\Omega_0 = 0.4$, respectively. For a scale-invariant spec-

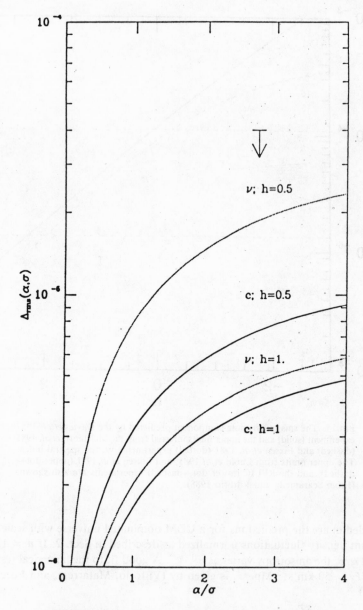

FIG. 6. CBR anisotropy expected in a CDM dominated universe for h = 0.5 (heavy continuous line) and for h = 1.0 (light continuous line); h is the Hubble constant in units of 100 km s^{-1}/Mpc. The arrow refers to the Melchiorri *et al.* 1981 upper limit. Dotted lines refer to a massive neutrino dominated universe. (From Vittorio and Silk 1985).

trum and $\Omega_0 > 0.4$ there is no conflict with the observed upper limit (see also Vittorio and Silk 1984).

The calculation of the intermediate scale anisotropy in Eq. (13) neglects the global space curvature. For the quadrupole anisotropy this approximation is not valid any longer, and, because of this, we restrict ourselves to the flat case. In a flat CDM dominated universe, where light traces the mass, the rms value of the quadrupole component is (Vittorio, Matarrese, and Lucchin 1988)

$$|a_2| = 1.3 \cdot 10^{-(4+1.285n)}. \tag{15}$$

This implies $n > 0$, in order to be consistent with the observational upper limit set by the Lubin et al. (1985) and Fixsen et al. (1981). Consistency with the RELIC upper limit requires $n > 0.35$. The amplitude of the higher multipoles are (Vittorio, Matarrese, and Lucchin 1988)

$$|a_l| = 1.3 \cdot 10^{-(4+1.285n)} \cdot \sqrt{\frac{\Gamma[l + (n-1)/2]}{\Gamma[(3+n)/2]}} \sqrt{\frac{\Gamma[(9-n)/2]}{\Gamma[l + (5-n)/2]}}. \tag{16}$$

The RELIC upper limit to the octupole anisotropy constrains the primordial spectral index to be greater than 0.15.

6. SIMULATING A MICROWAVE ANISOTROPY EXPERIMENT

The analysis of CBR anisotropy data with the likelihood method (see, e.g., Kaiser and Lasenby 1987) requires an explicit guess for the functional form of the CBR temperature correlation function. Davies et al. (1987) assumed a Gaussian functional form: $C(\alpha, \sigma) = C(0, \sigma) \exp \{-\alpha^2/[2(2\sigma^2 + \theta_c^2)]\}$, with a sky intrinsic coherence angle of $\theta_c \simeq 4^o$, and found $C(0, \sigma) = 3.7 \cdot 10^{-5}$ (corresponding to $\Delta T/T(8^o.2,3^o.5) = 3 \cdot 10^{-5}$).

The Davies et al. (1987) published data have also been analysed using the correlation function given in Eq. (7), for primordial spectral indices $-3 < n < 3$ (Vittorio et al. 1988). Figure 7 shows the value of the rms CBR temperature fluctuation, $C_{0M}^{1/2}$, obtained with the maximum likelihood method, and the likelihood ratio vs. the primordial spectral index n. For spectral indices between 0 and 1, the likelihood ratio shows a peak value ~ 10. A similar value was found by Davies et al. (1987) analysing the data with the Gaussian correlation function. So, in the framework of the likelihood analysis, white noise or scale-invariant density fluctuation spectra are also consistent with the Davies et al. (1987) data, producing best fit values of $C_{0M}^{1/2} = 10^{-4}$ and $C_{0M}^{1/2} = 5.9 \cdot 10^{-5}$, respectively.

Unfortunately, in the present case, the likelihood analysis provides only the best estimates for the parameters of the model, but does not provide a

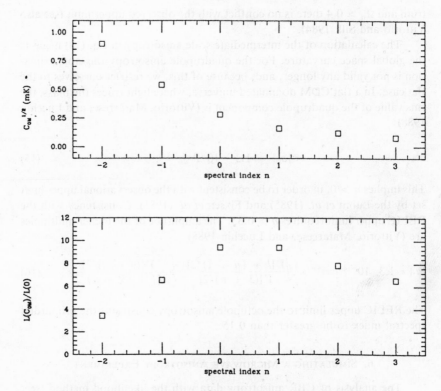

FIG. 7. The quantity $C_{0M}^{1/2}$, and the likelihood ratio vs. the perturbation spectral index assumed in analysing the Davies *et al.*1987 data. (From Vittorio, de Bernardis, Masi, and Scaramella 1988).

confidence interval for them. For evaluating the probability distribution for the parameters one can simulate the observations of a theoretical sky, including effects as, e.g., receiver noise, modulation geometry, beam pattern, etc. This seems also a promising method for fairly comparing theory and observations, since it takes into account all the important experimental effects otherwise neglected in a purely theoretical analysis.

The Davies *et al.* (1987) experiment has been simulated in the following way (Vittorio *et al.* 1988). Two hundred different realizations of a strip of the theoretical sky have been generated. Each of these strips is ~ 160° wide, 20° thick, and has a resolution of ~ 0°.8. They are realizations of the theoretical ensemble described by Eq. (7), where temperature fluctuations are determined by the Sachs and Wolfe effect: the density fluctuations are assumed to be Gaussian distributed, adiabatic, and scale-invariant. The rms temperature fluctuation of the theoretical ensemble is fixed to be $5.9 \cdot 10^{-5}$, as consistently found analysing the Davies *et al.* (1987) data for $n = 1$. Each strip is sampled as was done in the actual experiment, so a data set of 70 points per strip is generated. The receiver noise is simulated by adding to the theoretical temperature fluctuations a white noise of amplitude (rms) ~ 0.22 mK, the amplitude of error bars in the Davies *et al.* (1987) data.

Each data set has been analysed as the real data, obtaining values for $C_{0M}^{1/2}$ and likelihood ratios. Figure 8 shows the frequency of the results over the 200 realizations. If the temperature fluctuations are determined by scale invariant density fluctuations, the simulation of the experiment by Davies *et al.* (1987) shows that a likelihood ratio > 10 should be expected in 25% of the realizations. However the probability of having a likelihood ratio between 10 and 30 and $C_{0M}^{1/2} = 5.9$ (± 5.9) 10^{-5} is only 5%.

7. Summary

We presented CDM model predictions for the peculiar acceleration exerted on the Local Group by the material inside a galaxy sample of depth R. The convergence of the peculiar acceleration, exerted by the material contained

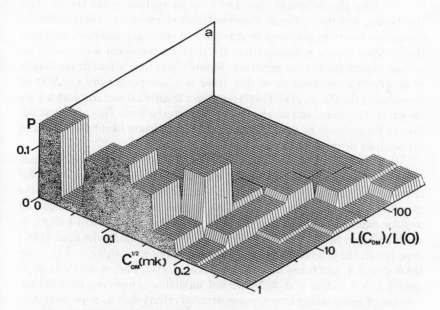

Fig. 8. Results of Monte-Carlo simulations of the Davies *et al.* 1987 experiment: the frequency of the results is plotted vs. the estimated $C_{0M}^{1/2}$ and the likelihood ratio, $L(C_{0M})/L(0)$. (From Vittorio, de Bernardis, Masi, and Scaramella 1988).

in nested subsamples, to a finite value is only a necessary condition for the absence of large-scale, large amplitude density inhomogeneities. To confidently assess this absence, we need to independently estimate the peculiar velocity of the considered sample relative to the comoving frame. We consider a flat CDM dominated universe, where light traces the mass, and we discuss the convergence of the peculiar acceleration. We also show that in this model the expected misalignment angle is less than 40° (95% confidence level) for a galaxy sample comparable with the IRAS sample.

Comparing among themselves different experiments can provide constraints on the primordial spectral index. In particular, if the Davies *et al.* (1987) experiment really provides a measure of the primordial density fluctuations and the RELIC result sets a realistic upper limit to the quadrupole anisotropy, then there is evidence for n to be greater than unity, if the density fluctuation spectrum on large scales is a single power law.

We have discussed the large scale CBR anisotropies expected in a CDM dominated universe and we have shown that the observational upper limits to the CBR anisotropy again constrain the value of the primordial spectral index. The tightest constraint comes from the RELIC upper limit to the quadrupole anisotropy which implies, for a flat CDM universe where light traces the mass, that $n > 0.35$.

We have also presented results of a recent analysis of the Davies *et al.* (1987) data, with the maximum likelihood method, assuming CBR temperature correlation functions predicted by gravitational instability scenarios. Although the likelihood analysis suggests that the data are consistent with a scale invariant density fluctuation spectrum, Monte Carlo simulations of the Davies *et al.* (1987) experiment show that there is a low probability ($\sim$ 5%) of reproducing the Davies *et al.* (1987) result in a theoretical scenario with a scale invariant, Gaussian, and adiabatic density perturbations. This of course addresses the question of the confidence of the maximum likelihood results in the analysis of the CBR data.

Finally, Monte Carlo simulations of the CBR anisotropy experiments can be used to define the experimental configuration for searching for CBR temperature anisotropy. This can be done as a first approach by studying the expected large-scale pattern of the CBR and by predicting abundances and angular diameters of the hot spots in the CBR temperature distribution, as we discussed in Sec. 3. For example, if $\sigma = 3^o.5$, as in the Davies *et al.* (1987) experiment, the expected angular diameter of a hot spot with $\nu \sim 2$ is $\sim 20^o$ for $0 \le n \le 1$, and is less than the Davies *et al.* (1987) beam switching angle only if $n > 1$ and/or $\nu \gtrsim 4$. Numerical simulations, however, have the advantage of easily taking into account detailed effects such as noise (detector, atmosphere, instrumentation, etc), sky coverage, modulation geometry, beam pattern, etc. Therefore, they seem to be extremely promising, from one side, for fairly comparing theory and observations and properly constraining theoretical models, and, from the other side, for assessing the best observational strategy for a positive detection of CBR temperature anisotropies.

ACKNOWLEDGEMENTS

The results presented at this meeting have been obtained in a collaborative effort with many friends. I am particularly indebted to Paolo de Bernardis, Roman Juszkiewicz, Francesco Lucchin, Silvia Masi, Sabino Matarrese, Roberto Scaramella, and Rosemary Wyse. I am also indebted to Alfonso Cavaliere for discussions and comments on this manuscript.

REFERENCES

Bardeen, J.M., Bond, J.R., Kaiser, N., Szalay, A.S. 1986. *Ap J.* **300**, 15.

Blumenthal, G., Faber, S., Primack, J., and Rees, M. 1984. *Nature.* **301**, 584.

Bond, J. and Efstathiou, G. 1987. *MN.* **226**, 655.

Davis, M., Efstathiou, G., Frenk, C., and White, S.D.M. 1985. *Ap J.* **292**, 371.

Davis, M. and Peebles, P.J.E. 1983. *Ap J.* **267**, 465.

Davies, R.D., Lasenby, A.N., Watson, R.A., Daintree, E.J., Hopkins, J., Beckman, J., Sanchez-Almeida, J., and Rebolo, R. 1987. *Nature.* **326**, 462.

Fabbri, R., Lucchin, F., and Matarrese, S., 1987. *Ap J.* **315**, 1.

Fixsen, D.J., Cheng, E.S., and Wilkinson, D.T. 1981. *Phys Rev Lett.* **44**, 1563.

Juszkiewicz, R. and Bertschinger, E. 1988, *Ap J Letters.* **344**, L59.

Juszkiewicz, R., Vittorio, N., and Wyse, R.M. 1988. in preparation.

Kaiser, N. 1988. preprint.

Kaiser, N., and Lasenby, A.N. 1987. preprint.

Klypin, A., Sazhin, M., Strukov, A., and Skulachev, D. 1986. *Pisma Astron. Zh.* in press.

Lubin, P. and Villela, T. 1987. in *Proceedings of the E. Fermi Summer School, Confrontation Bewteen Theories and Observations in Cosmology.* ed. F. Melchiorri. Varenna.

Lubin, P., Villela, T., Epstein, G., and Smoot, G. 1985. *Ap J Letters.* **298**, 1.

Meiksin, A., and Davis, M. 1985. *AJ.* **91**, 191.

Melchiorri, F., Melchiorri, B., Ceccarelli, C., and Pietranera, L. 1981. *Ap J.* **250**, L1.

Peebles, P.J.E. 1980. *Large-Scale Structure of the Universe.* Princeton: Princeton University Press.

_____. 1981. *Ap J. Letters.* **263**, L119.

_____. 1982. *Ap J.* **259**, 442.

Sachs, R.W. and Wolfe, A.M. 1967. *Ap J.* **147**, 73.

Sazhin, M.V., 1985. *MN.* **216**, 25p.

Scaramella, R. and Vittorio, N. 1988. *Ap J Letters.* in press.

Strauss, M. and Davis M. 1988. in *Large-Scale Structures of the Universe.* eds. J. Audouze, M.-C. Pelletan, and A. Szalay. Dordrecht: Kluwer Academic Publishers, p. 191.

Turner, M.S. 1987. *Lecture at the E. Fermi Summer School, Confrontation between Theories and Observations in Cosmology.* ed. F. Melchiorri. Varenna, Italy.

Villumsen, J. and Strauss, M. 1987. *Ap J.* **322**, 37.

Vittorio, N. and Silk, J. 1984. *Ap J Letters.* **285**, L39.

_____. J. 1985. *Ap J Letters.* **293**, L1.

Vittorio, N. and Juszkiewicz, R. 1987. *Ap J Letters*. **314,** L29.

Vittorio, N., Matarrese, S., and Lucchin, F. 1988. *Ap J*. **328,** 69.

Vittorio, N., Juszkiewicz, R., and Davis, M. 1986. *Nature*. **323,** 132.

Vittorio, N., de Bernardis, P., Masi, S., and Scaramella, R. 1988. *Ap J*. in press.

Yahil, A. 1985. in *The Virgo Cluster of Galaxies*. eds. O.G. Richter and B. Binggeli. Munich: ESO, p. 359.

Yahil, A., Walker, D., and Rowan-Robinson, M. 1986. *Ap J Letters*. **301,** L1.

PROBING COSMIC DENSITY FLUCTUATION SPECTRA

J. R. BOND

CIAR Cosmology Program, Canadian Institute for Theoretical Astrophysics, Toronto

ABSTRACT

The most popular approach to structure formation in the universe is to assume perturbations at early times form a homogeneous and isotropic Gaussian random field characterized by a density perturbation spectrum $\mathbf{P}_\varrho(k) \equiv (k^3/(2\pi^2)) \langle |(\delta\varrho/\varrho)(k)|^2 \rangle$. In this paper I summarize the status of a wide variety of tests based on the theory of Gaussian random fields and approximate analytic treatments of dynamics which probe different wavenumber ranges of the $\mathbf{P}_\varrho(k)$ function : essentially every region of k-space is covered by at least one test. Granted complete freedom in the choice of $\mathbf{P}_\varrho(k)$, there is currently no clear need for non-Gaussian statistics. If we regard the interpretation of ξ_{cc} and cluster patches determined from the Abell catalogue as biased, the Great Attractor hypothesis as being an oversimplification of a complex large-scale velocity field, and the Tenerife $\Delta T/T$ observation at $8°$ as non-primordial, then the required spectrum looks very promising for the *minimal* $\Omega = 1$ cold dark matter (CDM) model, with structure growing from scale invariant adiabatic initial conditions. If these interpretations of the observations stand, it is not clear whether the $\mathbf{P}_\varrho(k)$ suggested by *all* of the data can arise in any natural setting.

1. INTRODUCTION

There are currently two main techniques that can be applied to the study of structure formation on large scales in models based on Gaussian initial conditions. One can evolve realizations of the linear initial conditions by N-body (and/or hydrodynamic) techniques, measuring various observables in the realizations to confront with observations. The alternative is the approach adopted here: dynamical evolution is approximated by the Zeldovich solution, using it to map the Gaussian statistics of the filtered fluctuations in unperturbed Lagrangian space to non-Gaussian statistics in Eulerian space. Abundances and correlation functions of selected classes of points chosen to correspond to classes of cosmic objects are then directly computed by ensemble averaging various statistical 'operators'. The formalism to treat the statistics of density peaks in the unevolved Gaussian random field is given in Bardeen *et al.* (1986, hereafter BBKS) and is extended by Bardeen *et al.* (1988, hereafter BBRS). See also Bond (1986, 1987a,b,c, 1988) and Bardeen, Bond, and Efstathiou (1987, hereafter BBE). The inclusion of Zeldovich dynamics in this formalism was given by Bond and Couchman (1987, 1988).

For such calculations to be feasible it is necessary to identify the complex objects that we can observe with points conditioned by relatively simple criteria, *e.g.*, with peaks of the Gaussian random density field smoothed over a scale R_f roughly corresponding to the mass of the objects, with the selection function $P(\text{obj}|\nu, \nu_b)$ depending only upon simple parameters like the relative height ν of the peak (in units of $\sigma_\varrho(R_f)$, where $\sigma_\varrho(R_f)$ is the (linear) amplitude of the rms density fluctuations), and perhaps the height ν_b of a lower resolution 'background' (BBKS) field smoothed on $R_b > R_f$. For example, P might be unity if $\nu\sigma_\varrho(R_f)$ is above some critical value appropriate to the collapse of the R_f-scale peak *and* $\nu_b\sigma_\varrho(R_b)$ is below some threshold associated with merging, being zero otherwise. The $R_b \to R_f$ limit of this is similar to the Press and Schechter (1974) formalism for estimating the mass function of cosmic objects from a fluctuation spectrum. A variant of this approach appropriate to peaks is used in Section 2.1 and applied to the study of rare events in Section 2.2.

Estimations of correlation function evolution can be obtained by moving the peaks using the Zeldovich approximation: the position at time t of a peak initially at position t is $x(r, t) = r - a(t)[2H_0^{-2}/3] \nabla \Phi$ for $\Omega = 1$ universes, where Φ is the (linear and time independent) gravitational potential field, and a is the scale factor. The assumption in this model is that the peaks are primarily generated by short waves and follow the bulk flow of the mass described by $x(r, t)$, smoothed on some $R_b > R_f$ scale. The collapse of highly asymmetric peaks could also be followed using the Zeldovich approximation. However, if the peak is relatively spherically symmetric, as is the case if $\nu \gtrsim 2.5$, the Zeldovich approximation fails badly. In this case, a spherical approximation like the top hat model would do better: I adopt this in Section 2 to describe localized collapses.

In Section 2 I summarize the status of some of the quantities which can be calculated in the statistical theory. Table 1 lists a variety of tests of the fluctuation spectrum, some of which probe the shape of the spectrum over an extended range in k-space, and all of which probe the amplitude of the spectrum in some k-band. In this paper I emphasize calculations appropriate to Great Attractors: the velocity field in their presence and how probable they

TABLE 1
DIRECT PROBES OF THE FLUCTUATION SPECTRUM

PROBES	Wavenumber Range k^{-1} (h^{-1} Mpc)	Local Power Law Index	CDM Status
Shape and Amp Tests			
galaxy halos	0.2-1	$-2 \leq n \leq -1$	+
$\xi_{gg}(r)$, $w_{gg}(\theta)$ power law	0.5-10	$-2 \leq n \leq -1$	+
$w_{gg}(\theta)$ break	10-20	$n \gtrsim -1/2$	+
ξ_{cg}	5-20	$n \gtrsim -1/2$	+
ξ_{cc}	5-100	$n \sim -1$ (?)	−
ξ_{cg}, ξ_{cc} anisotropies	5-20	$n \leq -1/2$ (?)	+
Void/Wall Texture	5-15	$n \leq -1/2$	+ (?)
Local Flow	3-5		+
Large-Scale Flow	10-40 (wishful?)		?
Amplitude Tests			
Lyα clouds	0.04-0.1	$4 \leq \sigma_\rho(.05) \leq 6$	+
gal form. epoch z_{gf}	0.2-7	$3 \leq \sigma_\rho(.4) \leq 6?$	+
cl abund. $n_{cluster}$	4-7	$0.4 \leq \sigma_\rho(5) \leq 1$	+
rare gal events e.g., n_{GA}	8-20	$\sigma_\rho(15) \sim .2?$	−
rare cl events $n_{cl\,patch}$	~ 20+	$\sigma_\rho(25) \sim .1?$	−
CMB Test			
small $\theta \sim 2' - 20'$	3-30		+
interm $\theta \sim 2^o - 10^o$	200-1000		−
large θ (e.g., RELICT)	700-6000		+
BN distortion	0.001-0.01		− (?)

Notes: The range of wavenumbers listed above is a crude indication of where the tests based on Gaussian statistical techniques are applicable. N-body studies can go in closer to collapsed structures with accuracy. The Gaussian tests listed are direct probes only for a hierarchical theory for which dynamical evolution on the scales in question is relatively mild, and are not applicable, for example, to highly nonlinear portions of $n = 0$ spectra. Other interpretations of some 'tests' are certainly possible. In particular the models of Lyα clouds (Rees 1986, Bond, Szalay, and Silk 1988) and the high redshift dust source (Bond, Carr, and Hogan 1988) of the CMB spectral distortion (Matsumoto et al. 1987) are certainly controversial. While texture comparisons remain visual tests, Gaussian field methods are at best qualitative indicators. Although galaxy halo formation really requires N-body studies (Frenk et al. 1985, Quinn, Salmon, and Zurek 1986), a good indication of the final shape is found by considering the cross-correlation $\xi_{pk,\,\varrho}$, describing the statistically averaged shape of matter that can infall onto a galaxy. $n = 0$ spectra lead to halos with profiles that are too steep. The $n > -1/2$ limits from w_{cg} and the w_{gg} break are not very serious, but are meant to indicate that $n = -1$ will *not* do. I expect that the w_{gg} constraint on n is stronger. Similarly the $n \leq -1/2$ limits indicate $n \gtrsim 0$ will definitely not do.

are. I mainly discuss these tests within the context of the $\Omega = 1$ CDM model, with an amplitude parameterized by a biasing factor b_g (BBKS, Bond 1986). I take $b_g = 1.44$ following Bond (1987a,b); this compares with the value $b_g \approx 2.5$ used by Davis *et al.* (1985) and $b_g = 1.7$ used by BBE. Kaiser and Lahav (1988) give evidence for $b_g \approx 1.5$ from the streaming data in these proceedings; Kaiser and Cole (1988) have also used a similar value.

The data probing shorter scales than for clusters, $k^{-1} \lesssim 5 \, h^{-1}$ Mpc, seem to agree reasonably well with the predictions of this theory. Some of the large-scale tests suggest extra power might be required to explain them. A reasonable phenomenological approach is to add the extra power to a CDM base. This could arise physically if we were free to alter the initial conditions from the scale invariant form predicted in the standard inflation model. How difficult this is to arrange naturally is discussed by Bond, Salopek and Bardeen in these proceedings. There are three fiducial cases we consider: (1) CDM + plateau. with the initial spectrum having a ramp beginning at some wavenumber k_R which levels off to a scale invariant high amplitude plateau at some smaller wavenumber k_P. Such a spectrum could arise in 'double inflation'; (2) CDM + mountain, with the power falling down to either the standard CDM value for $k < k_P$, or even below it. A simple idealization of this which makes the calculations easy, CDM + spike, is to add a δ-function of extra power at a specific wavenumber: $P_\varrho = P_\varrho [CDM] + \sigma^2_{\varrho, P} k_P \delta(k - k_P)$: we show some of its effects in Section 2; (3) Initial spectra of fluctuations with arbitrary power laws. An example used in Section 2 is isocurvature baryon perturbations in an open universe with $n = -1$ fluctuations in the entropy-per-baryon ($n = -3$ is scale invariant), as advocated by Peebles (1987, 1988). The transfer function for such universes naturally imprints features at the horizon scale when the relativistic and non-relativistic matter have equal density. A mountain of power is generated followed by a precipitous $n = 3$ dropoff at low k. A fourth approach is to assume scale invariance but change the constituents of the universe, thereby modifying the transfer function which maps the initial fluctuation spectrum from the very early universe to the post-recombination one (BBE). None of these spectra can have power in the gravitational potential fluctuations above the power on very large scales: they cannot look like CDM + mountain; extra large-scale power looks more like the CDM + plateau model.

2. DIRECT PROBES OF THE FLUCTUATION SPECTRUM

2.1 - Abundances Using a Spherical Peak Model

With the statistical theory one can estimate the abundances of objects ranging from dwarf galaxies and Lyα clouds through galaxies, groups, and clusters, given models for the selection functions $P(\text{object}|\nu, R_f)$. Although these abundances qualitatively agree with observations for the CDM theory, for a more precise test we would ideally like an expression for the differential mass function $n(> \delta_*, M)d \ln M$ for objects with overdensity δ (relative to the background) above δ_* (i.e., belonging to a 'catalogue' of contrast δ_*) and with mass between M and $M + dM$. As we emphasized in BBKS, no satisfactory theoretical derivation of a semi-analytic form for $n(> \delta_*, M)d \ln M$ exists, and I believe this conclusion still remains valid, although many people, including myself, have explored the consequences of simple ansatzes for the mass function. A better approach is to calibrate $n(> \delta_*, M)$ for many different catalogue thresholds δ_* using N-body experiments. Although a beginning has been made to such a calibration (Efstathiou and Rees 1987), the high M (rare events) and low M (destruction through merging) ends have not been well determined.

The two main analytic techniques utilize either the Press-Schechter (1974, hereafter PS) ansatz or the theory of peaks of a Gaussian field. I prefer the latter (Bond 1987a) which I think is better motivated physically, although most authors seem to prefer the former (e.g., Schaeffer and Silk 1985, Coles and Kaiser 1988, Efstathiou and Rees 1988, Narayan and White 1988). Here I shall emphasize the peak formulation, and adopt Gaussian smoothing of the field. (Although top hat smoothing seems better motivated physically, problems arise since the short wavelength limit of the top hat filter of the density fluctuation spectrum only drops as a power $\sim (kR_{TH})^{-4}$; the top-hat-filtered CDM random density field is then only once differentiable and has an infinite density of peaks of tiny radius for a flicker noise unfiltered spectrum at high k). The ideal approach from the point of view of the theory of Gaussian fields is to consider peaks on a hierarchy of resolution scales $\{R_{f,s}\}$. In the limit in which the resolution scales become densely packed, a reasonable (although not rigorously derivable) expression for $n(> \delta_*, M)$ is

$$n(> \delta_*, M)d \ln M = N_{pk}(\nu\delta_*, R_f) \frac{d\nu_{\delta_*}}{d \ln M} d \ln M , \qquad (2.1)$$

the number density of peaks on the scale R_{fs} with overdensities in excess of δ_* which do not have overdensities when smoothed on the 'background' scale $R_{f,s+1} = R_{fs} + dR_f$ as large as δ_*. Here ν_{δ_*} is the height of R_f-scale peaks which have $\delta = \delta_*$ at the redshift at which $n(> \delta_*, M)$ is determined. $N_{pk}(\nu)$ is the differential peak density (BBKS).

To evaluate this expression the parameters of the smoothed Gaussian field ν and R_f must be related to the (possibly nonlinear) overdensity δ and the mass M by adopting a simplified model of the dynamics of collapse of the matter surrounding the peak. Here I shall use the spherical top hat model. For universes with $\Omega = \Omega_{nr} = 1$, where Ω_{nr} is the density in non-relativistic particles, the relation between $\nu\sigma_\varrho$ and the actual nonlinear overdensity is accurately given by

$$1 + \delta \approx (1 - v\sigma_\varrho/f_c)^{-f_c}, \, f_c = 1.68647, \, v\sigma_\varrho(z) < f_v \approx 1.606, \quad (2.2a)$$

$$X(r, t) = ar/(1 + \delta)^{1/3}, \quad (2.2b)$$

$$V(r, t) = Har(1 - v\sigma_\varrho/f_c)^{f_c/3-1} (1 - v\sigma_\varrho/f_{ta}), \, f_{ta} = f_c/(1 + f_c/3) = 1.08 \quad (2.2c)$$

The only failure in these expressions is the velocity V near turnaround, which occurs when $v\sigma_\varrho = 1.06$ in the exact model rather than at 1.08. Suppose collapse ceases once the overdensity rises to a value δ_v with a corresponding value of $f_v \equiv v\sigma_\varrho$ obtained from eq. (2.2a). After this redshift, the subsequent evolution of the overdensity is given by the $\sim (1 + z)^{-3}$ law:

$$1 + \delta = (1 + \delta_v) \, (v\sigma_\varrho(z)/f_v)^3, \, v\sigma_\varrho(z) > f_v. \quad (2.3)$$

The 'classical' value for the virialized density of the top hat model with $\Omega_{nr} = 1$ is $\delta_v = 170$, obtained with $f_v = 1.606$, the value quoted in eq. (2.2a).

The relation between the initial comoving top hat radius of the cloud required to determine the mass M, the height v of the peak and the filtering radius R_f is complex and indeed time dependent as infall occurs. In Bond (1987a), I adopted the mass associated with a Gaussian profile of comoving radius $r_s \equiv f_s(v, R_f)R_f$:

$$M = (2\pi)^{3-2} \bar\varrho_0 r_s^3(R_f, v), \quad \text{with} \quad r_s \sim 1.5R_f, \quad (2.4)$$

where $\bar\varrho_0$ is the current background density, and argued that $1 \lesssim r_s/R_f \lesssim 1.7$, with higher values appropriate for higher v peaks and smaller filtering radii. As a simple compromise I chose a v-independent value $f_s = 1.5$. The faint end of the PS fit to the numerical results for CDM given by Efstathiou and Rees (1987) indicate $f_s \sim 1.3$ might be a better choice. With these approximations for M and δ, eq.(2.1) becomes

$$n(> \delta_*, M)d \ln M = N_{pk}(v_{\delta_*}, R_f)v_{\delta_*} \frac{\gamma^2 R_f^2}{R_*^2} d \ln M, \, \langle k^2 \rangle \equiv 3\gamma^2/R_*^2. \quad (2.5)$$

The quantities γ and R_* introduced by BBKS are related to the average $\langle k^2 \rangle$ of the density spectrum as given in eq.(2.5). N_{Pk} is replaced by $2 \exp[-v_{\delta_*}^2/2]/(2\pi)^{1/2} [(2\pi)^{3/2}r_s^3]^{-1}$ in the PS theory, where f_s is typically chosen to be 1 rather than the 1.5 adopted here.

Fig. 1 illustrates how eq.(2.5) for 'virialized' structures with $\delta > 170$ compares with the data for the Gott-Turner groups and Abell clusters. The main

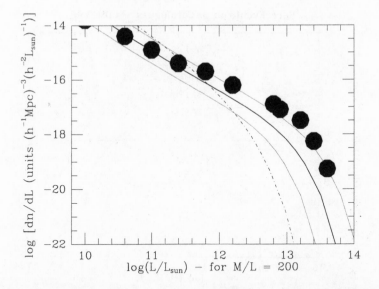

FIG. 1. The luminosity function computed for the CDM theory with biasing factor $b_g = 1.44$ and $f_s = 1.5$, assuming a fixed M/L ratio for the groups and clusters. The heavy solid line corresponds to $M/L = 200$, and the dotted lines have $M/L = 400$ (left) and 100 (right). The dot-dash line is the PS result (without the factor of 2 fudge usually included) with $f_s = 1$; $f_s = 1.5$ would shift it over toward the peak curve, although the shape would remain different. The black dots are data for the Gott-Turner groups and Abell clusters. The data for the groups is not particularly reliable.

conclusion to be drawn is that the CDM model does not obviously fail to reproduce the hierarchy of observed virialized structures in the universe. The more precise group analysis of Nolthenius and White (1987) using N-body realizations of the CDM model gives even better fits to the group abundances they determine from the CfA redshift survey.

2.2 - Abundances of Rare Events: Great Attractors and Cluster Patches

In this section, I estimate the abundances of very large-scale objects with relatively small overdensities $1 \lesssim \delta_* \lesssim 5$ which have not yet turned around. Equations (2.2) and (2.5) would still be applicable. Thus $\delta = 0.8$ if $\nu\sigma_\varrho = 0.5$, and $\delta = 3.5$ if $\nu\sigma_\varrho = 1$. Figure 2 illustrates how $n(> \delta_*, M)$ falls to high mass for varying δ_*. Note that since N_{pk} drops at low ν, for small M the mass function with $\delta > 1$ can fall below that with $\delta > 5$. The PS formalism suffers from this as well although not as severely. However, the high M values should provide a reasonable estimate of the rare event density; this will be true whether the PS or peak technique is used since the behaviour on

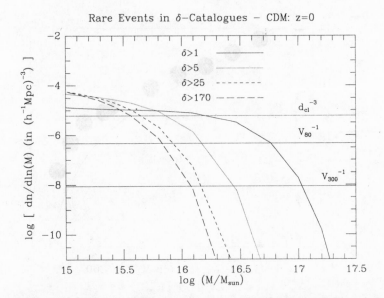

FIG. 2. The mass function $n(> \delta_*, M)$ derived from eq.(2.5) for the CDM model with $b_g = 1.44$ and $f_s = 1.5$ for the values of δ_* shown. For comparison the density of $(R \geq 1)$ Abell clusters, $d\,\bar{c}_l^3 = (55\ h^{-1}$ Mpc$)^{-3}$, and the densities corresponding to one object in volume limited samples of radius $R_{sample} = 80$ and $300\ h^{-1}$ Mpc are indicated, corresponding roughly to the elliptical survey region and to Tully's cluster sample.

the rare event side is dominated by the $\exp[-\nu_{\delta_*}^2/2]$ tail characteristic of Gaussians: when it begins to dominate the fall is dramatic. A difficulty of applying this technique is that abundances of objects of a given mass will then be very sensitive to the choice of f_s.

If a Great Attractor (GA) is identified with a peak of the Gaussian field with overdensity $\sim \delta_*$ and 'mass' in a logarithmic range $d \ln M$ about M, then the probability of finding a GA in a volume-limited sample of radius R_{sample} is

$$P(GA) = \frac{4\pi}{3} R_{sample}^3\, n(> \delta_*, M).$$

For the elliptical survey, I take $R_{sample} = 80\ h^{-1}$ Mpc. The mass excess is $\sim 5 \times 10^{16}\ M_\odot$ for the GA parameters presented in this conference by Faber and Burstein (1988). According to Figure 2, for the CDM model (with biasing factor $b_g = 1.44$) there would be about one $\nu\sigma_\varrho = 0.6$ object in the survey, but certainly no $\nu\sigma_\varrho = 1$ objects. In Section 2.6 I show that if the ~ 600 km s^{-1} flow experienced by Tully's Local (Coma-Sculptor) Cloud —

of which the Local Group is a small part — were entirely due to a GA, then $\nu\sigma_{\varrho} \approx 1$ would be required. However if about half of this velocity is due to other pullers and pushers (*e.g.*, the Local Void) which just happen to align the flow pattern in the Centaurus flow direction, then $\nu\sigma_{\varrho} \approx 0.5$ would suffice: the GA would not necessarily be too rare for the CDM model. This is sensitive to f_s: $f_s = 1$ would give a probability about a percent; hence to be more precise requires calibration of the $M - R_f$ relation. The $f_s = 1.5$ choice probably errs on the pro-CDM side.

These rare events arise from high ν peaks which are approximately spherical, justifying our use of the top hat model. With $f_s = 1.5$, 5×10^{16} M$_\odot$ corresponds to $R_f \sim 12$ h^{-1} Mpc for which $\sigma_\varrho = 0.17$, hence ν is 6 for $\nu\sigma_\varrho = 1$, and 3 for $\nu\sigma_\varrho = 0.5$. The three eigenvalues of the shear tensor for high ν are approximately (Bond 1987b) $\lambda_1 \approx (\nu\sigma_\varrho/3) (1 + 3e_\nu)$, $\lambda_2 \approx \nu\sigma_\varrho/3$ and $\lambda_3 \approx (\nu\sigma_\varrho/3) (1 - 3e_\nu)$. For given ν, the average value of e_ν is $\langle e_\nu | \nu \rangle \approx (3\pi/40)^{1/2}\nu^{-1}$; hence $3e_\nu \sim 0.25$ for $\nu = 6$ and 0.5 for $\nu = 3$, with non-negligible deviations from a spherical flow predicted in the latter case. However, the spherical $\delta - \nu$ relation still provides a reasonable estimate. In a pancake approximation (which will not be accurate for such small values of e_ν), $\delta \approx \nu\sigma_\varrho [1 - \nu\sigma_\varrho(1/3 + e_\nu)]^{-1}$: the overdensities predicted are not very disimilar from the spherical model.

The cluster patches advocated by Tully could also arise from Gaussian random fields. The large biasing factor of clusters implies that a great increase in the number of clusters will occur for relatively modest overdensities:

$$1 + \delta_{cl} \sim (1 + \delta_\varrho) \exp[(b_{cl} - 1)\nu\sigma_\varrho] \exp[-(b_{cl} - 1)^2 \sigma_\varrho^2/2]. \quad (2.6)$$

For clusters in the CDM model $b_{cl} \approx 6$, hence $\nu\sigma_\varrho = 0.5$ leads to $\delta_{cl} \approx 20$, and $\nu\sigma_\varrho = 1$ gives $\delta_{cl} \approx 670$. In a sample of radius $R_{sample} = 300$ h^{-1} Mpc there would be about one $\nu\sigma_\varrho = 0.5$ peak on scale $R_f \approx 20$ h^{-1} Mpc; over $R_f \approx 40$ h^{-1} Mpc, there would not even be one of these patches in the observable universe. By contrast consider the isocurvature baryon model with $\Omega = \Omega_B = 0.4$ normalized by mass traces light ($b_g = 1$ for J_3 normalization at 10 h^{-1} Mpc). In this case $b_{cl} \sim 4$. Over 300 h^{-1} Mpc there would be 3 regions with the cluster density enhanced by 7 in patches of $R_f = 40$ h^{-1} Mpc. This model would have quite a few $\nu\sigma_\varrho = 1$ regions within the elliptical sample volume as well.

2.3 - Two-point Correlation Functions

The correlation function for galaxies and clusters ξ_{gg} and ξ_{cc} and the cross correlation between clusters and galaxies ξ_{cg} can be computed assuming the above models for the objects as peaks of the smoothed Gaussian random field. (See BBKS, BBE and BBRS for details). The theoretical predictions for the

CDM theory are in agreement with the observed ξ_{gg} and the new determination of ξ_{cg} found by Lilje and Efstathiou (1988), but the CDM ξ_{cc} is definitely too low relative to the Bahcall and Soneira (1983) result. The shape of the correlation functions on large scales is very much conditioned by the shape of the fluctuation spectrum for CDM as it steepens towards the primordial Zeldovich shape as one passes from cluster scales where the local power law index of the fluctuation spectrum is $n \approx -1$ to $n = 1$ at large scales. The reported $\xi_{cc} \sim r^{-2}$ shape suggests a $n = -1$ ramp beyond cluster scale, delaying the drop to $n = 1$ to beyond $\sim 100\,h^{-1}$ Mpc. However, the Groth-Peebles (1977) break at $\theta \sim 3^o$ in the angular correlation function w_{gg} (θ) evaluated from the Shane-Wirtanen catalogue of the Lick survey, which has been confirmed by Maddox and Efstathiou (1988) in their Southern Sky Catalogue, implies such a power law extension is incompatible with w_{gg} (Bond and Couchman 1987). Further, the amplitude of ξ_{cg} would have to be more similar to the original Seldner and Peebles (1977) value than the smaller one determined by Lilje and Efstathiou (1988). Thus if ξ_{cc} persists in maintaining its current power law form to $\sim 50\,h^{-1}$ Mpc then it appears it cannot be compatible with a biasing picture based on Gaussian initial fluctuations.

I now consider the effect of extra power in a simplistic model which mimics the CDM + mountain spectrum and is somewhat similar to the isocurvature baryon spectrum. If a pulse of power is added at wavenumber k_P, then added to the usual scale invariant ξ_{ij} will be a term $b_i b_j \sigma_\varrho^2(k_P)\,\sin(k_P r)/(k_P r)$, where i and j can be c or g. The amplitude of this extra piece would then be approximately constant, of magnitude $\sim 0.3(\sigma_\varrho(k_P)/0.1)^2$ for clusters, out to $\sqrt{3}k_P^{-1}$. The addition of such a term could therefore mimic the large ξ_{cc} obtained, assuming we are not overly sold on the reliability of the r^{-2} power law. However such a term might already be ruled out. Lilje and Efstathiou (1988) find no evidence for such behaviour in ξ_{cg} whether or not they remove large-scale gradients from the Lick catalogue. Although if k_P^{-1} is large enough this contribution to w_{gg} would not have shown up in the Groth-Peebles determination since large-scale gradients are subtracted, the Maddox-Efstathiou determination does not require gradient subtraction: they find no evidence for such a term. These are both high precision tests and even the low $\xi_{cg} \sim 0.07$ and $\xi_{gg} \sim 0.02$ levels predicted could probably have been picked up.

2.4 - Two-point Function Anisotropies

The Binggeli (1982) effect (see also Struble and Peebles 1985, Rhee and Katgert 1987) that clusters are preferentially aligned along their long axis, and the Argyres et al. (1986) effect that galaxies are also preferentially found in the plane of the long axis of clusters out to $\sim 15 - 20\,h^{-1}$ Mpc is predicted by the theory of Gaussian statistics (Bond 1986, 1987c). The constraint imposed by requiring a cluster to be oriented along a specified axis forces the density waves that had to constructively interfere at the cluster peak to also preferentially constructively interfere along the long axis compared with other directions. For fluctuation spectra shallower than $n = 0$, long waves as well as short ones build the cluster, hence the effect can persist weakly out to large distances from the collapsed cluster. This coherent wave bunching will make the overdensities of clusters, of galaxies and of the mass field largest along the long axis and smallest along the short axis for a given radius. In addition the same effect will tend to align the long axes of the clusters so that they point toward each other. Further, the orientation of a central smaller scale peak will be correlated with the imposed cluster scale orientation, especially if the two scales are not inordinately disparate. A cD galaxy might correspond to Gaussian smoothing of $R_f \sim (1 - 2) \, h^{-1}$ Mpc, which is not very far from the $\sim 5 \, h^{-1}$ Mpc appropriate for clusters, suggesting a plausible origin of the Carter and Metcalfe (1981) effect that cD galaxies tend to be aligned with their host clusters.

The critical question is whether these effects are quantitatively large enough to explain the observed effects and whether they can be used to differentiate among possible fluctuation spectra. Little alignment will occur if the spectral index is $n > 0$: $n = 0$ spectra are uncorrelated and $n = 1$ spectra are anticorrelated; such indices can be ruled out over the range that alignments are observed.

If ξ for oriented clusters is expanded in spherical harmonics then, in the linear regime, the nonzero components will be monopoles ($\ell = 0$) for ξ_{cg} and ξ_{cc}, the usual correlation function when no account is taken of the orientation; a quadrupole ($\ell = 2$) for ξ_{cg}, and a quadrupole and octupole ($\ell = 4$) for ξ_{cc}. Nonlinear corrections to the statistics or the dynamics add all higher even multipoles, but these will be unimportant in the far field. For clusters the far field is $r \gtrsim 15 \, h^{-1}$ Mpc. If the mass correlation function is a power law $\xi \sim r^{-\gamma}$ in the far field then $\xi_{l=2} \sim r^{-(\gamma+2)}$ falls off faster, being negligible at distances where the monopole component can still be large. Nonetheless the anisotropy amplitude can be quite large to many times the filtering radius, similar to the alignment levels reported. (See BBRS for detailed calculations.)

2.5 - Large-Scale Streaming Velocities with Great Attractors and Repulsors

The most direct and probably the best method to relate the data on large-scale streaming velocities to theory is to average the velocity over some selection function, forming a bulk velocity v_{bulk}, which has a distribution $P(v_{bulk}|R_f)$ depending upon the specific choice of selection function, e.g., a Gaussian of scale R_f. The work discussed by Kaiser and Lahav (1988) and Szalay (1988) considerably extends this approach and places it on a firm theoretical foundation. The technique is unbiased in that our location is taken to be random. Alternatively one can use the data in an *a posteriori* fashion to recognize that we may not be in a typical region and condition the velocity field determination by the constraint that there may be large entities such as a GA surrounding us. One can compute conditional probabilities $P(v(r)|GA)$, $P(v(r)|GA, \text{Local Void})$ etc. The probability of the observed velocity field would then be $P(v(r)|GA) \times P(GA)$, shifting the question of large-scale streaming velocities to one of rare event probability $P(GA)$ for a given fluctuation spectrum, as described in Sec. 2.2.

Within linear theory, $P(v(r)|GA)$ is a Gaussian distribution. Assume that the GA can be modelled by a region smoothed with a Gaussian filter R_{ga} with linear perturbation amplitude $v_{ga}\sigma_{ga}$. Requiring the point to be a peak of the density field does not change the result shown here very much. The regions whose velocities we are interested in computing are assumed to be smoothed over a Gaussian scale R_c, which would typically be much smaller than R_{ga}. The mean velocity of such R_c-smoothed clouds at position r_2 subject to the constraint that a GA is located at position r_1 is given by (see BBKS, Appendix D, for the derivational technique)

$$\langle \mathbf{v}(\mathbf{r}_2)|v_{ga}(\mathbf{r}_1)\rangle = \hat{\mathbf{r}} \, v_{ga} \, \xi_v'(r; R_h)[H_D\sigma_{\varrho,ga}]^{-1} \qquad (2.7a)$$

$$= -\frac{H_D\mathbf{r}}{3} \langle \frac{\Delta M}{M} (< r; R_h)|v_{ga}\rangle, \ \mathbf{r} = \mathbf{r}_2 - \mathbf{r}_1, \qquad (2.7b)$$

$$\xi_v(r) \equiv \langle \mathbf{v}(\mathbf{r}_2) \cdot \mathbf{v}(\mathbf{r}_1)\rangle, \ \xi_v' \equiv d\xi_v/dr = -H_D^2 J_{3\varrho}(r; R_h)/r^2, (2.7c)$$

$$R_h \equiv [(R_{ga}^2 + R_c^2)/2]^{1/2}, \ \sigma_{\varrho,ga} \equiv \sigma_\varrho (R_{ga}), \ H_D \equiv \dot{D}/D. \qquad (2.7d)$$

The cross correlation of the velocity field with an asemble of Great Attractors is ξ_{vga}. Here, the velocity correlation function ξ_v and $J_{3\varrho}(r) \equiv \int_0^r \xi_v r^2 \, dr$ are to be smoothed over the intermediate filter R_h. $D(t)$ describes the linear growth of perturbations (Peebles 1980, Sec. II) and $= a(t)$ in a $\Omega = \Omega_{nr} = 1$ universe. Equation (2.7b) relates the mean velocity to the mean mass excess $\Delta M/M$ within the radius r from a the GA. The dispersion in the direction perpendicular and parallel to the line joining the point r_2 to GA are

$$\langle(\Delta v_\perp(r))^2\rangle^{1/2} = (2/3)^{1/2}\sigma_v, \ \Delta \mathbf{v} \equiv \mathbf{v} - \langle \mathbf{v}|v_{ga}\rangle, \ \sigma_v \equiv \sigma_v(R_c), \quad (2.8a)$$

$$\langle(\Delta v_\parallel(r))^2\rangle^{1/2} = [\sigma_v^2/3 - (\xi_v'(r)/(H_D\sigma_{\varrho,ga}))^2]^{1/2}. \qquad (2.8b)$$

The 3-dimensional velocity dispersion of the clouds in the absence of a GA is σ_v evaluated using R_c. It is not very sensitive to the specific value of R_c provided it is small enough, since the velocity spectrum peaks at $\sim 3 \ h^{-1}$ Mpc for CDM; this turns out to be similar to the coherence scale for velocities of field points smoothed on galactic scales, which defines the typical coherent flow unit. The behaviour of the mean velocities and the dispersions is shown in Fig. 3 for the CDM and isocurvature baryon models. Here I took $R_c = 3.2 \ h^{-1}$ Mpc, motivated by the above discussion. The choice $R_{ga} \sim 15 \ h^{-1}$ Mpc is arbitrary; $R_{ga} \sim 12 \ h^{-1}$ Mpc might be a better choice to get the mass excess to be $\sim 5 \times 10^{16} \ M_{\odot}$ if $f_s = 1.5$ as assumed in Sec. 2.2.

Note that eq. (2.7b) bears some similarity to the phenomenological formula for the GA velocity field adopted by Faber and Burstein (1988). For a power law correlation $\xi_\varrho \sim r^{-\gamma}$, $\gamma = 3 + n$ if $n \lesssim 0$, where n is the spectral index, $3J_{3\varrho}/r^3$ will be $(1 - \gamma/3)^{-1} \ \xi_\varrho(r)$, for $r \gg R_h$, with a 'core radius' of order R_h. Apparently $n = n_A - 2$, where their far field velocity falloff power is n_A, fit to be 1.7. Useful calibration of eq.(2.7) using their result is difficult as it is quite sensitive to the core radius which could be considerably shorter than the linear value if the GA has already turned around. A crude but simple estimate of how the perturbed (Eulerian) position x_2 has evolved from the unperturbed (Lagrangian) position r_2 is to adopt the spherical top hat model given by eq.(2.2): the distance contracts according to $x_2 = (r_2 - r_1)/(1 + \delta)^{1/3} + x_1$, where $\delta = \Delta M/M$ is the mean overdensity in the interior; e.g., $\delta = 3.5$ implies the significant contraction $x_2 - x_1 \approx 0.6(r_2 - r_1)$ over the core radius. At our distance there would not have been as much collapse of co-moving space.

The interpretation of Fig. 3a is the following: If the Local Group velocity toward the GA direction must reach 580 km s^{-1}, then it could be obtained coherently using $\langle v|GA \rangle$ alone, provided the GA has $\delta \sim 3.5$; or it could be only partially coherent if the GA has $\delta \sim 0.8$, with the GA contributing ~ 300 km s^{-1}, the remaining ~ 300 km s^{-1} being provided by the unconstrained part of the density field: such a value is within the 1σ upward fluctuation indicated by Fig. 3a. However, the objects around us which are doing the pulling and pushing should be identifiable. An example would be the Local Void, which lies above the plane of the Local Supercluster and in the direction opposite to the Centaurus cluster. The velocity field given by eq.(2.7) then points away from the void since $v < 0$ — voids are repulsive. As a crude estimate of the sort of amplitude we might expect, take our distance from the Local Void centre as 20 h^{-1} Mpc and let it have an 'overdensity' $v\sigma_\varrho = -0.5$, smoothed over 10 h^{-1} Mpc. With these parameters we would get a contribution of 320 km s^{-1} pointing away from the void, with a component directed below the supergalactic plane, and presumably a residual component in the GA direction.

2.6 - Are Mountains of Extra Power Called For?

In this paper we have seen that judicious interpretations of the observations (and the rather low b_g and high f_s choices) might make all of the data compatible with the CDM model. However this cannot be the case if the

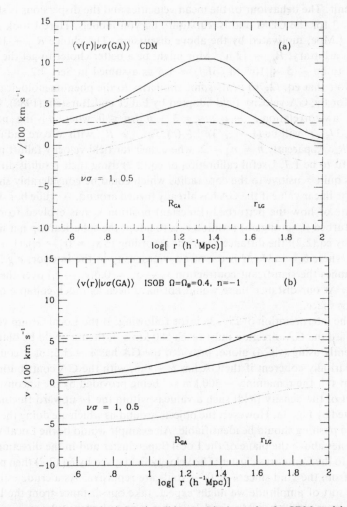

FIG. 3. The mean velocity $\langle v | GA \rangle$ (solid curves) is shown for the CDM model with $b_g = 1.44$ (3a) and the $\Omega = \Omega_B = 0.4$ isocurvature baryon model with $b_g = 1$ and $n = -1$ (3b) for two choices of height $\nu_{ga}\sigma_{ga} = 0.5$ and 1 (upper solid curve), corresponding to overdensities δ_{ga} in the spherical model of Sec. 2.2 of 0.8 and 3.5. The Gaussian smoothing radii are $R_{ga} = 15 \, h^{-1}$ Mpc for the GA and $R_c = 3.2 \, h^{-1}$ Mpc — the coherence length for velocities smoothed over galaxy scales, 0.35 h^{-1} Mpc, in the CDM model — for the clouds whose flow velocity we wish to determine; thus $R_h = 10.8 \, h^{-1}$ Mpc. With such a large R_{ga}, the top hat 'GA' would encompass the Centaurus cluster at $\sim 30 \, h^{-1}$ Mpc from us as well as the GA at 42 h^{-1} Mpc. The R_c scale corresponds roughly to the scale of our 'Coma-Sculptor' cloud, whose flow is relatively coherent as Faber, Burstein and Tully pointed out. The dotted curves are the ± 1σ variations about the mean velocity in the parallel direction (toward the GA) for the $\nu_{ga}\sigma_{ga} = 0.5$ case. The variations are similar for the $\nu_{ga}\sigma_{ga} = 1$ case. The dispersion in the perpendicular direction is indicated by the dot-dash curve.

Tenerife experiment (Davies *et al.* 1987) is really seeing primordial CMB anisotropies at the 4×10^{-5} level in a beam of size $\theta_{fwhm} = 8^o$. If we accept the need for extra power at large scales indicated by this experiment, yet still require compatibility with the large angle RELICT experiment of Strukov *et al.* (1988), a mountain of extra power is the required solution, the ramp plus plateau possibility being ruled out by RELICT. The idealized sharp mountain at wavenumber k_P — a spike — must be obtained by initial condition modification, not by transfer function modification of scale invariant initial conditions. Such a spike gives an enhanced ξ_{cc}, impressive cluster patches and increases the number of entities of the Great Attractor type, but might already be incompatible with w_{gg} and ξ_{cg}. It would add (in quadrature) a term to the 3D *rms* velocity dispersion given by $\sigma_{vP} \approx 1000 k_{P2}^{-1} \sigma_{\varrho P}/0.1$ km s^{-1}, where $k_{P2}^{-1} \equiv k_P^{-1}/100$ h^{-1} Mpc. If k_P^{-1} is large enough, the CMB anisotropy would just be due to the Sachs Wolfe effect. The spike would add (again in quadrature) an *rms* single beam temperature fluctuation, smoothed over a beam profile with Gaussian dispersion $\theta_s = 0.425\theta_{fwhm}$, given by (for $\Omega_{nr} = 1$ universes)

$$\frac{\Delta T}{T}(\theta_{fwhm}) \approx \frac{1}{2}(Hk_P^{-1})^2 \sigma_{\varrho P} e^{-(\ell\theta_s)^2/3} \tag{2.9}$$

$$\sim 6 \times 10^{-5} [k_{P2}^{-1}]^2 \, \frac{\sigma_{\varrho P}}{0.1} \, \exp\left[-k_{P2}^2 \left(\frac{\theta_{fwhm}}{3.9^o}\right)^2\right],$$

where $\ell_P = k_P \chi_0 \approx 60[k_{P2}^{-1}]^{-1} \approx (1^o)^{-1} k_{P2}.$

Here χ_0 is the comoving distance to the last scattering surface. (See Bond 1987a for the treatment of the beam smearing.) Choosing $k_P^{-1} \sim 300$ h^{-1} Mpc with $\sigma_{\varrho P} \sim 0.1$ could therefore give the reported Tenerife level. I would also guess that such a pulse might be compatible with the RELICT data, since there would be little pixel-to-pixel correlation, and the associated angular power spectrum for the spike (per logarithmic interval of angular wavenumber ℓ), $\ell^2 C_\ell = (\ell/\ell_P)^2 (1 - \ell^2/\ell_P^2)^{-1/2}$ for $\ell < \ell_P$, vanishing for larger ℓ, would have little power over the ℓ-range that the RELICT experiment is most sensitive to. (However, the effective beam smearing for RELICT is similar to that for the Tenerife experiment.) In conclusion, if we accept the need for large-scale power, a mountain of extra power from ~ 25 h^{-1} Mpc to ~ 300 h^{-1} Mpc might be compatible with all of the data as currently stated. However, there is apparently no natural physical mechanism to add such power.

Acknowledgement

This work was supported by a Canadian Institute for Advanced Research Fellowship, a Sloan Foundation Fellowship and the NSERC of Canada.

REFERENCES

Argyres, P.C., Groth, E.J., and Peebles, P.J.E. 1986. *AJ.* **91,** 471.

Bahcall, N. and Soneira, R. 1983. *Ap J.* **270,** 70.

Bardeen, J.M., Bond, J.R., Kaiser, N., and Szalay, A.S. 1986. *Ap J.* **304,** 15 [BBKS].

Bardeen, J.M. *et al.* 1988. in preparation. [BBRS].

Bardeen, J.M., Bond, J.R., and Efstathiou, G. 1987. *Ap J.* **321,** 28. [BBE].

Binggeli, B. 1982. *AA.* **107,** 338.

Bond, J.R. 1986. in *Galaxy Distances and Deviations from the Hubble Flow.* eds. B. F. Madore and R. B. Tully. Dordrecht: Reidel, p. 255.

_____. 1987a. in *The Early Universe.* ed. W. G. Unruh. Dordrecht: Reidel.

_____. 1987b. in *Cosmology Particle Physics.* ed. I. Hinchcliffe. Singapore: World Scientific.

_____. 1987c. In *Nearly Normal Galaxies.* ed. S. Faber. New York: Springer-Verlag.

_____. 1988. in *Large-Scale Structures of the Universe.* eds. J. Audouze, M.-C. Pelletan, and A. Szalay. Dordrecht: Kluwer Academic Publishers, p. 93.

Bond, J.R. and Couchman, H. 1987. *Proceedings of the Second Canadian Conference on General Relativity and Relativistic Astrophysics.* eds. A. Coley and C. Dyer. Singapore: World Scientific.

_____. 1988. in preparation.

Bond, J.R., Carr, B.J., and Hogan, C. 1988. Preprint.

Bond, J.R., Szalay, A.S., and Silk, J. 1988. *Ap J.* **324,** 627.

Carter, D. and Metcalfe, N. 1981. *MN.* **191,** 325.

Coles, S. and Kaiser, N. 1988. in preparation.

Davies, R.D., Lasenby, A.L., Watson, R.A., Daintree, E.J., Hopkins, J., Beckman, J., Sanchez-Almeida, J., and Rebolo, R. 1987. *Nature.* **326,** 462.

Davis, M., Efstahiou, G., Frenk, C.S., and White, S.D.M. 1985. *Ap J.* **292,** 371.

Efstathiou, G. and Rees, M. 1987. *MN.* **230,** 5P.

Efstathiou, G., Frenk, C.S., White, S.D.M. and Davis, M. 1988. preprint.

Faber, S., and Burstein, D. 1988, this volume.

Frenk, C.S., White, S.D.M., Davis, M., and Efstathiou, G. 1985. *Nature.* **317,** 595.

Groth, E.J., and Peebles, P.J.E. 1977. *Ap J.* **217,** 385.

Kaiser, N., and Lahav, O. 1988, this volume.

Lilje, P., and Efstathiou, G. 1988. *MN.* **231,** 635.

Maddox, S., Efstathiou, G. and Loveday, J. 1988. in *Large-Scale Structures of the Universe.* eds. J. Audouze, M.-C. Pelletan, and A. Szalay. Dordrecht: Kluwer Academic Publishers, p. 151.

Matsumoto, T., Hayakawa, S., Matsuo, H., Murakami, H., Sato, S., Lange, A.E. and Richards, P.L. 1987. preprint.

Narayan, R. and White, S.D.M. 1988. preprint.

Nolthenius, R. and White, S.D.M. 1987. *MN.* **225**, 505.

Peebles, P.J.E. 1980. *The Large Scale Structure of the Universe.* Princeton: Princeton University Press.

_____. 1987. *Ap J.* **277**, L1.

_____. 1988. this volume.

Press, W.H. and Schechter, P. 1974. *Ap J.* **187**, 425.

Quinn, P.J., Salmon, J.K. and Zurek, W.H. 1986. *Nature.* **322**, 392.

Rees, M. 1986. *MN.* **218**, 25P.

Rhee, G. and Katgert, P. 1987. *AA.* **183**, 217.

Schaeffer, R., and Silk, J. 1985. *Ap J.* **292**, 319.

Seldner, M., and Peebles, P.J.E. 1977. *Ap J.* **215**, 703.

Struble, M.F., and Peebles, P.J.E. 1985. *AJ.* **90**, 582.

Strukov, I.A., Skulachev, D.P. and Klypin, A.A. 1988, in *Large-Scale Structures of the Universe.* eds. J. Audouze, M.-C. Pelletan, and A. Szalay. Dordrecht: Kluwer Academic Publishers, p. 27.

Szalay, A. 1988, this volume.

VI

THEORY: PHENOMENOLOGICAL

N-BODY SIMULATIONS OF A UNIVERSE
DOMINATED BY COLD DARK MATTER

MARC DAVIS

Departments of Astronomy and Physics, University of California, Berkeley

and

GEORGE EFSTATHIOU

Institute of Astronomy, Cambridge

ABSTRACT

There exist a variety of observational constraints that can be used to test and reject theories of large-scale structure in the universe. Many of these tests require N-body simulations to follow the nonlinear clustering evolution of the models. In this review we describe how the standard cold dark matter (CDM) model compares to the available constraints. The theory is parameterized by two free parameters: the initial amplitude of the perturbation spectrum, and the horizon scale at the epoch of equality of the radiation and matter density. With judicious choice of these parameters (the latter of which is fixed by the Hubble constant), the CDM theory matches an impressive list of observations, but is inconsistent with the reports of clustering on scales $\gtrsim$ 5000 km/s. We describe recent simulations which show that the distribution of luminous galaxies in a CDM universe will be more strongly clustered than the mass distribution. The amplitude of the "bias" is sufficiently large to reconcile observations with the high cosmological density ($\Omega = 1$) assumed in the CDM model.

1. INTRODUCTION

Enormous progress has been made in the last several years toward understanding the development of large-scale structure in the universe. The theoretical underpinning of this progress is the presumption that the universe underwent an early inflationary episode, which provides the only known explanation for the near flatness of the universe, and also prescribes an initial spectrum of density perturbations. The current standard model is the minimal theory expected from any inflationary universe. The density perturbations arising from inflation are assumed to be random phase Gaussian noise with a scale invariant adiabatic spectrum. The universe is assumed to have the critical cosmological density parameter ($\Omega = 1$) and to be dominated by some form of cold dark matter (CDM), presumably a weakly interacting particle. This allows us to have a flat universe while still remaining compatible with the standard theory of primordial nucleosynthesis. The overall amplitude of the fluctuation spectrum is sensitive to unknown details of the field theory, so at present we adjust this free parameter to match observations. Subsequent evolution of these adiabatic density perturbations during the radiation dominated phase (Peebles 1982, Blumenthal and Primack 1983) leads to a unique spectrum at the epoch of recombination. These linear perturbations will evolve via gravitational instability into the nonlinear structure observed at present. The goal of N-body studies is to directly simulate the clustering into this nonlinear phase. Because the theory is so completely specified there is always the chance that it can be falsified if it fails to match observations.

We will describe the current effort of our "gang of four" collaboration (Davis, Efstathiou, Frenk and White), but will not attempt a review of all our past work (see White *et al.* 1987 and references therein). Most of our projects have been focused on direct comparison of the models to a variety of observational facts, which we summarize in Table 1. Any acceptable theory of large-scale structure should pass the tests listed in Table 1, although it must be remembered that some of the observations lack precision and others (*e.g.* filamentary structure) are rather qualitative and subjective. CDM has become the focus of attention because it passes most of the tests so well. Perhaps the model will eventually be falsified, but we have been unable to do so convincingly after four years of trying.

Setting the Hubble constant is equivalent to setting the horizon scale at the epoch of equality of radiation and matter density, which determines the peak in the perturbation spectrum. We presume that $H_0 = 50$ km s^{-1} Mpc^{-1}, because this value provides a good match to large-scale structure. The CDM model works well for smaller values of H_0, but fails to give sufficient large-scale structure for $H_0 = 100$. If in fact H_0 eventually settles in the range of 100 km s^{-1} Mpc^{-1}, the minimal CDM model is dead, quite apart from problems of the age constraints (see Bardeen *et al.* 1987, for a discussion of various non-standard CDM models). The amplitude of the initial perturbations is set by matching the observed strength of galaxy correlations $\xi_{gg}(r)$ (point 6 in Table 1). With no further adjustable parameters, the model must match the tests described in Table 1. The cosmological constant could be used as a further parameter, but we shall assume $\Lambda = 0$.

TABLE 1

WHAT A *GOOD* THEORY SHOULD EXPLAIN

On the scale of galaxies:

1. Galaxy abundance as a function of luminosity $\phi(L)$
2. Flatness of galaxy rotation curves at $v_{rot} = 150 - 250$ km/s
3. Luminosity-rotation correlation ($L \propto \sigma^4$)
4. Proper specific angular momentum λ and internal distribution
5. Morphology-environment correlation

On intermediate scales:

6. Amplitude and slope of $\xi_{gg}(r)$
7. Amplitude and slope of relative pair velocities $\sigma_{12}(r)$
8. Plausible bias mechanism

On larger scales:

9. Proper "filamentary frothiness" in galaxy clustering
10. Large voids, apparently empty
11. Correct abundance of rich clusters
12. Amplitude and shape of $\xi_{gc}(r)$
13. Sufficiently enhanced amplitude of $\xi_{cc}(r)$
14. Sufficient coherence in large-scale flows
15. Consistency with IRAS dipole direction $\Delta\theta(r)$

As a function of redshift:

16. Evolution in abundance of QSO's, AGN's, radio sources
17. Age of galactic disks
18. Change in populations of galaxies in clusters
19. Redshift of "galaxy" formation

2. SMALL-SCALE STRUCTURE

Let us discuss the tests one by one. The tests on galaxy size scales are detailed by Frenk *et al.* (1985, 1988). We evolved a series of simulations of comoving size 14 Mpc, so that galaxy halos are comprised of many particles. The halos in these models form with flat rotation curves (we measure $v_c(r) = (GM(r)/r)^{1/2}$) in exactly the range to compare to real galaxies, so item 2 falls out immediately. These halos extend for several hundred kpc, and the CDM model is quite inconsistent with halos truncated at ~ 50 kpc (*e.g.*

Little and Tremaine 1987). The model can thus be tested by measuring the extent of halos via radial velocity measurements of faint satellites or by the X-ray emission of hot gas.

From the models we can measure $N(v_c)$, the number density of halos with circular velocity $\gtrsim v_c$. From the observed galaxy luminosity distribution function (Felten 1985) we infer $N(L)$, the number density of galaxies with luminosity $\gtrsim L$. If we assume each halo to be associated with one galaxy and luminosity to be a monotonic function of circular velocity, then by matching $N(v_c) = N(L)$ we derive the functional form of $L(v_c)$, with results shown in Figure 1a. For comparison, we show the Tully-Fisher relation for spirals (dashed line), the Faber-Jackson relation for ellipticals (dotted line) and a data point for dwarf irregulars in Virgo from Bothun *et al.* (1984; see Frenk *et al.* 1988 for further details concerning these comparisons). The halos in our simulations must harbour galaxies of all types, so it is encouraging that our predicted $L - v_c$ relation lies between the observed relations for spirals and ellipticals and passes close to the data point for dwarfs. This agreement is especially noteworthy because we had no adjustable parameters in deriving our prediction.

Assuming that the 'light' associated with each halo may be computed as described above, we can determine a "mass-to-light" ratio for each halo. The results are shown in Figure 1b. Note that this is M/L measured at a surface

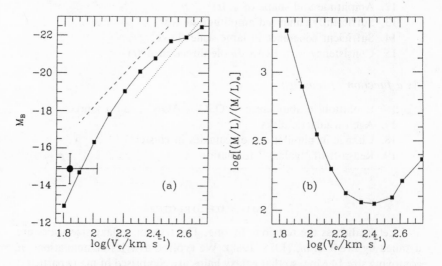

Fig. 1. (a) The Tully-Fisher relation for halos in the simulations. The "magnitude" of the "galaxy" associated with each halo is plotted against characteristic velocity. The dashed and dotted lines give the observed relations for spirals and ellipticals respectively. The point with error bars refers to dwarf irregulars in the Virgo cluster. (b) Predicted mass-to-light ratios in solar units for the "galaxies" associated with the halos in our simulations.

of density contrast of approximately 200. At this large radius, most halos with $V_c \simeq 200$ km/s have $M/L \simeq 100$ and for halos with shallower potential wells the M/L value is expected to be very large. It is not that the fainter galaxies have extraordinarily massive halos, but rather that their star formation efficiency is likely to be less efficient, as discussed by Dekel and Silk (1986). The upturn in M/L for larger V_c is a consequence of our assumption that each halo is associated with only one luminous galaxy; some of our halos have $v_c > 400$km/s and are uncharacteristic of even the brightest spirals. While some of these systems may correspond to the halos of bright ellipticals, others appear too massive to be associated with any single galaxy. Rather, they must be identified with the merged halos of groups and clusters of galaxies. This problem could be overcome by simulating the dissipative formation of galaxies, though we describe below an approximate prescription for assigning several galaxies to a common halo.

Angular momentum of galaxies has been discussed in detail by Barnes and Efstathiou (1987). Sufficient angular momentum is generated via tidal torques in the CDM model; the formation of centrifugally supported spiral galaxies would follow if gas collapsed dissipatively by a factor of ~ 10 in radius within the dark halos (Fall and Efstathiou 1980). The smaller specific angular momentum observed in the luminous portions of elliptical galaxies may result from angular momentum redistribution during the highly aspherical collapse and merging which occurs in the denser regions in the CDM models. In fact, in the simulations we find that halos grow both by slow accretion of diffuse material and by violent merging of subunits. Presumably a galactic disk will form under quiet conditions of gradual accretion, but disks are likely to be disrupted by strong mergers. The maximum ages of white dwarf stars in the disk of our own galaxy suggest it has not undergone a strong merger since $Z \sim 1$. The models demonstrate that strong merging for $Z < 1$ occurs mostly in the denser regions of the models, i.e. the group centers. It seems therefore that CDM provides a framework for the well established clustering differences of early versus late type galaxies (item 5 of Table 1). This statement can be better quantified by further analysis of our existing simulations.

3. INTERMEDIATE-SCALE CLUSTERING

We use the observed amplitude of galaxy clustering to set the initial amplitude of the perturbation spectrum. However, this does not guarantee that $\xi(r)$ will have a power law form with the observed slope. Furthermore we must determine the relationship of the galaxy distribution to the underlying mass distribution. It is well known that observations imply $\Omega \approx 0.2$ if galaxies trace the mass, so to meet the inflationary imperative that $\Omega = 1$, luminous galaxies must be a biased tracer of mass. Understanding the mechanism by which galaxies become a biased mass tracer is fundamental in any model of large-scale structure. The usual assumption is that bright galaxies are associated with rare peaks in the Gaussian noise field (Bardeen *et al.* 1986). In our initial studies we put the bias in "by hand", assuming galaxies were associated with the peaks of 2.5σ fluctuations in the initial density field. This procedure worked remarkably well but had no strong physical justification, although possible astrophysical biasing mechanisms have been discussed (e.g. Rees 1985, Dekel and Silk 1986). The absence of a compelling physical explanation of the needed bias has always been an unattractive feature of the CDM model.

However in our recent studies we have shown that halos, particularly those with $v_c \geq 200$ km/s are "naturally biased" by a purely gravitational effect. In the CDM model, deep potential wells are assembled late, when there are substantial fluctuations on larger scale. A protovoid will act as a section of an open uiverse and a protocluster will act as a section of a closed universe; the growth rate of linear density fluctuations will be enhanced in the dense regions and it will be depressed in the underdense regions. The net result is that denser regions are more efficient at condensing material into deep potential well objects, so a bias is guaranteed. The bias is expected to be less pronounced for smaller v_c because these objects are assembled earlier, when the amplitude of the large-scale structure is smaller and the growth rate of linear perturbations more nearly equal in the over and underdense regions.

This work is described by White *et al.* (1988). In this study we required more dynamic range than in our previous work because we needed to resolve individual galaxies as condensations of multiple points in a volume sufficiently large to measure galaxy clustering. We have begun therefore to use Cray supercomputers and can now evolve models of 262144 particles in potential grids of size 128^3. With the PPPM code our force softening length is 1/600th size of the periodic cube, and each model requires approximately 20 Cray hours to expand a factor of 8. In Figure 2 are presented slices of a snapshot of a model evolved in a cube of comoving size 3200 km/s. The mass of each particle is $7.0 \times 10^{10} M_\odot$ which is sufficient to resolve virialized halos with $v_c = 100$ km/s into approximately 10 particles.

The simulations are purely gravitational and so do not allow us to study the behavior of the dissipative gas from which galaxies must form. We have, therefore, included galaxy formation and merging in a way which, although plausible, remains somewhat *ad hoc*. At various stages during the evolution of a model we locate the most strongly bound particle in each dark matter halo. These are labelled "galaxies" and are assigned a circular speed at the radius of a sphere centered on each "galaxy" of mean overdensity 1000 times the present critical density. We then adopt simple algorithms to model galaxy mergers and to avoid multiple galaxy formation within each halo. The results are not especially sensitive to these procedures which are described in White *et al.* (1988). In Figures 2b and 2c we show the distribution of "galaxies" with $v_c > 100$ km/s and $v_c > 200$ km/s respectively. Note that the galaxies trace the ridges of high density in the matter distribution and are very deficient in the low density regions.

The spatial autocorrelation functions $\xi(r)$ for the mass distribution and for galaxies with $v_c > 100$ km/s and $v_c > 200$ km/s of this model are shown in Figure 3. The galaxy autocorrelation functions are steeper than that of the mass and have a larger amplitude. Over the range of separations shown in Figure 3 the mean correlation enhancement is ~ 1.3 for $v_c > 100$ km/s and ~ 2.5 for $v_c > 200$ km/s. When correlations are computed in redshift space with velocity broadening consistent with observed peculiar velocities, the slope of the correlations flattens to a power law of approximately $\gamma = 1.8$.

The amplitude of the correlations for "galaxies" with $v_c > 200$ km/s is a factor of 2 below that observed in magnitude limited galaxy catalogs; a sufficient level of natural bias occurs for "galaxies" with $v_c > 250$ km/s. The model predicts that the strength of clustering depends on the asymptotic circular velocity v_c of a galaxy, and, therefore, on the luminosity by the Tully-Fisher relation. There is little direct evidence for this in present samples, though good statistics are available for only a narrow range of luminosities. To some degree, however, galaxy morphology is correlated with v_c, and it is quite apparent that the strength of galaxy clustering is strongly dependent on morphology (e.g. Giovanelli et al. 1986, Davis and Geller 1976). Further study will hopefully provide a more quantitative measure of this point.

Previously the amplitude of the initial spectrum was adjusted to give the correct correlations of 2.5σ peaks, but now we must readjust the amplitude so that the naturally biased galaxy distribution has the correct correlation behavior. The correlation of the bias strength with the depth of the halo potential well makes this readjustment somewhat uncertain at present.

4. Large-Scale Clustering

N-body simulations in a box larger than 10000 km/s necessarily sacrifice small-scale resolution; the particle masses exceed the mass of a galaxy and it becomes necessary to use the statistical prescription to generate the bias. Large-scale clustering properties of a CDM universe are described in White et al. (1987). Items 9-11 of Table 1 were shown to be consistent with CDM, but items 12-14 presented problems.

We were able to reproduce the void in Boötes without difficulty if we sampled the galaxy distribution dilutely, as Kirshner et al. (1987) had sampled Boötes. However, we commented that in the CDM model Boötes sized voids would be very rare events, if completely devoid of galaxies. A new full survey of *IRAS* selected galaxies in the Boötes region (Strauss and Huchra 1988) found 3 galaxies within the void region where 11 were expected. These galaxies

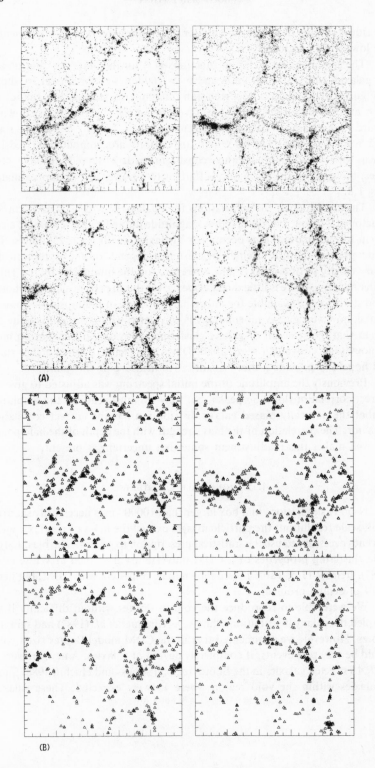

(A)

(B)

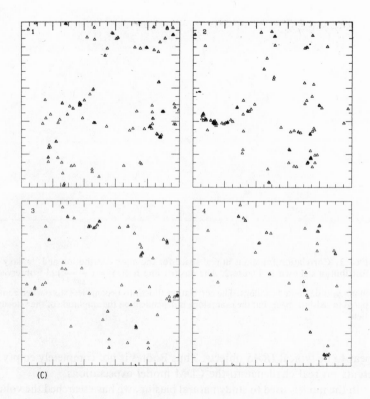

(C)

Fig. 2. (a) Four consecutive slices, each $1/8$ the thickness of the cube for a simulation of 262144 points in a volume 3200 km/s in length. The epoch is $z = 0$. All points are plotted. (b) The same four slices, but now only points at the center of potential wells of depth $\geq$ 100 km/s are plotted. (c) The same four slices, but now only points centered on potential wells of depth $\geq$ 200 km/s are plotted.

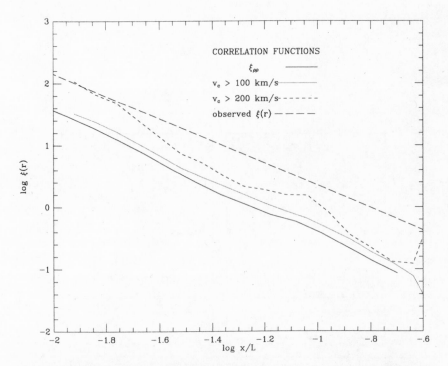

FIG. 3. Correlation functions in real space for the mass distribution and "galaxy" distributions shown in Figure 2. The dashed line is $\xi(r) = (\frac{r}{5h^{-1}})^{-1.8}$ observed in magnitude limited catalogs. The correlation functions become less steep when computed in redshift space, but the exact slope is dependent on the amplitude of the velocity field.

appear to be typical *IRAS* objects. Thus Boötes is not completely empty and presents no real challenge to the CDM model expectations.

In the models used to study natural biasing, we have searched the volumes of length size 2500 km/s to find the largest sphere devoid of all "galaxies" with $v_c > 100$ km/s. Plotted in Figure 4 is the average mass density as a function of distance from these void centers. The void radius was approximately 20% the size of the simulation, and the figure shows that the mass density within the voids is typically 20% of the mean. The voids are not empty, but they are certainly of low density, and they are not expected to be "filled" with faint galaxies. These voids are small compared to the size of those reported in the galaxy distribution, but they are as large as could be expected within simulations of this size.

White *et al.* (1987) demonstrated the observed abundance of the Abell clusters was compatible with CDM, but that the model cluster correlation func-

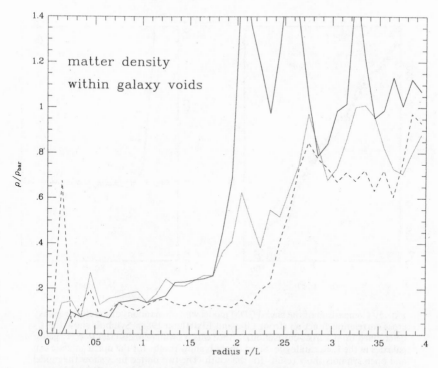

FIG. 4. The matter density distribution within the largest voids in each of 3 simulations of size $L = 2500$ km/s. We plot the density relative to the mean as a function of radius from the center of the void.

tions, $\xi_{cc}(r)$ and $\xi_{cg}(r)$, were of insufficient amplitude compared to observations (Bahcall and Soneira 1983, Seldner and Peebles 1977). This has always been the chief deficiency of the CDM model, and it is, therefore, of considerable interest to reconfirm the observational measures of cluster correlations.

Recently Lilje and Efstathiou (1988) have reanalyzed $\xi_{cg}(r)$ using redshifts for 204 Abell clusters. Using the redshift information and modern estimates of the galaxy luminosity function, they found a correlation amplitude a factor of two smaller than previously reported; furthermore, they showed that $\xi_{cg}(r)$ could not be reliably determined on scales $\gtrsim 20h^{-1}$ Mpc. Figure 5 shows a plot comparing $\xi_{cg}(r)$ from the models of White et al. (1987) and the recent measurements of Lilje and Efstathiou (1988). No parameters were used to scale one result to the other. The amplitude and shape are reasonably well reproduced by the model; there is no indication of any serious discrepancy.

The question of $\xi_{cc}(r)$ has been much discussed at this conference and the problem of projection effects, raised by Sutherland (1988) and presented here by Kaiser and Lahav (1988), must be studied further. The CDM predic-

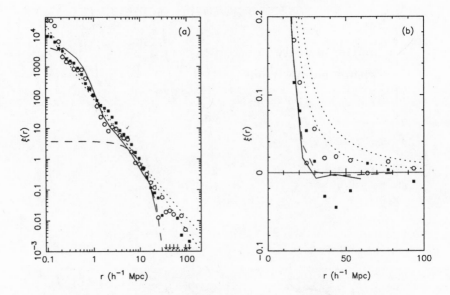

FIG. 5. Comparison of the biased CDM model with observations of the cluster-galaxy cross correlation function (from Lilje and Efstathiou 1988). Symbols show the ξ_{cg} determined from a cross-correlation of Abell clusters with richness class $R \geq 1$ against galaxies in the Lick catalogue. Solid symbols show results for the north galactic cap and open symbols show results for the south. The two dotted lines show the model $\xi_{cg}(r) = (r/8.8h^{-1} \text{ Mpc})^{-2.2}$ and $\xi_{cg}(r) = (r/6h^{-1} \text{ Mpc})^{-2.3} + (r/7h^{-1} \text{ Mpc})^{-1}$. The solid lines show ξ_{cg} for the CDM model as determined from N-body simulations while the dashed line shows an analytic prediction of ξ_{cg} using the statistical techniques described in Bardeen *et al.* (1987). Figure 5(a) shows these results on a log-log plot, while Figure 5(b) shows the results at large scales on a linear-linear plot.

tion is quite definite; there should not be substantial clustering of Abell clusters on scales larger than ~ 5000 km/s. The key issue here concerns reliability of the cluster catalogues. In our view, it would be premature to discard the CDM model on this account until more observational work has been done.

More recent considerations are related to the question of large-scale velocity flows in the universe. In principle, this is an easy calculation for models of structure with random phase initial conditions because linear theory should apply on the scales in question. CDM is not compatible with large amplitude flows with coherence lengths of order 5000 km/s. Superficially, it would seem that the coherence length of the velocity flows observed by Dressler *et al.* (1987) are considerably larger than predicted by the CDM model. However, it is important to assess the relative weights assigned to galaxies as a function of their distance from the observer; the more distant galaxies will carry much less weight than nearby galaxies so the effective "window function" for a bulk flow solution may be considerably smaller than the depth of the sample (Kaiser 1988).

Careful statistical comparisons of the elliptical galaxy data, together with results on the IRAS and optical dipoles have been described at this meeting by Kaiser and Lahav (1988). They find that a biased CDM model is compatible with these data, though there are indications of more power at large scales; their maximum likelihood analysis favours a larger amplitude for the CDM power spectrum than we have used in our simulations by a factor of about 2, but the data do not seem incompatible with our lower value.

Vittorio (1988) has shown that the misalignment angle between the microwave dipole direction and the IRAS gravity direction presents a strong test of the large-scale power spectrum, provided the IRAS galaxies at least approximately trace the mass distribution (item 15 in Table 1). The observed misalignment is small at modest distances, which is consistent with CDM but perhaps is a problem for the isocurvature baryon model.

5. CLUSTERING EVOLUTION

Based on the simulations we can certainly predict the evolutionary history of the correlation functions, but a suitable data set is not yet available for comparison. The evolutionary behavior of individual objects is much more difficult because we must include additional physics into simulations to begin to describe the dissipation of the baryons. The CDM halo models have been criticized as implying galaxy formation at too low a redshift, but as Baron and White (1987) have shown, galaxy formation could have continued to low redshifts ($z \sim 1 - 2$) without conflicting with observational searches for primeval galaxies. Constraints on the redshift of galaxy formation could offer an important constraint on the amplitude of the CDM fluctuation spectrum and could perhaps supply a decisive test of the theory, but at present the observational limits seem weak.

We also remark that there is no conflict between the standard CDM model and observations of bright quasars at $z \gtrsim 4$ (Efstathiou and Rees 1988). The comoving density of luminous quasars ($L > 10^{47}$ erg/s between $z = 2 - 4$) is small ($N_Q \simeq 10^{-6} h^{-3}$ Mpc^{-3}), and can be associated with proto-galactic nuclei in the CDM model if quasars are short lived and radiate at about the Eddington limit.

In some respects the kind of ongoing galaxy formation expected in the CDM model seems to us to be a positive virtue. For example, point 18 in Table 1 is another indicator that galaxies, as recently as $z = 0.4$, often have different stellar populations, indicative of more active star formation than at the present (*e.g.* Dressler and Gunn 1983). The CDM model predicts that galaxies at $z \sim 1$ should show an enhanced merger rate. More work clearly needs to be done on the expected evolution of galaxies, but certainly the view with the Space Telescope should be illuminating.

6. CONCLUSIONS

The CDM model is attractive because it fits in naturally with our current ideas of the early universe. The model has been tested against a large number of observations, yet there are no clear-cut discrepancies. The most serious problem comes from observations of the cluster-cluster correlation function, which may indicate a requirement for more power at large scales.

We have been motivated to investigate the CDM model in so much detail precisely because it contains so few parameters. Although we haven't been able to rule it out yet, we can at least point to the sort of observations and further investigations which need to be done. These include:

(i) *Tests of biased galaxy formation.* Our "natural biasing" mechanism makes a clear prediction that galaxies associated with halos of high v_c should be more strongly clustered than those with low v_c. This can be tested in a variety of ways, *e.g.* by comparing the correlation functions for faint and bright galaxies.

(ii) *Formation of galaxies.* With our choice of amplitude for the CDM spectrum, most halos with $v_c \sim 250$km/s form between redshifts $z \approx 3{-}1$. Such a low redshift of galaxy formation can be tested by primeval galaxy searches and by studies of spectral evolution (*e.g.* the amplitude of the 4000 Å break, Hamilton 1985). If massive protogalaxies were abundant at $z \gtrsim 3$, this would require either a revision of our choice of amplitude for the CDM spectrum, or perhaps indicate that the CDM theory is wrong.

(iii) *Clustering on large scales.* The CDM model makes a precise prediction of the amplitude of the cluster-cluster correlation function which is clearly at odds with the results of Bahcall and Soneira (1983). We have played down the discrepancy because we don't really know how reliable the Abell catalog is. However, in the near future, we should have a new X-ray selected cluster catalog from the ROSAT satellite which should be objective and complete. It will be fascinating to see whether the CDM model passes this test. The new deep galaxy survey constructed at Cambridge (Maddox, Efstathiou and Loveday 1988) also promises to shed some light on this problem as discussed at this meeting.

(iv) *Large-scale streaming motions.* As mentioned by several speakers at this meeting, the CDM model predicts a noisy velocity field with a relatively small coherence length; but this is not in obvious conflict with the available data. Further studies of large-scale streaming motions are necessary.

(v) *Extent of galaxy halos.* In the CDM model, we would expect galaxy halos to extend to large radii with approximately isothermal radial profiles. It should be possible to test this idea, either though X-ray observations or from kinematic studies.

(vi) *Anisotropies and spectral distortions of the cosmic background radiation.* Anisotropies in the cosmic microwave background would provide a clean test of the CDM model. Detailed predictions have been made by Bond and Efstathiou (1987). The results are within the accessible range for well designed experiments but are below the sensitivity levels of current experiments. The reported detection on scales of 8° discussed by Lasenby (1988) is too large to be compatible with the CDM model; we await with great interest to see whether this result is confirmed. We also mention the distortion of the backround radiation spectrum reported by Matsumoto *et al.* (1988). This perhaps indicates the presence of a substantial quantity of dust at redshifts $z \sim 40$, which would be difficult to reconcile with the CDM model.

Doubtless the reader will be able to add other items to this list. Even if the CDM model is eventually proved wrong, it is at least a good approximation to the truth!

Acknowledgements

We thank our collaborators, Carlos Frenk and Simon White, for allowing us to discuss our joint work. We have received financial support from NATO, the SERC, and the NSF.

REFERENCES

Bahcall, N. and Soneira, R. 1983. *Ap J.* **270**, 20.

Bardeen, J.M., Bond, J.R., Kaiser, N., and Szalay, A.S. 1986. *Ap J.* **304**, 15.

Bardeen, J.M., Bond, J.R., and Efstathiou, G. 1987. *Ap J.* **321**, 28.

Barnes, J. and Efstathiou., G. 1987. *Ap J.* **319**, 573.

Baron, E. and White, S.D.M. 1987. *Ap J.* **322**, 585.

Blumenthal, G.R. and Primack, J.R. 1983. in *Fourth Workshop on Grand Unification.* p. 256. ed. H.A. Weldon, P. Langacker, and P.J. Steinhardt. Boston: Birkhauser.

Bond, J.R. and Efstathiou, G. 1987. *MN.* **226**, 655.

Bothun, G.D., Aaronson, M., Schommer, R., Huchra, J., and Mould, J. 1984. *Ap J.* **278**, 475.

Davis M. and Geller, M.J. 1976. *Ap J.* **208**, 13.

Dekel, A. and Silk, J. 1986. *Ap J.* **303**, 39.

Dressler, A. and Gunn, J.E. 1983. *Ap J.* **270**, 7.

Dressler, A., Faber, S.M., Burstein, D., Davies, R.L., Lynden-Bell, D., Terlevich, R.J., and Wegner, G. 1987. *Ap J.* **313**, 42.

Efstathiou, G. and Rees, M.J. 1988. *MN.* **230**, 5p.

Fall, S.M. and Efstathiou, G. 1980. *MN.* **193**, 189.

Felten, J.E. 1985. *Comm Astr Sp Sci.* **11**, 53.

Frenk, C.S., White, S.D.M., Efstathiou, G., and Davis, M. 1985. *Nature.* **317**, 595.

Frenk, C.S., White, S.D.M., Davis, M., and Efstathiou, G. 1988. *Ap J.,* **327**, 507.

Giovanelli, R., Haynes, M.P., and Chincarini, G.L. 1986. *Ap J.* **300**, 77.

Hamilton, D. 1985. *Ap J.* **297**, 371.

Kaiser, N. 1988. *MN.* **231**, 149.

Kaiser, N. and Lahav, O. 1988, this volume.

Kirshner, R., Oemler, A. Schechter, P., and Shectman, S. 1987. *Ap J.* **314**, 493.

Lasenby, A. 1988, this volume.

Lilje, P. and Efstathiou, G. 1988. *MN.* **231**, 635.

Little, B. and Tremaine, S. 1987. *Ap J.* **320**, 493.

Maddox, S.J., Efstathiou, G., and Loveday, J. 1988. in *Large-Scale Structures of the Universe.* eds. J. Audouze, M.-C. Pelletan, and A. Szalay. Dordrecht: Kluwer Academic Publishers, p. 151.

Matsumoto, T., Hayakawa, S., Matsuo, H., Murakami, H., Sato, S., Lange, A.E., and Richards, P.L. 1988. *Ap J.,* **329**, 567.

Peebles, P.J.E. 1982. *Ap J Letters.* **263**, L1.

Rees, M.J. 1985. *MN.* **213**, 75p.

Seldner, M. and Peebles, P.J.E. 1977. *Ap J.* **215**, 703.

Strauss, M. and Huchra, J. 1988. *AJ., ***95**, 1602.

Sutherland, W. 1988. *MN.* **234**, 159.

Vittorio, N. 1988. this volume.

White, S.D.M., Frenk, C.S., Davis, M., and Efstathiou, G. 1987. *Ap J.* **313**, 505.

White, S.D.M., Davis, M., Efstathiou, G., and Frenk, C.S. 1988. *Nature.* **330**, 451.

THE STATISTICS OF THE DISTRIBUTIONS OF GALAXIES, MASS AND PECULIAR VELOCITIES

P. J. E. PEEBLES

Joseph Henry Laboratories, Princeton University

1. INTRODUCTION

The galaxy peculiar velocity field seems to have two noteworthy features: the rms value is large, perhaps $\sim$ 600 to 1000 km sec^{-1}, and the coherence length is large, $r_c \gtrsim 10$ h^{-1} Mpc (H $= 100$ h km sec^{-1} Mpc^{-1}). The first is based on the observation that the velocity of the Local Group relative to the rest frame defined by the microwave background radiation is 600 km sec^{-1} (Wilkinson 1986). Since we are in a relatively calm part of space, it seems likely that the rms velocity is no smaller than our motion. This velocity is considerably greater than the deviations from Hubble's law observed in our neighborhood (Sandage 1986): at distances hr $\sim$ 6 Mpc the scatter is no more than about 150 km sec^{-1} (Peebles 1988). The standard interpretation is that a large part of our peculiar motion is a component that varies only slowly with position. Possible examples of this large-scale component are seen in the Rubin-Ford effect (Rubin *et al.* 1976) and in the Great Attractor discussed in these proceedings by Faber and Burstein (1988).

A convenient measure of the length scales over which the peculiar velocity field, $\mathbf{v(r)}$, varies is provided by the statistic

$$< (v_1 - v_2)^2 > = \delta v(r)^2. \qquad (1)$$

The average is supposed to be taken over a fair sample of luminous (L $>$ L$_*$) pairs of galaxies with separations in a small range around r. We expect that the velocities, $\mathbf{v}_1$ and $\mathbf{v}_2$, of the two galaxies are uncorrelated in the limit of large separation so

$$\delta v(r \rightarrow \infty)^2 = 2 < v^2 >.$$

The observations lead us to suspect that $\delta v(r)$ decreases with decreasing r, reflecting the fact that the two galaxies tend to have a common component of velocity that cancels out the difference $\mathbf{v}_1$-$\mathbf{v}_2$. A coherence length for the velocity field is the separation r_c at which δv^2 is down from its value at $r \rightarrow \infty$ by a factor of 2:

$$\delta v(r_c) = \delta v(r \rightarrow \infty)/2^{1/2}. \tag{2}$$

At this separation the peculiar velocities of the two galaxies can be imagined to have on average equal contributions from a part common to the two galaxies and a part uncorrelated between the two galaxies.

At small r, $\delta v(r)$ is observed to be (Davis and Peebles 1983, Peebles 1984, Fig. 6)

$$\delta v = 3^{1/2}(340 \pm 40)\ (hr_{Mpc})^{0.13\ \pm 0.04}\ km\ sec^{-1},$$

$$0.1 \lesssim hr \lesssim 1\ Mpc. \tag{3}$$

The factor $3^{1/2}$ corrects from line-of-sight to three-dimensional velocities under the assumption of isotropy. The fact that δv in this equation is greater than the local noise in the Hubble flow is, I think, a reflection of the fact that we are in an unusually calm region. The fact that δv at $hr = 1$ Mpc is less than 600 km sec^{-1} indicates that the coherence length r_c is greater than $1\ h^{-1}$ Mpc, as we would expect from the Rubin-Ford and Great Attractor effects.

If the peculiar velocity field is produced by gravity then $\delta v(r)$ tells us something about the character of the mass distribution. I propose to consider here whether our constraints on $\delta v(r)$ are consistent with what would be expected if mass were distributed in the same way as galaxies. This question was first considered by Clutton-Brock and Peebles (1981); the discussion in the next section is an update. I will argue that the assumption that galaxies trace mass looks phenomenologically promising although, as is well known, it does force us to the inelegant assumption that the mean mass density is about 20% of the critical Einstein-de Sitter value. In Section 3 I present some generally negative comments on the biasing assumption required to reconcile the galaxy position and velocity observations with the Einstein-de Sitter density.

2. DYNAMICS UNDER THE ASSUMPTION THAT GALAXIES TRACE MASS

The statistic $\delta v(r)$ is determined by, among other things, the dimensionless mass autocorrelation function

$$\xi(r) = <(\varrho(\mathbf{x}) - <\varrho>)\ (\varrho(\mathbf{x} + \mathbf{r}) - <\varrho>) >/<\varrho>^2, \tag{4}$$

where $\varrho(\mathbf{x})$ is the mass density, $<\varrho>$ the mean value. The autocorrelation function is the Fourier transform of the power spectrum, $P(k)$:

$$\xi(r) = \int d^3k \ P(k) \ \sin kr/kr. \tag{5}$$

The rms velocity difference $\delta v(r)$ at small separations is determined by the condition that the relative acceleration $\sim v^2/r$ be balanced on average by gravity, for otherwise the galaxy clustering pattern would dissolve in less than a Hubble time. If the mass two—and three—point correlation functions agree with the galaxy functions then the balance condition is (Davis and Peebles 1983, eq. [44])

$$<(\mathbf{v}_1 - \mathbf{v}_2)^2> = KQ(Hr)^2 \ \xi(r) \ \Omega, \tag{6}$$

where (Davis and Peebles 1983; Groth and Peebles 1986)

$$\xi = (r_0/r)^\gamma, \ hr_0 = 5.4 \ \text{Mpc}, \ \gamma = 1.77; \tag{7}$$

the amplitude of the three-point correlation function is $Q \sim 0.7$; the mean mass density is $\varrho = \Omega$ times the critical Einstein-de Sitter value; the galaxy pair separation is r; and the dimensionless factor is $K(1.77) \cong 12$. Equation (6) agrees with the observations (eq. [3]) if the density parameter is

$$\Omega = 0.2. \tag{8}$$

This value is adopted in the following.

The energy equation (Irvine 1965) fixes the galaxy rms peculiar velocity in terms of the mass autocorrelation function, as was first discussed by Fall (1975). A convenient form of the energy equation is (Peebles 1980, Sec. 74)

$$<v^2> = \frac{6}{7} \ \Omega H^2 \ J_2, \ J_2 = \int_0^\infty rdr \ \xi(r). \tag{9}$$

The best estimate of J_2 seems to be that of Clutton-Brock and Peebles (1981) from the Lick catalog:

$$J_2 = 164 \ e^{\pm 0.15} \ h^{-2} \ \text{Mpc}^2. \tag{10}$$

With $\Omega = 0.2$ this gives

$$<v^2>^{1/2} = 530 \ e^{\pm 0.08} \ \text{km sec}^{-1}. \tag{11}$$

Since the motion of the Local Group relative to the microwave background is about 600 km sec^{-1} (Wilkinson 1986) equation (11) seems reasonable.

Equation (1) is

$$\delta v(r)^2 = 2 <v^2> - 2 <v_1 \cdot v_2>. \qquad (12)$$

At large r we can use perturbation theory to find $<v_1 \cdot v_2>$ (Peebles 1980, Sec. 72):

$$<v_1 \cdot v_2> = 4\pi(Hf(\Omega))^2 \int_0^\infty dk\ P(k)\ \sin kr/kr,\ f \cong \Omega^{0.6}. \qquad (13)$$

This with equation (5) for P(k) and equation (11) for $<v^2>$ shows how δv is expected to behave at large r.

To get a rough approximation to the integral in equation (13) let us note that on scales $20 \le hr \le 40$ Mpc $\xi(r)$ is small so the power spectrum is nearly flat, $P(k) \cong P_0$. In the approximation $P(k) = P_0$ equation (5) says

$$J_3 = \int_0^r r^2 dr\ \xi(r) \cong 2\pi^2\ P_0, \qquad (14)$$

and equation (13) is

$$<v_1 \cdot v_2> \cong (Hf)^2\ J_3/r. \qquad (15)$$

The estimate of J_3 from the Lick catalog is (Clutton-Brock and Peebles 1981)

$$J_3 = 596\ e^{\pm 0.2}\ h^{-3}\ Mpc^3. \qquad (16)$$

Equations (11), (12), (15) and (16) yield

$$\delta v^2 = (750)^2 - (1300)^2/hr_{Mpc}\ km^2\ sec^{-2}. \qquad (17)$$

In this approximation the coherence length (eq. [2]) is

$$r_c \sim 6\ h^{-1}\ Mpc. \qquad (18)$$

The coherence length may be underestimated because J_3 in equation (16) assumes $\xi = 0$ at $r \ge 25\ h^{-1}$ Mpc. Because of the volume factor $r^2 dr$ the integral J_3 is very sensitive to the value of ξ at large r (and the value of J_3 in eq. [9] is less sensitive), so that a small positive tail of ξ at larger r could appreciably increase r_c. Examples of this effect, which would be expected in the baryon isocurvature model, are shown in Peebles (1987).

We have no direct estimates of r_c from the observations, though it may be possible to get useful constraints from the sample of Burstein *et al.* (1987). Until such results are available we can get some feeling for the velocity field expected under the present assumptions by considering the mean velocity averaged through a spherical window of radius r:

$$\mathbf{v}_a = \int_r \mathbf{v} \, dV/V. \tag{19}$$

The mean square value of v_a is (eq. [15], or eqs. [13] and [16] of Clutton-Brock and Peebles 1981)

$$<v_a^2> = \frac{6}{5} (Hf(\Omega))^2 J_3/r. \tag{20}$$

With $r = 40 \, h^{-1}$ Mpc, $\Omega = 0.2$, and J_3 from equation (16) we get

$$<v_a^2>^{1/2} = 160 \text{ km sec}^{-1}, \ r = 40 \, h^{-1} \text{ Mpc}. \tag{21}$$

In an observation like that of Faber and Burstein (1988) one would move the sphere around to find a local maximum of $|\mathbf{v}_a|$; one would not have to look hard to find a 2 σ fluctuation so one would expect to see

$$v_a \sim 300 \text{ km sec}^{-1}, \tag{22}$$

at the depth $r = 40 \, h^{-1}$ Mpc of the Great Attractor. This is a factor of about two below what Faber *et al.* (1987) observe. Given the rather large uncertainties in the observations of the Great Attractor and of J_3, I consider this to be tolerably reasonable agreement. R. Juszkiewicz is working on a more detailed check of this point.

3. COMMENTS ON THE BIASING PICTURE

As is discussed in these proceedings by Kaiser and Lahav (1988), we could reconcile the velocity field observations with $\Omega = 1$ by assuming that mass clusters less strongly than do galaxies. This is a powerful concept and, as Efstathiou (1988) describes in these proceedings, it follows in a natural way in the scale-invariant cold dark matter model. There are of course problems. The most immediate, to my mind, is that the bias factor would have to be remarkably insensitive to the length scale. If equation (6) for δv^2 were biased because r_0^γ is overestimated by a factor $b^2 \sim 5$, then b^2 would have to be nearly constant at $0.01 \le hr \le 1$ Mpc, because the observed δv (eq. [3]) varies with r about as expected from equation (6). Since $\xi(r)$ varies by more than three orders of magnitude at $0.01 \lesssim hr \lesssim 1$ Mpc, it seems doubtful that one would have expected b^2 to be so nearly constant (unless $b^2 = 1$).

The statistic J_3 is mainly sensitive to $\xi(r)$ ar $r \sim 10\,h^{-1}$ Mpc. If $\xi(r)$ were correctly estimated from the galaxy distribution on this scale, and $\Omega = 1$, then we would expect, instead of equation (22), $v_a \sim 850$ km sec^{-1}. This would say that the Great Attractor is unusually weak, which would seem a little surprising. But the alternative assumption, that the bias factor has held nearly constant from 10 kpc to 10 Mpc separation, would also seem surprising.

4. CONCLUSION

The point of this discussion is that, under the assumption that mass clusters like galaxies, and with suitable adjustment of one parameter, the mean mass density, the predicted character of the galaxy peculiar velocity field is reasonably close to the observations. This assumption is not very popular because it would require that the mean mass density be less than the critical Einstein-de Sitter value. On the other hand, it seems to me that the fact that the assumption can be fitted to the observations in a simple and direct way is a considerable virtue, and that it therefore merits further close attention.

This research was supported in part by the National Science Foundation of the United States.

REFERENCES

Burstein, D., Davies, R., Dressler, A., Faber, S., Stone, R., Lynden-Bell, D., Terlevich, R., and Wegner, G. 1987. *Ap J.* **64**, 601.

Clutton-Brock, M. and Peebles, P.J.E. 1981. *AJ.* **86**, 1115.

Davis, M. and Peebles, P.J.E. 1983. *Ap J.* **267**, 465.

Efstathiou, G. 1988. this volume.

Faber, S. and Burstein, D. 1988. this volume.

Faber, S., Dressler, A., Davies, R., Burstein, D., Lynden-Bell, D., Terlevich, R., and Wegner, G. 1987. in *Nearly Normal Galaxies.* p. 175. ed. S.M. Faber. New York: Springer Verlag.

Fall, S.M. 1975. MN. **172**, 23p.

Groth, E.J. and Peebles, P.J.E. 1986. *Ap J.* **310**, 507.

Irvine, W.M. 1965. *Ann. Phys.* (N.Y.), **32**, 322.

Kaiser, N. and Lahav, O. 1988. this volume.

Peebles, P.J.E. 1980. *The Large-Scale Structure of the Universe.* Princeton: Princeton University Press.

_____. 1984. *Science.* **224**, 1385.

_____. 1987. *Nature.* **327**, 210.

_____. 1988. *Ap J.* in the press.

Rubin, V.C., Ford, W.K., Thonnard, N., Roberts M.S., and Graham, J.A. 1976. *AJ.* **81**, 687.

Sandage. A.R. 1986. *Ap J.* **307**, 1.

Wilkinson, D.T. 1986. *Science.* **232**, 1517.

THEORETICAL IMPLICATIONS OF SUPERCLUSTERING

AVISHAI DEKEL

Racah Institute of Physics, The Hebrew University of Jerusalem

ABSTRACT

I report here on three studies of the superclustering of clusters.

M. J. West, A. Oemler Jr. and myself have found, using N-body simulations, that the alignment of clusters of galaxies with the surrounding galaxy distribution is a useful discriminant between certain Gaussian cosmologies within the gravitational instability picture. The observed alignment of the Shane-Wirtanen galaxy counts with Abell clusters requires either a coherence length in the initial fluctuation spectrum on a scale of a few tens of megaparsecs, as in pancake scenarios or a flat ($n \leq -2$) power spectrum on such large scales. It is in conflict with hierarchical scenarios such as the Cold Dark Matter (CDM) model, which indicates a general weakness in 'filamentary structure' in such models.

G. R. Blumenthal, J. R. Primack and myself have estimated the effect of projection contamination in the Abell catalog on the cluster-cluster correlation function. A simple empirical test where pairs of small angular separations are excluded indicates only a $\sim 50\%$ effect in amplitude near the correlation length. A statistical estimate relating the cluster auto-correlation function to the cross-correlation function of clusters and galaxies suggests that at least one of these functions is inconsistent with the conventional models of Gaussian fluctuations in an $\Omega = 1$ universe, such as standard CDM or pancake scenarios.

D. Weinberg, J. P. Ostriker and myself have found that superclustering in a generic explosion scenario substantially exceeds that in flat, Gaussian models. The two points where three shells intersect are found to be the natural sites for the formation of rich clusters. A simple toy model is used to study the spatial distribution of rich clusters in an explosion picture with a random distribution of seeds. The shell topology gives rise to a correlation function which is close to the observed power-law, with richer clusters having stronger correlations. The correlation amplitude is larger than observed by a factor of ~ 2, but a natural anti-correlation of seeds reduces the superclustering to the observed level. Typical shell radii of a few tens of megaparsecs produce the observed number density of Abell clusters. Percolation superclustering analysis, cluster void probabilities, and topology tests confirm strong high-order correlations. Thus, the superclustering in certain explosion models is fairly consistent with the available observational data.

1. INTRODUCTION

This paper describes parts of three separate long-term projects. They address different implications of superclustering on the theory of the formation of large-scale structure in the universe.

The first project is an observational and theoretical effort by M. J. West, A. Oemler, Jr. and myself to study the structure and dynamics of rich clusters in N-body simulations of certain competing cosmological scenarios, in comparison with observed Abell clusters. Although clusters are non-linear systems, they might preserve some tracers of the initial conditions which led to their formation. What makes them appealing for such a study is that while they are relatively easy to identify and study, they are young dynamically and involve mostly gravity in their formation and evolution. We have found, contrary to what we had hoped for, that the light distibution in the observed central regions of clusters is similar in most aspects to the mass distribution of the simulated clusters in most of the scenarios. This general result tells us useful things, such as how efficient violent relaxation is in erasing the initial conditions, and how well the light profile traces the mass profile in clusters, but it means that the inner parts of clusters are not very helpful in trying to distinguish between cosmologies.

The outer regions of clusters are more promising, however. We find, for example, that subclustering at radii ~ 5 h^{-1} Mpc around clusters is sensitive to initial conditions and we suggest observations in that direction. Here, I will describe another property which we find useful for cosmology — the alignment of clusters with the surrounding galaxy distribution. We find that the observed large-scale alignment of the Shane-Wirtanen galaxy counts with Abell clusters (see Peebles 1988) indicates either a coherence length in the spectrum on a scale of a few tens of megaparsecs, as in pancake scenarios or a flat (n $\leq$ -2) power spectrum on such large scales. The observed alignment is in conflict with hierarchical scenarios such as the Cold Dark Matter (CDM) model, which seems to reflect a general deficiency of large-scale 'filamentary' structure in CDM.

My second contribution in collaboration with G. R. Blumenthal and J. R. Primack, attempts to evaluate the validity of the excessive two-point correlation function of Abell clusters, which provides constraints on the formation of very large-scale structure. We estimate in two different ways the effect of contamination of Abell clusters by foreground or background galaxies on the cluster-cluster correlations (an effect suggested by W. Sutherland and N. Kaiser). We find the effect to be weak. Relating the cluster-galaxy correlation function and the effect of contamination on the cluster-cluster correlation function, we conclude that the inconsistency with conventional Gaussian models, such as CDM in an $\Omega = 1$ universe, does not go way so easily.

We studied a possible scenario which is capable of reproducing the observed superclustering (and the streaming motions on very large scales) within the framework of Gaussian fluctuations with a Harrison-Zeldovich spectrum. This is an open ($\Omega < 1$) hybrid model of CDM and baryons, which carries more large-scale power in the spectrum of fluctuations but is not associated with high-order correlations in the linear regime.

Based on a non-Gaussian cosmological model a collaboration, which started with S. Saarinen and B. J. Carr, is going on with D. Weinberg and J. P. Ostriker. We study the formation of large-scale structure in the explosion picture where positive-energy perturbations sweep matter onto high-density shells. The pairs of knots where three shells intersect are found to be deep potential wells which are the natural sites for the formation of rich galaxy clusters. We use a simple toy model to study the spatial distribution of rich clusters in a generic type of explosion scenario, concentrating on statistical measures of superclustering and especially on the cluster-cluster correlation function. The toy model, parameterized by the distribution of shell radii and the filling factor, places spherical shells at random and identifies each intersection point as a cluster.

The two-point correlation function, which results from the shell topology despite the random distribution of seeds, is a power-law consistent with observations, with the richer clusters forming at the intersections of bigger shells and so having stronger correlations. However, the correlation amplitude in our simple models is larger than observed by a factor $\sim 2 - 3$. The superclustering must be reduced by an anti-correlation of the explosion seeds — an effect which has a natural origin. Typical shell radii $\sim 25 - 50h^{-1}$ Mpc and filling factors $\sim 0.3 - 0.6$ are required to produce the observed number density of clusters. A toy model with a power-law radius distribution $\propto R^{-4}$ for $R < R_{max}$, can reproduce the richness distribution of clusters in the Abell catalog. Percolation supercluster statistics, cluster void probabilities, and topology tests confirm the presence of strong high-order correlations, reflecting the non-Gaussian nature of the superclustering. Thus, superclustering in the explosion scenario substantially exceeds that in the $\Omega = 1$ Gaussian models and is quite distinguishable from them. We conclude that certain explosion models are consistent with the available observational data.

2. ALIGNMENT OF CLUSTERS WITH THEIR SURROUNDINGS: A PROBLEM FOR CDM?

2.1 - Introduction

As an attempt to constrain the range of viable cosmogonic scenarios, rich clusters of galaxies have been studied in a series of papers by Michael J. West, Augustus Oemler Jr. and myself. We have compared N-body simulations of clusters formed from different cosmological initial conditions with observations. In paper I (West, Dekel, and Oemler 1987), we studied density profiles and velocity dispersion profiles. We found good agreement between the simulated mass profiles and the observed light profiles, indicating that the radial distribution of light and mass inside clusters follow each other, but the profiles were found to be poor discriminants between cosmological scenarios. In Paper II (West, Oemler, and Dekel 1988), we focused on substructures in and around clusters. We found the subclustering in the inner $\sim 3\ h^{-1}\ Mpc$ to be poorly correlated with the initial conditions, probably because of violent relaxation, but subclustering in the vicinity of rich clusters is a discriminant which might be useful. In Paper III (West, Dekel and Oemler 1988) we study alignments and ellipticities. I will describe here only one part of this work — an effect which has been found to provide a useful discriminant between competing cosmological scenarios. This is the alignment of clusters with the surrounding galaxy distribution.

Most clusters of galaxies are elongated to a degree which allows a relatively unambiguous definition of a major axis for each cluster. The position angles of first ranked cluster galaxies show a strong tendency to be aligned with the major axis of their parent cluster (Sastry 1986, Dressler 1980, Carter and Metcalfe 1980, Binggeli 1982, Struble 1987, Tucker and Peterson 1987). On larger scales, up to $\sim 30\ h^{-1}\ Mpc$, the clusters themselves exhibit a statistical tendency to be aligned with the lines connecting them to neighboring clusters (Binggeli 1982). Although a subsequent study (Struble and Peebles 1985) found the evidence for such an alignment to be weaker, more recent work (Rhee and Katgert 1987) gives more support to the effect.

In a recent study of a possibly related effect on intermediate scales, Argyres *et al.* (1986, hereafter AGPS) and Lambas, Groth, and Peebles (1988, hereafter LGP) have found that Shane-Wirtanen galaxy counts in the regions surrounding Abell clusters tend to be systematically higher along the direction defined by the cluster major axis (or by the position angle of its first ranked galaxy). This suggests a correlation between the orientation of rich clusters and the large-scale galaxy distribution which extends to at least $15h^{-1}$ *Mpc*. Taken together, these various observations indicate the alignment of structure on scales from several tens of kiloparsecs on up to several tens of megaparsecs. These alignments are one aspect of what is sometimes referred to as 'filamentary structure' in the distribution of galaxies. Accounting for these observations presents a formidable challenge to any theory of the formation of structure in the universe.

The cosmogonic scenarios which we address in this work all assume pure gravitational clustering from small-amplitude, Gaussian density fluctuations. Depending on the exact form of the initial spectrum of fluctuations, the nature of the dark matter and the cosmological model, structure may have formed either by clustering bottom-up from small to large scales, or via fragmentation top-down from large to small scales. In CDM, for instance, the sequence proceeds bottom-up, with nearly simultaneous collapse of systems on galactic scales (Peebles 1982, Blumental and Primack 1983). In the case of hot dark matter, like neutrinos, small-scale fluctuations would have been erased and thus large-scale perturbations would have been the first to collapse, resulting in the formation of flattened superclusters ('pancakes'), followed by subsequent fragmentation to galaxies and clusters (Zeldovich 1970; Doroshkevich, Shandarin, and Saar 1978). Hybrids of these two extreme senarios are also possible, as might result from the presence of different types of perturbations or different types of dark matter (e.g. Dekel 1983, 1984a, 1984b; Dekel and Aarseth 1984), or if the universe underwent more than one inflationary phase (Silk and Turner 1987, Turner *et al.* 1987).

Qualitative arguments suggest that elongated and aligned clusters could arise in various cosmogonies. The flattening cannot be accounted for by rotation (e.g. Rood *et al.* 1972, Dressler 1980, Noonan 1980); anisotropic velocity dispersion is required. In general, there is high probability for any initial perturbation to be aspherical (Doroshkevich 1970, Bardeen *et al.* 1986). Any initial asphericity is amplified during gravitational collapse (Lin, Mestel and Shu 1965) and some of it is preserved later on (e.g. Aarseth and Binney 1978). It would seem natural to expect flattened clusters which tend to be correlated with one another in the pancake scenario, where large-scale collapse to sheets and filaments is expected to occur first. Hierarchical scenarios are expected in general to show weaker large-scale alignments, with decreasing strength for steeper spectra (larger power index n). The CDM spectrum, for example, is expected to yield certain filaments on galactic scales, where the spectrum is flat ($n \simeq -2$) and objects of different scales collapse almost simultaneously, but not on scales of $\sim 10\ h^{-1}$ *Mpc* and up, where the spectrum is steeper ($n \simeq 0$) (Nusser and Dekel 1988). Nevertheless, the situation might be more complex. For instance, elongation and alignment might arise in hierarchical clustering as a result of tidal interactions between neighboring protoclusters (Binney and Silk 1979, Palmer 1981, 1983). Even though this has been found by N-body simulations (Dekel, West, and Aarseth 1984, Dekel 1984b) not to be enough to explain the Binggeli alignment, it might still produce some alignments on smaller scales. Also, simulations of cluster formation in hierarchical clustering (e.g. White 1976, Cavaliere *et al.* 1986) have shown that subclustering during the cluster collapse is also capable of producing an overall flattening of the galaxy distribution which persists long after the collapse. These effects on the resultant alignments have to be quantified.

The simulations used in this study are described brifly in Sec. 2.2. In Sec. 2.3 the alignment is measured in the simulated clusters of the different scenarios and compared with the observations of AGPS. Our conclusions are in Sec. 2.4.

2.2 - The Simulations

In order to generate high-resolution cluster simulations beginning from realistic cosmological initial conditions, a novel approach for stretching the dynamical range was used, in which the clusters were simulated in two separate stages. In step I, low-resolution, large-scale cosmological simulations were performed in order to find the locations where protoclusters formed for a given random realization of the initial conditions. These results were then used to generate the initial conditions for step II, in which high-resolution simulations of individual clusters and their surroundings were performed. Such an approach makes it possible to study the detailed properties of rich clusters with sufficient resolution to detect systematic differences which may arise due to the different initial conditions.

The desired initial fluctuation spectra for the different cosmologies were generated using a method based on the approximation of Zeldovich (1970) for describing the linear evolution of fluctuations. The spectra considered here have the general power-law form

$$\langle |\delta_{\vec{k}}|^2 \rangle \propto k^n \qquad (2.1)$$

over certain intervals of wave number k. Simulations were performed for the following five initial spectra: a) a pancake scenario with $n = 0$ on large scales and no fluctuations below a critical coherence wavelength, b) a hybrid scenario where a large-scale component ($n = 0$) with a coherence length as in the pancake scenario is combined with a small-scale component ($n = 0$) whose amplitude was one-half that of the large-scale component at the coherence length, and c), d), and e) three hierarchical clustering scenarios with power spectrum indices $n = 0$, -1, and -2 respectively. The CDM spectrum can be approximated near the relevant scales for rich clusters by either the $n = 0$ or $n = -1$ models, while $n = -2$ is more appropriate for CDM on the scale of galaxies. All of these simulations assumed an Einstein-de Sitter universe

(Ω = 1). (The n = 0 simulations were also repeated for the case of an open universe, Ω_0 = 0.15, but the clusters there ended up relatively isolated so that there were not enough particles in the regions surrounding the clusters to allow meaningful alignment analysis.) The large-scale simulations of step I were performed with ~ 4000 equal-mass particles using a comoving version of a direct N-body code (Aarseth 1985). Four random realizations were performed for each of the different theoretical scenarios. The stages of the simulations that correspond to the present epoch were determined by matching the slope of the two-point correlation function, $\xi(r)$, with the slope γ = 1.8 of the observed galaxy correlation function, ignoring biasing. Equating the correlation length of the simulations with the claimed value for galaxies, $r_0 \simeq 5h^{-1}$ Mpc (Davis and Peebles 1983; although see Oemler et al. 1988), then sets the scaling from simulation to physical units. With this scaling, the diameter of the simulated volumes corresponds to ~ $100h^{-1}$ Mpc, and the coherence length in the pancake and hybrid scenarios results in superclusters of roughly $30h^{-1}$ Mpc in diameter. Rich clusters were then identified in these large-scale simulations using a simple group-finding algorithm described in Paper I. The five richest clusters found by this procedure in each simulation, with a mean number density of $\simeq 6 \times 10^{-6} (h^{-1}$ Mpc$)^{-3}$, were assumed to correspond to Abell clusters of richness $R \geq 1$.

Having identified the rich clusters in each simulation of step I, new simulations were then performed using the same initial fluctuations, but now modeling smaller volumes centered on the locations of each of the clusters (a total of 20 clusters per cosmogonic scenario). The radius of these volumes was 45% that of the large-scale simulations. These high-resolution simulations of individual clusters were run using a non-comoving version of the Aarseth code, with ~ 1000 equal-mass particles. With the adopted scaling each particle in these simulations should correspond roughly to an L_* galaxy.

In order to study the simulated clusters in a manner similar to that used by observers, three orthogonal projected views of each cluster have been used. A cubic volume with origin at the center of the simulated volume was cut to ensure equal depth along the line of sight. The cluster center in each case was then taken as the location of the density maximum of the projected distribution of particles within this cube, as determined by an iterative count procedure using square grid cells. Several representative clusters formed in the different cosmological scenarios are shown in Figure 2.1.

2.3 - Statistical Analysis of Alignment

In order to compare the simulations with available observational results, the same test used by AGPS to analyse the observational catalogs has been applied here to the simulated clusters. Each projected view of each simulated cluster was assigned a redshift drawn at random from the cluster redshifts in

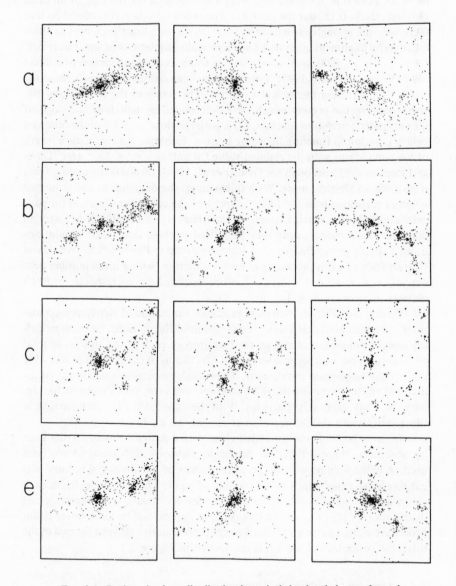

FIG. 2.1. Projected galaxy distribution in typical simulated clusters formed in the different cosmological scenarios. Labels denote: (a) pancake scenario (b) hybrid scenario; (c), and (e) hierarchical clustering scenarios with $n = 0$ and -2 respectively. Each box is 30 h^{-1} *Mpc* on a side. Each particle roughly represents an L_* galaxy. The large-scale fluctuations were chosen from the same set of random numbers so that the clusters in the different scenarios are directly comparable.

the sample of 137 clusters with $z \leq 0.1$ and $b \geq 40$ which comprised the data set used in the study of AGPS (83 of the clusters in their sample were taken from the earlier study of Struble and Peebles 1985). Whereas AGPS determined the major axis of each cluster by eye, from either the galaxy distribution on the palomar Sky Survey prints or from the position angle of the brightest cluster galaxy, the major axis of each of the simulated clusters was found here from the moments of inertia of all particles within a distance corresponding to a projected angular separation $\Theta = 0.25°$ from the cluster center. Then, the mean surface density of particles in different angular and radial bins around each cluster was measured. Following AGPS, four circular rings of radius Θ centered on each cluster were used: $(0.25°, 0.5°)$, $(0.5°, 1°)$, $(1°, 2°)$ and $(2°, 4°)$. [AGPS also used a bin $(4°, 8°)$ which was not possible to use here because of the limited size of our simulations. They detected no clear signal of alignments in that bin anyway.] For comparison, an angle of $\Theta = 4°$ subtends a distance of $\sim 17\ h^{-1}\ Mpc$ at a redshift of $z = 0.1$ (for $\Omega = 1$). The relative position angle ϕ of each particle was then defined as the angle between the radius vector to the particle and the cluster major axis. Next, for each radial bin, Θ, the excess surface number density of particles, $\eta(\phi)$, was determined by first counting particles in bins of $10°$ in ϕ, and then subtracting the mean surface density of particles within the given radial bin. Once this was done for all simulated clusters, the means and standard deviations of $\eta(\phi)$ were computed for all clusters of a given cosmological scenario. The results for some of our scenarios are presented in Fig. 2.2. For comparison the results of AGPS are reproduced in Fig. 2.3.

For the pancake scenario (Fig. 2.2a), the particle counts show a clear tendency to be systematically higher along the direction defined by the cluster major axis; this effect extending to $\Theta = 4°$ and perhaps more. Moderate alignment is detected in the hybrid scenario and in the case $n = -2$. No alignment is detected for $n = 0$ or $n = 1$.

To provide a more quantitative measure AGPS fit the $\eta(\phi)$ data for each Θ bin to a function of the form $a\ cos(2\phi)$, by defining for each cluster, i,

$$a_i(\Theta) = \frac{2}{9} \sum_{k=1}^{9} \eta_i(\Theta, \phi_k) cos(2\phi_k), \qquad (2.2)$$

where $\phi_k = \pi(2k - 1)/36$ radians. Then, the mean value for all clusters, $a(\Theta)$, was computed, along with its standard deviation $\sigma(\Theta)$. As a measure of the significance of the alignment, a χ^2 is defined by

$$\chi^2 = \sum_{j=1}^{m} \left(\frac{a(\Theta_j)}{\sigma(\Theta_j)} \right)^2, \qquad (2.3)$$

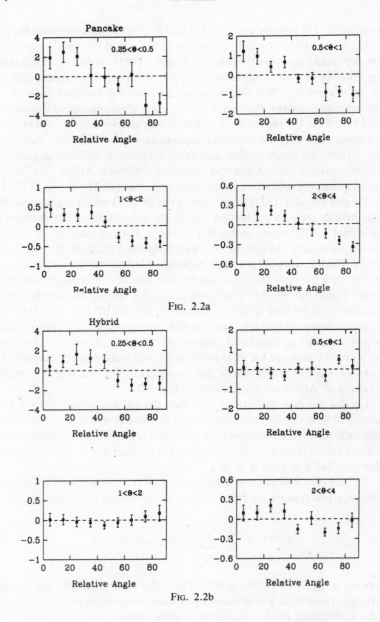

FIG. 2.2a

FIG. 2.2b

FIG. 2.2. Alignment in the simulated clusters of the different scenarios. Shown is the surface density of galaxies, η, as a function of the azimuthal angle ϕ as measured from the cluster major axis ($\phi = 0$), for different radial bins (Θ): (a) pancake scenario; (c), (d) and (e) hierarchical scenarios with $n = 0$, -1, and -2 respectively.

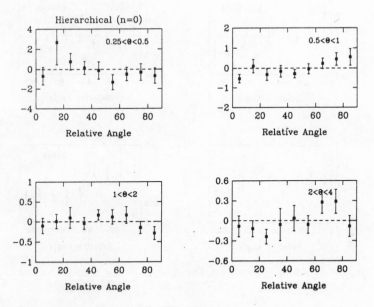

FIG. 2.2c

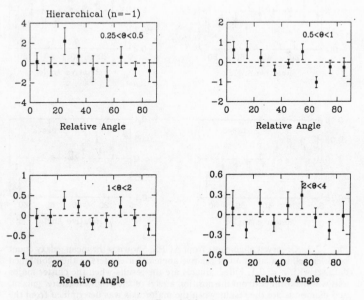

FIG. 2.2d

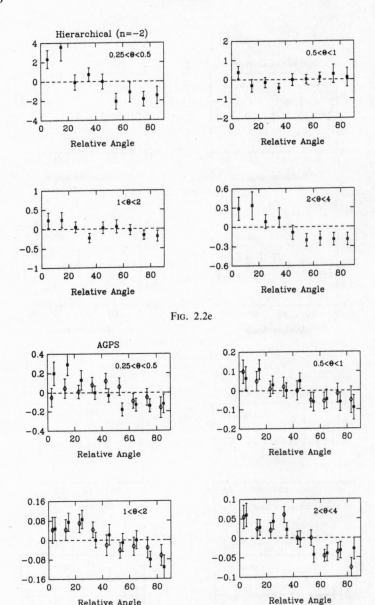

FIG. 2.2e

FIG. 2.3. Observed alignments from AGPS. Shown is the mean galaxy count per Lick cell as a function of relative angle ϕ for clusters in their combined data set with $z < 0.1$. Filled squares are the results when the cluster major axis was determined from the position angles of the brightest cluster galaxy, and diamonds are the results when the major axis was determined from the galaxy distribution within the inner regions of the cluster.

where m is the number of Θ bins. From their sample of 137 clusters (the combined data sets 1 + 2 in their paper), AGPS find $\chi^2 = 21.3$ when the cluster major axes were determined from their brightest galaxies, and $\chi^2 = 13.9$ when the galaxy distribution in the inner regions of the cluster was used. LGP find $\chi^2 = 11.5$ and 17.1, respectively. This statistic should approximate the usual χ^2 distribution with m degrees of freedom (in this case $m = 5$), i.e. $\chi^2 \simeq 5$ for no significant alignment. Because the probability of having measured the above values of χ^2 in the absence of any true alignments is exceedingly small, this is a strong signal of large-scale alignment of clusters with the surrounding galaxy distribution.

This same analysis has been applied here to the results of the simulated clusters shown in Fig. 2.2. Values of χ^2 found for each of the different scenarios are listed in Table 1, along with the corresponding probabilities of having measured such a value of χ^2 in the absence of any true alignments (here $m = 4$ instead of 5). The *only* scenarios which yield results comparable or more significant than those observed are the panacake, hybrid, and $n = -2$ scenarios.

TABLE 1
χ^2 VALUES AND CORRESPONDING PROBABILITIES

Scenario	χ^2	$P(\chi^2)$
pancake	43.77	< 0.001
hybrid	11.47	0.002
hierarchical n = 0	6.44	0.184
hierarchical n = −1	5.03	0.294
hierarchical n = −2	18.44	<0.001
AGPS (observed north)		
brightest galaxy	21.3	0.0008
clusters core	13.9	0.02
LGP (observed south)		
z ≤ 0.055	11.5	0.042
z ≤ 0.075	17.1	0.0042

2.4 - Conclusion

The orientations of clusters with respect to the galaxy distribution on larger scales were found to provide a very interesting test. A very pronounced tendency for alignments is found in the pancake scenario, on scales up to at least $10h^{-1}$ *Mpc*. Alignments were seen in the $n = -2$ case and in the hybrid case. No detectable alignments were found for $n = 0$ or $n = -1$ hierarchical clustering scenarios, which approximate CDM on cluster scales. (The ellipticities of the main bodies of the simulated clusters, on the other hand, were found to be independent of the initial conditions and therefore not very useful for cosmology.)

That the pancake scenario leads to alignments between clusters and the large-scale galaxy distribution is not too surprising, since the coherence length in the initial fluctuation spectrum results in the formation of large-scale filaments and sheets with which the clusters are associated. Likewise, it would seem that the lack of such alignments in the $n = 0$ and $n = -1$ hierarchical clustering scenarios reflects the dominance of small-scale fluctuations which results in a different sequence of formation in which clusters collapse and virialize before supercluster formation, and are not much correlated with them. These results are in agreement with the earlier results of Dekel, West, and Aarseth (1984) concerning cluster-cluster alignment, and indicate that tidal forces do not have as significant an effect on the final shapes and orientations of clusters as naive estimates might suggest. That a positive, though weaker, signal of alignment is detected in the $n = -2$ case and in the hybrid model reflects the important contribution of large-scale fluctuations there, which both induce tidal effects and make the collapses of clusters and superclusters follow each other in quicker succession so that there is a cross-talk between them.

The observational evidence for large-scale alignments of the galaxy distribution around Abell clusters found by AGPS and LGP is found to be consistent with that predicted by the pancake scenarios or the $n \leq -2$ hierarchical models. The hierarchical models with $n = 0$ and $n = -1$, which approximate CDM on the relevant scales, reproduce no detectable alignments, in clear conflict with the observations. [The fact that analytic estimates based on Gaussian fluctuations (J.R. Bond, private communication) do seem to predict some sort of alignments in CDM might be explained by the different definition used for 'alignment' which is not directly comparable to the observed property.]

I would like to argue that this alignment is a measure of one aspect of the general concept sometimes referred to as 'filamentary' structure. Our results indicate that standard CDM does have difficulties in reproducing the filamentary distribution of galaxies on scales beyond $10h^{-1}$ Mpc (see also Dekel 1984b; Dekel, West and Aarseth 1984). It warns us that the large-scale filaments seen in some of the N-body simulations of CDM (White et al. 1987) might be artifacts of the numerical procedure, as has been argued recently by P.J.E. Peebles (private communication).

More work is needed to determine whether or not similar alignments could also be produced by models which are not based solely on gravitational instability; for example, we are in a process of studying alignments in the explosion scenario.

3. Contamination of the Cluster Correlation Function: is the Superclustering Problem Real?

3.1 - Introduction

Perhaps the strongest clue for the presence of very large-scale structure in the universe is provided by the spatial distribution of rich clusters of galaxies. The two-point correlation function of Abell clusters (Abell 1958), $\xi_{cc}(r)$, is about twenty times stronger than the galaxy-galaxy correlation function, and it remains positive out to about 50 h^{-1} *Mpc* (see Bahcall 1988a for a review). These strong correlations are in clear conflict with the traditional cosmogonic scenario which assume Gaussian initial density fluctuations in an Einstein-deSitter ($\Omega = 1$) universe (e.g. Barnes *et al.* 1985). In particular, the 'standard' cold dark matter scenario (CDM), when normalized to fit either the distribution of galaxies or the isotropy of the microwave background, fails to reproduce the observed ξ_{cc} (Bardeen, Bond, and Efstathiou 1987; Batuski, Melott, and Burns 1987; White *et al.* 1987). CDM predicts that ξ_{cc} should become negative beyond $\sim 20\ h^{-2}$ *Mpc*, because if clusters form at high density peaks then $\xi_{cc}(r) \propto \xi(r)$ at large separations (Kaiser 1984), and the matter two-point correlation function, $\xi(r)$, is predicted to go negative at $\sim 20\ (\Omega h^2)^{-1}$ *Mpc* in a CDM universe.

The important theoretical implications of the apparently strong cluster correlations, which I will partly address in the next section, motivate first a careful evaluation of the reality of the observed result. The limited number of clusters in the redshift surveys leads to quite large uncertainties in ξ_{cc} (e.g. Ling, Frenk, and Barrow 1986). But the most serious questions concerning the validity of the measured ξ_{cc} as an indicator for real large-scale clustering arise because of possible systematic selection effects. Sutherland (1988), whose work has been discussed in this meeting by Nick Kaiser, has raised the possibility that the high amplitude of ξ_{cc} is mostly due to the effect of contamination intrinsic to the Abell catalog (or similar cluster catalogs). The effect is over-counting pairs of small angular separations as a result of mutual contamination of the galaxy counts in the one cluster by galaxies of the other. The general contamination in the Abell catalog by foreground and background galaxies is a well-known problem; it has previously been estimated, for example, by Lucey (1983) using Monte Carlo realizations. This effect, Sutherland argues, is responsible for the anisotropy detected by Bahcall, Soneira and Burgett (1986) in the distribution of cluster pair separations in redshift space. When they plot separation in redshift against angular separation (i.e. projected on the sky), they find a strong excess of pairs with small angular separations, mostly below 10 h^{-1} *Mpc*. While these authors blame the effect on cluster peculiar velocities on the order of 2000 km s^{-1}, namely the so called 'fingers of God', Sutherland argues that the anisotropy extends to too high redshift separations, and is a natural result of the projection effect as estimated by him.

We are investigating the contamination effect on ξ_{cc} of the northern Abell sample. I describe below preliminary results which ignore the effect of cluster correlation functions of order ≥ 2.

First, a brief description of the Abell procedure in classifying the clusters in his catalog. For any cluster which he detected on the Palomar Sky Survey plates, he estimated the distance based on the magnitude, m_{10}, of the 10th brightest galaxy, which is assumed to be a standard candle. Using this distance he counted galaxies within a circle of a radius which corresponds to $r_A = 1.5\ h^{-1}\ Mpc$ about the cluster center. Galaxies are included only if their magnitude falls in the band $(m_3, m_3 + 2)$, where m_3 is the magnitude of the third brightest galaxy. The background count in a reference field on each plate, chosen to be away from any cluster, has been subtracted out. The clusters were classified into richness classes $R = 0,1, \ldots 5$ if their net count fell in the range 30-49, 50-79, $\ldots$ 300 - respectively. The clusters were also assigned a distance class $D = 1,2, \ldots 6$ according to m_{10}. The outer boundary of $D = 4$ is at about 300 $h^{-1}\ Mpc$ $(z \simeq 0.1)$ and the $D = 6$ clusters extend to $\simeq 600\ h^{-1}\ Mpc$. The richness classification is assumed to be independent of distance out to $D = 6$, and the catalog is assumed to be complete for $R \geq 1$. There are 102 clusters in the $R \geq 1$, $D \leq 4$ sample, which, together with the cluster redshifts, is the main source for the current direct estimates of ξ_{cc}.

3.2 - Simple Empirical Estimate

To obtain the most simple estimate we compare the correlation function of the 102 clusters in the redshift sample as is, with that obtained eliminating the pairs with small projected separations — those which are blamed for the correlation excess.

We first calculate $\xi_{cc}(r)$ as in Bahcall and Soneira (1983, hereafter BS). This is shown in Figure 3.1. It can be approximated by

$$\xi_{cc}(r) = (r/r_o)^{-\gamma},\ r_o = 25\ h^{-1}\ Mpc,\ \gamma = 1.8 \qquad (3.1)$$

(also Klypin and Kopylov 1983). Then, we eliminate from the pair counts, both in the Abell sample and in the appropriate Poissonian sample which we use as a reference, all pairs that are of projected separation less than 10 $h^{-1}\ Mpc$. The resultant correlation function is also shown in Fig. 3.1. The difference between the two correlation functions is small; less than a factor of 25% in amplitude on any scale.

This empirical test might suffer from over-simplification, but, nevertheless, it does indicate that the projection effect is quite small, unlike what is expected based on the estimate of Sutherland.

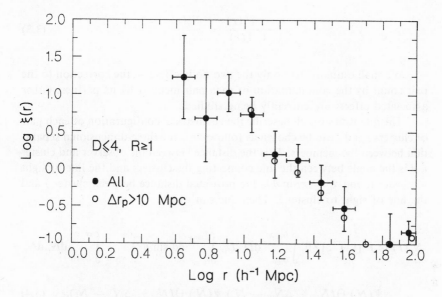

FIG. 3.1. Empirical test: The cluster-cluster correlation function in redshift space of the 102 $R \geq 1$, $D \leq 4$ Abell clusters (error bars) is compared with the correlation function of the same sample from which all pairs of angular separations less than 10 h^{-1} *Mpc* were excluded.

3.3 - Statistical Estimate of the Two-Body Effect

We now wish to estimate the expected effect of contamination on ξ_{cc}, for assumed number density profiles of galaxies in extended halos around the clusters.

Write the desired cluster two-point correlation function of the "pure" clusters using the integral definition

$$1 + \xi(r) = N_p(r)/N_{pp}(r), \tag{3.2}$$

where $N_p(r)$ is the number of cluster pairs with separation in the interval $(r, r + dr)$, and $N_{pp}(r)$ is the expected number of pairs in an equivalent Poisson distribution. Denote the analogous quantities as derived from the contaminated catalog by $\xi'(r)$, $N_p'(r)$ and $N_{pp}'(r)$ respectively. We wish to calculate $[1 + \xi'(r)]/[1 + \xi(r)]$.

The Poisson pair counts scale with the total number of pairs within the volume. Most pairs are of large angular separation, on the order of the volume size R, which are not expected to be affected by contamination. Thus, the ratio $N_{pp}'(r)/N_{pp}(r)$ is much closer to unity than the ratio $N_p'(r)/N_p(r)$ for $r \ll R$, and the desired effect is therefore given, to first order, by

$$\frac{1 + \xi'(r)}{1 + \xi(r)} = \frac{N_p'(r)}{N_p(r)} . \tag{3.3}$$

We shall estimate here only the two-body effect — the correction to the pair count by the contamination of each pair member by its partner. Other associated effects are currently being studied.

The quantities which describe the geometrical configuration of each pair of clusters i and j can be chosen as follows: r is the three-dimensional separation between the members, D_i is the distance between the observer and cluster i, θ is the angle between the line connecting the clusters and the line of sight to cluster i, and $x_{ij} = r \sin \theta$ is the projected distance between cluster j and the line of sight to cluster i. Then one can write

$$N_p'(r) \propto r^2 [1 + \xi(r)] \int_{D_{\min}}^{D_{\max}} dD_i \, D_i^2 \int_{\theta_{\min}}^{\theta_{\max}} d\theta \, \sin \theta \int_{N_i=0}^{\infty} \int_{N_j=0}^{\infty} dN_i \, dN_j$$

$$\Psi(N_i) \, \Theta(N_i + \Delta N_{ij} - N_c) \, \Psi(N_j) \, \Theta(N_j + \Delta N_{ji} - N_c). \tag{3.4}$$

The r-dependent term in front of the integrals is the same for $N_p(r)$ and it therefore drops out from the ratio $N_p'(r)/N_p(r)$. The function $\Psi(N)$ is the multiplicity function of Abell clusters such that $\Psi(N)dN$ is the number density of clusters with Abell count of galaxies in the range $(N, N + dN)$. $N_c = 50$ is the critical count which borders between richness classes $R = 0$ and $R = 1$. The contribution to the counts in cluster i by galaxies of cluster j is given by the correction term ΔN_{ij}. $\Theta(N)$ is the usual step function; only pairs in which the "contaminated" richness of each cluster, $N + \Delta N$, is larger than N_c, qualify to be included in the pair count.

The correction term can be written as a product of three terms:

$$\Delta N_{ij}(D_{ij}, x_{ij}, N_j) = D(D_{ij}) \, \sigma(x_{ij}) \, N_j. \tag{3.5}$$

The distance dependence is given as a function of D_{ij}, the ratio of the distance to cluster j projected on the line of sight to i and D_i. We assume a universal shape for the projected density profile around each cluster, $\sigma(x)$, which is weighted by the Abell number count N.

The distance dependence of the correction term is calculated from

$$D(D_{ij}) = D_{ij}^2 \, \pi r_A^2 \int_{L_2 \, D_{ij}^2}^{L_3 \, D_{ij}^2} dL \, \Phi(L)/\bar{n}, \tag{3.6}$$

where $\Phi(L)$ is the galaxy luminosity function and L_3 and L_2 are the intrinsic luminosities corresponding to the magnitudes m_3 and $m_3 + 2$ which define the Abell band. The varying lower limit of the integral takes into account the fact that some foreground galaxies, which are intrinsically fainter than L_2, are included in the Abell counts as if they were brighter than L_2, because their distance is overestimated when they are assumed to belong to the cluster under counting.

The projected number density profile of each cluster, assuming spherical symmetry, is related to the three-dimensional number density profile, $n(r)$, via

$$\sigma(x)N = \int_x^{R_{max}} dr\, n(r)\, r\, (r^2 - x^2)^{-1/2}. \tag{3.7}$$

For the profile $n(r)$ we use a most crucial observational input: the cluster-galaxy correlation function, $\xi_{cg}(r)$, of Abell clusters of richness $R \geq 1$ and galaxies from the Lick counts. A possible realization of ξ_{cg} is an ensemble of randomly distributed clusters of richness $R \geq 1$, with density profiles weighted by their richness,

$$n(N, r) = \bar{n}\, \xi_{cg}(r)\, N/\langle N \rangle, \tag{3.8}$$

out to a large halo radius, R_{max}. Here $\langle N \rangle$ is the mean richness

$$\langle N \rangle = \int_{N_c}^{\infty} dN\, \Psi(N)\, N \Big/ \int_{N_c}^{\infty} dN\, \Psi(N), \tag{3.9}$$

and $\bar{n}$ is the mean number density of galaxies with luminosities in the Abell band,

$$\bar{n} = \int_{L_2}^{L_3} dL\, \Phi(L). \tag{3.10}$$

Thus, we need three observational inputs: $\Phi(L)$, $\Psi(N)$, and $\xi_{cg}(r)$ with R_{max}. For the galaxies we adopt a Schechter Luminosity function,

$$\Phi(L) = \frac{\Phi_*}{L_*} \left(\frac{L}{L_*} \right)^{-\alpha} e^{-L/L_*}. \tag{3.11}$$

Following the Abell definition of the luminosity band one obtains for $\alpha = 1.3$

$$L_2 = 0.28\, L_*, \quad L_3 = 1.75\, L_*. \tag{3.12}$$

Then, in Eq. (3.10), $\bar{n} = 1.07\Phi_*$. We adopt the Schechter parameters

$$\alpha = 1.3, \quad \Phi_* = 0.01\ (h^{-1}\ Mpc)^{-3}. \tag{3.13}$$

A more uncertain input is the Abell cluster multiplicity function $\Psi(N)$ in the vicinity of $N = N_c$. The steepness of this function just below N_c determines how many clusters have real richnesses just below N_c, that can easily be upgraded by contamination into richness class $R = 1$. Figure 3.2 shows the distribution of cluster richnesses in the Abell catalog, binned in intervals of $\Delta N = 5$. There is an obvious bend near N_c which may reflect the onset of the catalog incompleteness below N_c. We use different fits to $\Psi(N)$ of the $D \leq 4$ sample, using power-laws of the form

$$\Psi(N) \propto (N/N_c)^{-\psi}. \tag{3.14}$$

A power-law with $\psi = 3.5$ is a reasonable fit in the range $40 \leq N \leq 60$ about N_c. It serves as an upper limit on smaller scales, where the distribution of groups in the catalog of Gott and Turner (1977) can be fitted by $\psi = 2$ (see the next section). A power of $\psi = 5$ provides an upper limit on the possible power below N_c. We consider only clusters of $N \geq 30$ (i.e. $R \geq 0$).

For the cluster-galaxy correlation function we test two alternative observational results. The old result of Seldner and Peebles (1977, hereafter SP)

$$\xi_{cg}(r) = (r/7h^{-1}\ Mpc)^{-2.5} + (r/12h^{-1}\ Mpc)^{-1.7}, \tag{3.15}$$

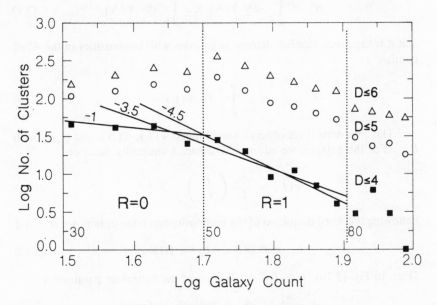

FIG. 3.2. Multiplicity function of Abell clusters in bins of 5 galaxies.

and the recent result of Lilje and Efstathiou (1988, hereafter LE)

$$\xi_{cg}(r) = (r/8.8h^{-1} Mpc)^{-2.2}. \qquad (3.16)$$

The ξ_{cg} of SP has strong power on large scales, indicating extended halos of galaxies around clusters. This ξ_{cg} is in conflict with 'standard' CDM. The LE result, which is different mostly because of a difference in the adopted luminosity function for the Lick galaxy counts, indicates more limited cluster halos, in better agreement with CDM. In each case we tried two alternative maximum radii for the halos around the clusters, $R_{max} = 25\ h^{-1} Mpc$ and $50\ h^{-1} Mpc$. Recall that the mean separation between $R \geq 1$ neighboring Abell clusters is $55\ h^{-1} Mpc$, so the assumed values for R_{max} should lead to an overestimate of the contamination effect.

The correction to the observed $\xi_{cc}(r)$ of the contaminated catalog (BS) is calculated by evaluating the multiple integral (3.4) under the assumed $\xi_{cg}(r)$, $\Psi(N)$ and $\Phi(L)$. The resultant 'pure' $\xi_{cc}(r)$ are shown in Figure 3.3 for the various choices of parameters.

If the ξ_{cg} of LE is adopted, the contamination hardly affects ξ_{cc}. Even with the maximal cluster multiplicity function ($\psi = 5$) and halo extent ($R_{max} = 50\ h^{-1} Mpc$), the effect on ξ_{cc} at $r = 25\ h^{-1} Mpc$ is weaker than the 1σ upper limit of 25% - effect indicated by our empirical test of the previous section. If we adopt the ξ_{cg} of SP instead, the effect is somewhat stronger. For the reasonable choice of parameters ($\psi \leq 3.5$ below N_c and $R_{max} = 25\ h^{-1} Mpc$) the correction is still on the order of 25%. The effect becomes large only when the extreme parameters are used; for $\psi = 5$ and $R_{max} = 50\ h^{-1} Mpc$, the correction factor in ξ_{cc} is of order 4 at $25\ h^{-1} Mpc$. This latter result, which we view as an overstimate of the two-body effect, is comparable to the estimate claimed by Sutherland.

3.4 - Conclusion

The contamination of the Abell catalog by foreground and background galaxies systematically enhances the cluster-cluster correlation by preferentially introducing cluster pairs of small angular separations. It is very important to estimate this effect because of the strong implications of the high ξ_{cc} on the conventional theories of structure formation, and in particular on standard CDM.

Our empirical test, where we simply eliminated pairs of projected separation less than $10\ h^{-1} Mpc$, showed a weak effect; $\xi_{cc}/\xi_{cc}' \simeq 0.75$ at $25\ h^{-1} Mpc$.

The analytic estimate of the effect depends mostly on the assumed cluster-galaxy correlation function. The recent ξ_{cg} estimate of Lilje and Efstathiou (1988) yields $\xi_{cc}/\xi_{cc}' \simeq 0.75$ in agreement with the empirical test. The old ξ_{cg}

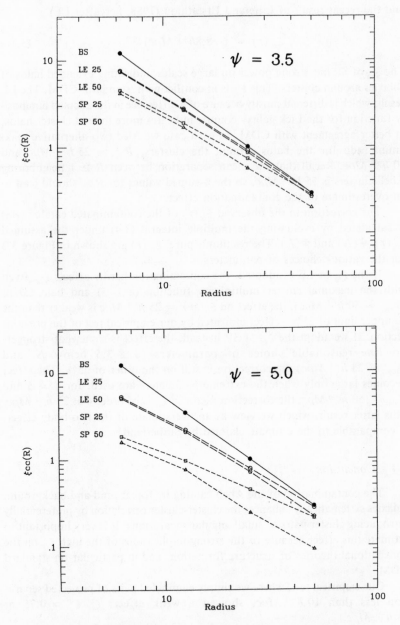

FIG. 3.3. Corrected cluster-cluster correlation function for the cluster-galaxy correlation function of Lilje and Efstathiou (1988, LE) or of Seldner and Peebles (1977, SP), with R_{max} = 25 or 50 h^{-1} Mpc. The assumed cluster multiplicity function is a power-law with ψ = 3.5 (top) and ψ = 5 (bottom).

estimate of Seldner and Peebles (1977) can give rise to corrections of $\xi_{cc}/\xi_{cc}' \lesssim 0.5$, but only if the cluster halos are assumed to extend to $50 \ h^{-1} \ Mpc$ (!) and the cluster multiplicity function is assumed to be steep ($\psi = 5$) for $R = 0$ clusters. The former is an over-estimate because it assumes that the clusters extend beyond the half mean separation between them. The latter means that clusters are missing from the catalog in a rapidly growing rate as N decreases, starting just below $N_c = 50$.

If the result of the empirical test for ξ_{cc} is to be taken as is, the extreme result based on the ξ_{cg} of SP, with very extended halos and a very steep cluster multiplicity function, can be excluded based on the analytic estimate.

The theoretical implication is that ξ_{cc} and ξ_{cg} cannot simultaneously be compatible with the predictions of CDM (and similar traditional scenarios). If one of these functions is low enough in amplitude, the other must be too high. The high ξ_{cg} of SP is by itself incompatible with CDM (independently of whether the associated ξ_{cc} may or may not be compatible with CDM). The lower ξ_{cg} of LE is in better agreement with CDM, but the corrected ξ_{cc} is still in conflict with the theory. However, projection effects from three-point correlations might be stronger.

4. SUPERCLUSTERING IN THE EXPLOSION SCENARIO: ONE POSSIBLE SOLUTION?

4.1 - Introduction

Since all the cluster samples studied so far yield a similar correlation function, and there is no convincing evidence that it has been severely overestimated, it would be worth while to adopt the observed excess of ξ_{cc} as a working hypothesis and seek a theoretical explanation.

Within the framework of Gaussian fluctuations, the attemps to get more power on large scales have baryons playing an important role in an open universe (e.g. Dekel 1984a) or in a universe with a large cosmological constant. The hybrid scenario of baryons and CDM (Bardeen, Bond, and Efstathiou 1987; Blumenthal, Dekel, and Primack 1988) and the baryon isocurvature scenario (Peebles 1987), both have substantial power on scales of $50 \ h^{-1} \ Mpc$. However, in addition to giving up the simplicity of the Einstein-deSitter cosmology, these scenarios may run into problems with the observed isotropy of the microwave background, so they require either a finely tuned cosmological constant or ad hoc reionization at $z \sim 100$.

The difficulties of the Gaussian models on large scales motivate a serious consideration of non-Gaussian scenarios. One possibility is the cosmic string model, where clusters form by accretion onto string loops, which have non-Gaussian correlations because they are chopped from the same parent loop (Turok 1985; Primack, Blumenthal, and Dekel 1986; Scherrer 1987). This picture has some promising features, but detailed predictions await more accurate studies of string-loop fragmentation and loop velocities, since the current results from numerical simulations are in conflict with one another (compare, for example, Albrecht and Turok 1985 with Bennet and Bouchet 1988).

Here I will focus on the explosion scenario. I will describe below the surprising result that a *random* distribution of shells could produce the observed clustering of clusters which form at the vertices where three shells intersect. This is a part of a study of the formation of large-scale structure in generic explosion models, which is being carried out by David Weinberg, Jeremiah P. Ostriker and myself [Weinberg, Dekel and Ostriker 1988 (Paper I); Weinberg, Ostriker and Dekel 1988 (Paper II)]. I wish to refer to Bahcall (1988b) and to Geller (M. Geller, private comm.), who have referred in general to a similar model of cluster formation primarily on the basis of observational considerations.

In this picture, positive energy perturbations drive material away from "seeds", sweeping gas into dense, expanding shells that cool and fragment into galaxies (Ostriker and Cowie 1981; Ikeuchi 1981). Explosions of supermassive stars could act as seeds (also Carr and Ikeuchi 1985). Alternatively, the long-wavelength radiation produced by superconducting cosmic strings could sweep plasma into shells which could be a few tens of Mpc's in size (Ostriker, Thompson, and Witten 1987; 1988). In fact, the scenario does not require explosive energy at all — negative density fluctuations grow by gravity alone into expanding voids that have structure similar to other cosmological blast waves (Bertschinger 1983, 1985; Ostriker and McKee 1988).

The explosion scenario naturally accounts for the "bubbles" in the galaxy distribution (e.g. de Lapparent, Geller, and Huchra 1986). Galaxy redshift surveys suggest that shells of radius 10 to 30 h^{-1} *Mpc* with a filling factor of order unity may dominate the structure, such that the dynamical interactions of shells must play an important role in the development of clusters, superclusters, and voids. Saarinen, Dekel, and Carr (1987) have simulated clustering in a universe of interacting shells, and found that the explosion scenario may be able to account for the observed galaxy distribution. In Paper I, we extend this study to a detailed investigation of shell interactions, focusing on two- and three-shell collisions. The work of Paper II, some of which I will describe here, grew out of this dynamical investigation and it explores one of the explosion scenario's most striking predictions — a distribution of clusters with strong signatures of high-order superclustering on scales of ~ 50 h^{-1} *Mpc*.

The explosion seeds produce shells that expand, enclosing an ever-increasing fraction of space, and eventually overlap. Gravitational instabilities grow slowly on isolated shells (White and Ostriker 1988), so shell interactions are essential to the formation of massive galaxy clusters. When two shells cross they interesect in an overdense ring, which accretes matter from the shells;

the two voids "push" matter into the circular wall where the shells overlap and from this wall into the ring (Figure 4.1). Rich clusters then form when a *third* shell intersects this ring in two points — "knots". (When three shells overlap they intersect in two, and only two, points, independent of their relative sizes!) The knots are the sites of deep potential minima that accumulate the surrounding matter (Figure 4.2). The process is somewhat analogous to the formation of structure in the pancake scenario (Zeldovich 1970), where the matter collapses to flat "sheets", then flows towards the lines of intersection, and along those to their points of intersection.

These results suggest a simple, geometrical toy model for the spatial distribution of clusters in the explosion scenario. For a given distribution of

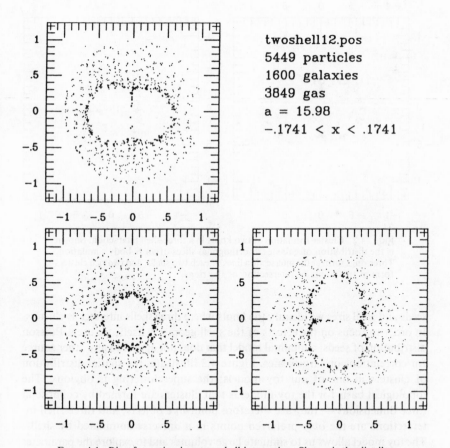

twoshell12.pos
5449 particles
1600 galaxies
3849 gas
a = 15.98
−.1741 < x < .1741

FIG. 4.1. Ring formation in a two-shell interaction. Shown is the distribution of galaxies (and background gas) in orthogonal slices of an N-body simulations. The overlapping wall has been evacutaed into a dense ring.

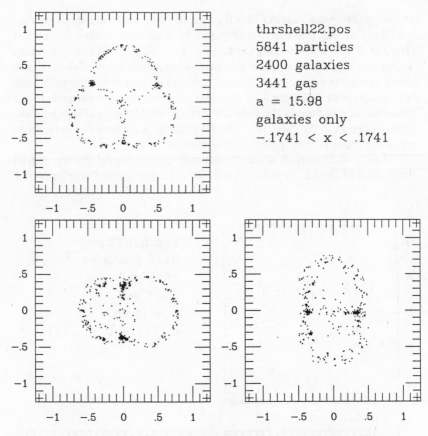

thrshell22.pos
5841 particles
2400 galaxies
3441 gas
a = 15.98
galaxies only
−.1741 < x < .1741

FIG. 4.2. Cluster formation at two knots in a three-shell interaction. Shown is the distribution of galaxies in orthogonal slices of an N-body simulation. Three rings form at the intersection lines of each two shells, and two rich clusters form at the points of intersection of the rings.

seeds and shell sizes we identify the knots where three shells intersect as clusters. In order to focus on the effects of the shell geometry itself, we use a Poisson distribution of seeds, a minimal model that does not appeal to any other source for correlations on large scales. Figure 4.3 illustrates the spatial distribution of clusters in one of our toy models; the superclustering is obvious. The topological basis for the toy model is that clusters are formed by collapse in three dimensions — they are therefore point-like objects, and three-shell intersections are the *only* preferred points in a universe dominated by shells. The toy model allows us to simulate large volumes and to explore the parameter space of the explosion scenario in order to identify the range of models that seems most promising.

cluster projections: power law model

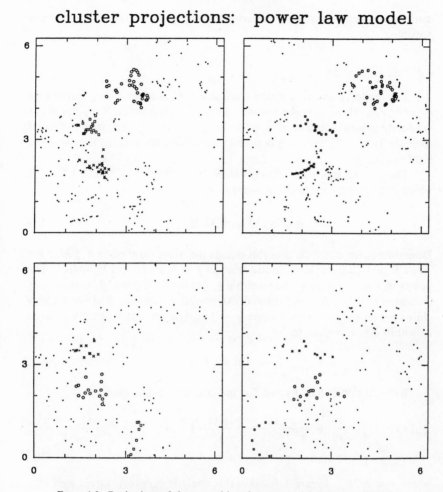

FIG. 4.3. Projections of cluster positions in a toy model. Top: a realization of the power law model. Bottom: the merged version of the same realization. The models have $\beta = -4.5$, $f = 0.3$, and the merger radius is $R_{mer} = R_{max}/5$. Left and right panels are projected along different axes. Hexagons, crosses, and squares indicate members of the three largest superclusters, identified by percolation at an overdensity of 20. Tick marks are in units of R_{max}.

I will discuss here only some highlights of Paper II, which includes a very detailed investigation of several toy models. In Sec. 4.2 I describe the simplest version of equal-size shells, focusing on the number density of clusters and ξ_{cc}. In Sec. 4.3 I address one of our more realistic models that assumes a spectrum of shell sizes, where we also compute the cluster mass function and look for trends of correlation strength with richness. Finally, in Sec. 4.4 I summarize other measures of superclustering in comparison with observations when possible.

4.2 - Equal Size Shells

The model is specified by the shell radius R_{sh} and the filling factor $f \equiv n_s(4\pi R_{sh}^3/3)$, where n_s is the number density of shell seeds. We place shells randomly in a cubical box of side $50R_{sh}/3$, and identify each intersection of three shells as a cluster, using periodic boundaries. We average the results of eight realizations.

The number density of clusters is (Kulsrud 1988) $n_c = (9\pi/16) f^3 R_{sh}^{-3}$, which implies a mean neighbor separation

$$\bar{d} \equiv n_c^{-1/3} \simeq 0.83 \, R_{sh} \, f^{-1}. \tag{4.1}$$

Demanding, for example, that this equals the observed $\bar{d}$ of 55 h^{-1} Mpc for Abell $R \geq 1$ clusters, leads to the relation $f = 0.45(R_{sh}/30 \, h^{-1} \, Mpc)$. The points in Figure 4.4 show the resulting $\xi_{cc}(r)$ in our toy model. It is sharply truncated at $r = 2 \, R_{sh}$, as expected. At smaller separations it is remarkably close to a power law, with a slope $\gamma \sim 1.5$. The correlation length is given to within a few percent by

$$r_0 \simeq 1.6 \, R_{sh} \, f^{-0.85}. \tag{4.2}$$

Kulsrud (1988) has derived for the constant radius model

$$\xi_{cc}(y) = \frac{3}{4f}\frac{1}{y} + \frac{2}{\pi^2 f^2}\frac{(1-y^2)(1+3y^2)}{y^2} + \frac{1}{3\pi^2 f^3}\frac{(1-y^2)^2}{y}, \tag{4.3}$$

where $y \equiv r/2R_{sh}$. Figure 4.4 shows this analytic ξ_{cc} which agrees well with the numerical results. The probability of finding a cluster in a randomly chosen volume is $\propto f^3$, because it requires three shells. Given a cluster, a formation of a cluster elsewhere on one of the parent shells requires two additional shells, so it contributes a term $\propto f^2/f^3$. This term is $\propto r^{-2}$ at $r < R_{sh}$ because of the two-dimensional nature of the parent shell where the other cluster forms. Similarly, formation of a cluster on one of the three rings requires only one additional shell and it therefore contributes a term $\propto f/f^3$, which is $\propto r^{-2}$ at small r. The guaranteed presence of the twin cluster contributes a term $\propto 1/f^3$, which dominates only at very low f. Thus, the three terms in eq. 4.3 come from five-, four- and three-shell configurations respectively.

The two-point correlations have a maximum range of $2R_{sh}$. Higher-order correlations, however, can extend beyond $2R_{sh}$. For example, a connected

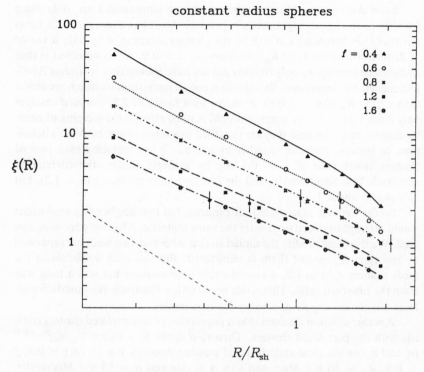

FIG. 4.4. Correlation function of the constant radius model. Points are average results of eight realizations for each of the indicated filling factors f. Curves show the analytic formula of Kulsrud (1988) for the same filling factors. Dashed line at lower left is an $r^{-1.8}$ power law. Small points with vertical bars mark the mean neighbor separations d.

chain of shells can produce clusters with positive N-point correlations out to $2(N-1)\,R_{sh}$.

How does the model compare with observations? The power law form is encouraging, and while the slope $\gamma \simeq 1.5$ is somewhat shallow, it is acceptable given the uncertainties. The correlation amplitude, however, is too high. In all cases the ratio $r_0/\bar{d}$ exceeds the observed value of 0.5 by at least a factor of three, as can be seen in Figure 4.4, where vertical bars mark $\bar{d}$. Take the fact that the observed ξ_{cc} is still a power law (and certainly positive) at r_0, which requires $r_0 < 2R_{sh}$ because correlations vanish beyond the shell diameter. By equation (4.2), this condition implies $f > 0.77$. The lower limit on f, by equations (4.1) and (4.2), forces

$$r_0/\bar{d} = 1.93\,f^{0.15} > 1.85, \qquad (4.4)$$

in *clear conflict* with the observed value of 0.5.

Several effects could reduce the correlation amplitude. First, only some fraction of the knots might actually produce Abell-like clusters. But it turns out that even throwing out 7/8 of the clusters increases $\bar{d}$ by only a factor of 2, so constraining $r_0 < R_{sh}$ still leaves $r_0/\bar{d} > 0.9$. A second effect is that shells which overlapped only recently did not have enough time to accrete Abell-like clusters by the present. By throwing out all intrinsic pairs which are closer than $s_{min} = R_{sh}$ (for $f = 0.6$), n_c drops by a factor of 2.3, but $r_0/\bar{d}$ changes only from 1.7 to 1.6, not nearly enough. A third effect is the merging of nearby clusters, either because they are classified as a single cluster by Abell's definition, or because they fall together by gravity. When replacing each pair of clusters closer than $R_{mer} = 0.2 \ R_{sh}$ by a single cluster, the correlation amplitude falls significantly, and the function becomes flatter ($\gamma \simeq 1.2$), but $r_0/\bar{d}$ is still about 1.3.

So far we have placed seeds at random, but two neighboring explosions would attempt to sweep up mostly the same material, effectively forming one shell. We therefore modify the model so that whenever two seeds are separated by less than R_{sh} one of them is eliminated. But this *anticorrelation* of the seeds reduces $r_0/\bar{d}$ to 1.2, a considerable improvement but still a long way from the observed ratio. This crude prescription illustrates the qualitative effects one would expect in a more realistic model.

Another solution assumes that a population of uncorrelated clusters coexists with the correlated clusters. Then $r_0/\bar{d}$ drops by a factor $(n_c/n_t)^{(6-\gamma)/3\gamma}$ (n_t and n_c are the total and clustered number density). For $(n_c/n_t) = 0.3$, $f = 0.3$, $R_{sh} = 30 \ h^{-1} \ Mpc$, and $\gamma = 1.5$, one gets $\bar{d} = 55 \ h^{-1} \ Mpc$ and $r_0 = 27 \ h^{-1} \ Mpc$, consistent with the observed ξ_{cc}. But note that the random clusters must outnumber the correlated clusters, which would require the *ad hoc* addition of large density fluctuations to the model.

In summary, while the simplest model is attractive in magically producing correlated clusters from uncorrelated shells with approximately the correct functional form, it is clearly incorrect, and no simple modification that we have considered can rescue it. The model fails at low filling factors because ξ_{cc} tends to run into the cutoff at $2R_{sh}$ while its amplitude is still too high, and it fails at filling factors above the limit imposed by the positivity of ξ_{cc} at r_0 because the dimensionless correlation length is too strong.

4.3 - A Range of Shell Sizes

The presence of very large shells allows ξ_{cc} to continue beyond the typical shell diameter, giving hope for improvement. The models require new parameters, but they produce a spectrum of cluster masses that can be tested against additional observational constraints. The surface density of mass swept up by the shell is proportional to its radius, so the cluster mass is roughly proportional to the multiple of the three radii of the interacting shells. Assuming that M/L is constant, we can compare the form of the model mass function to the observed luminosity (or multiplicity) function of clusters, and study the dependence of correlation strength on cluster richness.

I will focus here on a model with a power-law distribution for shell radii, with a sharp cutoff at some maximum. The superconducting cosmic string model predicts such a distribution, with an upper cutoff due to the last generation of shells that can cool and form galaxies before Compton cooling becomes ineffective (Ostriker and Thompson 1987). Something similar might also arise in a model where density fluctuations with a scale-free power spectrum collapse, cool, and convert some fraction of their energy into explosive blast waves.

The probability for a shell to have a radius R is taken to be $P(R) \propto R^{\beta}$ for $R \leq R_{max}$. When $\beta = -4$, shells in equal logarithmic bins of radius occupy an equal volume. The value $\beta = -4.5$ is predicted by the superconducting cosmic model for shells of radius $R \geq 5$ Mpc (Ostriker, Thompson and Witten 1988). In computational realizations we must have a lower cutoff R_{min} as well. The total filling factor, for $\beta \leq -4$, diverges as R_{min} goes to zero, so here f corresponds to shells with $1/2 \, R_{max} < R < R_{max}$ only. We eliminate shells that happen to be placed entirely within larger shells. The model is thus specified by β, f, and R_{max}. The dynamic range is $R_{max}/R_{min} = 8$. Because clusters of mass $M/M_{max} = R_1 R_2 R_3/R_{max}^3 > 1/8$ cannot have parent shells smaller than R_{min}, such a simulation correctly represents the cluster distribution for masses greater than $M_{max}/8$.

4.3.1 - Mass-Luminosity Function

Bahcall (1979) finds that the distribution of cluster richnesses in the Abell catalog (see Fig. 3.2) and the group catalog of Gott and Turner (1977), is well described by the function

$$\Psi_c(L) = \Psi_0(L/L_0)^{-2} \exp(-L/L_0) \qquad (4.5)$$

$$L_0 = 2.5 \cdot 10^{12} \, h^{-2} \, L_{\odot}, \qquad \Psi_0 = 5.3 \cdot 10^{-6} \, (h^{-1} \, Mpc)^{-3} \, L_0^{-1}.$$

Here L represents the total luminosity within the Abell radius, and $\Psi_c(L)dL$ is the number density of clusters in the range $(L, L + dL)$. The minimum luminosity of an $R = 1$ cluster is $1.9 \cdot 10^{12} \, h^{-2} \, L_{\odot}$, about L_0. The complete redshift sample of $D \leq 4$ clusters in the northern Abell catalog (Hoessel, Gunn, and Thuan 1980, HGT) contains 102 clusters of richness $R \geq 1$, implying a mean number density $n_c = 6 \cdot 10^{-6} \, (h^{-1} \, Mpc)^{-3}$. The corresponding mean neighbor separation is $d = 55 \, h^{-1} \, Mpc$.

Figure 4.5 shows the mass function of model clusters. The cutoff in radius at R_{max} induces the exponential decline at high masses. As long as the filling

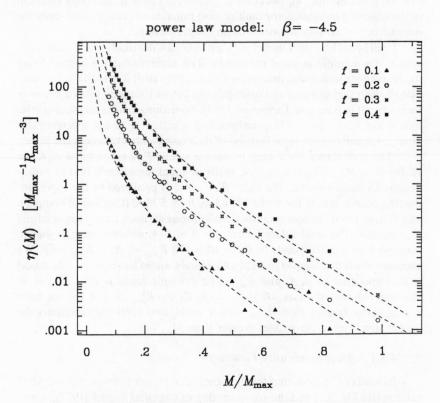

FIG. 4.5. Mass function of the power law model with $\beta = -4.5$, filling factors of 0.1, 0.2, 0.3, and 0.4. Points are average results of eight realizations. Dashed lines are Schechter-like functions fit to the numerical results, with characteristic $M_* = 0.14\, M_{max}$.

factor is low, n scales with f^3 just as in the constant radius case. The dashed lines show the fit of a function $\Psi(M) = f^3 A(M/M_*)^{-2} exp(-M/M_*) M_{max}^{-1} R_{max}^{-3}$. We allowed A and M_* to assume their best fit values, but the curves differ only by the f^3 factor. The best fit parameters for $\beta = -4.5$ are $A = 2066$ and $M_* = 0.14\, M_{max}$. The value of M_* is insensitive to the assumed behavior at small masses because most of the data points lie in the exponential cutoff region. The range $-4.5 < \beta < -3.5$ gives an acceptable mass function within the uncertainties. The mass function near M_* is insensitive to our prescription for eliminating shells because only a few shells of this size are eliminated. Shells with small radii begin to fill space, and the mass function depends on our approximation to the action of sweeping, so it is not appropriate to compare the model with poor groups. However, the power law model naturally accounts for the observation that the cluster $\Psi(L)$ smoothly extends the luminosity function of rich groups, since the transition from three-shell processes to ring and shell fragmentation that give rise to smaller groups is a gradual one.

We can identify objects of mass $M > M_*$ as Abell clusters of richness $R \geq 1$. For the above mass function, $\bar{d}_* = 0.29\, R_{max}\, f^{-1}$ for $\beta = -4.5$. This is a direct generalization of equation (4.1). A similar integration of the cluster luminosity function yields $\bar{d}_0 = 58\, h^{-1}\, Mpc$ for the mean separation of clusters with $L > L_0$. Using the best fit values of A and M_*, the filling factor required to make $\bar{d}_* = \bar{d}_0$ is $f = 0.25\, (R_{max}/50\, h^{-1}\, Mpc)$.

4.3.2 - Correlation Function

Figure 4.6 shows ξ_{cc} for $f = 0.3$, and its richness dependence. The correlation function at small distances is again nearly a power law. Instead of cutting off sharply at $2R_{sh}$, it rolls over gradually towards $2R_{max}$. We determine γ by a least squares fit of $(r/r_0)^{-\gamma}$ in the range $10 > \xi_{cc} > 0.5$. The correlation length r_0 determined in this way typically exceeds by 5 - 10% the radius at which ξ_{cc} itself is unity. The amplitude of ξ_{cc} is higher for richer clusters, in qualitative agreement with the observed trend, but in our simulation $r_0/\bar{d}$ becomes somewhat smaller for more massive clusters. The dependence of ξ_{cc} on f is similar to that in the constant radius model; only for $f \geq 0.3$ does the power law extend relatively unchanged to $\xi_{cc} = 1$. The constraint that γ should be $\sim 1.5 - 2$ imposes $f \geq 0.3$. But the constraint that r_0 should be $\approx 0.5\bar{d}$ is not satisfied for any $f \geq 0.1$. For $f \geq 0.3$ the model r_0 is a factor of two or more too large. For small f, on the other hand, r_0 becomes greater than R_{max}, which is unacceptable because ξ_{cc} is observed to be still positive at $2r_0 \simeq 50\, h^{-1}\, Mpc$.

On balance, the parameter values $\beta = -4.5$ and $f = 0.3$ come closest to meeting the observational constraints on ξ_{cc}. The best fit slope for $M \geq M_*$ clusters is -1.75. ξ_{cc} does fall below the power law for $\xi_{cc} \leq 1$, but the observations are uncertain enough in this regime for the drop to be acceptable. The correlation length is $1.5\, R_{max}$, in some disagreement with the positivity condition $r_0 < R_{max}$. The most serious discrepancy is that $r_0/\bar{d} = 1.5$, about a factor of three higher than observed.

4.3.3 - Modified Power Law Models

Merging clusters considerably improves the agreement with observations (Figure 4.7). With $R_{mer} = R_{max}/5$, $r_0/\bar{d}$ drops to 0.95 and $\gamma \simeq 1.5$. $r_0/\bar{d}$ thus remains about a factor of two too large, and r_0 still slightly exceeds R_{max}. We have tested various extreme prescriptions for assigning mass to the merged clusters and found negligible effects.

Finally, we anticorrelate the seeds as before. One can turn strongly overlapping shells into a single shell that conserves their combined energy ($\propto R^5$),

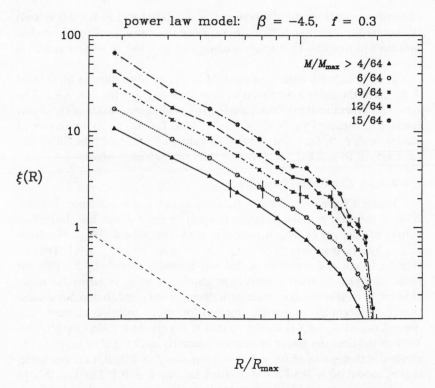

FIG. 4.6. Correlation function of the power law model, with $\beta = -4.5$ and
$f = 0.3$. Different symbols indicate different mass cutoffs — the correlation
function is evaluated for all clusters exceeding the specified mass. Abell clusters
of $R \geq 1$ correspond to $M/M_{max} \geq 0.14$. Vertical bars mark the mean
neighbor separations.

or volume, or surface area. But our simpler prescription of eliminating seeds
that lie inside larger shells differs from more elaborate merger procedures only at the 10 — 20% level. Now $\gamma \simeq 2.2$, $r_0 \simeq 1.4\ R_{sh}$, n_c drops by nearly a
factor of five, so $r_0/\bar{d} = 0.81$ — a great improvement over the standard
model. The mass function retains the same form and characteristic mass, just
reduced by a constant factor. Mergers turn out to have only a minimal impact
on the anticorrelated model; $r_0/\bar{d} = 0.74$. In all three mdified versions of the
power law model, $r_0/\bar{d}$ is nearly independent of mass, in agreement with the
observed trend.

   In summary, although none of the power law models accurately satisfies
all of the observational constraints, they are an improvement over models with
a single shell size. The mass function of models with $\beta \sim -4$ agrees well with
the form of the observed luminosity function, and the values of f and R_{max}

modified power law: $\beta = -4.5$, $f = 0.3$

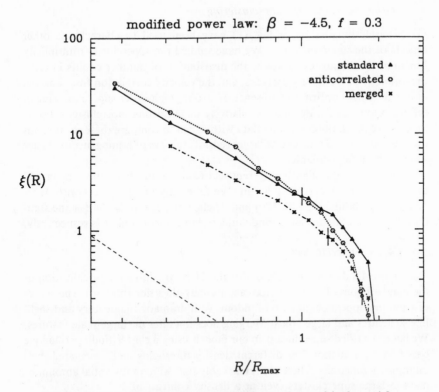

FIG. 4.7. Correlation functions of modified power law models, $\beta = -4.5$, $f = 0.3$ and clusters with $M/M_{max} > 9/64$. Triangles (solid line) show the standard model. Hexagons show the anticorrelated model, where seeds lying inside larger shells are eliminated. Crosses show the merged model, in which cluster groups smaller than $R_{mer} = R_{max}/5$ are merged into single clusters. Vertical bars mark the mean neighbor separations (near the lower right corner for the anticorrelated model).

needed to reproduce the observed number density of clusters are quite plausible. In the constant radius case even the minimal constraint $r_0 < 2R_{sh}$ forces $r_0/\bar{d} \geq 1.85$. For power law models the separation at which correlations vanish, $2R_{max}$, significantly exceeds the diameter of a typical cluster's parent shells, so the effects of the cutoff are less harsh. The most successful model, $\beta = -4.5$, $f = 0.3$, with anticorrelated seeds and mergers of close groups, has $r_0/\bar{d} \approx 0.75$, only about 50% larger than observed.

We note in passing that superconducting cosmic strings might produce elongated ellipsoidal bubbles rather than shells because the strings have substantial peculiar velocities. Since elongated structures create correlations at large distances, changing shells to ellipsoids might reduce the remaining problems.

4.4 - Other Measures of Superclustering

Several statistics beyond $\xi_{cc}(r)$ have been used to characterize other aspects of the superclustering. We have studied the supercluster multiplicity function, the frequency of voids, the distribution of number counts in cells, the topology of isodensity surfaces, and the velocity correlation function. All of our statistics confirm the presence of strong, high-order superclustering in the explosion models. Beyond the relatively small existing supercluster catalogs there is not much observational data with which to compare these results, but the situation should improve as larger redshift surveys of homogeneous cluster samples become available.

We have used a slightly different set of simulations from those described before, with fewer clusters per run. We take only those clusters with $M > M_*$, leaving about 230, 130, 120, and 95 clusters per simulation for the standard, merged, anticorrelated, and anticorrelated/merged models respectively.

4.4.1 - Superclusters

Figure 4.3 displays projections of the $M > M_*$ clusters in a realization of the standard power law model (top) and its merged version (bottom). The cluster distribution appears highly non-random, with dramatic filamentary and shell-like structures and large voids. Merging does not alter the large-scale features. We have identified superclusters in our models using a cluster-finding technique based on "percolation" at different density thresholds, and calculated their multiplicity functions. These are remarkably flat, with a substantial amount of mass in large superclusters even at a density contrast of 100.

In the analysis of Batuski, Melott and Burns (1987, BMB) of a sample of 225 $R \geq 0$ Abell clusters of $\varrho/\bar{\varrho} = 2.8$, two superclusters, with 38 and 36 members respectively, contain one third of all the clusters in the sample. The explosion models are clearly capable of producing such large superclusters. All of the models appear roughly consistent with the BMB data, although the constant radius and the standard power law models show perhaps too mucn superclustering. Bahcall and Soneira (1984, BS4) provide four separate catalogs of superclusters from the sample of 104 Abell clusters, at overdensities of 20, 40, 100, and 400. At an overdensity of 20 a single supercluster, whose central concentration is like the Corona Borealis supercluster, contains nearly 15% of all the clusters in the survey. Again the explosion models can easily produce such superclusters. In fact the three modified power law models all predict that ~ 15% of the clusters should be in groups of multiplicity 9 or larger, in agreement with the BSA result. Given the small size of the observational sample, the merged power law model seems quite consistent with the BS4 multiplicity function.

BMB compare observational results to numerical simulations of several initially Gaussian models. The ξ_{cc} of all these Gaussian models are weaker than observed, the opposite problem from our explosion models. BMB then find less large-scale superclustering in their models than in the observations, with only 20 - 30% of clusters in groups larger than 10 at $\varrho/\bar{\varrho} = 2.8$. The multiplicity functions of the Gaussian models are declining at this point — (there are fewer clusters in the > 10 bin than in the $5 — 10$ bin) whereas the multiplicity functions of the explosion models are still rising. It seems that the differences between the explosion models and the Gaussian models would become still more evident at larger multiplicities and higher density contrasts. The only Gaussian models that can reproduce the observed ξ_{cc} appeal to an open cosmology dominated by baryons or hot dark matter, but these would also fail in reproducing very big superclusters.

Superclustering, as measured by the multiplicity function, depends on high-order correlations in addition to the two-point correlations. We expect strong high-order correlations in the explosion models, as a result of multiple-shell interactions; four shells, for example, give rise to n-point correlation functions up to $n = 6$. Elongated superclusters can form along chains of overlapping shells. While ξ_{cc} vanishes beyond $2R_{sh}$, higher-order correlations can extend beyond the shell diameter! This is in contrast to the Gaussian models whose statistical properties are completely specified by their power spectrum or by its Fourier transform, the two-point correlation function; all higher order correlations vanish except where they are created by non-linear evolution.

4.4.2 - Voids

The void probability function $P_0(V)$, the probability that a randomly placed volume V is empty, depends on correlation functions of all orders. A Poisson distribution with number density n, for example, has no other correlations, yielding the familiar $P_0 = \exp(-nV)$. For a Gaussian distribution only the one and two-point correlations are non-zero, so $P_0 = \exp(-nV + n^2V^2\langle\xi\rangle/2)$.

In Figure 4.8 we plot P_0 as a function of nV, the expected number of clusters in a sphere of volume V. There are big differences between the various explosion models — we show here only the results for the merged power law model, which has the lowest P_0 of all. The differences between the explosion models and a Poisson distribution are enormous, an order of magnitude by $nV = 4$, and 3—4 orders of magnitude by $nV = 8$. We have analyzed the simulated cluster sample of BMB for CDM and neutrinos in the same way. The most weakly clustered of our models has void probabilities that are substantially higher than the most strongly clustered Gaussian models, an open universe dominated by CDM.

The void probability function is therefore a potentially powerful tool for testing theoretical models. One needs large samples to measure low values of P_0, however. A sample of N clusters contains N/nV independent volumes of size V, so one can measure probabilities down to $\sim nV/N$. Bahcall and Soneira

A. DEKEL

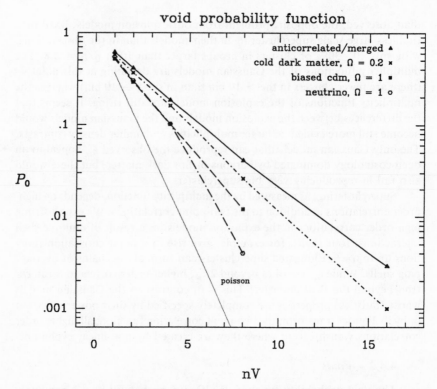

FIG. 4.8. Void probability function for the anticorrelated/merged explosion model in comparison with Gaussian models (as simulated by Melott).

(1982) have reported the existence of a very large void in the Abell cluster distribution, with a volume $\sim 10^6\,h^{-3}$ Mpc^3 and $nV \sim 6$. The probability of finding such a void is about 0.001, 0.01 and 0.1 in the Poisson, CDM and explosion models respectively. However, because Bahcall and Soneira search for a large empty volume rather than place randomly centered spheres, the statistical significance of this void is unclear.

We also generalize P_0 to the number count distribution, the probability $P_N(nV)$ that a randomly placed volume V contains N particles, and we make the appropriate predictions (paper II).

4.4.3 - Topology

The method was introduced by Gott, Melott, and Dicknson (1986) and discussed at length by Weinberg, Gott, and Melott (1987, hereafter WGM). Gott, Weinberg, and Melott (1987), and Melott, Weinberg, and Gott (1988) have applied this technique to the galaxy distributions of Gaussian models. After smoothing, we draw contour surfaces at a range of threshold densities

(ν = the number of standard deviations from the mean) and measure the genus of the surface. The genus G_s is the difference between the number of holes and the number of isolated regions. The number of "holes", in the sense of donut holes, is the maximum number of closed curves that can be drawn on the surface without dividing it further. At high or low ν contours typically break into isolated bags surrounding local peaks or valleys, so they have a negative genus. At intermediate thresholds contours tend to be multiply connected, "sponge-like", and therefore have positive genus.

For random phase distributions the genus is symmetric in ν because high and low density regions are statistically indistinguishable (e.g. Hamilton, Gott, and Weinberg 1986). The transition between isolated clusters or voids and sponge-like contours occurs at $\nu = \pm 1$. Non-random phase models generally produce an asymmetric $G_s(\nu)$ (WGM); this asymmetry can serve as an important distinguishing feature between Gaussian and non-Gaussian models. The amplitude of the $G_s(\nu)$ curve reflects the amount of structure present in the cluster distribution on scales of the smoothing length, and we can use this amplitude to define an "effective index" n_{eff} that can be compared to other models with a power law spectrum.

All the explosion models display definite asymmetry in $G_s(\nu)$. Isolated superclusters appear over an expanded range of ν with G_s crossing zero at $\nu \sim 0.7$; isolated voids begin to appear only at $\nu \leq -1.2$, and for the constant-radius models the curve peaks below the median density. The minimum at positive ν tends to be deeper than that at negative ν, indicating more isolated superclusters than isolated voids. The asymmetry is opposite to that of "bubble" models in which mass points lie uniformly on walls surrounding empty voids (Figure 14 of WGM). Rather, these cluster distributions show the "meatball" topology seen in isolated cluster distributions (WGM, Figure 19). The shape of the $G_s(\nu)$ curves is very similar for all the models, but the amplitudes are different; they indicate for the modified power law models n_{eff} = 0.1, 0.25, and 0.4 for the merged, anticorrelated, and anticorrelated/merged cases respectively. The other models show a negative index; the $f = 1.0$ model, in particular, has very few holes, with $n_{eff} = -1.5$.

Preliminary results from an analysis of 155 Abell clusters by Gott and collaborators are consistent with a random phase distribution and an index $n_{eff} \sim 0.5$ (Gott, private communication), but the sample contains only 4 "holes", so the results are not strong enough to rule out any of these models.

4.4.4 - Velocities

Because clusters form on expanding shells, they should acquire peculiar velocities. We estimate the peculiar velocities by making the simplifying assumptions that the peculiar velocity of each shell is a fixed fraction of its Hubble velocity, each shell contributes an equal amount of mass to the cluster, and the cluster conserves the net momentum of this matter. For shells that overlap substantially, the velocity calculated in this way approaches the velocity of the intersection point itself. We calculate the peculiar velocity correlation function $\langle \vec{v}(\vec{x}) \cdot \vec{v}(\vec{x} + \vec{s}) \rangle$. It is positive and decreasing at separations s smaller than the typical shell radius R, and it turns negative at $s \simeq R$. Nearby clusters tend to share the peculiar velocity of their common shells, so their velocity vectors point in the same direction. This turns to an anticorrelation at $s > R$ because many pairs in this separation range lie on opposite sides of the same shell. More cluster pairs share common shells when the filling factor is low, hence a deeper anticorrelation shows up. Any correlations beyond the shell diameter disappear in the noise. This pattern of positive velocity correlations at small separations turning to *anticorrelations* at larger separations should be a distinctive signature of the explosion models. If clusters formed instead by gravitational collapse of density fluctuations, we would more likely find nearby clusters falling towards each other, producing an anticorrelation at small s.

The rms peculiar velocity is about 2/3 the shell peculiar velocity for the constant radius models and 1/3 the R_{max} peculiar velocity for the power law model. The peculiar expansion velocities of shells depend on details of the cosmological scenario. Shells that sweep up all matter in a flat universe approach a self-similar state (Bertschinger 1983, Ikeuchi, Tomisaka, and Ostriker 1983) where the peculiar expansion velocity is 20% of the Hubble velocity. However, peculiar velocities will eventually decay in an open universe, and in a universe dominated by collisionless dark matter whose gravity will slow a shell down until the dark matter inside catches up and crosses it. It appears that the typical cluster velocities will be too low to produce the $\sim$ 1000 km s^{-1} broadening suggested by the anisotropy of pair separations in the redshift sample (Bahcall, Soneira, and Burgett 1986). The large velocities on large scales is currently the most severe problem for the explosion picture.

In summary, all our statistical measures demonstrate the impressive degree of large-scale structure in the explosion models. Each of these superclustering statistics depends on the various orders of correlation functions in a different way, so they can provide at least partially independent tests of the models. All of the explosion models show substantially more superclustering than the with traditional Gaussian models. The merged power law model agrees quite well with the observed supercluster multiplicity function. The other models are reasonably consistent with the multiplicity data, except that they tend to have fewer isolated clusters at the higher density thresholds. No observational analyses are available for the void probability function or number counts, though these are not difficult to measure in principle. The supercluster-void topology of all the explosion models displays a "meatball" asymmetry. This asymmetry does not appear in the current observational data, but larger cluster samples will be needed before this becomes a statistically significant discrepancy. Cluster peculiar velocities are expected to be fairly small, but they should show a systematic pattern of correlation at close separations and an anticorrelation at separations between R_{sh} and $2R_{sh}$.

5. CONCLUSION

The three different projects described above are meant to demonstrate how unique the system of rich clusters of galaxies is in constraining the theories of the formation of large-scale structure. These objects could be detective out to cosmological distances of order $z \sim 1$, and their clustering process must be dominated by gravity, so they should make ideal tracers of large-scale initial conditions and dynamics. Although the study of internal properties of clusters yielded little evidence for any specific initial conditions, the immediate vicinity of clusters, on scales of $1 - 10 \ h^{-1} \ Mpc$, and its relationship to the central parts, does contain very interesting cosmological signatures. But of crucial importance is the superclustering of clusters, which indicates non-trivial structure on scales of $10 - 100 \ h^{-1} \ Mpc$. While the galaxy two-point correlation function vanishes in the noise at about $10 \ h^{-1} \ Mpc$, the cluster correlations provide strong statistical evidence for non-trivial structure on larger scales. Combined with the presence of filamentary structure and big voids in the distribution of galaxies on scales of a few tens of megaparsecs, and the intriguing coherent motions on similar scales, the cluster distribution indicates detectable dynamical evolution of structure on very large scales, on the order of 5% of the present horizon, and it has the potential of probing the structure on even larger scales.

Our analysis of the systematic effect of contamination in the Abell catalog indicates that although the effect is interesting the superclustering is not a fluke — it might be real and should be taken seriously.

The very large-scale structure is not a natural outcome of the standard theories which are based on Gaussian, scale-invariant initial fluctuations in an Einstein-de Sitter universe. The repair of the Gaussian models requires either an open cosmological model, or a non-zero cosmological constant. Alternatively, the explosion scenario offers a natural source of non-Gaussian fluctuations, which can reproduce and easily over-produce the observed superclustering in a generic way.

What makes the cluster distribution in the explosion scenario so powerful is that it emerges straightforwardly from generic topological considerations. This allows the use of simple toy models which provide strong constraints on the possible nature of the model. For example, the required sizes of ~ 30 $h^{-1} \ Mpc$ for the biggest shells indicate explosions which are non-trivial energetically. Such explosions could not be seeded by supernovae in a single galaxy; they must originate in something like detonations, or superconducting cosmic strings. The model provides quantitative predictions which will be testable with larger redshift samples of clusters.

The moral is that we are in great need for larger, homogeneous, redshift samples of rich clusters of galaxies. The Abell catalog to distance class $D = 6$, and its southern counterpart are there, waiting for cluster redshifts to be measured. Cluster catalogs based on their X-ray emission will also be very useful; they would suffer even less from possible contamination because the cluster cores are smaller in X-rays, and they will provide independent evidence for the degree of superclustering. On smaller scales, more detailed studies of the galaxy distribution a few megaparsecs away from the centers of nearby clusters also have the promise of providing very useful cosmological information.

ACKNOWLEDGEMENTS

I am deeply indebted to my collaborators on the various projects described here: G.R. Blumenthal, A. Oemler Jr., J.P. Ostriker, J.R. Primack, D. Weinberg and M.J. West.

REFERENCES

Aarseth, S.J. 1985. In *Multiple Time Scales*. ed. J.U. Brackbill and B.I. Cohen. p. 377. New York: Academic.

Abell, G. 1958. *Ap J Suppl.* **3**, 211.

Albrecht, A. and Turok, N. 1985. *Phys Rev Letters.* **54**, 1868.

Argyres, P.C., Groth, E.J., Peebles, P.J.E., and Struble, M.F. 1986. *AJ.* **91**, 471. (AGPS).

Bahcall, N.A. 1979. *Ap J.* **232**, 689.

_____. 1988. in *Large-Scale Structures of the Universe,* eds. J. Audouze, M.C. Pelletan, and A. Szalay. Dordrecht: Kluwer Academic Publishers, p. 229.

Bahcall, N.A. and Soneira, R.M. 1982 *Ap J.* **262**, 419.

_____. 1983. *Ap J.* **270**, 20 (BS).

_____. 1984. *Ap J.* **277**, 27 (BS4).

Bahcall, N.A., Soneira, R.M., and Burgett, W.S. 1986 *Ap J.* **311**, 15.

Bardeen, J.M., Bond, J.R., and Efstathiou, G. 1987. *Ap J.* **321**, 28.

Bardeen, J.M., Bond, J.R., Kaiser, N., and Szalay, A.S. 1986. *Ap J.* **304**, 15.

Barnes, J., Dekel, A., Efstathiou, G. and Frenk, C.S. 1985. *Ap J.* **295**, 368.

Batuski, D.J. and Burns, J.O. 1985. *AJ.* **90**, 1413.

Batuski, D.J., Melott, A.L., and Burns, J.O. 1987 *Ap J.* **322**, 48. (BMB).

Bennet, D. and Bouchet, F. 1988. *Phys Rev Letters.* **60**, 257.

Bertschinger, E.W. 1983. *Ap J.* **268**, 17.

_____. 1985. *Ap J.* **295**, 1.

Binggeli, B. *AA.* **107**, 338.

Binney, J. and Silk, J. 1979. *MN.* **188**, 273.

Blumenthal, G.R. and Primack, J.R. 1983. in *Fourth Workshop on Grand Unification.* ed. H.A. Weldon, P. Langacker, and P.J. Steinhardt. p. 256. Boston: Birkhauser.

Blumenthal, G.R., Dekel, A. and Primack, J.R. 1988. *Ap J.* **326**, 539.

Carr, B. and Ikeuchi, S. 1985. *MN.* **213**, 497.

Carter, D. and Metcalfe, N. 1981. *MN.* **191**, 325.

Cavaliere, A., Santangelo, P., Tarquini, G., and Vittorio, N. 1986. *Ap J.* **305**, 651.

Davis, M. and Peebles, P.J.E. 1983. *Ap J.* **267**, 465.

Dekel, A. 1983. *Ap J.* **264**, 373.

_____. 1984a. *Ap J.* **284**, 445.

_____. 1984b. in *Eighth Johns Hopkins Workshop on Current Problems in Particle Theory.* eds. G. Domokos and S. Koveski-Domokos. p. 191. Singapore: World Scientific.

Dekel, A. and Aarseth, S.J. 1984. *Ap J.* **238**, 1.

Dekel, A., West, M.J., and Aarseth, S.J. 1984. *Ap J.* **279**, 1.

de Lapparent, V., Geller, M., and Huchra, J. 1986. *Ap J Letters.* **302**, L1.

Doroshkevich, A.G. 1970. *Astrophysica.* **6**, 320.

Doroshkevich, A.G., Shandarin, S.F., and Saar, E. 1978. *MN.* **184**, 643.

Dressler, A. 1980. *Ap J.* **236**, 351.

Gott, J.R. and Turner, E.L. 1977. *Ap J.* **216**, 357.

Gott, J.R., Melott, A.L., and Dickinson, M. 1986. *Ap J.* **306**, 341.

Gott, J.R., Weinberg, D.W., and Melott, A.L. 1987. *Ap J.* **319**, 1.

Hamilton, A.J.S., Gott, J.R., and Weinberg, D.H. 1986. *Ap J.* **309**, 1.

Hoessel, J., Gunn, J.E., and Thuan, T.X. 1980. *Ap J.* **241**, 486. (HGT).

Ikeuchi, S. 1981. *PASJ.* **33**, 221.

Ikeuchi, S., Tomisaka, K., and Ostriker, J.P. 1983. *Ap J.* **265**, 583.

Klypin, A.A. and Kopylov, A.I. 1983. *Sov Astron Letters.* **9**, 41.

Kulsrud, R. 1988, in preparation.

Lambas, D.G., Groth, E.J. and Peebles, P.J.E. 1988, *AJ.* **95**, 975 (LGP).

Lilje, P.B. and Efstathiou, G. 1988. *MN.* **231**, 635 (LE).

Lin, C.C., Mestel, L., and Shu, F.H. 1965. *Ap J.* **142**, 1431.

Ling, E.N., Frenk, C.S., and Barrow, J.D. 1986. *MN.* **223**, 21p.

Lukey, J.R. 1983. *MN.* **204**, 33.

Melott, A.L., Weinberg, D.W., and Gott, J.R. 1988. *Ap J.,* in press.

Noonan, T. 1980. *Ap J.* **238**, 793.

Nusser, A. and Dekel, A. 1988. in preparation.

Oemler, A., Schechter, P.L., Shectman, S.A., and Kirshner, R.P. 1988, in preparation.

Ostriker, J.P. and Cowie, L.L. 1981. *Ap J Letters.* **243**, L127.

Ostriker, J.P., and McKee, C.M. 1988. *Rev Mod Phys.* **60**, 1.

Ostriker. J.P., Thompson, C., and Witten, E. 1986. *Phys Lett B.* **280**, 231.

_____. 1988. *Rev Mod Phys.,* in preparation.

Palmer, P.L. 1981. *MN.* **197**, 721.

_____. 1983. *MN.* **202**, 561.

Peebles P.J.E. 1982. *Ap J Letters.* **263**, L1.

_____. 1987. *Ap J Letters.* **315**, L73.

_____. 1988. this volume.

Primack, J.R., Blumenthal, G.R. and Dekel, A. 1986. in *Galaxy Distances and Deviations from Universal Expansion.* eds. B.F. Madore and R.B. Tully. Dordrecht: Reidel, p. 265.

Rhee, G.F.R.N. and Katgert, P. 1987. *AA.* **183**, 217.

Rood, H.J., Page, T.L., Kintner, E.C., and King, I.R. 1972. *Ap J.* **175**, 627.

Saarinen, S., Dekel, A. and Carr, B. 1987. *Nature.* **325**, 598.

Sastry, G.N. 1986. *PASP.* **80**, 252.

Scherrer, R. 1987, preprint.

Seldner, M. and Peebles, P.J.E. 1977. *Ap J.* **215**, 703 (SE).

Silk, J. and Turner, M.S. 1987, preprint.

Struble, M.F. 1987. *Ap J.* **323**, 468.

Struble, M.F. and Peebles, P.J.E. 1985. *AJ.* **90**, 582.

Sutherland, W. 1988. *MN.* **234**, 159.

Tucker, G.S. and Peterson, J.B. 1987, preprint.

Turner, M.S., Villumsen, J.V., Vittorio, N., and Silk, J. 1987, preprint.

Turok, N. 1985. *Phys Rev Letters.* **55**, 1801.

Weinberg, D.H., Dekel, A., and Ostriker, J.P. 1988, in preparation. (Paper I).

Weinberg, D.H., Ostriker, J.P., and Dekel, A. 1988, in preparation. (Paper II).

Weinberg, D.H., Gott, J.R., and Melott, A.L. 1987. *Ap J.* **321**, 2. (WGM).

West, M.J., Dekel, A., and Oemler, A.Jr. 1987. *Ap J.* **316**, 1. (Paper I).

_____. 1988. *Ap J.* in preparation. (Paper III).

West, M.J., Oemler, A., and Dekel, A. 1988. *Ap J.* in press. (Paper II).

White, S.D.M. 1976. *MN.* **177**, 717.

White, S.D.M., Frenk, C., Davis, M., and Efstathiou, G. 1987. *Ap J.* **313**, 505.

White, S.D.M. and Ostriker, J.P. 1988, in preparation.

Zeldovich, Ya.B. 1970. *AA.* **5**, 84.

VII

PROPERTIES OF GALAXIES AT HIGH Z AND LOW Z

RECENT OBSERVATIONS OF DISTANT MATTER: DIRECT CLUES TO BIRTH AND EVOLUTION

DAVID C. KOO

Space Telescope Science Institute, Baltimore

and

Lick Observatory, Board of Studies in Astronomy and Astrophysics

ABSTRACT

Highlights of recent deep observations of field galaxies, clusters of galaxies, radio galaxies, quasar absorption lines, and quasars are used to illustrate our progress since the 1981 Vatican Conference on Astrophysical Cosmology and to review the current status of evidence for evolution in their intrinsic properties and large-scale clustering. The birth and ages of galaxies can be explored directly by exploiting these classes of objects to search for primeval galaxies.

1. INTRODUCION

Except for the microwave background measurements, most of the other observations presented in this volume are equivalent to snapshots, many quite detailed, of the universe as it appears now, and reach distances that only encompass less than 1% of the accessible volume. In contrast, although the data become much coarser as we peer farther into the past to higher redshifts and fainter limits, the resultant glimpses of the early history of the universe provide, in principle, direct evidence for changes in the contents, properties, and distribution of matter. Even the birth process may be visible.

These faint observations are also important as powerful constraints to various cosmological scenarios, including the popular Cold Dark Matter (CDM) theory, which has been so successful in explaining many of the known correlations among a variety of galaxy properties and their clustering and motions, all with a critical density of $\Omega = 1$ for the universe. Of interest for faint observations, this theory predicts that the bulk of star formation should have occurred at small redshifts, typically from $z = 0.5$ to perhaps 2 or so, a range that is quite accessible within an $\Omega = 1$ universe. For example, assuming little correction for the shape of the spectra (i.e., a spectral index of $- 1$) and no evolution in luminosity for most galaxies (i.e., those similar to spiral galaxies with active star formation), the predicted brightness of a typical galaxy of $M_v = - 22$ (Hubble constant of 50 km sec^{-1} Mpc^{-1} assumed, so the universe is about 13 Gyr old) should be around V = 20.6 at z = 0.5, 22.2 at z = 1, and 23.9 at z = 2. A 4-m telescope equipped with modern CCD detectors would image to these limits easily and should even yield redshifts in several hours, if emission lines are strong, as expected with active star formation.

Since the last Vatican Conference on Astrophysical Cosmology in 1981 (Brück *et al.* 1982, henceforth VAC) when CCDs were still a novelty, tremendous progress has been made. A handful of the more recent and exciting (at least tantalizing) discoveries will be highlighted here for five classes of objects receiving the most attention in observations of distant matter; field galaxies, clusters of galaxies, radio galaxies, quasar absorption lines, and quasars themselves. For each I will touch on the current status of our view of their evolution, on the presence and perhaps evolution of large-scale clustering at lookback times that are a significant fraction of the age of the universe (z > 0.1), and on new candidates for primeval galaxies.

2. FIELD GALAXIES

2.1 - Evolution

At the VAC Gunn in his review on this subject wrote as follows: (1) on counts and colors: "There is no evidence for evolution at J^+ mag brighter than about 23" and "the agreement among various workers is not very good", where J^+ refers to a photographic broadband blue; (2) on redshifts: he discussed his pioneering survey of 58 optically selected field galaxies to B ~ 20 that had a median redshift consistent with no evolution, but also an unexplained high redshift tail; (3) on the 4000 Å break: "It is quite clear that one does see the expected stellar evolution in ellipticals as one looks to larger... redshifts," demonstrated with an example at z = 0.75.

Today the data quantity is vastly superior, but our picture is still far from final: (1) on counts: the depth has reached almost 10 times fainter (Tyson 1988), with divergence of the no-evolution model and data by about a factor of 10 (Fig. 1); even if the claim for no significant field galaxy evolution in luminosity and color to z ~ 0.8 from the multicolor work of Loh and Spillar (1986) is not accepted, the nature of the strong color evolution seen in the multicolor data of Tyson (1988) and Koo (1986b) remains ambiguous but suggestive of

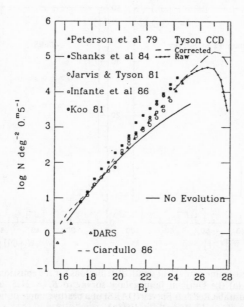

FIG. 1. Recent published galaxy number counts reduced to a photographic blue band. The solid curve represents the best estimate no-evolution model (see Ellis 1988b for details and references).

more extensive star formation at moderate redshifts $z = 0.4$ to 2; (2) on redshifts: nearly 600 redshifts to limits $\sim$ 5x fainter than that mentioned above are now available from two independent surveys. Both are consistent in suggesting that luminosity evolution has indeed been slight, if any, since $z \sim 0.4$, but that many galaxies do show signs of extensive star formation, either through very ultraviolet colors (Koo and Kron 1988b) or from the overall increase in the strengths of [O II] emission lines (Ellis 1988a), as seen in Figure 2. The high redshift tail mentioned by Gunn above is not confirmed (Koo 1985); (3) on the 4000 Å break: among four groups all working to limits of $z \sim 0.8$, Hamilton (1985) and Oke(1983) both claim no evidence for evolution while Spinrad (1986) and Dressler and Gunn (1988) see a decrease in the amplitude of the break consistent with "expected evolution" (see Figure 3).

2.2 - Clustering Evolution

This area of research remains largely unexplored and among existing data, inconclusive if not inconsistent. Based on 4-m photographic surveys to B $\sim$ 24, Koo and Szalay (1984) and Pritchet and Infante (1986) show consistent amplitudes for the two-point autocorrelation function that suggest little cluster-

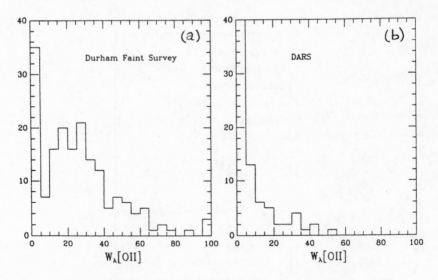

FIG. 2. Rest-frame equivalent width distributions for the emission line [O II]
3727 Å in (a) the Durham faint galaxy survey to B_j = 21.5, and (b) the
Durham-Australia Redshift Survey (DARS) of a nearby sample of field galaxies
complete to B_J = 17 (from Ellis 1988a).

ing evolution, whereas Stevenson *et al.* (1985) measure amplitudes nearly a
factor of two lower, implying that clustering has increased significantly since
larger redshifts. Based on field-to-field fluctuations being twice Poisson in the
CCD data of Tyson and 5 times Poisson as reported by Ellis (1987) for
photographic data, Koo (1988) notes that these values are entirely consistent
with a simple extrapolation of the − 0.8 power-law slope found for bright
galaxies and of the no-evolution extension of the scaling by number density
of counts. Using spectral redshifts (Ellis 1987) or multicolor estimates of redshifts (Loh 1988) to measure the spatial correlation amplitude at z ~ 0.5, both
report values consistent with lower amplitudes and hence clustering evolution.
On larger scales (~ 100 Mpc), simple histograms of the redshifts in some fields
suggest the presence of large-scale voids and perhaps sheet-like superclusters
(Ellis 1987; Koo and Kron 1988b; see Fig. 4), but before we accapt their reality
the significance of these fluctuations needs to be carefully compared to realistic
simulations, since even the CDM model may naturally produce such features
(White *et al.* 1987) in a fraction of independent samples.

2.3 - Primeval Galaxies

The prospect of detecting primeval galaxies, i.e. those undergoing their
initial star formation, is an area of research pioneered nearly two decades ago

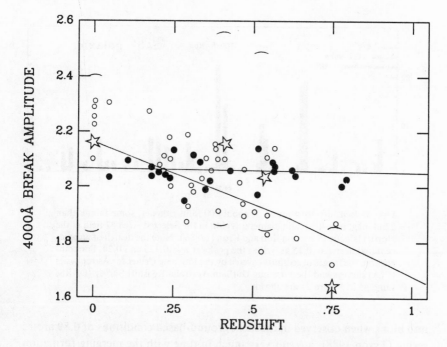

FIG. 3. Observations of the 4000 Å break amplitude versus redshift from Spinrad (1986) in open circles, Hamilton (1985) in solid circles, and Dressler (1987) in stars that represent the median and brackets that enclose the central two-thirds of each sample. The sloping solid line is the prediction for evolving ellipticals; the horizontal line is for reference (see Spinrad 1986 for more details).

by Partridge and Peebles (1967), but it has received much attention in the last few years, though virtually ignored at the VAC. In addition to the largely negative results reviewed by Davis (1980) and Koo (1986a), three recent surveys for field primeval galaxies deserve attention: (1) an extremely deep survey for Lyman-α to R $\sim$ 26 limits has been completed by Pritchet and Hartwick (1987), with negative results that begin to severely constrain traditional models (see Figure 5); more recent explorations of the extended formation times and perhaps gradual merging of subcomponents may, however, allow the CDM model to survive (Baron and White 1987); (2) preliminary results from comparing a very deep ultraviolet CCD image (see Majewski 1988) covering a 10 square arcmin field to a blue photographic image reaching B $\sim$ 24 show no candidates with z $>$ 3.5. Any such candidate would be recognized by its invisibility in the optical UV, due to the expected drop of flux at the Lyman continuum break at 912 Å; (3) about 20% of the B $>$ 26 objects break up

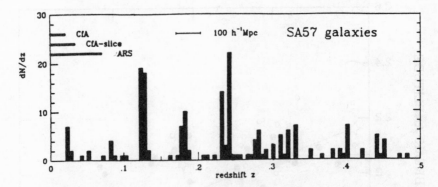

Fig. 4. Redshift distribution of about 200 faint galaxies, some fainter than 22nd mag, in a 40 arcmin diameter region in the Selected Area 57 field at the North Galactic Pole from Koo and Kron (1988b). Note the dearth of galaxies centered near z = 0.15 as well as the peaks at z = 0.125 and 0.24. For comparison, the ranges of redshifts covered by the 15th mag Center for Astrophysics (CfA) surveys and the 17th mag Durham Australia Redshift Survey (DARS, same as in Figure 2) are shown.

into blobs when observed in excellent ground-based conditions of 0.89 arcsec seeing (Tyson 1988), a result very much in tune with the merging formation scenarios of at least the CDM model.

3. CLUSTERS OF GALAXIES

3.1 - Evolution

Gunn in his VAC review mentions the initiation with Dressler of spectroscopic investigations of the Butcher-Oemler effect, in which distant (z ~ 0.4) compact rich clusters exhibit a larger fraction of blue galaxies in their cores than their present-day counterparts (Butcher and Oemler 1978). By now, several groups have contributed to the accumulation of several hundred spectra for perhaps a dozen or so clusters (see Dressler and Gunn 1988 for a recent review) with important results. First, these data largely support the membership of enough blue galaxies to confirm the photometric effect, as redefined (Butcher and Oemler 1984; see Figure 6). Of more interest, unlike most cluster spirals today which have spectra consistent with reasonably continuous star formation, spectra of distant cluster galaxies often show strong enough Balmer absorption lines, with weak, if any, accompanying emission lines, to suggest bursts of star formation (about 1% to 10% by mass) a billion years or so prior to the epoch being observed (see Figure 7). Moreover, some

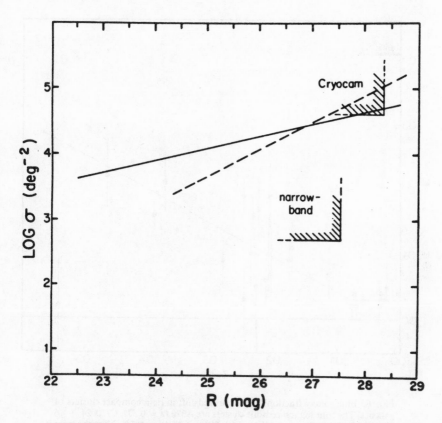

FIG. 5. Comparison of observational limits on primeval galaxies (PG) set by Pritchet and Hartwick (1987) with theoretical predictions. The solid line represents the predictions of a variety of models by Meier (1976). The dashed line represents models of PGs at z = 5 by Davis (1980), modified as described by Koo (1986a). Their observed limits correspond to less than one PG per CCD field at 6800 Å for the narrow-band filter observations, and less than one object per 272 arcsec² for the Cryocam observations. The sense of the limits is to exclude PGs whose properties lie to the upper left of the plotted points. The limits are plotted for pure Lyman-α sources with angular extent less than approximately 2."5 (Cryocam) and less than approximately 1."5 (narrowband). No correction for the sampling in redshift space has been made in this diagram (taken from Pritchet and Hartwick 1987; see their text for details).

evidence for activity and evolution in otherwise very red cluster galaxies at these redshifts is also found, either by changes in the 4000 Å break (Gunn and Dressler 1988) or by ultraviolet excesses (MacLaren *et al.* 1988, Couch and Sharples 1987). The underlying physical mechanisms remain uncertain, but range from cluster-gas ram-pressure induced star-formation to galaxy

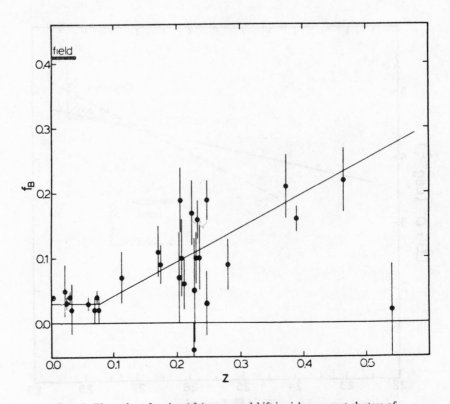

Fig. 6. Blue galaxy fraction (f_B) versus redshift in rich, compact clusters of galaxies. The four highest redshift clusters are A370 (z = 0.37), CL 0024 + 16 (z = 0.39), 3C 295 (z = 0.46), and CL 0016 + 16 (z = 0.54). Note the large dispersion of f_B at high redshifts. The "field" refers to local field galaxies; the solid line is an eye-drawn trend of increasing f_B with redshift claimed by Butcher and Oemler (1984). This figure is a simplified version of their Figure 3.

interactions; whatever the cause, different clusters at the same redshift may show different average properties, hinting that the evolutionary clocks are set cluster by cluster rather than by universal effects. As previously mentioned, field galaxies also show excess star formation activity at the same redshifts, but this vital relationship of evolution and environment has yet to be studied in detail.

3.2 - Cluster-Cluster Evolution

This area of research is still in its infancy but is likely to blossom over the next five years, as more systematic searches for distant clusters are undertaken (Gunn *et al.* 1986, Ellis 1988*b*). These will be among the first surveys to yield

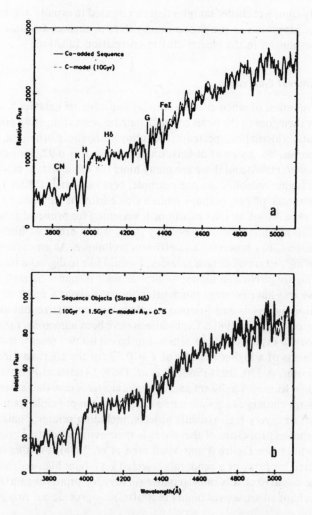

FIG. 7. (a) The coadded spectrum of the 62 normal sequence (red) galaxies identified in AC 103, AC 114, and AC 118, plotted on a rest-wavelength scale. Prominent absorption features are identified. For comparison, the spectrum of a standard old population model (10 Gyr c-model with a prompt 1Gyr initial burst) is superimposed upon the observed spectrum. The agreement between the two is seen to be excellent; (b) the coadded spectrum of the 11 sequence galaxies with noticeably stronger Hδ absorption. The comparison spectrum (light line) is the combination of a burst model spectrum and the old population model spectrum shown in (a) added together so as to contribute equal amounts of light at λ_{rest} = 4000 Å and reddened assuming A_V = 0.5 mag. Note that the poor match between the observed and model spectrum in the vicinity of Hβ (λ4861) is due to a strong night-sky emission line falling on this feature in the observed spectra of many of the Hδ-strong objects (from Couch and Sharples 1987).

reasonably complete cluster samples that can be used to explore not only evolution in the properties of the clusters themselves or their constituent galaxies, but also evolution in the cluster-cluster correlation function.

3.3 - Primeval Galaxies

The question of when clusters formed and whether galaxies had already existed by then goes to the heart of the debate between competing possibilities for the initial fluctuation spectrum, including adiabatic, isothermal, and now CDM theories. We know of at least one cluster at $z = 0.92$, found optically by Gunn *et al.* (1986) and there are many hints for more distant clusters from objects at higher redshifts. As one example, bent radio tails (Miley 1987) suggest the presence of gas, perhaps from a rich cluster of galaxies, to redshifts of nearly two. As will be later mentioned, searching for primeval galaxies near known high redshift objects, especially those which are most likely to exist within clusters, is a powerful and efficient technique. As an excuse to show an interesting picture of distant galaxies, I would like to digress a bit and mention the highly publicized luminous arcs around distant clusters.

These arcs have received much attention, mainly as a result of the AAS presentation by Lynds and Petrosian (1986), but few realize that an arc was first reported by Hoag (1981). Explanations have been numerous, ranging from light echoes to explosion shock shells, but based on two recent, independent measurements of a higher redshift of $z = 0.72$ for the arc than the $z = 0.37$ for the cluster, A 370, itself (Soucail *et al.* 1987, Lynds and Petrosian 1988), gravitational lensing of a distant galaxy by the cluster is now the favored theory. Using distant clusters as gigantic telescopes, we can in principle magnify otherwise very faint, very high redshift objects, including primeval galaxies. One of the common properties of these arcs is their extreme relative brightness in the ultraviolet (see Figure 8 and MacLaren *et al.* 1988), but this is perhaps the expected property of a randomly selected very-faint high-redshift galaxy. Given the negative results of systematic surveys for primeval galaxies, serendipity, perhaps an arc seen at higher redshifts, may provide our first good case.

4. Radio Galaxies

4.1 - Evolution

At the VAC the claim was made that "the fact, ... that strong extragalactic radio sources ... if they are not QSOs, are uniquely and without exception ... ellipticals and S0's." Based on this assumption, van der Laan and Windhorst not only suggested an increase of a factor of 100 in density by $z = 0.8$ in radio galaxies, but also suggested the detection of color evolution, since many

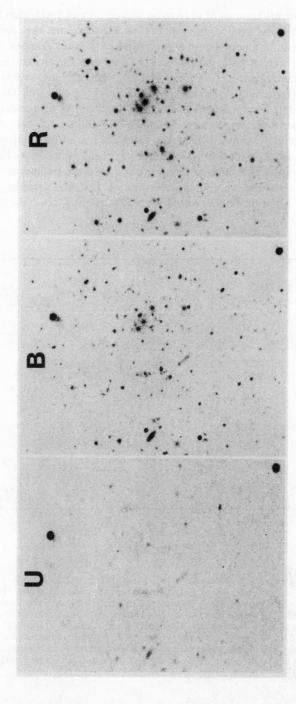

FIG. 8. Pictures of cluster 0024 + 16 at z = 0.39 showing a possible three piece "luminous arc". North is to the top, east to the left. The size of the images is ∼ 130 arc sec E-W and ∼ 160 arc sec N-S; at 6.3 Mpc per arc sec at the cluster redshift and $H_0 = 50$ km sec^{-1} Mpc^{-1} and $q_0 = 0.5$, the arc is ∼ 240 kpc from the cluster center. The images are produced from plates taken with the 4 m telescope at Kitt Peak National Observatory. The red (R), blue (B), and ultraviolet (U) plates were exposed for 60m, 60m and 150m, respectively; taken in 1975, 1980, and 1983, respectively; and correspond to rest-frame 4400 Å, 3350 Å, and 2600 Å, respectively, if the arc is at the cluster redshift. Recent observations of an arc in the cluster A370 favor the theory that an arc is a background galaxy gravitationally lensed by the rich cluster.

of the optical identifications of mJy radio sources, about 1000 times fainter than the 3CR Jy level catalog, were quite blue. Furthermore, they suggested a cutoff at redshifts z between 3 to 5 for these mJy radio galaxies.

Today our view of radio galaxies has changed dramatically and in many ways complicates any picture for evolution. First of all, the radio source counts show a distinct flattening of the slope at low fluxes (Figure 9), which is inconsistent with simple extrapolations of the traditional classes of quasars, elliptical galaxies, and a few very weak radio spirals (Windhorst *et al.* 1985, Danese *et al.* 1987). Secondly, the view that radio galaxies form a homogeneous class has been demolished. With over 100 redshifts of the mJy sources, we find that many of the brighter blue radio galaxies are not star-forming ellipticals at high redshifts z > 1, but rather morphologically peculiar, perhaps interacting, galaxies (see Figure 10) at moderate redshifts typically z ~ 0.2 to 0.6 (Kron *et al.*

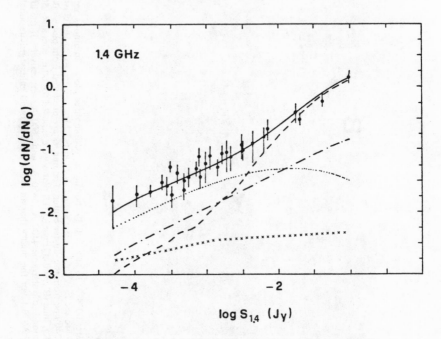

FIG. 9. Deep counts at 1.4 GHz in relative differential form ($dN_0 = 150S_{1.4}^{-2.5}$ sr^{-1} J$_y^{-1}$ such that an Euclidian rise would be a horizontal line. The data points are those listed by Condon and Mitchell (1984) and Windhorst *et al.* (1985). The dashed and the dot-dashed lines show, respectively, the contributions of steep-spectrum and "flat"-spectrum ellipticals, S0's, and QSOs; the dotted line displays the expected counts from evolving starbust/interacting galaxies. The crosses display the counts of unevolving spirals and irregulars. The solid line is the sum of all the above contributions (taken from Danese *et al.* 1987).

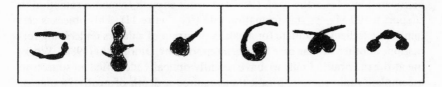

FIG. 10. Hand-drawn pictures of several optical identifications of mJy radio galaxies displaying diverse morphologies suggestive of close interactions (see Kron *et al.* 1985 for more details).

1985, Windhorst *et al.* 1987). This view of radio galaxies belonging to different classes occurs even at the 3CR level, where vastly improved CCD imaging by, e.g., Hutchings (1987) or Heckman *et al.* (1986) shows that strong radio galaxies also frequently possess peculiar morphologies, including the mysterious tendency for alignment between optical images and radio jets among distant sources (McCarthy *et al.* 1987*b*; Chambers *et al.* 1987). Until this diversity is well studied and understood, any claims for evolution, either of the underlying optical or infrared hosts in luminosity and color (see e.g. Spinrad 1986 or Lilly and Longair 1984), or of the radio population (Peacock 1985; Windhorst 1984), should be treated with caution.

4.2 - Clustering Evolution

Since radio sources can be detected to very high redshifts without the usual optical bias, they may also serve as powerful tracers of large-scale clustering with time, assuming of course that their typical environments are not changing at different epochs. Until one reaches at least the mJy level of around 50 sources or so per square degree (where typical separations are around 10 arcmin or a few Mpc at $z = 0.5$ and higher), any clustering amplitudes extrapolated from that of galaxy-galaxy correlations found today would be difficult to detect. Of course, most high redshift radio sources might reside in clusters and thus possess higher correlation amplitudes, as found among clusters today. At present several similar redshifts, hinting of large-scale clustering, are already quite common among these deep samples (Kron *et al.* 1985), but a more detailed analysis awaits improved redshift coverage.

4.3 - Primeval Galaxies

Despite the disappointing results of not finding the 20th mag blue mJy radio galaxies to be ellipticals undergoing extensive star formation at high redshift, as suggested by Katgert *et al.* (1979), radio galaxies have recently provided some of the best primeval galaxy candidates. Most exciting is the giant (10″ or 100 kpc) gas cloud at $z = 1.82$, identified with a 3CR source, and discovered by McCarthy *et al.* (1987*a*) to emit enough Lyman-α radiation to support a 100 M$_\odot$/yr star formation rate (see Figure 11). This object is certainly an excellent candidate for a galaxy or group of galaxies undergoing formation; another similar case has been reported by Djorgovski (1988). Working at the mJy radio limit, we have recently optically identified all sources in a complete radio sample, a task that required a depth of nearly 26 mag or so (Windhorst *et al.* 1987). What is tantalizing are the dozen or more faintest candidates, for they appear to be higher-redshift counterparts to the confirmed 3CR gas clouds. So far, spectroscopy has not been successful in yielding any redshifts on any of these "fuzzballs", but this and other samples should soon provide more examples of these giant Lyman-α gas clouds. In contrast to these radio sources Lilly (1988) has discovered a radio galaxy at z ~ 3.4 and argues that it contains evolved stars and is ~ 2 Gyr old. If so, at least some galaxies not associated with quasars formed at z ~ 4.5 or earlier.

5. QSO Absorption Lines

5.1 - Evolution

These features in the spectra of QSOs, which occur in various classes with the narrow Lyman-α lines most studied, were barely mentioned at the VAC but are now standard topics at many meetings; recently an entire conference was devoted to this very active area of research (Blades *et al.* 1987). These

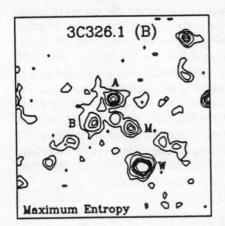
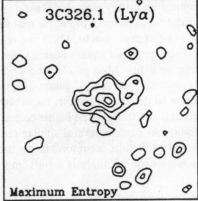

Fig. 11. Contour maps of a 100 kpc gas cloud primeval galaxy candidate, 3C 326.1 (z = 1.825): left panel represents the broad-band blue (B), which is free of any line emission; right panel a 60 Å narrow-band filter, centered on the redshifted Lyman-α. Both fields are 37.4 arcsec on each side, with North to the top, East to the left. The data were obtained with the KPNO 4-m telescope and a CCD, and processed with a Maximum Entropy algorithm (taken from Djorgovski 1988).

lines, observable to the highest redshifts, provide unique probes of the intergalactic medium as well as otherwise invisible gaseous structures that may not even be associated with luminous matter. Along with the counts of radio sources and QSOs, these absorption features give some of the least ambiguous pieces of evidence for evolution in the universe (at least with standard cosmologies). The increase with redshift per comoving volume of the number of narrow Lyman-α lines has been put on more solid footing (greater than 3σ above no-evolution predictions) with, e.g., the high precision work of Murdoch et al. (1986), as seen in Figure 12; using observations with IUE, this

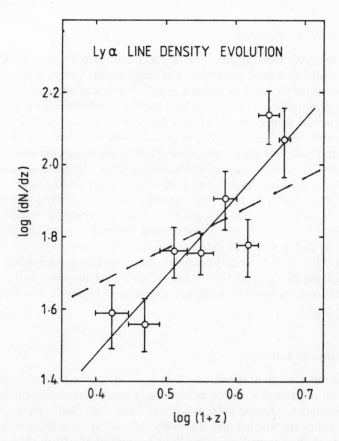

FIG. 12. Log dN/dz plotted as function of log $(1 + z)$ for a Lyman-α absorption line sample with equivalent widths $\gtrsim 0.32$ Å. The vertical bars are 1σ errors, and the horizontal bars represent chosen ranges of Δz. The solid line drawn is the Maximum Likelihood fit to the 277 lines of the Lyman-α sample (taken from Murdoch et al. 1986); for $q_0 \lesssim 0.5$, the no-evolution predictions are shallower than the dashed line.

evolution has been found to continue to redshifts $z \sim 1$ (Jenkins *et al.* 1987). An interesting result from the Murdoch *et al.* (1986) work is the confirmation of previous claims that the density of lines decrease near the QSO; this "inverse effect" or "proximity effect" is presumably the result of the increased ionization in vast volumes (radii of several Mpc) surrounding bright QSOs (but see Tytler 1987 for a contrary view). This effect places constraints on the properties of the IGM at high redshifts (Bajtlik *et al.* 1988). In addition, a recent 15 times improvement of the Gunn and Peterson (1965) measurement of the lack of neutral hydrogen in the IGM at high redshifts by Steidel and Sargent (1987) shows that the IGM must still be highly ionized beyond $z = 3$.

5.2 - Clustering Evolution

At the VAC, Oort mentions that he had recently received a preprint from Sargent claiming that no correlation was found among Lyman-α lines in two QSOs separated by 3 arcmin and at $z \sim 2.5$. This line of attack on clustering at high redshifts remains important (and the results still not settled; see Webb 1987), since the number density of these lines are far greater than that of other high redshift objects, namely QSOs or radio sources. Moreover, these gas clouds may well uniformly permeate most of space (except perhaps regions around clusters with hot gas and bright ionizing QSOs), including areas devoid of luminous galaxies. An interesting new result is the finding that "voids" of such lines roughly on 5 h^{-1} Mpc scale do exist (Crotts 1987), with maybe a filling factor of about 5% by redshifts of 3, but this result has recently been contested by Ostriker *et al.* (1988); the original note by Rees and Carswell (1987) claimed no such voids. With the advent of Space Telescope and its capability to reach below 3000 Å, we may not only secure new clues to the physical properties of these lines and the IGM at low redshifts, but also trace their clustering evolution from today to the highest redshifts. In principle, this diagnostic can yield q_o.

5.3 - Primeval Galaxies

The relationship of the narrow Lyman-α lines to the formation and environment of galaxies at high redshift is being investigated by several groups (e.g., Ikeuchi and Ostriker 1986, Bond *et al.* 1988, Rees 1986) with no general consensus on even whether they are clouds confined by the IGM, as originally proposed in the comprehensive study by Sargent *et al.* (1980). The broad, damped Lyman-α absorption lines and those from heavy metals, on the other hand, are proposed to be directly related to galaxies, probably as gaseous extensions of galaxy disks. Since such features are found to cover 20% of the sky in one sample, whereas simple predictions suggest 4%, these disks were either larger in the past or more numerous (Smith *et al.* 1986, Wolfe *et al.* 1986). Using narrow band filters tuned to Lyman-α at the redshift of the damped lines, Smith *et al.* (1987) have searched for the underlying or nearby galaxies. Their results constrain the star formation rate of these high redshift, possibly primeval spiral galaxies, to less than 1 $M_\odot$/yr, far less than the 100 $M_\odot$/yr values of the giant clouds mentioned in the previous section and perhaps even less than that from our Milky Way. This approach to finding primeval galaxies is attractive in that neutral hydrogen gas is known to exist in the search area at a specific and very high redshift.

6. QSOs

6.1 - Evolution

In their review of QSO evolution, Schmidt and Green (VAC) in the first sentence say: "The space density of quasars increases steeply with distance out to a redshift of three at least." Later, based mainly on the failure to find high redshift QSOs by Osmer (1982), they state "the number of observable quasars beyond a redshift of 3.5 is an order of magnitude lower than that predicted on the basis of a smooth extrapolation from lower redshifts", i.e., a high redshift cutoff exists that may signal an important epoch in the history of galaxy and QSO formation. At that time, only one useable QSO fainter than B = 20 had a firm redshift, and although the picture drawn by Schmidt and Green (1983), in which more luminous QSOs were evolving faster in density (luminosity-dependent density evolution), did fit the available data well, radically different physical scenarios, such as one in which the total volume density of QSOs has been approximately constant but their luminosities were brighter in the past (pure luminosity evolution) could not be distinguished (Mathez 1976, Cheny and Rowan-Robinson 1981), as demonstrated in Figure 13.

Today about 250 redshifts or more of such faint QSOs are available, mainly from the fiber-optic spectroscopic survey of the Durham group (Boyle *et al.* 1987). Combined with data from several other groups, a simple scenario is emerging that describes the overall changes in the luminosity function to the highest redshifts (see Figure 14). Using the break in the shape of the luminosity function as a fiducial marker of evolution, we find that QSOs evolved, as an ensemble, mainly in luminosity with a *gradual decline* in density. In other words, there was not a much larger density of QSO objects (or events) in the past and there is no evidence for an abrupt cutoff at redshifts near 3 to 4. Moreover, there are insufficient numbers of observed QSOs to ionize the IGM as required by the Gunn-Peterson test (Shapiro 1986). Although one

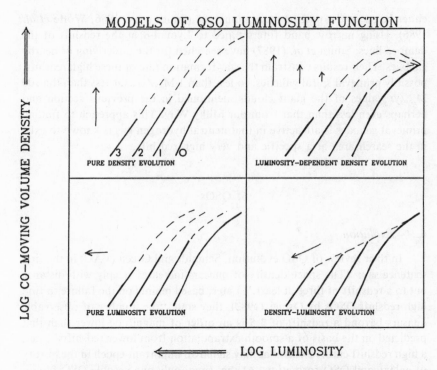

LOG CO–MOVING VOLUME DENSITY

MODELS OF QSO LUMINOSITY FUNCTION

PURE DENSITY EVOLUTION

LUMINOSITY–DEPENDENT DENSITY EVOLUTION

PURE LUMINOSITY EVOLUTION

DENSITY–LUMINOSITY EVOLUTION

LOG LUMINOSITY

FIG. 13. Rough representations of several proposed models of QSO luminosity-function evolution. The solid lines for redshifts z = 0, 1, 2, and 3 show the observed luminosity function of QSOs (and Seyfert 1s) to a limit of B ~ 20; note the lack of discrimination among the models. The dashed lines show the predictions for samples complete to magnitudes fainter than 20. The arrows show the direction of evolution of the luminosity function towards higher redshift; the luminosity-dependent density evolution model of Schmidt and Green (1983) predicts larger density increases for brighter objects.

interpretation is that brighter QSOs die off faster after their birth at high red-shifts (Koo and Kron 1988a), alternatives which modify birth and death rates and luminosity history are equally acceptable to the extent that the continuity equation is satisfied (Cavaliere *et al.* 1982). Both the finding that many low-redshift QSOs appear to be activated by interactions (Stockton 1982, Hutchings 1987) and the claim that QSOs change from lower to higher density environments with redshift (Yee and Green 1987) argue against any picture in which QSOs were all formed in the distant past.

6.2 - Clustering Evolution

Although the clustering history of QSOs may be complicated by evolution in their sites of formation, QSOs remain one of the few available probes

QSO LUMINOSITY FUNCTION

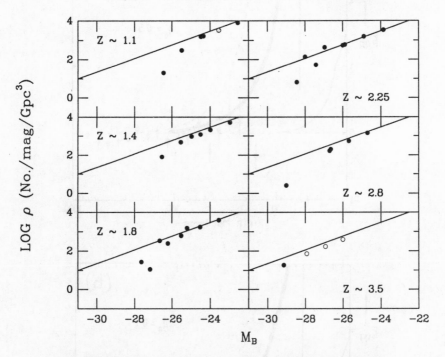

FIG. 14. Each panel shows the observed comoving volume density of QSOs at different redshifts near 1 and larger. A Hubble constant of 50 km sec⁻¹ Mpc⁻¹, a critical $q_0 = 0.5$ and a spectral index of $\alpha = -1$ have been assumed in the calculation of absolute magnitudes. Open circles show values corresponding to the detection of one QSO, though none were found. The solid lines serve as fiducials. Note that the main trend in the evolution of the apparent break of the luminosity function is for higher redshift QSOs to be brighter and if anything, systematically fewer with no abrupt cutoff at even high redshifts. Details of the data from several groups and a density-luminosity model that fits can be found in Koo and Kron (1988a).

at high redshift. Except for a handful of pairs and triplets of QSOs that suggest possible association in superclusters (Oort *et al.* 1981), Woltjer and Setti at the VAC had little else and thus stated: "the data are certainly inadequate to establish fully clustering of quasars." Today the situation has improved but we are still at the 3σ level, though we can expect substantial gains in this area of research as more surveys achieve spectroscopic completeness. Two recent examples of positive detection include a 170 QSO sample that shows strong correlations at less than 10 Mpc (Shanks *et al.* 1987) as seen in Figure 15 and

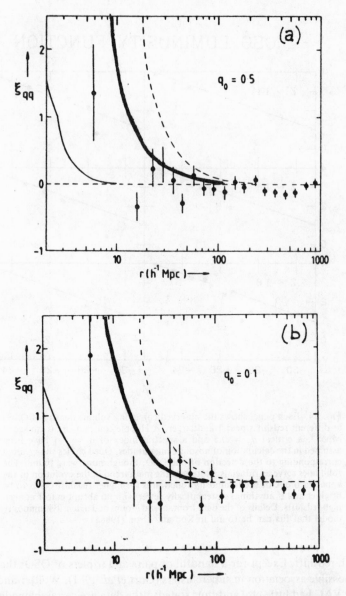

FIG. 15. (a) The QSO two-point correlation function ξ_{qq} for a complete sample of 172 QSOs assuming a value of $q_0 = 0.5$. QSO separation, r, is measured in comoving coordinates. The thin solid line is the expected galaxy correlation function at $z = 1.5$ based on the 'stable' clustering model. The thick solid line and dashed line are the predicted rich cluster correlation functions at $z = 1.5$ based on the 'stable' and 'comoving' models respectively. (b) As (a) for the $q_0 = 0.1$ model (from Shanks *et al.* 1987).

the claim by Shaver (1987) for a significant excess of QSOs on scales larger than superclusters at redshifts z < 0.5 in the direction of the Local Group motion.

6.3 - Primeval Galaxy Candidates

Although the redshift 4 barrier was first toppled by Warren *et al.* (1987*a*) nearly a year ago with a z = 4.01 QSO, a flurry of five new cases has recently appeared, including the record holder of z = 4.43 by Warren *et al.* (1987*b*) and one at z = 4.40 found accidentally by McCarthy *et al.* (1988), while observing a z = 0.71 radio galaxy. These push the epoch of formation, for at least some objects, beyond z = 4 and support the QSO evolution picture drawn above in which no abrupt cutoff of QSOs has yet been found. Moreover, these QSOs provide new high-redshift beacons to possible primeval candidates as have already been found in two lower redshift cases. On their first attempt using narrow-band imaging Djorgovski *et al.* (1985) found a z = 3.22 Lyman-α fuzzy object, presumably a galaxy (though possibly one with some non-thermal activity or interaction with the nearby QSO) (see Figure 16). Another one at z = 3.27 was found near the interesting QSO gravitational lens pair, MG 2016 + 112, by Schneider *et al.* (1986, 1987). In its perversity, the universe has not yielded another one despite deeper surveys of over four dozen other high redshift QSOs (see Djorgovski 1988)!

7. SUMMARY

7.1 - Evolution

Beyond redshifts of z ~ 0.4, evolution in number density, luminosity, or spectral properties has been observed with reasonable confidence in all classes of high redshift objects. The coherent picture that emerges is the importance of gas in this evolution, as the fuel for active nuclei that produces radio galaxies and QSOs, as the source itself in the case of the QSO absorption lines, and as revealed by increased star formation activity in field galaxies and cluster galaxies. The physical mechanisms for the observed evolution remain poorly understood, largely because the major theoretical advancements have been results from N-body simulations using different fluctuation spectra with perhaps some biasing. Admittedly, dynamics that rely solely on gravity may play an important role in controlling evolution through interactions and mergers, but complex inelastic gas processes are likely to dominate more and more as we look further back in time when less gas had already undergone a conversion into stars. The explosive galaxy formation theory of Ostriker and Cowie (1981) and Ikeuchi (1981) shows how gas may even affect large-scale

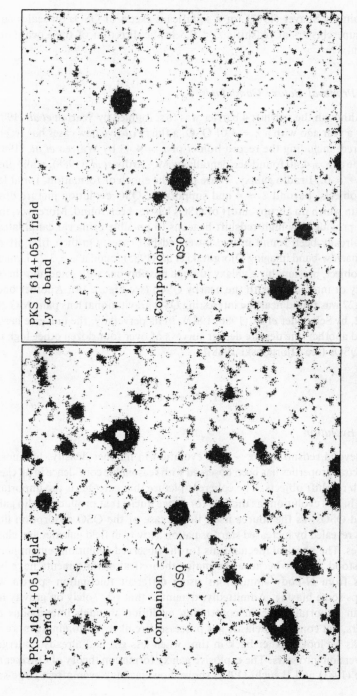

FIG. 16. CCD pictures (North at top; East to left; 73 arcsec each side) of the region around the radio QSO, PKS 1614 + 051 at z = 3.218, taken with a broadband red filter (r) and with a narrow band filter that brackets Lyman-α redshifted by z ~ 3.22. A possible bridge of Lyman-α emission appears to connect the QSO and its companion, presumably a galaxy interacting with it (from Djorgovski *et al.* 1987).

structure; similarly, bright QSOs can ionize vast volumes of the surrounding IGM. The incorporation of gas into some of the N-body codes has already begun (Quinn, private communication) and will undoubtedly provide new insights and understanding of the rapid influx of new observational data in the years to come.

7.2 - Clustering

Several 3σ level detections of clustering at large redshifts have already been reported for QSOs, Lyman-α lines, and field galaxies. Despite its importance as a probe of various cosmological scenarios and perhaps even q_o itself (assuming the comoving clustering scale is approximately invariant over most of the later life of the universe), the research area of tracking the evolution of clustering on all scales is very much in its infancy. With the advent of several fiber-optic spectrograph systems, especially those designed for use in dedicated redshift programs of faint objects, we can expect tremendous progress in this field over the next decade.

7.3 - Primeval Galaxies

Starting with the Golden Age of primeval galaxy searches in the 60's, when Partridge and Peebles (1967) proposed that galaxies at birth would be red (high redshift) and fuzzy (with star-formation occurring at maximum extent of the protogalactic gas cloud), continuing to a Renaissance era in the 70's when the dissipative galaxy formation models of Larson (1974) inspired Meier (1976) to suggest that primeval galaxies might instead be blue (low redshift) and compact (star formation occurring after collapse), we are now in the 80's with CDM models predicting that they might be blue (low redshift) but fuzzy (with blobs continually merging) or with recent IRAS observations hinting that they might even be red (due to dust obscuration) and compact (as galaxies undergoing bursts of star formation). We appear to be on the verge of a new era, when primeval galaxies will finally be discovered. Several good candidates have recently been found among high redshift radio galaxies and QSOs, with many promising avenues yet to be explored. Overall, the clues point to the oldest galaxies being formed at redshifts beyond 3 to 4.

These searches are motivated by the hope of pinning down the age of galaxies and the epoch of galaxy formation, as well as constraining the subsequent processes that control the evolution of star formation rate, initial mass function, chemical enrichment, gas dynamics, structure formation, etc. A reasonable concern is that our underlying assumption of coeval formation of all galaxies is wrong. If so, we can no longer rely on special classes of objects, such as the reddest galaxies, strong emission line galaxies, cluster galaxies, QSO absorption lines, or QSOs to reveal the history of "normal" galaxy forma-

tion. In the worst case, perhaps even a direct attack through a redshift study
of extremely faint field galaxies will not yield definitive answers. On the other
hand, much larger ground based telescopes will soon be a reality; larger CCDs
should be forthcoming; deep infrared imaging and spectroscopy are progressing rapidly; Hubble Space telescope may yet fly; etc. Even if we do not reach
the Holy Grail of watching galaxies being born, the search itself will be exciting and full of surprises. "Look afar and see the end from the beginning"
(Fortune Cookie 1983).

REFERENCES

Bajtlik, S., Duncan, R.C., and Ostriker, J.P. 1988. *Ap J.* **327**, 570.

Baron, E. and White, S.D.M. 1987. *Ap J.* **322**, 585.

Blades, J.C., Norman, C., and Turnshek, D.A. 1987. *Proceedings of STScI Workshop on QSO Absorption Lines: Probing the Universe.*

Bond, J.R., Szalay, A., and Silk, J. 1988. *Ap J.***324**, 627.

Boyle, B.J., Fong, R., Shanks, T., and Peterson, B.A. 1987. *MN.* **227**, 717.

Brück, H.A., Coyne, G.V., and Longair, M.S. (eds.) 1982. *Astrophysical Cosmology.* Vatican City State: Pontifical Academy of Sciences. (VAC).

Butcher, H. and Oemler, A. 1978. *Ap J.* **219**, 18.

_____. 1984. *Ap J.* **285**, 426.

Cavaliere, A., Giallongo, E., Messina, A., and Vagnetti, F. 1982. *AA.* **114**, L1.

Chambers, K., Miley, G., and van Breugel, W. 1987. *Nature,* **329**, 604.

Cheny, J.E. and Rowan-Robinson, M. 1981. *MN.* **195**, 497.

Condon, J.J. and Mitchell, K.J. 1984. AJ. **89**, 610.

Couch, W.J. and Sharples, R.M. 1987. *MN.* **229**, 423.

Crotts, A.P.S. 1987. *MN.* **228**, 41p.

Danese, L., DeZotti, G., Franceschini, A., and Toffolatti, L. 1987. *Ap J Letters.* **318**, L15.

Davis, M. 1980. in *IAU 92, Objects of High Redshift.* eds. G.O. Abell and P.J.E. Peebles. p. 57. Dordrecht: Reidel.

Djorgovski, S. 1988. in *Towards Understanding Galaxies at Large Redshifts.* eds. R.G. Kron and A. Renzini. p. 259. Dordrecht: Reidel.

Djorgovski, S., Spinrad, H., McCarthy, P., and Strauss, M. 1985. *Ap J Letters.* **299**, L1.

Djorgovski, S., Strauss, M., Perly, R., Spinrad, H., and McCarthy, P. 1987. *AJ.* **93**, 1318.

Dressler, A. 1987. in *Nearly Normal Galaxies.* ed. S.M. Faber. p. 265. New York: Springer-Verlag.

Dressler, A. and Gunn, J.E. 1988. in *Large-Scale Structures of the Universe,* eds. J. Audouze, M.-C. Pelletan, and A. Szalay. Dordrecht: Kluwer Academic Pubblisher, p. 311.

Ellis, R.S. 1987. in *IAU Symp. 124, Observational Cosmology.* eds. A. Hewitt, G. Burbidge, and L.Z. Fang., p. 367. Dordrecht: Reidel.

_____. 1988a. in *Towards Understanding Galaxies at large Redshifts.* eds. R.G. Kron and A. Renzini, p. 147. Dordrecht: Reidel.

_____. 1988b. in *High Redshift and Primeval Galaxies.* eds. J. Bergeron, D. Kunth, and B. Rocca-Volmerange.

Fortune Cookie. 1983, eaten with inspiration by the author.

Gunn, J.E. and Dressler, A. 1988. in *Towards Understanding Galaxies at Large Redshifts*. eds. R.G. Kron and A. Renzini. p. 227. Dordrecht: Reidel.

Gunn, J.E. and Peterson, B.A. 1965. *Ap J.* **142**, 1633.

Gunn, J.E., Hoessel, J., and Oke, J.B. 1986. *Ap J.* **306**, 30.

Hamilton, D. 1985. *Ap J.* **297**, 371.

Heckman, T.M., Smith, E.P., Baum, S.A., van Breugel, W.J.M., Miley, G.K., Illingworth, G.D., Bothun, G.D., and Balick, B. 1986. *Ap J.* **311**, 526.

Hoag, A.A. 1981. *BAAS.* **13**, 799.

Hutchings, J.B. 1987. *Ap J.* **320**, 122.

Ikeuchi, S. 1981. *PASJ.* **33**, 211.

Ikeuchi, S. and Ostriker, J. 1986. *Ap J.* **301**, 522.

Jenkins, E.B., Caulet, A., Wamstecker, W., Blades, J.C., Morton, D.C., and York, D.G. 1987. in *QSO Absorption Lines: Probing the Universe*. eds. J.C. Blades, C. Norman, and D.A. Turnshek. p. 34. (STScI).

Katgert, P., de Ruiter, H.R., and van der Laan, H. 1979. *Nature.* **280**, 20.

Koo, D.C. 1985. *AJ.* **90**, 418.

———. 1986a. in *The Spectral Evolution of Galaxies*. eds. C. Chiosi and A. Renzini. p. 419. Dordrecht: Reidel.

———. 1986b. *Ap J.* **311**, 651.

———. 1988. in *Large-Scale Structures of the Universe,* eds. J. Audouze, M.-C. Pelletan, and A. Szalay. Dordrecht: Kluwer Academic Publishers, p. 221.

Koo, D.C. and Kron, R.G. 1988a. *Ap J.* **325**, 92.

———. 1988b. in *Towards Understanding Galaxies at Large Redshifts*. eds. R.G. Kron and A. Renzini. p. 209. Dordrecht: Reidel.

Koo, D.C. and Szalay, A.S. 1984. *Ap J.* **282**, 390.

Kron, R.G., Koo, D.C., and Windhorst, R.A. 1985. *AA.* **146**, 38.

Larson, R.B. 1974. *MN.* **166**, 585.

Lilly, S.J. 1988. *Ap J.* **333**, 161.

Lilly, S.J. and Longair, M.S. 1984. *MN.* **211**, 833.

Loh, E.D. 1988. in *Large-Scale Structures of the Universe,* eds. J. Audouze, M.-C. Pelletan, and A. Szalay. Dordrecht: Kluwer Academic Publishers, p. 529.

Loh, E.D. and Spillar, E.J. 1986. *Ap J Letters.* **307**, L1.

Lynds, R. and Petrosian, V. 1986. *BAAS.* **18**, 1014.

———. 1988, preprint.

MacLaren, I., Ellis, R.S., and Couch, W.J. 1988. *MN.* **230**, 249.

Majewski, S.R. 1988. in *Towards Understanding Galaxies at Large Redshifts*. eds. R.G. Kron and A. Renzini. p. 203. Dordrecht: Reidel.

Mathez, G. 1976. *AA.* **53**, 15.

McCarthy, P.J., Spinrad, H., Djorgovski, S., Strauss, M.A., van Breugel, W., and Liebert, J. 1987a. *Ap J Letters.* **319**, L39.

McCarthy, P.J., van Breugel, W., Spinrad, H., and Djorgovski, S. 1987b. *Ap J Letters.* **321**, L29.

McCarthy, P.J., Dickinson, M., Filippenko, A.V., Spinrad, H., and van Breugel, W.J.M. 1988. *Ap J Letters.* **328**, L29.

Meier, D.L. 1976. *Ap J.* **207**, 343.

Miley, G. 1987. in *IAU Symp. 124 Observational Cosmology.* eds. A. Hewitt, G. Burbidge, and L.Z. Fang. p. 267. Dordrecht: Reidel.

Murdoch, H.S., Hunstead, R.W., Pettini, M., and Blades, J.C. 1986. *Ap J.* **309**, 19.

Oke, J.B. 1983. in *Clusters and Groups of Galaxies.* eds. F. Mardiorossian, G. Giuricin, and M. Mesetti. p. 99. Dordrecht: Reidel.

Oort, J.H., Arp, H., and de Ruiter, H. 1981. *AA.* **95**, 7.

Osmer, P.S. 1982. *Ap J.* **253**, 28.

Ostriker, J.P., Bajtlik, S., and Duncan, R.C. 1988. *Ap J Letters.* **327**, L35.

Ostriker, J.P. and Cowie, L.L. 1981. *Ap J Letters.* **243**, L127.

Partridge , R.B. and Peebles, P.J.E. 1967. *Ap. J.* **147**, 868.

Peacock, J.A. 1985. *MN.* **217**, 601.

Pritchet, C.J. and Hartwick, F.D.A. 1987. *Ap. J.* **320**, , 464.

Pritchet, C. and Infante, L. 1986. *AJ.* **91**, 1.

Rees, M. 1986. *MN.* **218**, 25p.

Rees, M. and Carswell, R. 1987. *MN.* **224**, 13p.

Sargent, W.L.W., Young, P.J., Boksenberg, A., and Tytler, D. 1980. *Ap J.* **42**, 41.

Schmidt, M. and Green, R.F. 1983. *Ap J.* **269**, 352.

Schneider, D., Gunn, J.E., Turner, E., Lawrence, C., Hewitt, J., Schmidt, M., and Burke, B. 1986. *AJ.* **91**, 991.

Schneider, D., Gunn, J.E. Turner, E., Lawrence, C., Schmidt, M., and Burke, B. 1987. *AJ.* **94**, 12.

Shanks, T., Fong, R., Boyle, B.J., and Peterson, B.A. 1987. *MN.* **227**, 739.

Shapiro, P.R. 1986. *PASP.* **98**, 1014.

Shaver, P.A. 1987. *Nature.* **326**, 773.

Smith, H.E., Cohen, R.D., and Bradley, S. 1986. *Ap J.* **310**, 583.

Smith, H.E., Cohen, R.D., and Burns, J.E. 1987. in *QSO Absorption Lines: Probing the Universe.* eds. J.C. Blades, C. Norman, and D.A. Turnshek. p. 148. (STScI).

Soucail, G., Mellier, Y., Fort, B., Mathez, G., and d'Odorico, S. 1987. *IAU Circular No. 4482.*

Spinrad, H. 1986. *PASP.* **98**, 269.

Steidel, C.C. and Sargent, W.L.W. 1987. *Ap J Letters.* **318**, L11.

Stevenson, P.R.F., Shanks, T., Fong, R., and MacGillivray, H.T. 1985. *MN.* **213**, 953.

Stockton, A. 1982. *Ap J.* **257**, 33.

Tyson, J.A. 1988, preprint.

Tytler, D. 1987. *Ap J.* **321**, 69.

Warren, S.J., Hewett, P.C., Irwin, M.J., McMahon, R.G., Bridgeland, M.T., Bunclark, P.S., and Kibblewhite, E.J. 1987a. *Nature.* **325**, 131.

Warren, S.J., Hewett, P.C., Osmer, P.S., and Irwin, M.J. 1987b. *Nature.* **330**, 453.

Webb, J.K. 1987. in *IAU Symp. 124 Observational Cosmology.* eds. A. Hewitt, G. Burbidge, and L.Z. Fang. p. 803. Dordrecht: Reidel.

White, S.D.M., Frenk, C.S., Davis, M., and Efstathiou, G. 1987. *Ap J.* **313**, 505.

Windhorst, R.A. 1984. *Ph. D. thesis.* Leiden University.

Windhorst, R.A., Dressler, A., and Koo, D.C. 1987. *BAAS.* **18**, 1006.

Windhorst, R.A., Miley, G.A., Owen, F.N., Kron, R.G., and Koo, D.C. 1985. *Ap J.* **289**, 494.

Wolfe, A., Turnshek, D., Smith, H.E., and Cohen, R.D. 1986. *Ap J Suppl.* **61**, 249.

Yee, H.K.C. and Green, R.F. 1987. *Ap J.* **319**, 28.

FIELD AND CLUSTER GALAXIES:
DO THEY DIFFER DYNAMICALLY?

VERA C. RUBIN

*Department of Terrestrial Magnetism,
Carnegie Institution of Washington*

1. INTRODUCTION

Astronomers are optimists by nature. To decipher the large-scale structure of the universe from observations of bits and pieces of galaxies made at vast distances demands optimism. And the results discussed in this volume, the initial successes in mapping large-scale motions in the universe, show that our optimism is warranted.

However, we should not be so impressed by our success that we fail to assess our procedures critically. Between the observations and the conclusions lie many assumptions, some of which we continually examine: Malmquist biases, extinction corrections, sampling effects. I examine a different question here: Do galaxies of similar morphology have similar dynamical properties, regardless of their local environments and their evolutionary histories? Over fifty years ago Hubble and Humason (1931) noted that galaxies are segregated by types, depending upon their location. More recently Dressler (1980a) quantified this separation by Hubble types: 80% of field galaxies are spirals, but as few as 15% of cluster galaxies are spirals, and these are most often found in the cluster periphery. Depending predominantly on the local density, clustering properties differ (Davis and Geller 1976), angular correlations differ (Haynes and Giovannelli 1988), luminosity functions differ (Sandage *et al.* 1985), and dwarf fraction differs (Sharp, Jones, and Jones 1988), among other properties.

Is it then reasonable to expect that the Tully-Fisher and the Faber-Jackson relations have identical zero points and identical slopes at all locations across the sky and in depth? Are the rotation curves the same whether we study galaxies in the field or in clusters or with optical telescopes or with radio telescopes?

For parts of these questions answers exist. It is now well-established that some spiral galaxies in clusters have HI disks which are deficient in neutral hydrogen (Chamaraux *et al.* 1980, Haynes *et al.* 1985) compared to their cohorts in the field, and that in Virgo this deficiency decreases as a function of distance from M87 (van Gorkom and Kotanyi 1985), decreases as a function of later Hubble type, is a function of orbital parameters (Dressler 1986), and is accompanied by a truncation and an asymmetry related to the direction to M87 (Warmels 1985). Surprisingly, the CO diameters, normalized by HI diameters, are larger near the core of the cluster (Kenney and Young 1986), making it possible that the HI profile width can be artificially reduced if the HI disk does not extend as far as the peak of the rotation curve (Stauffer, Kenney, and Young 1986). Distributions and fluxes of HII regions (Kennecutt *et al.* 1984) of cluster and field spirals are generally indistinguishable. Optical spectra of the inner parts (for the few galaxies with measured velocities) appear normal (Chincarini and de Souza 1985), but no sample prior to the one discussed here has rotation curves for cluster galaxies which extend to the outer regions of spiral disks. We do not yet know if the maximum rotation velocity of the optical disk is never, sometimes, often, or always located beyond the limits of the neutral hydrogen disk in HI deficient galaxies. This is a large ignorance.

It is fair to assume, I think, that dynamical differences between field and cluster spirals are not large, or variations would already be known. But even subtle differences are important if we misinterpret small variations among galaxies as large-scale velocities. There are numerous occasions in the history of astronomy when small differences which were overlooked have led to large misunderstandings. All studies of large-scale motions are based on the premise that we understand galaxy properties so well that we can define parameters or correlations that remain unchanged as a function of environment, and that we can use these correlations to estimate the distance and hence the expected velocity of recession for any galaxy. In this contribution, I examine the rotation properties of galaxies in spiral-rich clusters, and in the compact Hickson groups, and ask if they differ as a function of position in the cluster, and from field spirals.

2. Rotation Curves for Galaxies in Spiral-Rich Clusters

2.1 - Forms of Rotation Curves

Although over 100 spiral galaxies now have accurate emission line rotation curves extending virtually to the limits of the optical disk, few cluster galaxies have comparable rotation curves. Data necessary to carry out a comparison of optical rotation curves of field and cluster galaxies are woefully incomplete. Over the past few years Rubin *et al.* (1988a) have observed optical rotation curves for about 20 Sa, Sb, Sc, and Irr galaxies in four spiral-rich clusters:

Cancer, Peg I, Hercules, and DC 1842-63 (Dressler 1980b). While individually many of the rotation curves appear normal, a study of their mass forms [i.e., log(V^2R) interior to R versus log R] by Burstein *et al.* (1986) indicates that statistically they differ subtly from the mass forms defined by field galaxies. Particularly, rotation curves with steeply rising inner velocities and rising outer velocities (mass types I) are absent from the cluster sample.

Of the cluster galaxies, 13 have morphology that can be called normal. Ten have rotation curves of normal form; only seven have rotation curves of normal amplitude. Rotation curves for some cluster galaxies are shown in Figure 1. Two-thirds of the galaxies of normal morphology have rotation curves of normal form; most of the galaxies of peculiar morphology have rotation curves which are peculiar. Thus, while we can generalize that normal galaxies have normal rotation curves and peculiar galaxies have peculiar rotation curves, the counter examples are equally important. The most common abnormality is a falling rotation curve, which is observed for about 1/3 of the cluster galaxies, of which about half have normal morphology.

A good correlation is found (Whitmore *et al.* 1988) between the outer gradients of the rotation curves and the distances from the centers of the clusters, in the sense that galaxies near the cluster core tend to have falling rotation curves, while galaxies farther from the cluster center [and the field galaxies studied earlier (Rubin *et al.* 1985)] have flat or rising rotation curves (Fig. 2). The outer gradient is defined (Whitmore 1984) as the percentage increase in the rotation velocity from $R = 0.4R_{25}$ to $R = 0.8R_{25}$, normalized to Vmax. While the sample is small, especially when divided into Hubble types, and the scatter large, gradients which are zero or negative (i.e., falling rotation curves) occur only in galaxies observed at small central distances, while large gradients are seen only in galaxies at large central distances. It is interesting that, although the *range* of values for the outer gradient in the field sample is as large as the range of values observed in the cluster sample, 6 of the 16 cluster galaxies (with well determined rotation curves) have zero or negative outer gradients, compared with only 7 of the 50 or so field spirals. Moreover, while for the field sample the outer gradients clump about zero or slightly positive values, for the cluster galaxies the values spread uniformly over the entire range. Many cluster spirals have rotation properties which resemble those of field spirals, but there is the additional small population of spirals near the cluster cores whose dynamics are unlike field spirals. The cluster galaxies must have had a more diverse evolutionary history than the field galaxies to establish these differences.

The ratio of mass-to-red luminosity ($M/L = V^2R/L_R$) interior to R is shown as a function of radial distance in the galaxy in Figure 3 for several cluster galaxies. The galaxies are arranged in order of the distance from the cluster centers, $R_{cluster}$. All cluster galaxies, except the few closest to the cluster centers, exhibit integral M/L ratios which increase with galaxy radius. A

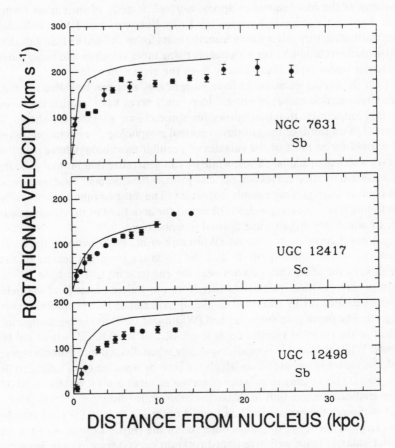

Fɪɢ. 1a. Rotation velocities in the plane of the galaxies as a function of nuclear distance for 3 galaxies in the Peg I cluster, compared with the rotation curves predicted by field spirals of corresponding morphology and luminosity. The mean velocity and the 1σ error bars are indicated for all measurements in each radial bin. The cluster rotation curves we have observed are generally low compared to their field cohorts.

luminous disk of constant M/L ratio, the assumption made in models which deconvolve the mass into disk and halo, is a horizontal line on Figure 3. This is not an acceptable fit to the observed M/L variation, and hence is not a suitable mass model for any program galaxy beyond about 0.3 Mpc from the cluster center. Cluster galaxies beyond this distance require a dark halo.

I show in Figure 4 the gradient in M/L [defined as:

$$M/L(0.8R/R_{25})/M/L(0.1R/R_{25})$$

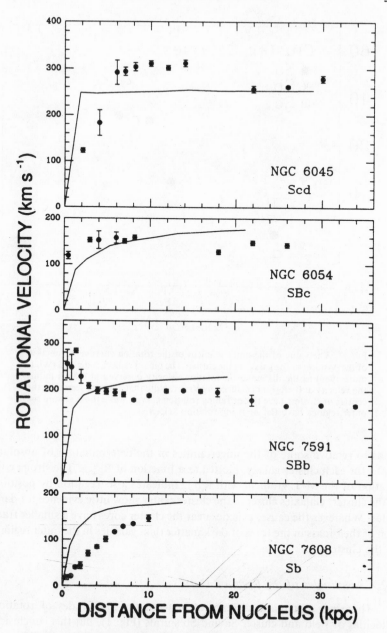

FIG. 1b. Rotation velocities in the plane of the galaxies as a function of nuclear distance for 2 galaxies in Hercules (NGC 6045 and NGC 6054) and 2 galaxies in the Peg I cluster, compared with the rotation curves predicted by field spirals of corresponding morphology and luminosity. Note the peculiar forms for some of the rotation curves.

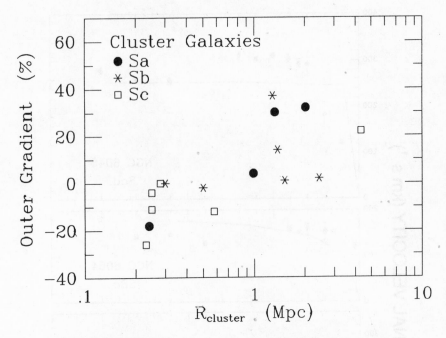

Fig. 2. The value of the outer gradient of the rotation curve as a function of the position of the galaxy in the cluster. The outer gradient is defined (Whitmore 1984) as the difference in rotation velocity observed at $0.8R_{25}$ minus the velocity at $0.4R_{25}$, normalized by Vmax. Note that galaxies seen near the centers of clusters have flat or falling rotation curves, while those seen in the outer regions have flat or rising rotation velocities.

so as to remove some of the uncertainties in the determination of absolute M/L] for each cluster galaxy, plotted as a function of $R_{cluster}$. The strong correlation of the M/L gradient with $R_{cluster}$ indicates a segregation by position in the cluster. Galaxies closer to the cluster center show little evidence for dark halos. Whatever the cause, galaxies near the cluster cores have a smaller fraction of their mass in the form of dark matter than galaxies in the outer regions of the clusters.

2.2 - Amplitudes of the Rotation Curves

There are systematic differences between the amplitudes of rotation velocities of field and cluster Sa and Sb spirals (Fig. 1), but this conclusion is of low statistical weight due to the small sample size and the possibility of systematic differences in magnitude scale compared with the field galaxies. We place each cluster galaxy at its mean cluster distance, and use its observed Vmax to predict an absolute magnitude, based on field galaxies of the same

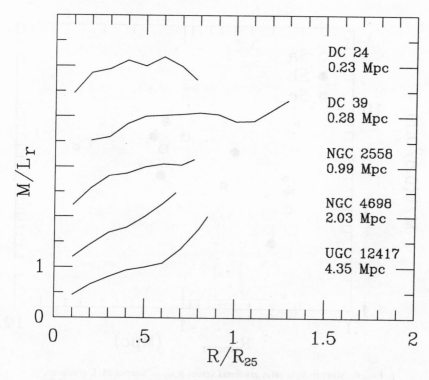

FIG. 3. Mass-to-red light ratios (normalized to 1 at $0.5R_{25}$) versus the distance from the center of the galaxy; galaxies are arranged in order of increasing distances from the centers of the clusters.

Hubble type; alternatively, we use its photometric absolute magnitude to predict a rotation curve and the value of Vmax. All Sa and Sb cluster galaxies have rotation curves which fall below the rotation curve predicted (Rubin *et al.* 1985) for a field galaxy of equivalent Hubble type and luminosity. An equivalent statement is that the Tully-Fisher relation has a different zero-point for the cluster and field samples.

This result, if supported by more extensive rotation curve data, has implications for the evaluation of the Hubble constant and for large-scale motions. However, we are aware of the numerous systematic effects which can enter the cluster analysis (Rubin *et al.* 1988a). Different selection procedures for the field and cluster samples may make the field synthetic rotation curves inappropriate for the cluster sample. Cluster galaxy apparent magnitudes, most from our CCD photometry, may be systematically too bright compared with the Zwicky values for field spirals. Our data indicate that internal extinction in cluster spirals may be smaller than internal extinction in the field counter-

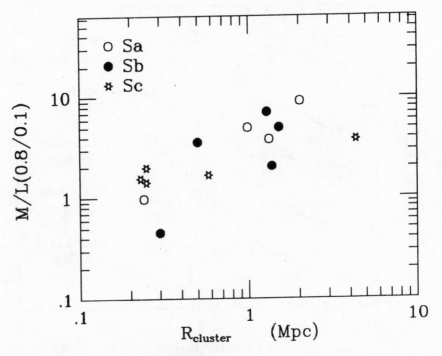

FIG. 4. Mass-to-light ratio gradients versus $R_{cluster}$. Steeper M/L gradients, evidence of massive halos, are found in galaxies in the outer regions of the clusters.

parts, a suggestion previously advanced by van den Bergh (1984) for cluster spirals and by Sandage and Tammann (1981) for Sa galaxies. Or each cluster may have a motion of 600 to 800 km/s away from the observer with respect to the smooth Hubble flow. This is expected for DC 1852-63, which is located near the apex of the microwave background dipole and not far on the sky or in velocity from the Great Attractor. However, velocities of approach, not recession, are predicted on this model for the Cancer and Peg I clusters. This is discussed further in Rubin (1988).

2.3 - Comparison of Optical Studies and 21-cm Observations of Clusters Spirals

In many observations and discussions of the Tully-Fisher relation, optical observations of field spirals are combined with 21-cm observations of cluster spirals. Many calibrators are field spirals. It is an unsolved puzzle why the optical data, generally for field galaxies, generally with blue magnitudes, sometimes show a steeper slope and a separation by Hubble type in a plot of luminosity versus log Vmax, compared with the 21-cm data, which are generally for cluster galaxies, often with H magnitudes.

The determination of the slope for the Tully-Fisher relation has been the subject of many studies (Tully and Fisher 1977, Roberts 1978, de Vaucouleurs *et al.* 1982, Richter and Huchtmeier 1984, Rubin *et al.* 1985, Kraan-Korteweg *et al.* 1987). While the Rubin field sample gives a slope of 10, other samples, predominantly cluster spirals, show slopes generally in the range from 5 to 8. Our small cluster sample also indicates shallower slopes, 6.1 for Sb's, 8.8 for Sc's. Perhaps the difference is simply the result of mixing two populations; field spirals with a steep slope and cluster spirals with a shallower slope .

Kraan-Korteweg *et al.* (1987) suggest that the large number of very bright galaxies in the Rubin field sample may increase the slope in the Tully-Fisher relation. But removing those with luminosities greater than $10^{11}L_\odot$ (i.e., the top 30%) reduces the slope for the field sample insignificantly.

To compare Vmax optical with the 21-cm profile width, we need a galaxy-by-galaxy comparison for HI deficient galaxies. Only then can we answer the question, To the limits of the optical disks what are the rotation properties of galaxies which have tidally truncated HI disks? In a pioneering effort Chincarini and de Souza (1985) obtained optical rotation curves for 8 spirals in Virgo known to be deficient in HI, and showed them to be normal. However, the limited extent of the optical spectra are not adequate to answer the above question.

Data do not yet exist to investigate directly such an effect, but I attempt a first look. I plot in Figure 5 the Fisher-Tully relation for Virgo spirals from the recent study of Kraan-Korteweg (1987), along with the best fit she determines. Vmax values come from integrated 21-cm profiles. Only galaxies with measured deficiency values are shown; this leaves 6 galaxies classified Sa or Sab, and 25 classified Sbc, Sc, or Scd. Although the sample suffers from the segregation usual in samples with only a small range in luminosity for each Hubble type, with Sd and Im at lowest magnitudes and Sa and Sab at highest, all galaxies define the same broad Tully-Fisher line. However, of the 8 galaxies with deficiencies over 0.80 (i.e., HI deficient by $\log^{-1} 0.8$ compared with field galaxies), 5 are classified Sab, 3 are Sc. All Sa's lie far from the mean line, as they will if Vmax for these galaxies is too small. The case is far from proven, but is suggestive and points a direction for future observations. It will be important to learn the optical rotation properties of the 5 Sab galaxies.

Only eight cluster galaxies in our sample have measured HI deficiencies (Giovanelli and Haynes 1985). The sample size is disappointingly small, yet there is a trend of HI deficiency to increase with increasing depression of the cluster rotation curves from the template rotation curves (Figure 6). That is, cluster spirals whose rotation velocities lie lowest when compared with field galaxies of the same morphology and luminosity have the largest measured HI deficiencies. Following the discussion at the Study Week, and in an effort to examine

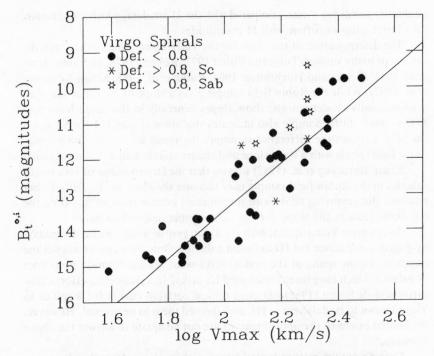

FIG. 5. The Tully-Fisher diagram for spirals in the Virgo cluster (from Kraan-Korteweg *et al.* 1987) which have a measured deficiency (Giovanelli and Haynes 1985). Although the scatter is large, there is no separation by Hubble type. However, Sab galaxies with the largest HI deficiencies lie highest with respect to the mean line copied from Kraan-Korteweg *et al.* See text for details.

this effect for a larger sample, Burstein and Rubin are presently calculating deficiency values for the field and cluster spirals in the Aaronson *et al.* (1986) sample, to see if these deficiency values correlate with the inferred deviations from a smooth Hubble flow. If these two parameters correlate, (and there is as yet no evidence that they do), this would offer evidence that deviant dynamical properties are an alternative explanation for large-scale motions.

3. DYNAMICS OF SPIRALS IN DENSER ENVIRONMENTS

Spiral galaxies located in large clusters, especially late type spirals, could be anomalies: they may inhabit outlying regions, even if seen near the centers by projection effects, or they may represent a population which has maintained a normal spiral morphology by lack of interaction with other cluster members. In order to resolve questions concerning differences between spiral galaxies isolated in the field and those located in truly crowded regions, we have ex-

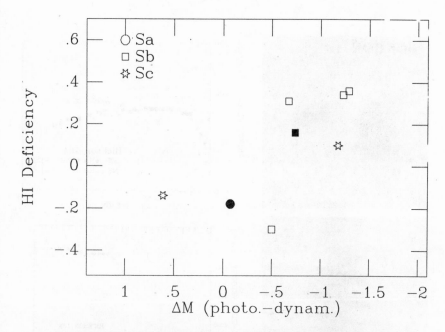

FIG. 6. The HI deficiency for 8 galaxies in the Cancer and Peg I clusters, plotted against $\Delta M_{photo-dyn}$ (i.e., the difference between the photometrically determined absolute magnitude and that determined from the rotation curve). A positive value of ΔM means that the cluster rotation velocity is low compared with that of a field galaxy of corresponding Hubble type and luminosity. Open Symbols, Peg I, filled symbols, Cancer.

tended our observations to study the dynamical properties of spirals located in the compact groups of galaxies catalogued by Hickson (1982). If these are bound groups, then the spirals inhabit regions of density as high as those near the cores of large clusters.

Compact groups pose an intriguing question for galaxy environmentalists. Crossing times are short; mergers and tidal effects would destroy the galaxies in times much less than a Hubble time. Yet morphological and spectral evidence indicate that many of these groups are dynamical entities. Rubin *et al.* (1988b) are currently completing a study of 40 spirals and 10 elliptical and S0 galaxies in 12 Hickson groups. Although many of the galaxies have a relatively normal morphology, most of them have rotation patterns which are at least moderately abnormal. Characteristic abnormalities are shown in Figures 7 and 8. Hickson 16a, an SBab spiral, and Hickson 100a, an Sb galaxy, have rotational velocities which are dissimilar on the two sides of the major axis, although each side separately would probably not be recognized as abnormal. A continuous warping might produce such an effect but, especially in H100a, the

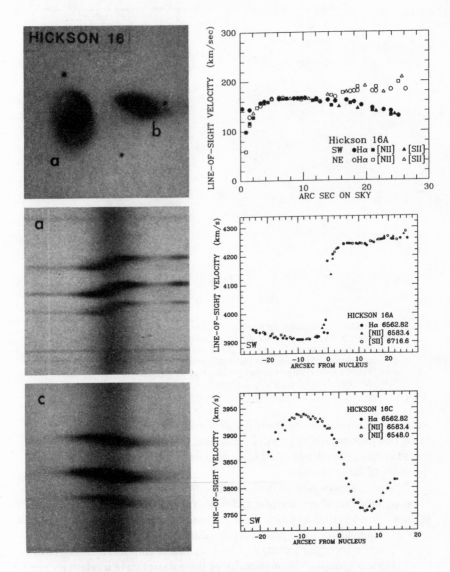

FIG. 7. Observations of Hickson 16 (Arp 316). The red image, showing galaxies a and b, comes from a Kitt Peak 36-inch CCD frame taken by J. Young. Major axis spectra of H16a (SBab) and H16c (Im) were taken with the Palomar 200-inch double spectrograph; the scale and dispersion are 0.8A/pix and 0.59''/pix. Integration times were 60 and 70 minutes, respectively. Note the lack of agreement between the SW and the NE rotation velocities in H16a beyond 14'', corresponding to the interarm region, and the sinusoidal rotation curve for H16c.

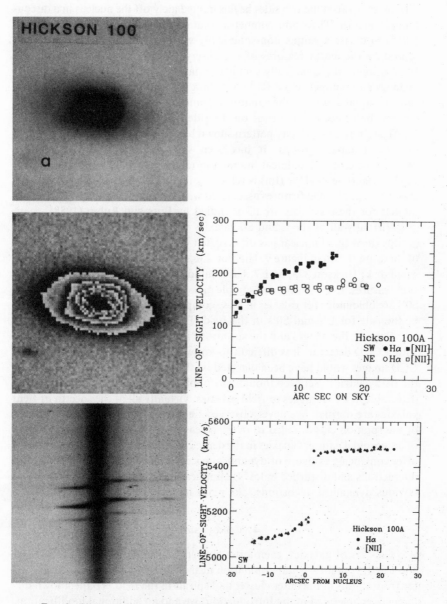

FIG. 8. Observations of Hickson 100a (Sb), N is up and E to the left. The red image comes from a KPNO 36-inch CCD frame taken by J. Young. Note the curious structure to the E, and the boxy intermediate isophotes. Velocities obtained from the Palomar 200-inch spectrum (scale and dispersion as in Fig. 7; integration time 60 min.) show a strong SW/NE asymmetry starting just beyond the nucleus.

difference between the two sides begins immediately off the nucleus in a discontinuous manner. These non-symmetric rotation curves have led Boss and Rubin (1988) to devise a simple non-spherically symmetric mass model which can reproduce the major features of the observed asymmetrical rotation curve. In addition to the spherically symmetric dark halo with $\varrho \propto R^{-2}$, this model includes an asymmetric $\varrho \propto R^{-1} \cos \theta$ dark halo. In the adopted model, contours of equal density in the equatorial plane are non-concentric approximate circles; the rotation curve rises on the side where the density is higher.

The sinusoidal velocity pattern shown by H16c is also often seen in spirals in these compact groups. It has been suggested that this pattern is a characteristic of a dynamical interaction (Schweizer 1982, Rubin and Ford 1983), even though H16c (Im) is relatively featureless and exhibits few of the major morphological features associated with a recent interaction. Halo mass models for these systems are also studied by Boss and Rubin (1988).

Virtually all of the E and S0 galaxies we have observed in the Hickson groups show small nuclear gas disks, often with high rotation velocities. Two of these are shown in Figure 9. Hickson 23c, an S0, contains a counter-rotating small disk; Hickson 37a, an E7, has a rapidly rotating disk with strong emission features. Even this limited sample makes it clear that gas disks in E's and S0's are ubiquitous for galaxies in these compact groups. Gas disks may equally be the rule for E's and S0's in the field (Phillips *et al.* 1986).

Because the images and the spectra of the Hickson galaxies are obtained with a CCD detector, it is difficult to know if the weak features seen on the CCD images would have been detected on the earlier photographic images of the field galaxies. Yet even though morphological peculiarities might now be detectable in CCD images of field galaxies, virtually all of the spectra of field galaxies are normal. We never observed field galaxies in which the two sides of the rotation curves failed to overlap reasonably well, nor rotation curves of a sinusoidal nature. Galaxies in the dense environment of the compact groups offer convincing evidence of dynamical peculiarities, probably related to tidal interactions and disturbed halos. By implication, it is likely that smaller, less obvious dynamical distortions take place in galaxies located in clusters.

4. Conclusions

Clusters of galaxies, even loose clusters like Cancer, Hercules (A2151), and Peg I, contain a spiral population near their cores, whose properties differ from spirals in the outer cluster and spirals in the field. To the known differences enumerated in the Introduction, we add an additional modification. Spiral galaxies located near the cluster cores have falling rotation curves. This suggests that only in the dense cluster cores are properties sufficiently extreme that they can modify the mass distributions of the halos. Halos of spirals in cluster cores may have been stripped by gravitational interactions with other

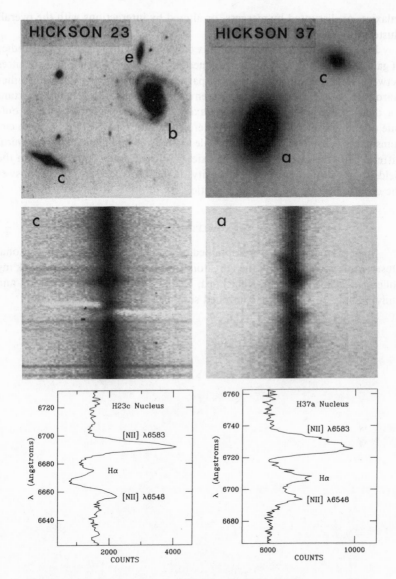

FIG. 9. Red images and spectra of Hickson 23c (S0) and Hickson 37a (E7). The images were taken with the KPNO 36-inch telescope (H23 by J. Young), and the spectra with the Palomar 200-inch double spectrograph (dispersion and scale as in Fig. 7; integration times 60 min. and 33 min.). The spectra are displayed after removal of 80% of the galaxy continuum radiation. Note strong Hα absorption, and the counter-rotating [NII] along the major axis of H23c. In H37a, the rapid rotation in Hα and [NII] is observed in position angle 312°, the position angle joining H37a and H37c, but 36° from the measured major axis of H37a.

galaxies, or disrupted by mergers, or altered by interactions with the overall cluster gravitational field.

It is not difficult to reconcile this conclusion with recent 21-cm studies of galaxies in clusters which have produced little evidence for major differences between field and cluster spirals. In particular, from their extensive studies Aaronson *et al.* (1986) point out that environmental effects are not important for their conclusions because the galaxies studied are generally located outside of the cluster cores, and are generally of late type. Although these circumstances will aid studies of large-scale motions by minimizing any dynamical differences which exist between galaxies in cluster cores and galaxies in the field, they will not satisfy the curiosity of those of us who want to answer the question posed by this paper's title.

ACKNOWLEDGEMENTS

Spectra and images were obtained at Palomar and Kitt Peak National Observatories, and I thank the Directors for observing time. I also thank my colleagues: Dave Burstein, Kent Ford, Deidre Hunter, Brad Whitmore, and Judy Young, for for their continued support and collaboration.

REFERENCES

Aaronson, M., Bothun, G., Mould, J., Huchra, J., Schommer, R. A., and Cornell, M. E. 1986. *Ap J.* **303**, 536.

Boss, A. P. and Rubin, V. C. 1988, in preparation.

Burstein, D., Rubin, V. C., Ford, W. K., and Whitmore, B. C. 1986. *Ap. J Letters.* **305**, L11.

Chamaraux, P., Balkowski, C., and Gerard, E. 1980. *Ap J.* **83**, 38.

Chincarini, G. and de Souza, R. 1985. *AA.* **153**, 218.

Davis, M. and Geller, M. J. 1976. *Ap J.* **208**, 13.

de Vaucouleurs, G., Buta, R., Bottinelli, L., Gouguenheim, L., and Paturel, G. 1982. *Ap J.* **254**, 8.

Dressler, A. 1980a. *Ap J.* **236**, 351.

_____. 1980b. *Ap J.* **42**, 569.

_____. 1986. *Ap J.* **301**, 35.

Giovanelli, R. and Haynes, M. P. 1985. *Ap J.* **292**, 404.

Haynes, M. P. and Giovanelli, R. 1988, this volume.

Haynes, M. P., Giovanelli, R., and Chincarini, G. L. 1985. *Ann Rev Astr Ap.* **22**, 445.

Hickson, P. 1982. *Ap J.* **255**, 382.

Hubble, E. and Humason, M. L. 1931. *Ap J.* **74**, 43.

Kennecutt, R. C., Bothun, G. D., and Schommer, R. A. 1984. *AJ.* **89**, 1279.

Kenney, J. D. and Young, J. 1986. *Ap J Letters.* **301**, L13.

Kraan-Korteweg, R. C., Cameron, L. C., and Tammann, G. A. 1987, preprint.

Phillips, M. M., Jenkins, C. R., Dopita, M. A., Sadler, E. M., and Binette, L. 1986. *A J.* **91**, 1062

Richter, O. G. and Huchtmeier, W. K. 1984. *AA.* **132**, 253.

Roberts, M. S. 1978. *AJ.* **83**, 1026.

Rubin, V. C. 1988, this volume.

Rubin, V. C., Burstein, D., Ford, W. K., and Thonnard, N. 1985. *Ap J.* **289**, 81.

Rubin, V. C. and Ford, W. K. 1983. *Ap J.* **271**, 556.

Rubin, V. C., Hunter, D., and Ford, W. K. 1988a, in preparation.

Rubin, V. C. Whitmore, B. C., and Ford, W. K. 1988b. *Ap J.* **333**, 522.

Sandage, A., Binggeli, B., and Tammann, G. A. 1985. *AJ.* **90**, 1759.

Sandage, A. and Tammann, G. A. 1981. *A Revised Shapley-Ames Catalog of Bright Galaxies,* Washington, D.C.: Carnegie Institution of Washington.

Schweizer, F. 1982. *AJ.* **252**, 455.

Sharp, N. A., Jones, B. J. T., and Jones, J. E. 1988. *MN.* **185**, 457.

Stauffer, J. B., Kenney, J. D., and Young, J. S. 1986. *AJ.* **91**, 1286.

Tully, R. B. and Fisher, J. R. 1977. *AA*. **54**, 661.

van den Bergh, S. 1984. *AJ*. **89**, 608.

van Gorkom, J. and Kotanyi, C. 1985. in *The Virgo Cluster*. p. 61. ed. O.G. Richter and Binggeli, B. Munich: ESO.

Warmels, R. H. 1985. in *The Virgo Cluster*. p. 51. ed. O.G. Richter and Binggeli, B. Munch: ESO.

Whitmore, B. C. 1984. *Ap J*. **278**, 61.

Whitmore, B. C., Forbes, D. A., and Rubin, V. C. 1988. *Ap J*. **333**, 542.

VIII

PROSPECTS FOR THE FUTURE

PROSPECTS FOR THE FUTURE
OR THE OPTIMISTIC COSMOLOGIST

AVISHAI DEKEL

Racah Institute of Physics, The Hebrew University of Jerusalem

Believe me, I did not invent this title. The organizing committee is to be blamed. I find it almost impossible to say something intelligent summarizing a workshop of such a team of leading experts. Moreover, how can one talk of prospects for the future if we can't even predict trivia such as whether we are doomed to fall into the Great Attractor or not. We might avoid it, of course, if $\Omega < 1$. (We asked a taxi driver in Rome how should we spell Great Attractor in Italian and he replied: "You mean Grande Attrattore? Like Sofia Loren?"). And then, what does 'optimism' mean in this context? Do we really hope to reach the ultimate 'theory of everything'? Understanding everything might make it all very boring for us at the end. Perhaps more exciting is the intellectual struggle associated with the challenges set by conflicting evidence for our theories?

Trying to meet the challenge anyway, I consulted some of my friends. I must quote George Blumenthal, who advised me as follows: "I can think of only three reasonable approaches: (1) Speak doubletalk, or Yiddish. No one will understand and nobody will cite you for stupidity; (2) Discuss upcoming observations and where they may lead theory. This is the safe choice; (3) Make an outrageous prediction. People will laugh and think you are joking, but this is your chance to be the Zwicky of the 1980s if you are right". Unfortunately, I can speak neither doubletalk nor Yiddish, but I will try to follow his other two suggestions. Since every participant is going to contribute to the summary of this meeting, I will be brief and will not make any attempt to be complete or even objective. This is, by all considerations, my own biased view.

The safe way to make a prediction is by extrapolation from the past into the future. When I look into the evolution of our field in the recent past, the one thing which strikes me most is the poor causal connection between theory and observation. This is especially surprising in a field which we all regard as very phenomenological. Let me demonstrate my point with a few examples. Consider first some of the *observational* results which were the major issues of discussion in this workshop, and ask in each case whether the result was predicted by theory, or even whether the observational effort was motivated *a priori* by theory.

Most of results concerning the *spatial distribution* of galaxies were neither predicted nor motivated by theory. The extensive study of galaxy correlations by Peebles and co-workers was certainly guided by general theoretical considerations, but I do not think that the discovery of a power-law correlation function has been predicted *a priori* by any specific theory. Superclusters and voids were found independently of theory. Some exceptions were the discoveries of 'filaments' and 'bubbles', which were predicted by theories of pancakes and explosions, respectively. The correlation of galaxy type with environment had not been predicted *a priori* either (and it is not well understood even today).

The discovery of large-scale *streaming motions* — the theme of this meeting which has been thoroughly disscussed by Burstein, Faber, Lynden-Bell, Mould, and Rubin — was (and still is) a big surprise for all theories.

The discovery of *superclustering* of clusters of galaxies, as measured for example by the cluster-cluster correlation function discussed here by Neta Bahcall and myself, was neither predicted nor motivated by theory.

The study of angular *temperature fluctuations* in the microwave background radiation as described by Anthony Lasenby was motivated by theory, but the current demanding upper limits were certainly not expected *a priori*.

Consider, in turn, some of the major *theoretical* ideas concerning the formation of large-scale structure, and ask whether they were motivated by observations. Causality is somewhat more apparent here.

The theory of *inflation* was motivated by very general observational evidence such as the causality indicated by the large-scale isotropy of the microwave background.

The idea of *non-baryonic* dark matter was partly motivated by observations. The fact that the observed isotropy of the microwave background is hard to explain in a baryonic universe provided some motivation. But the main argument for it is still wishful thinking. The theory of primordial nucleosynthesis and the measured abundances of He and D put an upper limit on the mean density of baryons, $\Omega_b \leq 0.2$. Then, the theoretical desire to have a flat universe with $\Omega = 1$ and $\Lambda = 0$, motivated by Inflation, suggests non-baryonic dark matter. As for the existence of candidates, the neutrino-dominated scenario had been somewhat motivated by the false alarm concerning the detection of $\sim 30 \, eV$ mass for the electron neutrino in the early eighties, but our current 'standard' Cold Dark Matter Model (CDM) was (and still is) a purely theoretical speculation (ask Jim Peebles, George Blumenthal, or Joel Primack).

Neither the idea of *Cosmic Explosions* nor the galaxy formation scenario based on *Cosmic Strings* were motivated by observations.

The newly revived scenarios of an open universe where *baryons* play an important role, both the isocurvature version which has recently been suggested by Jim Peebles and the hybrid versions which I have been working on during the years, were motivated by the evidence for very large-scale structure; first the cluster-cluster correlations and then the streaming velocities.

The *Decaying Particle Cosmology* was motivated by the wish to reconcile the dynamical evidence for $\Omega \sim 0.2$ with the theoretical desire for $\Omega = 1$. Certainly not by experimental particle physics.

The popular idea of *biased galaxy formation* was motivated by the same desire. But recall that this idea first emerged in the context of CDM, trying to reconcile the predicted correlation length with the correlation length observed for galaxies.

The study of *Gaussian Processes*, which was the basis for some of what Dick Bond and Alex Szalay told us here, has been initiated by the classical work of Doroshkevich, and has been pursued in order to provide a semi-analytic tool for comparing theories with observations. The same is true of other efforts made in developing various statistics to quantify this comparison, such as the maximum likelihood method discussed here by Nick Kaiser.

The theoretical study of the *microwave* fluctuations, some of which has been described here by Nicola Vittorio, is guided, to a certain extent, by the null discovery of such fluctuations on any angular scale.

Going through the above examples, it is perhaps not a big surprise that our field is led by observations, but it is quite surprising to realize what a weak effect the theoretical ideas actually have on the observational discoveries. There is some influence in the opposite direction, but even here it is surprising to notice that some of the major theoretical efforts were not motivated by observations.

The prospects for the future are easy to predict in this case. More of the same!

Let me summarize first some of the *observational developments* which are expected in the near future, and the major questions which they hopefully answer.

Galaxy redshift surveys. I am told by Margaret Geller and John Huchra that we should expect complete wide-angle surveys like the slices which they have shown us here, i.e. out to a magnitude limit of 15.5 (an effective depth of $\sim 100\ h^{-1}\ Mpc$ for L_* galaxies), in seven years. Deeper surveys to $m \leq 17.5$ ($\sim 250\ h^{-1}\ Mpc$ for L_*) in cones of $\sim 100\ deg^2$ are expected in five years, and 20th magnitude surveys will come next. The number of IRAS galaxies for which redshifts are measured is expected to be doubled soon. One can pose at least three important questions which should motivate such surveys: (a) Does the galaxy-galaxy correlation function go negative at $\sim 20\ h^{-1}\ Mpc$

or beyond? This is crucial for CDM and more generally for the standard assumption of scale-invariant Zeldovich spectrum of initial fluctuations; (b) Can one quantify the spatial distribution of galaxies in terms of statistically meaningful measures of filamentary structure, alignments, topology, etc.? (c) Are we approaching a 'fair sample'?

Velocities and redshift-independent distances. Martha Haynes promises that Tully-Fisher distances to spirals will be extended from the current ~ 150 h^{-1} *Mpc* in the Perseus-Pisces region to ~ 300 h^{-1} *Mpc* within one to three years. What one really needs though is a radio relescope like Arecibo in the southern hemisphere, and lots of optical time. Distances to ellipticals using the revised Faber-Jackson method can be, and should be, extended beyond the current 60 h^{-1} *Mpc* limit. A common desire we all share, of course, is to refine our current distance indicators, discover new ones, and hopefully understand the internal physical correlations of galaxy properties on which they are based, but I am quite skeptical about the chances of making significant progress on this front in the near future. On the other hand, the ability to extract the large-scale patterns of the velocity field by applying linear-theory iterations to a uniform redshift sample, as demonstrated so impressively at this meeting by Amos Yahil and Marc Davis using the IRAS sample, is a very promising new approach. After exploring the motion towards the Great Attractor, one can expect detailed mapping of the velocity field in the Perseus-Pisces region, and in larger volumes.

I share the worry expressed here by some of us that we cannot regard the current samples as "fair" samples as long as the significant features we detect, like superclusters, voids and streaming motions, are on scales comparable to the volume sampled itself. I can see how the recent evolution from the old notion of 'Virgo Infall' to the current idea of 'Grande Attrattore' will lead next to a discovery of 'Grande Repulsore' in the opposite side of the sky, and then to something which we will probably call in the next Vatican Workshop 'Attrattore di Tutti Gli Attrattori'' (see Figure 1), and so on.

After pursuing so impressively the 'geographical approach' of mapping the velocity field and modeling it with a simple model, we should proceed to *a priori* statistical analysis which would enable quantitative comparison with theory. The use of multi-parameter models, even if they are very descriptive and they provide a great fit to the data, is limited. The difference is analogous to the difference between discovering America and understanding plate tectonics. A simple example of a useful measurable quantity is the mean ('bulk') velocity in a given volume(s). Most interesting is the coherence length of the velocity field, and efforts will be made to analyse the velocity-velocity correlation function. This analysis is complicated by the fact that only the radial velocities are available, but I expect that several techniques will be developed to make its application useful.

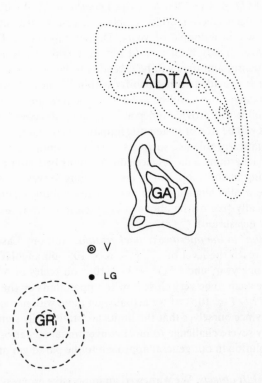

Fig. 1. Do we have a fair sample? The current geographical notion of *Grande Attrattore*, which has replaced the old model of *Virgo Infall*, is probably not the final word. Next, one might expect the discovery of an analogous *Grande Repulsore* — a negative perturbation somewhere in the opposite side of the sky. Then, when our samples expand further, we shall naturally discover something bigger and more distant, which we might naturally name *Attrattore di Tutti Gli Attrattori*, and so on.

Cluster-clustering. The 64 h^{-1} thousand dollar question is whether the cluster-cluster correlation function is, indeed, as high as it seems to be based on the studied sub-samples of the Abell catalog. Is the correlation length as large as half the mean separation between neighbors ($r_0/\bar{d} \simeq 0.5$), as argued by Neta Bahcall, or is it significantly smaller? The current large correlation length is in clear conflict with all current theories that are based on Gaussian, scale-invariant fluctuations in an Einstein-deSitter flat universe, but a value of $r_0/\bar{d} \simeq 0.1 - 0.2$ would be compatible with 'standard' CDM, for example. Significant progress is expected on this issue. The southern Abell catalog has been completed recently and it allows a new, immediate analysis of the angular cluster correlation function, $w(\theta)$. One can expect to have redshifts for all $R \geq 1$, $D \leq 4$ ($\leq 300 \; h^{-1}$ Mpc) southern Abell clusters, a sample comparable to the northern redshift sample which has been analysed by Bahcall and Soneira, within a couple of years. Deeper surveys, to $D \leq 6$ (≤ 600 h^{-1} Mpc), have already been carried out in narrow cones as described by Geller and Huchra, and they will be extended to larger areas until the whole Abell catalog is provided with redshifts within the next decade. Automatic cluster-finding algorithms will be applied to the new digitized sky survey of Cambridge and it will provide independent, more objective cluster catalogs. Such catalogs will be free of systematic human biases, but they will still suffer from systematic problems such as the contamination by projected foreground/background galaxies (see contributions by Kaiser and by myself). Most promising in the short run is an upcoming survey of X-ray clusters. Because of the smaller angular extent of the X-ray emitting regions, this catalog will be practically free of the above contamination effect, and will provide more reliable measure of $\xi_{cc}(r)$.

Fluctuations in the microwave background. Anthony Lasenby promises that we shall reach the level of $\delta T/T \sim 6 \times 10^{-6}$ on angular scales of $30''$ — $4'$ within one year, and $\delta T/T \sim 3 \times 10^{-6}$ on scales of $5'$ — $2°$ within five years. We seem to be very close now to upper limits on the various scales at the level of $\delta T/T < 10^{-5}$. If we actually get there with no believable detection, and convince ourselves that the limits are statistically significant, it will become a very severe challenge to most conventional scenarios, and it might require a revolution in our general approach to the paradigm of gravitational instability.

High redshift objects. We witnessed an impressive progress in the last few years, as described by David Koo, and we can expect more in the future. The homogeneous surveys of quasars and absorption clouds have the potential of providing a new dimension to our view of the formation of large scale structure — the time evolution. The interpretation of the data, however, is not trivial. A study of the time evolution of clustering, for example, will require a decomposition of the effects of luminosity/density evolution of the sources themselves and the gravitational clustering process. I do believe that if the data improve, and if we are clever enough, we will be able to extract useful constraints separately on the physics of the sources and on the cosmology which they are embedded in.

Predicting observations is relatively easy. Let's proceed now to the harder part of the prospects for the future, namely, trying to predict how the various *theories* will develop. This is going to be almost pure speculation. In these days of crisis in the stock exchange markets, let me use an analog — the 'Dow Jones' index of popularity for cosmological scenarios. Figure 2 shows first my view of the evolution of this index in the last decade.

We ended the seventies with only *baryons* in mind, but realized around the turn of the decade, facing the isotropy of the microwave background and

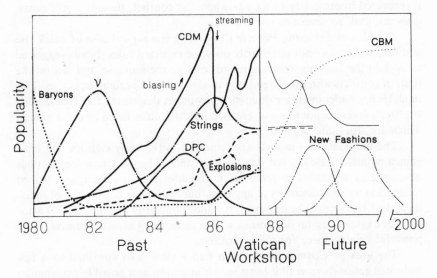

FIG. 2. Evolution of the 'Dow Jones' popularity index for theories of the formation of structure in the universe; before the Vatican Workshop and after it. See the text for details.

nucleosynthesis constraints, that non-baryonic dark matter may be required. Only a small group of non-baryonic skeptics were left to keep the baryonic models on low fire.

Neutrinos became the highlight of the early eighties, associated with the revival of the Zeldovich pancake picture, but then this scenario lost popularity drastically because of the difficulties it has in forming galaxies in time without over-clustering on larger scales. Two little positive jumps in the popularity of neutrinos occurred recently: the idea of anti-biasing helped easing some of the difficulties, and the discovery of large-scale streaming velocities reminded us that we need more power on large scales in the spectrum of fluctuations. The neutrino-dominated cosmology might provide the required large-scale power, especially if $\Omega < 1$.

Cold Dark Matter took over in about 1983 and has been the 'hot'theory since. Its main appeal is in its predictive power, which has been pursued very impressively by the N-body simulations, some of which were described by Marc Davis. Among the major milestones on its way to become the current 'standard' model one can mention the continuous great success in explaining many properties of galaxies, the problem of too weak clustering which was answered by the natural idea of 'biased galaxy formation', and recently the discovery of streaming motions on large scales. This discovery was at first interpreted as a fatal problem for CDM, and caused a drastic drop in its popularity, but it recovered impressively to a stage where the conflict, though it still exists, does not look so severe to some of us.

The idea of *Decaying Particle Cosmology* is a typical case of quick rise and fall, which does not necessarily obey the expected rules. It was suggested to explain the 'missing mass' required to close the universe, but fell on the basis of conflicts with the Virgo-infall model and the flat rotation curves. This model is not under intensive discussion any more, despite the fact that the difficulties it faces are not more severe than the difficulties faced by other models which are still quite popular.

The non-Gaussian models also emerged in the early eighties and have gained popularity slowly but steadily ever since. The *explosion* scenario has developed as an analog to processes in the interstellar medium, and gained recognition with the discovery of apparent 'bubbles' in the distribution of galaxies in the CfA 'slice of the universe'. It gained popularity among theorists when the idea of super-conducting cosmic strings came along as an alternative, more powerful driving force for the explosions.

The idea of *Cosmic Strings* also had a slow start confined to a few dedicated researchers until it burst to our attention as a possible explanation for non-Gaussian clustering of galaxies and clusters. It gained popularity in some circles at an outrageous rate until Jim Peebles brought the community back to its senses with two privately circulated 'screeds' of embarrassing questions concerning galaxy formation in the string picture as compared to observations.

This is my evaluation of where the various thoeries stand at present. How will they do in the future? This brings me to the *outrageous predictions*.

I take seriously the difficulties of CDM to cope with the following problems: (a) the very large-scale structure reflected by the cluster clustering and the large coherence length of the streaming towards the Great Attractor; (b) the filamentary structure on scales beyond $10 \ h^{-1} \ Mpc$; (c) the late galaxy formation and the fragility of disks; (d) the finite extent of the halo of the Galaxy; (e) the lack of actual, existing candidate particle. I therefore predict a steady decrease in popularity for CDM. This is an outrageous prediction because the current popularity gradient, as reflected by the general attitude of the participants of this workshop, is certainly still positive.

Neutrinos will probably maintain their moderately-low level of popularity. The shrinking experimental upper limits on the mass of the electron neutrino leave the tau neutrino as the only viable neutrino candidate which can still close the universe (need $m_\nu \sim 30 \ eV$). The predicted clusters in the neutrino scenario are somewhat too big and too hot even if galaxy formation in them was subject to anti-biasing. Massive halos around dwarf galaxies, if confirmed, pose another argument against neutrinos as the dark matter candidate there.

Cosmic Strings, I am afraid, might go down in popularity too. In addition to the difficulties raised by Jim Peebles concerning galaxy formation seeded by strings, I am confident that it will be realized soon that our understanding of the basic processes involved in the string scenario, and in particular the fragmentation of strings, is only poorly understood. The fragmentation pattern in the string simulations is extremely sensitive to the way the intercommuting process, and the resultant nasty kinks that form and propagate along the strings, are treated numerically. The different simulations by the different groups show very different results. This will add to the complexity of the string scenario, and will cause a continuous damage to its popularity. Enthusiasm is expected, at least for a while, with regard to a new type of strings — *Fundamental Strings* — which is an attempt to relate a cosmological string scenario to the concept of super-strings. The parameters of this model seem to be predicted by theory (e.g. $G_\mu \sim 10^{-3}$) but its ability to make a reasonable galaxy formation scenario is still to be investigated.

The *Explosion* model, on the other hand, is in a sense more immune against loss of popularity because it is backed by powerful forces that can match almost any observation with an ingenious theoretical idea. It has been found, as I briefly described in my earlier contribution, that clusters supercluster naturally in a generic explosion scenario, significantly more than in any of the other scenarios under discussion, and in better agreement with observations. It remains to be seen, however, how one can explain the large streaming motions in the explosion picture.

The existing reservoir of theoretical ingenuity, which has given rise to fancy ideas like decaying particles, cosmic strings, cosmic explosions and cosmic explosions generated by super-conducting cosmic strings etc. etc., promises that we will see more new theories which will emerge in order to face new conflicting evidence, attract attention for a while, and then decay the way the decaying particles did, perhaps to burst again when needed.

My outrageous prediction, which is not so outrageous for me if you consider my work in the past, is that the dark matter will eventually prove to be *baryonic*. After all, aren't baryons the only dark matter candidate which is known to exist? Take for example planets as the dark matter building blocks. Baryons, in an open universe, guarantee suffcent power on large scales to explain both the streaming motions and the superclustering (or 'super-pancaking'). The baryonic universe has started regaining popularity these days since Jim Peebles proposed the isocurvature model. I predict that baryonic models will go strong in the near future. Still, one should not give up so easily the nice features of CDM in explaining galaxies, which is basically an argument for an effective logarithmic slope of $n \simeq -2$ for the spectrum of fluctuations on galactic scales. The ideal scenario might, therefore, be a baryonic universe with a CDM-like spectrum of fluctuations on galactic scales. The isocurvature model comes close. Also, baryonic matter would behave as 'cold' matter if the gas had collapsed into objects (e.g. black holes) before galactic scales entered the horizon, giving rise to an exciting scenario of 'Cold Baryonic Matter'.

Despite my attempts to indicate the prospects for the future, I do not recommend we rush to buy stocks based on my speculations. Judging from the rapid evolution of ideas in the recent past and the current state of relative confusion in our field, one must admit that the future is mostly unpredictable. Nevertheless, this might be a good reason for 'optimism'.

FIG. 3. I must admit that I did something useful during the long discussions we had. I tried to sketch amateur cartoons of some of the participants at their best moments. It matches the high spirit of the workshop, but I have mixed feelings about how it would fit in these serious proceedings. Anyway, since the editors encouraged me to put in at least one example and they decided to take the responsibility, here it is. It is left for the reader to decide who this character is.

COMMENT

In order to recapture, in a somewhat systematic way, the lively and creative exchanges that took place during the Study Week and to preserve the personal enthusiasm that characterized much of the week's work, the Scientific Organizing Committee decided that at the end of the week, following the summary by Avishai Dekel, each participant would be asked to present for a period of time, not to exceed five minutes, whatever they wished in terms of final thoughts concerning the Study Week. Since the organizers had designed the meeting as a research week with the hope of opening new horizons of discovery or, at least, defining more accurately the questions to be addressed, a great deal of importance was attached to this final exercise of the week. From the recording a written transcript was provided to each participant for personal editing and the results are presented in the following pages.

The order of speakers was determined by drawing names from a hat, — John Huchra's best western sombrero, to be precise. This session was presided over by George V. Coyne, whose remarks to each as they began their small discourse have been, except for fragments retraceable from their remarks, discretely suppressed.

GEORGE V. COYNE, S.J.

SUMMARY COMMENTS

DAVID BURSTEIN

I would like to make two points. The first is that the prospects for getting peculiar velocities in the future seem somewhat limited in the following sense. We are currently stuck with distance predictors that give us, at best, accuracies of about 15%. This means that, for an individual object at a distance of 10,000 km s^{-1}, you can easily measure peculiar velocities of 3000 km s^{-1} or more. That doesn't get us very far. It seems to me that the only prospect for getting peculiar velocities on much larger scales is to look at groups of galaxies, groups of groups, and to try to beat down the statistics in that region. Even that game will not get us much past about 10,000 or 15,000 km s^{-1}, and when I looked to see if we could measure peculiar velocities at, say, 50,000 km s^{-1}, the future actually looks pretty dim with present techniques. We simply do not have the accuracy and we do not have the number of galaxies that will be necessary to measure peculiar velocities on the very largest scales. That is my first point.

My second point is that we should come away from this meeting with a clear desideratum to map out the plane of our Galaxy. It is apparent that, for whatever reason, our universe is created in such a way that, by our current view, much of that matter that is nearby (within 5,000-6,000 km s^{-1}) is obscured by galactic extinction. I do not think it is a coincidence that Perseus-Pisces dives into the Galactic Plane at a longitude of 120°, and the Great Attractor dives into the plane at a longitude of 310°, situated on opposite parts of the sky. The chances that galactic obscuration could do this are not negligible. We could have just as well been oriented so that our Galactic Plane would have totally obscured the large-scale structure that Brent Tully has identified.

So, the question in my mind is to what extent are the large-scale structures we see molded by this large dark band, and Zone of Avoidance that exists in our Galaxy?

NETA BAHCALL

The discussions held this week reveal the great progress made over the last several years in the field of the large-scale structure and motion in the universe. The existence of large-scale structure as a common phenomenon is now clearly established. Observations of the spatial distribution of galaxies, clusters, and even quasars, reveal a consistent picture of the existence of large-scale structure to scales of $\sim 50\text{-}100h^{-1}$ Mpc. Three dimensional maps of the galaxy and cluster distribution show the existence of large overdense regions surrounding voids, or underdense regions, that may extend to $50h^{-1}$ Mpc or more. Specific quantitative tools such as the correlation function and the "sponginess" measure of the distribution have been applied to describe the nature of the structure. Some of these results were discussed here this week. The correlation function analysis yields an average strength and scale of the correlated structure, showing positive galaxy correlations to $\sim 10h^{-1}$ Mpc, and considerably stronger correlations to larger scales for galaxy clusters. While the correlation analysis provides important information regarding the structure, additional quantitative measures are needed in order to describe the shape, extent, and nature of the distribution on large scales.

Observations over the last decade preceded theories in revealing structure and motion in the universe. New theories quickly followed and, as we heard summarized, are still in the process of development. It is difficult to predict at this point which of the theories — baryonic models, cold dark matter, explosion, or something else — will provide a consistent picture of the universe. Additional observations of structure and motion, combined with a quantitative description of the results, will be available over the next several years. These will surely provide new limits and ideas on theoretical models.

The use of clusters of galaxies as a tracer of the large-scale structure has proved to be effective, revealing the largest-scale structures yet observed. The cluster correlation function, with its large amplitude and scale-length, has stimulated new ideas such as biased galaxy formation and large structures surrounding clusters, as well as encouraging suggestions of cosmic strings as a possible source of cosmic fluctuations. However, the strong cluster correlation function has also created a problem for some models that cannot produce as strong a clustering and as much power on large scales as suggested by the cluster observations. While the correlation amplitude may be uncertain by about 30% due to various effects (see article above), it is still an order of magnitude stronger than the galaxy correlations, and stronger than estimated for example by cold dark matter simulations. New and improved cluster samples such as will be available in the future from digitized sky surveys and from X-ray cluster surveys should be able to reduce the uncertainty on the amplitude and scale of the cluster correlation, as well as determine whether weak correlations exist on scales $\gtrsim 50h^{-1}$ Mpc.

The use of quasars as tracers of the large-scale structure was discussed only briefly this week. I expect that this new tool will yield important results, including the dependence of the correlations, or structure, on redshift. Current results indicate strong quasar correlations, close to the cluster correlation strength, that also extend to nearly ~ 100h^{-1} Mpc (for $z \lesssim 2$). (No significant correlations are detected at $z > 2$ in the current samples.) Larger complete samples are expected to be analyzed in the next few years; these should yield a better understanding of the evolution of the large-scale structure and the role of quasars in these early structures.

Peculiar velocities on large scales have been discussed in detail during the week. It is important to continue the investigation of the possible motion of clusters in superclusters, summarized earlier, by using accurate distance indicators to these clusters. If large motions are indeed detected, the quantitative nature of the motion will yield important information regarding the masses of superclusters and/or possible non-gravitational effects such as explosions.

GEORGE EFSTATHIOU

I think that perhaps one of the reasons that we're having this session has to do with the Church's love for confession. If I had arranged it, each of us in turn would have confessed our biases and prejudices to Father Coyne at the start of the meeting. During the meeting, he would assess our performance and at the end he could hand out our penance in the form of camera ready sheets. The more "sins" you'd committed, the more camera ready sheets you would get!

We have been asked for our personal view of the meeting. It isn't sensible to confess in public, but I will admit to a couple of things that have bothered me. I have been frustrated by Jim Peebles' remarks of filaments in the N-body simulations. It goes against his intuition that a cold dark matter simulation should show fairly large-scale filaments. Is there a mistake in the initial conditions, or in the N-body dynamics? I understand Jim's point, but it isn't clear to me that anything is wrong. I am frustrated because I can't answer a query like this until I have interrogated my computer. Jim's remarks are qualitative so it isn't even clear what would constitute a satisfactory answer. In any case, there is an interesting problem to investigate. The second point concerns comparisons with the "standard cold dark matter" model. Several of us have compared the CDM model to observations and concluded that it is O.K. But I have noticed that at least three different values have been assumed for the amplitude of the power-spectrum; so we aren't all discussing the same CDM model after all, although it appears that way to others. Let's be precise about the amplitude and make sure that everyone knows the value that we are using.

One of the nicest things about coming to a meeting like this is discussing work before it has been published. That's very valuable. You show people graphs, you say "look at this" and "look at that". You get a good idea of what is going on. It has been really useful to discuss the results of the big galaxy survey that my students are working on in Cambridge. Hopefully, we should be able to tackle some of the important issues raised at this meeting.

I thank the organisers of this very successful workshop and I am especially grateful to Father Coyne for granting me absolution.

RICHARD BOND

As someone who inhabits the space of theories, and only sometimes the space of reality my interest in what I've been hearing this week is to learn how I should limit the regions of theory space I regularly haunt. In Avishai Dekel's stock market language I would avoid spreading my portfolio over too many models, and invest heavily in a winner. There is one model that we've all invested heavily in and that's the biased cold dark matter theory. We should emphasize strongly that it is not only just a pretty theory but it is a minimal theory, in the sense that we start with an inflationary model and one of the simplest assumptions for the nature of the dark matter. There has been *no other* viable mechanisms proposed but inflation to smooth out the universe. I'm sure that you all came away with the strong belief that inflation implies scale invariance after my talk in which I twisted and turned to try to break scale invariance. This means that, within the context of the theoretical models based upon a quantum origin for fluctuations, the isocurvature baryon models with the non-scale-invariant power laws advocated by Jim Peebles are extremely difficult to arrange, and are at least as unlikely as the extra power deviants that I was talking about.

We should, therefore, adopt the view that the biased cold dark matter model requires very firm evidence against it for us to completely reject it. What we've seen this week is a fantastic development of cosmography, the subject of making maps of the universe, and the maps are extending out from us to far distant lands. But it is only the most distant and the most hidden that gives the standard biased CDM model the most difficulty: the most hidden being the Great Attractor and the most distant being the cluster patches that Brent Tully tells us we should take very seriously. If so, we must have power which is of an enormous amplitude on large scales ($\delta \sim 0.1$ at ~ 50—$100h^{-1}$ Mpc), and the cluster-cluster correlation function is just one aspect of that. I think the Henry-Huchra-Postman cluster survey provides even more compelling evidence for very large-scale patchiness of the clusters in redshift space. It is hard to imagine how Abell could have biased his cluster selection in the redshift direction. Making sure that these cluster islands really exist is one of the most important things for us to show in the next number of years. On the other hand, I would say that, since we don't know how much of what Brent Tully is seeing is due to a biasing problem in the cluster catalogue, then we should not be surprised if these objects do not survive in new cluster catalogues independent of Abell's eyes.

Given that there are all sorts of exciting possibilities out there indicating very large-scale structure, it is still important for us to keep exploring theoretical space in order to find those new rising theories that Avishai pointed to in his forecast of trends in theory. Perhaps hints of very large-scale structures at high redshift from the distribution of Lyα clouds may force us into the very radical. However, we should bear in mind that, as with the space of theories, the space of current observational interpretations is not always synonymous with the space of reality.

As a final point: The most important thing that I'll be looking for over the next few years is confirmation of Anthony Lasenby's large angle microwave background anisotropy at a few degrees, which would destroy the scale invariant cold dark matter theory if the anisotropies are shown to be primordial.

SANDRA FABER

Since I was assigned to be one of the conference summarizers, I have tried to think out a more encompassing set of remarks. Actually, I think Avishai's summary was very much in the flavour of my own thoughts. I thought it was a brilliant summary, Avishai.

The great advances at this meeting, for me, were three. First, there are George Efstathiou's new galaxy-galaxy and galaxy-cluster correlation functions, which I had not seen before, and the implications that they have for cold dark matter. Secondly, I commend the magnificent work that has been done by Amos Yahil and Mark Davis and their collaborators in interpreting the IRAS survey. When I first heard about the IRAS work — I think it was at the Princeton cold dark matter meeting over a year ago — I was nery negative about it for all the reasons that various people have raised. I've now completely turned around on this, and I think it has been well worth all the effort that has been invested. It might not be the ultimate answer, but it's certainly opened our eyes to a new way of studying the universe. I see three things about it that are really significant.

Looking at those velocity flow maps, one realizes the importance of being freed from over-simplified infall flow models. This has been obvious to our group as we have tried to map our own velocity information and been forced to put in so many extra parameters. We're very conscious of the inelegance of this. It really is wonderful to have maps that don't have parameters like that and can be compared one-for-one with every observed galaxy.

Secondly, I think that the IRAS results as they now stand are on the edge of confirming once and for all that the velocities we see *are* due to density inhomogeneities. This is something we have all believed, but it's very nice to see it actually demonstrated. I think the demonstration will go down as one of the fundamental experiments in astrophysics, because it's the crucial information that we needed for the gravitational instability picture.

Thirdly, I think that the IRAS approach still needs work. Right now, the predicted motion of the Local Group is too close to Virgo, and in fact, I think that's exactly the direction one would have picked from just looking at the *visible* galaxies in Brent's catalogue of nearby galaxies. We know though that the Local Group doesn't move quite in that direction — it moves more toward the Great Attractor. So it's clear to me that something as yet is still missing from the data base; there needs to be more mass in the Great Attractor. My hope is that filling in the galactic plane better by interpolating rationally from one side of the plane to the other will fill in the missing link. In the longer term we need to survey the plane better, perhaps with 21 cm or X-rays.

The third advance for me at this meeting was the more detailed N-body simulations for biased galaxy formation and cold dark matter. I would say a gratifying point was seeing for the first time all the correlations between different kinds of galaxies, their velocities of rotation, masses, and so on, as a function of density and environment. Qualitatively, these effects were all predicted before. We knew, just from simple linear statistics, what kinds of galaxies would form and where in the CDM picture. Still, it was nice to see that demonstrated more precisely and more quantitatively.

A major question, though, is the question of exactly how good the biasing is. I would have liked to have heard more discussion about this. I asked Mark Davis and George Efstathiou the following question. I said, "if you observe your simulations the way observers observe the universe and measure the virial masses of little groups, attach a certain mass per galaxy that way and then infer the mass of the entire volume — this is the way we measure the mass density of the universe traditionally — how far short of the total does that inferred mass actually fall?" And the answer was, it falls short by a factor of 5, they said, which is exactly the factor we need in order to make Ω equal to unity. So that's very reassuring.

In this model now you can look and see where that missing mass is, and the answer they gave was that some of it is in the voids; and then we have these very high mass-to-light ratios of dwarf galaxies. That's where the rest of it is. The voids I can understand, obviously, and we've got to do something drastically different from what we're now doing in order to find the matter that's in the voids that never made galaxies. But I am puzzled about the matter around the dwarf galaxies. Anything that's around a galaxy to me is potentially fair game for discovery. Why haven't we discovered it so far? Where are the baryons that are associated with that matter? Have they simply not fallen into those galaxies yet? Perhaps one of these gentlemen will reply to that, but let me summarize by saying simply that looking at these simulations in more detail and trying to figure out clever ways of observing the missing mass might be very informative.

I will briefly enumerate three observational questions for the future. First, I believe that the future of large-scale motion work lies in two regions: in Perseus-Pisces, where we see very interesting structures and there is a hope of correlating motions with the morphology; and in the Great Attractor, where we badly need to figure out what the backside of this entity looks like and whether all the drama is merely on the front side. If so, we've already seen all there is, in which case it's not such a large or interesting structure. Second, I believe we badly need an unbiased and believable cluster catalogue to help us answer whether or not any of the cluster correlations remain positive beyond 3,000 kilometers per second. Third, I believe we are on the verge of seeing cosmic microwave background fluctuations. Even the foreground patch of the Great Attractor would have produced an observable fluctuation in the microwave background. I'd like to see emphasis on actually making maps and finding real features instead of being content with just RMS measurements. However, I'm sure the people in this field already feel this way.

My overall assessment of theory at this meeting is that cold dark matter is still alive and well, and I would not have put the gradient at zero as Avishai Dekel did; I would have put it positive.

JEREMY MOULD

Astronomers are divided into many tribes and there are three main tribes represented here. The first tribe is the tribe of observers. It's a very primitive tribe. We are given to gathering on mountain tops and worshipping the stars and the moon, and recently we were told that we should also worship the Great Attractor. There was some concern about this because it was difficult for some of us to see it. The second tribe is the tribe of model makers, and in this context the function of the model makers is to determine the peculiar velocities, from magnitudes and velocities that the observers measure. The third tribe is the tribe of theorists and the function of the theorists is to deduce the initial fluctuation spectrum from the peculiar velocities that we find from the models.

So what is there left that hasn't been talked about for each of the tribes to take home to their own camps? Well, the observers' tribe learned that there were good and poor data, and clearly we want to concentrate on the good material. Of course, this is not a fixed data set. We cannot simply confine ourselves to the galaxies in the Second Reference Catalogue; we have to measure those discrepant erroneous diameters and get them right.

A lesson for the parameterized model maker to take home, I think, is that now that the number of parameters is getting large, we need to know what the error bars are on the ten parameters. When you minimize chi-squared with ten parameters, you get out a ten by ten error matrix and you want to know what the covariance between, for example, the Virgo infall and the Great Attractor infall is, so that we know how much freedom there is to trade off one against the other.

I cannot offer any suggestions as to what the theorist tribe might take home, apart from noting that Avishai Dekel thought that they should wait and see what we came up with next.

An impression I have from the unparameterized models which, I agree with Sandy Faber, were the big new advance at this meeting is that, apart from Virgo, Hydra-Centaurus was probably the largest entity responsible for our local acceleration. The reason I got that impression was that apparently the acceleration vector has converged early as you move out in that direction towards Virgo and Hydra-Cen; that was also the appearance, I thought, of Amos Yahil's models.

At this meeting I think the question, is there a Great Attractor?, has been replaced by the question, do the IRAS density maps predict the observed velocities? In other words, does light trace mass? And if not, do we need to add extra attractors? In addition, we should, as Nick Kaiser said, worry about the uniqueness of these models. Perhaps we do need to add more biasing to the model, put more mass or less in the peaks.

Based on Marc Davis' results I think a tentative answer to the question: do the IRAS density maps predict the observed velocities? is yes. Remember, most of the outflowing test particles that we've seen in our own Tully-Fisher work are in the foreground of the Hydra-Cen mass distribution, at 1.8 Virgo distances, to introduce a new unit of distance. One is actually in it, at 2.8 Virgo distances. That's the Hydra cluster and we see zero peculiar velocity for it, plus or minus a large number. Clearly, as has been said before and was said at IAU Symposium 130 in Hungary, we have to try harder to get behind the Hydra-Cen mass distribution where inflow is predicted. Remember too, that it is not sufficient to observe galaxies in the 4500 kilometer per second peak discovered in Dressler's survey, because we found the Cen 45 subcluster peak was mostly peculiar velocity. We have to step out further in distance and that's particularly a problem for the Parkes Tully-Fisher observations, because we are right at the limit of the telescope. But if we had a few more distant clusters, the picture would be a lot clearer. I join the plea for a southern hemisphere Arecibo.

So, finally, to reiterate a bit, I think the future in this particular subfield lies in perturbing the IRAS map that Amos Yahil and Marc Davis have shown us, and in minimizing chi-squared between the observed and predicted velocity distributions. If we have to stick with the parameterized models, I think we should try pulling the Great Attractor in, placing it between the 3000 km/s and 4500 km/s peaks, and spreading it out, moving it over the zone of avoidance in the Pavo-Indus region. In the meantime, we will try to provide a lot more data to fully constrain the problem.

BRENT TULLY

I would like to dwell on the issue of the morphology of large-scale structure. My approach has been rather phenomenological, more so than that of almost anyone else in this room. George Efstathiou gives me a hard time because he is obstreperous, but I contend that the approach is a valid one in an area where we are so incredibly ignorant.

Jim Peebles delights in warning us about connecting the dots. We could be grossly misled. But a few pictures may be pushing us in surprising directions: a lizard in Perseus, a stick man centered on the Coma Cluster, a vast plane in Pisces-Cetus. These phenomena could not have been found by statistical investigations, because we would never have asked the right questions.

Consider that possible plane in Pisces-Cetus extending over 0.1c. There are reasonable reservations about whether it exists. I am relatively confident that it is real because the one-dimensional two-point correlation test provides a strong confirmation. But the situation will be clarified soon enough, when the new redshift surveys that extend to this region are analyzed.

There *can* be more to connect-the-dot patterns than a pretty picture. If truth is beauty, might beauty portend truth?

JOHN HUCHRA

What has impressed me the most at this meeting, a situation which is especially apparent at conferences discussing problems at the frontiers of science, is the basic conflict between theory and observations which seems to continue no matter how advanced the theory is nor how complete the observations are. At this meeting the theorists seems to be running a little rampant again and it's about time for those of us who are observers to rein them in a little and bring them at least a little closer to reality.

One annoying problem I want to point out, also mentioned earlier by Sandy Faber and George Efstathiou, is that we keep forgetting to lay out properly our ground rules and definitions. I don't mean definitions on viewgraphs but rather the definitions of those things that we are trying to actually go out and look at or measure. A good example for the observers is the definition of a cluster of galaxies. What exactly is a cluster? We had better have a good definition of what one is before we really start talking about the distribution of these somethat messy things (or complaining about the inability of theories to describe the statistics of their spatial distribution)! We need to get our house in order before either the theorists or observers can make any progress on such problems.

Now I want to play a role that would have been called a few weeks ago "snake-oil salesman," but to continue the analogy used by Avishai Dekel and Nick Kaiser this morning, this role might now be more likened to "stockbroker." I think the time has come for the observers to start concentrating on the big project needed to make any headway on large-scale-structure problems: a Digital Sky Survey. Let me not say too many bad things about photographic plates, but those of you in this room who have taken some know that they are not exactly the greatest detectors in the world. We *now* have much better detectors available and even better ones are coming in the next year or so. It is not beyond our reach to start projects to map the sky with galaxies to 19th or 20th magnitude over the next 4 or 5 years.

I would like to make a strong plea that we consider a national or even international effort to do a digital sky survey with CCD's and 1 arc sec pixels. For those of you who are not observers, 1 arc sec pixels are about what is needed to identify galaxies to 20th magnitude. The integration time should be set to produce ~ a few percent photometry at 20th. Such a project could cover a hemisphere, would contain roughly a terrabyte of data, and would occupy the photometric dark time on a 4-meter class telescope for about 5 years. If you wish to go a little less faint or do somewhat poorer photometry, you could go faster, or even better yet, cover the sky in more than one color.

People have objected to the amount of storage such a survey might take, but there are answers to that problem. With current technology a terrabyte will fit on roughly 50 laser disks, but that number will come down. I have seen a preliminary announcement from a German company for optical disk technology with sub-micron spot sizes; this could bring the storage requirement down to ten disks or even a single disk. Imagine having a digital version of the Palomar Sky Survey in your briefcase!

Such a CCD survey has amazing advantages. National centers and computer networks could allow almost instant access to the digital data to astronomers all over the world. CCD's can work into the red or near infrared and thus minimize the galactic extinction problem. You can easily do all those things we all like to do to galaxies — get their profiles, diameter, inclinations, position angles, you name it, from a homogeneous data base. You could identify x-ray or infrared sources. You could also (I hope!) figure out how to pick out galaxy clusters in a reproducible fashion. That's what I'd like to sell to you. Let's do it!

ANTHONY LASENBY

As a member of the primitive tribe of observers, I must say I've learnt a lot over this past week. It has been very beneficial for me. I am particularly glad to have learnt about the three-dimensional structure in some detail. I now think I have some feeling as to where the various clusters and superclusters are and am able to some extent to picture them in my mind. But of course as one continues with this basic programme of finding out the real details of what we have arounds us, it continues to look more and more complex, as one should have expected it to. One worry I have on large-scale structure is that perhaps we are not yet using the right statistic for evaluating it. The two-point correlation function is good since we understand (or are beginning to understand) how to put errors on it, and it is independent of the density of the fields, but I am not sure that it is adequate to quantify what we are seeing in terms of complex structure. Perhaps a new statistic — multifractals perhaps, though doubt has already been expressed about these — will be necessary to encode the higher order correlation functions which contain the information we need.

On the question of velocities, it is clear that there do exist large-scale motions, and I think the agreement between the data sets which Dave Burstien has been showing is very impressive. However, perhaps it is rather too early as yet to say that they are definitely caused by a single object — the Great Attractor. It would be nice to see some exploration of the range of models which could give rise to a similar quality of fit, and of course a definitive test will be possible when we are able to look *beyond* the Great Attractor, and see if infall is occuring there.

I really liked seeing an attempt to determine all the components of the velocity field. As several people have said, the diagrams shown by Amos Yahil and Mark Davis were very impressive, particularly Amos' diagrams of the self-consistent velocity and density fields. Even if one cannot believe all the details, because of problems with shot noise for example, this is still an extremely nice blend of observations and dynamical theory.

One detail I'd like to see cleared up is what Brent Tully was saying about H_0 and Malmquist bias, which is a subject we don't seem to have returned to very much this week. If H_0 is really as high as 95 km s^{-1} Mpc^{-1}, it seems it would represent a crushing blow for cold dark matter theory, so that for this and many other reasons the accurate determination of H_0 is still of course a top priority. On this topic, let me put in another quick 'plug' for microwave background observations. The determination of H_0 via observations of the Sunyaev-Zeldovich effect in clusters of galaxies is a possibility that has been discussed for some years now. With the advent of the enhanced Cambridge 5-km telescope, due to start operation in late 1988, for the first time it will be possible to make accurate maps of the Sunyaev-Zeldovich decrement in several galaxy clusters, with reasonable integration times. Thus useful bounds on H_0, set via this technique, will be possible within a few years.

Now H_0 is critical for CDM. On this question of critical numbers I'd like to support what Avishai Dekel was saying with respect to the tendency to take a single number and perhaps too quickly treat it as having demolished a theory (or provided great support for it). I think our theories now, like nature itself, are becoming complex enough that in some cases we don't immediately know what they are telling us. Time has to be spent working out consequences. For instance, in the meeting this week we have seen that the bulk velocities observed are not inconsistent with cold dark matter after all. The two can be reconciled. As a theory I think that cold dark matter is looking more complete and impressive all the time, and one is particularly struck by its predictive power from only a few assumptions.

On the microwave background, as Nick Kaiser was saying during the week, there has been a sort of 'golden age' up to now, where one could represent the confrontation between theory and experiment in terms of a single number — the $\Delta T/T$ on a given scale. However, if the Tenerife 'detection' is really of intrinsic anisotropy then detailed comparison on a 'morphological' level will start becoming possible (and necessary). Again a 'critical number' approach may be too simplistic. However, if one is still talking about critical numbers, and our detection turns out *not* to be of intrinsic anisotropy, so that one has to dig to still smaller fluctuation levels, then I'm glad to report that 10^{-5} will be enough. I was speaking with Marc Davis yesterday, and he said that if we could get the $\Delta T/T$ limits below 10^{-5}, then we'll have eliminated everything but cold dark matter, and he promised not to invent any alternative theories. So I'd like that written into the record!

To be realistic, in summary, on our microwave background fluctuations, we, Rod Davies, myself and colleagues, are the first observers to find fluctuations in a situation where we can actually go back and look in detail at what we've found. There is already some evidence for the reality of the fluctuations in the agreement of two independent experiments on different angular scales (our 5° and 8° experiments). However, the 'clincher' has obviously got to be to prove that the frequency dependence of what we have found has the form appropriate to a black body perturbation. Thus experiments at higher (and lower) frequencies will be vital, and we fully intend to carry these out and discover definitively the nature of the fluctuations we see.

MARTHA HAYNES

I'd like to return to the analogy that Avishai made about plate tectonics and cartography. I believe that we are at a very exciting point, and I think we will be giving Avishai Dekel a lot more information in the next few years that will help him develop his theory of plate tectonics.

One of the important points that we have learned recently, since the IAU Symposium 130 in Hungary in particular, is that there seems to be continuing convergence of ideas among us even when we look at very different data sets and pursue them in different ways. Although perhaps the details are not all ironed out yet, I find it encouraging that a lot of the pictures of large-scale structure that are emerging are similar. For example, when we take the southern part of Pisces-Perseus and the CfA slice, the structures are remarkably similar when they are displayed in a similar manner. When we consider how our ideas have modified in the last five or ten years, I think this convergence of thought means that we are certainly on the right track. That's true not only for observations but for theory. It seems to me that while we haven't ruled out all theories and we haven't ruled in some of them, cold dark matter makes some nice preditions that we can now go out and test. Maybe we'll rule cold dark matter theories out, but at least we have to be on some of the right tracks.

At the same time, I think we've also shown the importance of pursuing the subject from a variety of approaches. I can say that sometimes similarity in structure exists, but in some places, Pisces-Perseus just doesn't look like anything else. We have also to worry, therefore, that our little part of the universe is not like everywhere else, and locally at least we certainly have evidence now that we inhabit an anomaly. We should really wonder then whether our view of the universe would be different if we were sitting on the outskirts of Perseus, for example.

It is unfortunate at this point that while we have pictures of superclusters like Pisces-Perseus, Coma-A1367, and Hercules, we really can't measure distances to better than about 15%, and that just is not good enough. But, within the next few years, we ought to be able to do better.

I'd like to bring us back to the cartography to take a look at an analogy that we might keep in mind, and that's the story of the first "discovery" of the Pacific Ocean. Verrazano was looking to sail to India, not to discover a continent. While exploring the East Coast of the United States, he sailed past the outer banks of North Carolina Cape Hatteras, and he discovered the Pacific Ocean; at least, he thought he did. He looked over the narrow cape, saw another body of water, and said, "Ah, we can follow that to India. I don't have to do anything else. I can sail back". Now, he should have seen the land on the other side of the spit and the bay, but perhaps it was foggy that day, or perhaps he had to finish in time for a coffee break (as I am supposed to).

This little story does teach us something, that we should worry when the scales that we are studying are about the same size as our sample, and so I'll just leave with that warning. But, on the other hand, I know that, as Brent Tully has said, we will be continuing to extend our scales and we'll probably find more interesting structures and motions on larger scales too.

AMOS YAHIL

I want to concentrate on the rather narrow area that I myself have worked on, and to try and discuss some prospects for the future. I tend to go along with what Jeremy Mould has said, that the initial comparison between the peculiar velocity field predicted from the IRAS density distribution and the observed velocity field, as shown by various people here, is encouraging, but that, in detail, there is a lot more work left to be done. I think there is work to be done on both the side of density and velocity, and I want briefly to outline what I think are the major tasks in the next, two, three, or five years.

Let me start with things that we can already do, but have not yet done, on the density side. One objective is to put in the contribution of the early-type galaxies in a more realistic way. You have seen the sort of simple-minded, quick-fix algorithm of double counting which Marc Davis showed. We really need to supplement, indeed to replace it, by a careful analysis of the ratio of optical to infrared galaxies. Perhaps then we can get a better idea of the total mass distribution, and improve the calculation of gravity. That can be done with existing data.

The second thing we can do with existing data is to put in nonlinear corrections. This has so far not been done, but we have some ideas, and hopefully can report on them within a short time. I think this is also a doable problem.

One can always improve by obtaining more observations. We are ourselves pushing the IRAS measurements down to five degress from the galactic plane. I also agree with the suggestion of Sandy Faber, that we not replace what is left—whether it is five or ten degrees—by a simple homogeneous distribution, but by a more sophisticated interpolation. That is a nice idea, and it can improve our predicted velocities considerably.

I am hoping that another observational handle on density will come if ROSAT gives us a list of elliptical galaxies which we might use in a similar way to the IRAS spiral galaxies, but that is into the future.

On the side of the velocity distribution, we need to expand on what has already been done. One of the nice things about that is that many people can get those peculiar velocities in many different ways. My prediction is that many observers, both present here and ones not at this meeting, will now be scrambling to get as many peculiar velocities as they can. The body of data on peculiar velocities will then begin to grow very, very rapidly. I agree with what Martha Haynes said earlier, that the direction of Perseus-Pisces is an extremely interesting first shot, because we have relatively little information in that direction, and the density perturbation there is comparable to the Great Attractor in Hydra-Centaurus. So that is a place to go first.

Now, given that we are going to make all those improvements in the density determination, and obtain more peculiar velocities, we can begin to do a careful one-on-one comparison between the prediction of the theory and the observed peculiar velocity field. This can be done in a very, very detailed way, and will give us primarily two things.

First, we will have a quantitative test of biasing. I am stretching my neck out here saying that we will overcome all the observational difficulties, and actually be able to confront theory. This goes back to a point which Jim Gunn made in his summary talk at IAU Symposium 117 on *Dark Matter in the Universe* at Princeton two years ago, after the initial IRAS results were presented. He said that, if you have biasing, you violate the principle of superposition, and therefore vectors add up in different ways, and velocity fields get distorted. So, if you find good agreement between the predicted and observed velocity fields, you are fairly confident that biasing is minimal. On the other hand, if you do not find agreement, you may be able to find a prescription by which you re-scale the density perturbations, so as to obtain a better predicted velocity field, and then you have discovered biasing. I think that is a project that we should continue to pursue, and hopefully, when the data improve, will be able to accomplish.

Secondly, if we do find that everything fits together, we can realistically seek the cosmological density parameter, Ω_0. Again, I will be optimistic, and say that perhaps we can do that in a reliable way. So, my hope is that by the time of the next Vatican meeting, in another five or six years, they can shut us all up in the Sistine Chapel, and not let us out until we have decided on the value of Ω_0.

VERA RUBIN

To capture the spirit of the meeting, and in the fashion of the times, I present a List. *The Prettiest Pictures:* Yahil's galaxy plots showing gravitationally induced velocity vectors in the supergalactic plane, Tully's tinker-toy clusters, and Koo's faint UV arc. *The Biggest Failure:* using the microphones. *The Biggest IF:* IF light traces mass. *The Most Overworked Words:* toy, tophat, generic. *Questions I Wish I Had Asked, to Burstein, Faber, Mould, and Lynden-Bell:* Assuming a motion due to the Great Attractor, why is the decrease in dispersion in the Aaronson, Huchra, and Mould spiral sample so much more dramatic than for the Faber-Burstein elliptical sample? Does it arise from different sky coverage, or do spirals better map the large-scale flow than do ellipticals?

First Disappointing Statement: Geller's opening remarks that photometric surveys are not accurate enough to do meaningful statistics. *Most Beautiful Doodles:* Alex Szalay's. *Most Trivial Statistic:* every left-handed participant sat next to a left-handed participant. *Best Food:* every meal. *Second Disappointment:* unease concerning possible biases in the Abell catalogue. *What I Was Pleased To Learn:* Martha Hayne's impressive evidence for luminosity function differences in regions of different densities and how well Marc Davis's N-body simulations show such properties.

What I Was Not Really Surprised To Learn: questions concerning the reality of a Virgo infall. *What I Would Most Like to Learn:* the cause of the motion with respect to the microwave background and its relation to the inferred motion toward the Great Attractor; is it only gravity? *Least Esoteric Question:* how should we properly define groups, clusters, and superclusters, for they will surely in the future play the role that galaxies now play in defining large-scale structure and motions.

Most Optimistic View of The Great Attractor: probably at present the best detailed description of our motion, but unfortunate in: 1) being located close to the zone of avoidance, the zone made unobservable by the Milky Way; and 2) precariously near the velocity limit of the observed sample. *Big Question:* how will more observations change the model of the Great Attractor?

Most Thanks: to all the participants for their thoughtful contributions and spontaneous contributions, their agreements and disagreements all carried out cheerfully and in the spirit of advancing our understanding, even though we did not answer the 31 questions asked at the outset of the meeting. *Overall Conclusion:* it should be no surprise to the participants to hear that I support the concept that large-scale motions exist. A variety of observations of several samples now show that large-scale motions do exist. The interpretation of velocity structure may be affected by large-scale differences among galaxies. These must be sorted out by future observations.

I opened the meeting with the 16th century sky map of Centaurus. Perhaps a map of the world at that time would have been equally appropriate. When we compare 16th to 20th century world maps, and 16th and 20th century sky maps, we see an enormous change. I presume that the cosmography of the universe will change equally much in the next few hundred years.

JIM PEEBLES

There are several points in our discussion where it seems to me that we have reached a crisis point, such that relatively small further advances in theory and observation might be expected to yield a considerable improvement in our understanding. Here is my list of the hottest crises.

The main topic of our meeting is the large-scale peculiar velocity field. I remind you that the big news from the Burstein *et al.* study was not the magnitude of the velocity but rather the suggestion that the coherence length over which it varies may be as large as 4000 km s^{-1}. If this were so then as Dick Bond, Marc Davis and others have noted, it would be a considerable embarrassment for the scale-invariant cold dark matter model. Faber and Burstein have presented a thorough and closely reasoned case for their interpretation, which we certainly must take seriously. But the case would be stronger still, if a model with a small coherence length consistent with inflation had the benefit of a similar degree of attention and adjustment and been found to be wanting. An important start in this direction has been taken by Kaiser, as is described in these proceedings, but I have the impression that his method of weighting reduces sensitivity to the Burstein *et al.* picture, that the peculiar velocity has an appreciable component that varies over scales ~ 4000 km sec^{-1}. I think we understand the issues here, and that we may even be converging on a resolution, but am not so sure which way the resolution will go, large coherence length or small.

The second crisis point is the interpretation of the tendency of the galaxy space distribution to show linear features. There was a time when I was skeptical of this effect, but there is no believer like a reformed sinner. I am convinced by the redshift maps, and by the statistical tendency of elliptical galaxies to line up with the large-scale galaxy distribution, that the effect is real and crying for an explanation.

The numerical simulations of the scale-invariant cold dark matter model shown by George Efstathiou tend to have remarkable linear structures that are strikingly similar to what is observed. If this impression is confirmed by closer checks it will be a dramatic triumph for this model (and for the general class of gravitational instability pictures with roughly similar power spectra). I am a little uneasy about the effect, however, because we seem to have in these simulations an example of pattern formation without a source for the pattern. In Benard convection rolls one can see the effect of the container. In pancaking in hot dark matter one can see the effect of the hard high frequency cutoff of the power spectrum. Before we buy the linear structures seen in the N-body model simulations with slowly varying power spectra (flatter than k^{-3}) we are owed an explanation of where these patterns come from.

The explosion picture predicted the existence of linear features in the galaxy space distribution, as the remnants of the ridges piled up by the explosions. That is an impressive success. However, I am also impressed by an apparent problem for a simple explosion picture. We know that the local sheet of galaxies, at distances less than about 900 km sec^{-1}, is moving very nearly uniformly at 600 km sec^{-1} relative to the rest frame defined by the microwave background. The local galaxies off the sheet are moving at very nearly the same peculiar velocity. How does one explain this in an explosion picture?

My third crisis has to do with the shape of the galaxy two-point correlation function, ξ. At small separations ξ is well approximated as a power law, $\xi \propto r^{-\gamma}$, $\gamma = 1.77$. At separation, $r \sim 10h^{-1}$ Mpc (H = 100 h km sec^{-1} Mpc^{-1}), γ has a feature, a local excess over the power law, followed by an increase of slope. (Beyond that, at hr $\gtrsim$ 15 Mpc, ξ has not yet been reliably measured.) This feature was discovered by Ed Groth and myself some ten years ago, but the evidence for it has been critically debated only recently as a result of stimulating discussions by Margaret Geller and her colleagues. The outcome is that I am not convinced of the reality of the feature in the galaxy distribution at the Lick depth. Even so, it was with more than a little relief that I learned of the preliminary results of George Efstathiou and his colleagues, that seem to reveal the same feature in the deeper Cambridge APM survey.

What is the significance of this feature in ξ at hr $\sim$ 10 Mpc? A clue might be that the feature appears where $\xi(r)$ passes through unity. This suggests to me that the feature may have something to do with the transition from the highly non-linear character of the density fluctuations on small scales to the linear character of the density fluctuations observed on large scales; but I know of no convincing application of this idea (including my own).

N-body models with power law initial power spectra tend to make the slope of the mass autocorrelation function, $\xi_{\varrho\varrho}$, at $\xi_{\varrho\varrho} \sim 1$ much steeper than the slope of the galaxy two-point correlation function at $\xi \sim 1$. The unwanted steep slope is the result of the tendency of newly forming levels of the clustering hierarchy to collapse in a nearly radial way. I do not think that this radial collapse describes what is happening in the Local Supercluster. One can think of several ways to avoid radial collapse: assume mass does not cluster like galaxies, or assume the initial power spectrum is not a pure power law or, perhaps, assume the initial density fluctuations are not Gaussian. The scale-invariant cold dark matter model is an example of the first possibility. I gather that the N-body model simulations of this model described by George Efstathiou could be used to predict the expected shape of the galaxy two-point correlation function at $\xi \sim 1$ in this picture. We will await the results with considerable interest.

Another possibility is that the initial power spectrum of the mass distribution is not well approximated as a pure power law. An example is the baryonic isocurvature model, where the power spectrum of the baryon distribution develops a spike at a wavelength $\lambda \sim 100$ Mpc. The spike adds to the mass autocorrelation function a broad shoulder, with width $\sim \lambda$, which is at least in the wanted direction. Perhaps when we next meet I will be able to give an assessment of the prospects for this idea.

The final crisis is the value of the density parameter, $\Omega = \varrho / \varrho_{crit}$. The elegant answer is $\Omega = 1$. The dynamical estimates pretty consistently indicate $\Omega \sim 0.1$ to 0.3. As the quality and variety of the dynamical tests have improved I have started to take seriously the possibility that the universe knows something we do not know, that Ω may be less than unity. The tension between our belief in the reasonableness of the argument for $\Omega = 1$ and our wish to accept the observations will be an interesting case study for the sociology of science.

If $\Omega = 1$ then galaxies have to be more strongly clustered than is mass. Also, if the dark mass has negligible pressure, then there is an upper bound on the redshift at which the sites of galaxy formation can have been determined, because mass concentrations exist and these concentrations grow by gravitational instability, drawing in galaxies and mass and so tending to erase the original segregation of mass from galaxies. The redshift bound seems to be particularly tight in the scale-invariant cold dark matter model. My impression of the numerical simulations of this model is that galaxies are being assembled at redshifts less than unity. If so, I think this is a crisis for the model because, as David Koo described for us, there already is pretty good observational evidence for the existence of well-developed galaxies at redshifts greater than unity. It's reasonable to hope that we will see a resolution of the conflict between early and late galaxy formation in the not too distant future, because the observations of galaxies at redshifts $z \gtrsim 1$ are improving rapidly. This is a result of the dramatic improvement of optical and, very recently, panoramic infrared detectors, and of the very impressive work of the observers. The Ω puzzle is more complicated, in many ways; it is likely to be with us for some time.

MARC DAVIS

During the course of several long sleepless nights this week I perused the volume from the last Vatican Study Week, *Astrophysical Cosmology,* and I was intrigued to read Martin Rees' opening introduction. He addressed seven points for consideration at that workshop. They start with such questions as whether we know the Hubble constant to a precision better than 25%, and they end with the question of whether comprehensible physical processes at very early times can account for the overall homogeneity but small-scale roughness of the universe.

That week Steven Weinberg told us about the fantastic notion of inflation, baryogenesis and other exotica that were currently happening in the world
of grand unification physics. It was all very new, very young, and the notion
of fluctuation generation was unresolved. Also at that meeting Sandy Faber
argued strongly and very impressively that the flat rotation curves and the
observed scaling behavior of galaxies implied that we could understand quite
a few properties of galaxies if the effective spectral index of perturbations was
$n = -2$. I remember giving her a hard time about that, because I was con-
cerned particularly with the problems such a spectral index would cause on
large scales.

Inflation, of course, had just been invented. Cold dark matter wasn't invented until a year later. After cold dark matter was invented it was obvious
that spectral index $n_{eff} = -2$ was going to be the case on galaxy scales, but
that we wouldn't have a problem with large scales. So Sandy Faber's contribution was really prescient. Massive neutrinos were in vogue at that meeting;
Dennis Sciama spoke about them, but even then I note in the proceedings that
we had serious questions about whether that model would possibly work.
Massive neutrino models basically died the following year.

Now cold dark matter is the standard model and it's a matter of taste
whether you like the model. I happen to like it because it is so specific and
so minimal, as Dick Bond has emphasized, and therefore we can make all types
of predictions and it's a great model to try to shoot down. Now, it is really
all a house of cards, because not only does the model fail if we have 600 km/s
coherent bulk flows on a large-scale or if the cluster-custer correlation function is sufficiently large, but it fails just as well if we can find convincing
evidence of rotation curves that fall at large radii. If we start to see edges of
galactic halos, the theory dies. The rotation curves in our models extend hundreds of kiloparsecs and are essentially flat over this range. There's no way
around that in the simulations. The theory will also die if the Hubble constant settles in at 100 km s^{-1} Mpc^{-1}. I think the proponents of CDM got away
awfully easily this week as there was little discussion of the actual value of
H_o. To teach myself a little bit more about the Hubble constant, I've been
teaching a seminar to undergraduates this semester and I've learned considerably more about the situation. There are many remaining problems but
there is a lot of consistency in the different measurements of the calibrators
of H_o, and I don't think the calibrators are off by a factor of 2. I think Brent
Tully is right. 95 is a lot better estimate than 50. So I don't know what to
do at the moment.

I would like to mention one thing about Avishai Dekel's Dow-Jones prediction. I can see that he is a careful market watcher, but I think he missed one
blip in the CDM history. In particular, the biased CDM model was issued
simultaneously with the original model, because the theory didn't work at all
without the bias; the first announcement of success for the theory came with
the bias. However, as Gary Steigman and others emphasized, this invoked the
tooth fairy too many times because the bias had no physical motivation.
Without a physical understanding of the bias, the CDM stock was oversold,
and the theory was not all that convincing. I personally feel that it was the

natural bias, invented only last year, that really raised the stock of CDM up to its present level, and I agree that it is still climbing (I recommend 'buy').

Six years from now, at perhaps the next Vatican Study Week, my prediction, for what it is worth, is that CDM will have passed the observational tests on large scale. The clusters-cluster correlations will not be insurmountable and the drift velocities will be acceptable, but H_o is still going to be a problem. At that time, we are going to wonder how nature could be so perverse as to make a theory seem so consistent with the observations and yet clearly be wrong.

DONALD LYNDEN-BELL

As someone who believes in the importance of conventions I also think that society needs people who are prepared to break them; and Brent Tully's cosmography and bravery in exploring the forbidden regions, beset by the ogres of observational selection, I both admire and deprecate.

Now, Avishai Dekel told us about the future of the stock that is most closely conventional. As a conventionalist, I would like to back it. I think the stock is very down at present. It's very depressed, like my voice. The ordinary matter is the only matter we know of and of course it's very bad to predict from the only thing you know, but I like to do it nevertheless. I think we'll need a guillotine for all those who predicted the cosmological abundance. I think something is very wrong there, because omega is just a little greater than 1, and all matter is baryonic.

Now, I thought one of the objects of these brief remarks was to explain what we were going to do when we went away. As usual, I stand criticized by Sandy Faber, so I'll go away and allow for the mass associated with the seen mass and not just the seen mass in trying to calculate the optical dipole. I'll also extrapolate over the galactic plane as she suggests, and I'll calculate the nearby part of the dipole from actual luminosity rather than from angular-diameter squared, because if it's true that a lot of the dipole comes from nearby; then we know about all those galaxies.

I have been greatly impressed, as I was when we wrote that paper on the Great Attractor, that the velocity field is well-delineated by the Great Attractor by and large. I have also been very impressed, as have others, by the IRAS density maps. Their gravity field, I think, is at the present time not well-represented by the Great Attractor model, though the velocity field *is* and that raises a doubt. If you correct the picture from Huchra's catalogue for the incompleteness, you find that between Virgo and the Great Attractor there is an offset of the center of mass. This includes material only within 15 degrees of the supergalactic plane. If you now go off the supergalactic plane and try and plot density following those regions which seem to be the strongest on the plane of the sky, you find even further complications. I think, for instance, that the region in Capricorn may in practice be as important or even more important than the Great Attractor.

NICOLA VITTORIO

Many ideas have been discussed during this week and I feel I am going back home with a lot of homework to do. A crucial issue for comparing theory and observations concerns having a fair sample of the universe. As Peebles commented, the fairness of the sample depends on the quantity to be measured. In the case of the large-scale peculiar velocity field it is important to be careful, since, as we heard, the accelerating material seems to be just outside the Seven Samurai galaxy sample.

In my opinion the consistency of the CDM model with the large-scale flow is not a settled issue. The problem is the scale at which the flow occurs. Only if the Seven Samurai sample really shows that the drift occurs on scales as small as $15h^{-1}$ Mpc, will a CDM model, where light traces the mass, fulfill the observational constraints.

Undoubtedly, cold dark matter provides a very specific model for the formation and evolution of the large-scale structure. The natural biasing that Marc Davis discussed here seems to be an automatic way of introducing a bias in the galaxy distribution. Also, a natural biasing seems to be quite model independent, as it rests mainly on the gravitational growth of structure in a hierarchical scenario. Unfortunately, I must add, I am not convinced that a biased cold dark matter model satisfies the large-scale drifts constraints.

For the cosmic microwave background it is, of course, crucial to confirm the Davies *et al.* result and to provide tighter bounds on the temperature anisotropy. The RELIC II satellite is very promising in this respect.

BILL STOEGER

Well, I have been listening from the fringe, so to speak. I am not sure where the fringe is. It's somewhere between the outer edge of the Great Attractor and the last scattering surface. I have been amazed at the progress that has been made in mapping what I would call our cosmological neighborhood, the region of intermediate scale to which we have access with precise astronomical technique. However, I have a few worries, which have been reflected upon already by other people here which are superimposed on my amazement at what has been accomplished. One of them is that the structures that we are finding are just about the size, or a little bit bigger perhaps, than the sample volume we have been using. Another concern is selection effects. We are, as a matter of fact, coming to understand the important selection effects much better than before, and developing much more uniform ways of dealing with galaxy samples. But there still seems to be a lot more to be done in terms of dealing with this crucial issue.

Then, there is the whole problem which Nicola Vittorio and Donald Lynden-Bell mentioned: the fair sampling hypothesis. At what distances are we going to be able to say that we have a fair sample of the galaxy distribution in the universe? This is a question which worries me and a number of the people I work with. As we extend our more precise astronomical observations to larger and larger distances, we find structures, and possibly even coherent flows, on those larger scales. Where will this end? Are we approaching distance scales, as maybe Tony Tyson's deep galaxy count work and radio source counts would indicate, at which we shall be able to say that we do have spatial homogeneity on scales above between 200 and 500 Mpc? I hope so!

That leads me to the second half of my remarks, which is related more closely to the last scattering surface than it is to the Great Attractor. The issue is: How does what we have been talking about here in this Study Week fit in with our use of cosmological models, or *a* cosmological model, on larger scales? When we employ a Friedmann-Robertson-Walker (FRW) model to characterize or describe the universe — or an almost-FRW model — we must always specify the particular length scale below which it does not work. We already have a good idea about what the lower limit for that length scale is. Certainly, from what we have seen, FRW or almost-FRW models, do not adequately describe our universe on scales below 200 Mpc. On such scales, the universe is still too inhomogeneous. Above what length scale can we be sure that an almost-FRW model fits the universe? We do not yet know.

Some of the work that I have been doing with George Ellis and with others is aimed at giving precision to what is an almost-FRW model and what is not. We still do not have a workable criterion. But I think that the one thing that gives us hope is in the confrontation between the data provided by the more precise observational measurements of traditional astronomy (with which we can map our cosmological neighborhood) and the data we obtain from microwave background observations, which tell us about the epochs at last scattering and before. If we really believe that the microwave background, in its orthodox interpretation, is smooth enough to indicate that on some length scale in the past there was almost-homogeneity, then surely we can construct an argument to demonstrate that there is almost-homogeneity now on some very large length scale, even though it is much larger than we can deal with astronomically at present.

Relevant to this particular point, I believe that it behooves theorists and cosmological background observers alike to show that the smoothness of the background we have seen up to now is smooth enough to support an almost-FRW model of the universe on scales above some large value. That still has not been demonstrated. If you go back into the literature to see why people think using an FRW model is valid in the first place, you find that the strongest arguments rest on what is known as the Ehlers-Geren-Sachs theorem, which is outlined in its most accessible form in Hawking and Ellis: *The Large Scale Structure of Space-Time*. We really have to go back and re-examine (and certainly strengthen) that theorem (by proving it in its "almost form") in order to confirm the view that at least on some very large scale the almost-smoothness of the microwave background justified our use of FRW, or almost-FRW. These large scales on which we may have spatial homogeneity are linked to the intermediate scales we have been talking about, through models of galaxy formation, and the consistency they impose between these two types of observations — the astronomical and the microwave-background. What are the connections between the progenitors of the intermediate-scale structures we are seeing, like the Great Attractor and the surrounding galaxies affected by it, and the fluctuations (or lack of fluctuations) in the microwave background?

NICK KAISER

I would like to say a few words about the present status of the cold dark matter model in the light of what we have heard at this meeting.

The biggest problem is the Hubble constant. Most theorists want H of 50 (or less) in order to get the most pronounced large-scale structure, yet we have heard that most data point to a larger value. The problem of what are the real uncertainties here is very involved; however, one cannot help but feel somewhat ostrich-like in simply hoping that this problem will just go away.

If we brush this problem aside then the predictions of cold dark matter (CDM) on intermediate scales (2-10 Mpc/h say) look good, but given that one has the initial amplitude and bias as free parameters these successes are less impressive. In order to really test the models we must go to much larger scales. Fortunately, models like CDM are very cooperative here. If we stick to linear theory, we can simply write down the probability distribution for counts of galaxies in flux, redshift and angle space, and for peculiar velocities. Perhaps the most pressing questions here are; Is the divergence of the velocity field proportional to the underdensity of galaxies? This is expected if the structure is growing by gravitational instability - even if the galaxies are biased. What is the bias factor b which appears as the constant of proportionality? Are the fluctuations consistent with the cold dark matter power spectrum?

My feeling is that we are now seeing a fairly consistent picture emerging if all the different observations are taken with a pinch of salt. The comparison of peculiar velocities and the various dipole moments at least suggest a positive answer to the first question, and seem to give consistent and reasonable values for the bias parameter. When we come to the question of whether the spectrum of fluctuations agrees, things are more complicated. Provided we only look at these very 'broad brush' statistics, the distribution with depth looks quite compatible with the CDM prediction, and at least with the normalisation I prefer, the amplitude looks compatible. It seems that rumours of the death of CDM at the hands of these data were greatly exaggerated. Whether this will remain true when we look at the data in more detail remains a very challenging open question.

MARGARET GELLER

It will come as no surprise that I've been hearing many others mention the points on my list. I can thus be very brief.

I, like many others, have been very much impressed by the IRAS results. They demonstrate the power of a nearly uniform all-sky survey, where the selection biases are perhaps better understood than they are in the optical. In fact, I was struck by the difficulty of comparing the IRAS results with current optical catalogues. This problem brings me to the need John Huchra emphasized: we're going to have to obtain well-calibrated photometric catalogues in the optical. There is already the promise of progress toward this goal with the APM survey for the South and with the scanning of the new Palomar Sky Survey in the North. In the long run, we will probably need a digital survey to address all the problems.

One issue which particularly concerns me is the comparison between the scale of the largest structures and the extent of our redshift surveys. It's sobering that every survey we do seems to contain a structure as big as the survey. Thus it's rather difficult to derive statistics we can trust. The errors and uncertainties in the statistics we generally use are probably larger than we have thought.

We have to know the frequency of structures like the void in Boötes. Voids of this size appear to be common; most surveys which can contain them do. From redshift surveys, we don't have very good (or any) limits on structures which are intermediate in scale between Boötes ($6,000$ km s^{-1}/H$_o$) and gigaparsec scales. We would like to know the distribution of these very large structures and to have some statistics which are sensitive to the high order properties of the galaxy distribution. Perhaps we should be measuring the distribution of sizes of voids or the properties of the sheets. Of course, the first is a tall order when the surveys aren't big enough to contain more than one of the largest known voids!

Another statistical issue, also affected by problems in catalogues, is the relationship between the distribution of rich clusters and the distribution of individual galaxies. The cold dark matter models cannot match the current results. Certainly the new digitized sky surveys with objectively derived cluster catalogues will help us to make some progress in understanding clusters of galaxies as tracers of the large-scale matter distribution. However, I think that we will not really understand the relationship until redshift surveys for clusters and for individual galaxies overlap sufficiently that we can examine the relationship directly. There's hope that within ten years we will actually see such measures. We will be able to compare what we expect physically with what we see.

One certainty is that none of us are going to have any problems finding something to do. We'll all probably be surprised at some of the things we find.

DAVID KOO

Well, my worst fear has been borne out. I was the last one. Fortunately, not all my points have been expressed. Let me start by mentioning the conclusions of the 1981 Vatican Study Week, in which Longair emphasized the need for systematic surveys and here I quote: "...particularly profitable, I would suggest: further detailed studies of nearby galaxies, deep surveys of large areas of sky, the velocities and spectra of large samples of galaxies, the relation of active galaxies to the properties of the stellar and gaseous component of their parent galaxies and to their environments, measurement of fluctuations in the microwave background radiation with high precision" and "studies of how the properties of galaxies and the large-scale distribution of galaxies and diffuse matter change with cosmic epoch''. Except for perhaps the relationship of AGNs to host galaxies, I was extremely impressed how the relatively small number of observers here have covered so well almost all these areas with great advances; and probably just as remarkable has been the work of the theorists in terms of their energy, creativity, and sophistication in handling the large amounts of new data. Although I had come fully expecting to be convinced of the Great Attractor, I leave wondering whether some of the observed peculiar motions result more from repulsion by large pockets of under-densities, locally and perhaps beyond 5,000 km per second.

On the theoretical side I came expecting to see convincing evidence that the standard biased cold dark matter theory is inadequate to explain all this new data but I leave somewhat amazed by how well it actually does work. On the other hand, I am looking forward to future improvements by my next door neighbour here (A. Lasenby) in exploring the microwave background and am hoping that his detection really is positive on the large scales.

I also came hoping to see much discussion on different and quantitative diagnostics of the large-scale structure, especially of the voids, but I think that despite the mathematical rigorousness of PC's (visually blind alien from the Pisces-Cetus supercluster introduced by G. Efstathiou in the discussions) view of the world, the universe is really more complex and beautiful, or ugly if you're a theorist, than we can imagine with numbers. We should not forget the power of Eccle's human eye-brain pattern recognizer, with which we notice arcs around the clusters, bubbles or filaments in the galaxy distribution, different Hubble types of galaxies, maybe spectral features not much discussed here, and unusual classes of objects, such as quasars, blasars, AGNs, etc... Basically, much of what we learn results from recognizing deviations from the norm. Along these lines, I, like very many others, was extremely impressed by the flow diagrams of Amos Yahil. I guess that's because I am an observer, but, unlike others who were impressed by the attraction seen in certain areas, I was really amazed by where I thought I could see outflows from the voids. A simple question is: are we actually seeing the voids pushing? This point is important, for if they are pushing away, they must be true underdensities, and not just regions with lots of hidden matter.

Along the theme of cartography and maps that have been expressed by many here, I felt like an outsider exploring the frontiers, instead of mapping all the little cities on the East Coast. Our domain encompasses the more distant but 99% of the Universe that was largely ignored, perhaps appropriately so for this workshop. Besides penetrating new continents or even the poles, our approach, though crude, may even include drilling holes into the earth, probing the ocean bottom, or sending up balloons and rockets.

One point that I'd like to emphasize is the role of gas. Although gravity is very easy to model, gas is where the action is, especially in much of the detected evolution beyond redshift of 0.1. Gas is intimately linked to much of what is observable; to the morphology, physical parameters, and formation of galaxies, active galactic nuclei, the intergalactic medium, and quasar absorption lines; and to even the large-scale motions discussed in this workshop.

After all, what would be the effects of bulk flow among huge volumes of gas, either today or especially at early epochs before stars and galaxies were formed (and before simple dynamics are applicable)?

The future is very promising for us explorers. We'll soon have space telescope, large 10 meter or maybe 16 meter telescopes on the ground, vast imaging and spectroscopic surveys with large-area CCD's, and new adaptive and active optics that provide high resolution. Infrared imaging with arrays has already begun, giving us new eyes to explore the distant past.

In closing, since I am the last one, I wish to offer great thanks to the Vatican for a delightful week, more busy than free, but full of wonderful food, interesting science, great company, and very warm hospitality.

INDEX